An Agenda of Science for Environment and Development into the 21st Century

An Agenda of Science for Environment and Development into the 21st Century

Based on a Conference held in Vienna, Austria in November 1991

Edited by

J.C.I. Dooge *(Centre for Water Resources Research, University College, Dublin)*

G.T. Goodman *(Stockholm Environment Institute)*

J.W.M. la Rivière *(International Institute for Hydraulic and Environmental Engineering, Delft)*

J. Marton-Lefèvre *(International Council of Scientific Unions)*

T. O'Riordan *(School of Environmental Sciences, University of East Anglia)*

F. Praderie *(International Council of Scientific Unions)*

Compiled by

M. Brennan *(International Council of Scientific Unions)*

International Conference on An Agenda of Science for Environment and Development into the 21st Century (ASCEND 21)

Organized by the International Council of Scientific Unions (ICSU)

In cooperation with the Third World Academy of Sciences (TWAS)

Hosted by the Federal Government of Austria

Co-sponsored by:
European Science Foundation (ESF)
International Institute for Applied Systems Analysis (IIASA)
International Social Science Council (ISSC)
Norwegian Research Council for Science and the Humanities (NAVF) with the Norwegian Academy of Science and Letters
Stockholm Environment Institute (SEI)

The contents of the individual chapters, of the General Introduction, and of the three Section Introductions are the
responsibility of the separate authors concerned and do not necessarily represent the opinion of any of the sponsoring
organizations.

Acknowledgements
Financial Support:
The Government of Austria
The John D. and Catherine T. MacArthur Foundation
The United Nations Development Program
The co-sponsoring organizations

Equipment: IBM, Austria

Contents

APPENDICES

General Introduction

J.C.I. Dooge and J. Marton-Lefèvre

BACKGROUND

The International Council of Scientific Unions, ICSU, which celebrated its 60th anniversary in 1991, has long been involved in environment and development issues through its wide-ranging network of national Scientific Members, international Scientific Unions and international interdisciplinary Scientific Committees. This experience, combined with ICSU's long-standing status as a non-governmental, non-political organization, stood it in good stead when Maurice Strong, Secretary General of the United Nations Conference on Environment and Development (UNCED), speaking at ICSU's Global Change Forum in Paris in September 1990, asked ICSU to act as principal scientific adviser in the preparation of the Rio UNCED, or "Earth Summit" (June 1992).

ICSU responded to this invitation by mobilizing the broad world community of scientists with which it is in touch in a number of ways, including the encouragement of its National Scientific Members to take part in the national preparations for the Rio Conference so that the views of the scientific community in their country were taken into consideration; the involvement of scientists in the reviewing of UNCED documents of a scientific content and the joining in a number of the Working Parties set up to address the UNCED issues. ICSU Officers were closely involved in the UNCED Working Party on Science for Sustainable Development and in the drafting of the relevant chapter of the UNCED "Agenda 21" document.

Finally, ICSU decided to organize an International Conference on the contribution which science could and should make to the analysis and solution of problems of development and environment. The Conference theme, "An Agenda of Science for Environment and Development into the 21st Century" gave it its acronym, ASCEND 21. The Conference was held during the last week of November 1991 in Vienna.

ORGANIZATION OF CONFERENCE

ICSU, primarily concerned with natural sciences, wished to involve in the organization of this Conference, not only its own members, but also co-sponsors with expertise in other areas of the sciences. These co-sponsors were: the Third World Academy of Sciences (TWAS), the European Science Foundation (ESF), the International Institute for Applied Systems Analysis (IIASA), the International Social Science Council (ISSC), the Norwegian Council for Science and the Humanities (NAVF) with the Norwegian Academy of Science and Letters, and the Stockholm Environment Institute (SEI). The Government of Austria kindly offered to host the meeting. The Conference and its extensive preparations were financed by generous grants from the Government of Austria, the John D. and Catherine T. MacArthur Foundation, the United Nations Development Programme and the co-sponsors.

A Conference Advisory Committee was set up, consisting of ICSU Officers, members of the ICSU Committee on Science and Technology in Developing Countries (COSTED) and ICSU's Advisory Committee on the Environment (ACE), together with representatives of all co-sponsors and other experts. The list of members of the Advisory Committee is found in an Appendix to this volume.

The objectives of the ASCEND 21 Conference were to bring together the understanding and judgment of the world scientific community (encompassing natural, social, engineering and health scientists), on the issues of highest priority for the future of the environment and development, so as to define the agenda of science in these areas for the next 10 to 20 years. The outcome was to provide both a consolidated contribution to the Rio Conference and a perspective for the future of international science in these and related areas. The results of the ASCEND 21 Conference as reflected in this volume are intended to serve as a basis for future action in environment and development by the scientific community.

From the very beginning of the planning, it was recognized that ASCEND 21 had to look beyond the state of the art, to formulate the environment and development research agenda and to help identify the scientific knowledge base which would be needed for rational policy decisions during the coming decades. Furthermore, in order for the all-important interlinkages between environment and development, and their relationships to science, to be properly addressed, the entire Conference needed to be conceived as a dynamic building process aimed at identifying these interlinkages.

The participants of the ASCEND 21 Conference were invited individually after selection by the Advisory Committee from over a thousand proposals made by the co-sponsors, UN bodies, and other relevant organizations. Out of some 300 invitations sent out, some 275 responded positively. ASCEND 21 was finally attended by some 250 experts from nearly 70 countries. Approximately half came from developing countries in Africa, the Arab region, Asia and Central and South America. Each person, whatever their professional position, attended the Conference strictly on a personal basis. The full list of participants is found in an Appendix.

This Introduction is followed by the Conference Statement. The subsequent parts of the volume contain the key-note addresses given on the first and last days of the Conference and the sixteen Conference theme chapters which were substantially revised in the light of the intensive discussions which occurred in Vienna. Each section of chapters is introduced taking into consideration the specific discussions in the working groups and plenary sessions.

CONFERENCE THEMES AND DYNAMICS

The issues in front of the Conference were divided, for convenience, into three sections: the problems of environment and development, the scientific understanding of the Earth System, and contributions of science to environment and development strategies. Teams of authors were invited by the Conference Advisory Committee to prepare a chapter on each of sixteen themes. Co-authors were selected from different parts of the world (in most cases one from a developed and another from a developing country) with complementary expertise and experience. These teams of authors were brought together at the headquarters of IIASA in July 1991 to discuss their own chapters and also to discuss the interconnections between the themes and ensure that the broad-range of highly complex issues were dealt with as fully as possible.

The authors were asked to produce forward-looking chapters, based on an informed judgment of the present state of the science – what is known for certain, what can be anticipated with a high degree of probability, what is not known and what needs to be known – and then to identify future priority issues and suggest how science can help to resolve them. Authors were given the task of covering their respective themes in the space of twenty pages, and their chapters were prepared in advance of the meeting, so that they could be sent to participants early for study. Immediately after the end of the Conference these chapters were revised in the light of the discussions and it is these revised chapters which are found in this volume.

Participants were also stimulated in their thinking about the issues of the Conference by two events which were scheduled on the first day of the meeting: a series of **key-note addresses** which are reproduced in full in this volume, and an evening **Panel Session on Science and Policy** at which seven invited distinguished persons spoke of their own experiences in bringing science closer to the policy process. The panel, which was open to the public, was moderated by Alvaro Umaña, former Minister of Environment of Costa Rica, and other panelists were: Bert Bolin, Chairman of the Intergovernmental Panel on Climate Change (IPCC); Robert Corell, Assistant Director for Geosciences, National Science Foundation, USA; Maneka Gandhi, former Minister of Environment of India; Nay Htun, Director of Programmes and Special Adviser to UNCED; F. Kreutzer, former Minister of Environment of Austria; and Djibril Sène, former Minister of Higher Education and Scientific Research of Senegal.

Panelists agreed that scientists have a responsibility to continue to pursue their scientific work but also to seek an increased dialogue with policy-makers in order to provide to them independent and objective assessments on the problems of environment and development. Such assessment cannot be done by one scientist alone, but must be the result of a collective effort which should also provide advice on establishing priorities for science. The importance of involving a wide community of scientists (from the natural and social sciences) in partnership with others, particularly UN bodies, governments and industry, was stressed. Panelists highlighted the responsibility of the scientific communities in both the North and the South to work together in efforts to find solutions to environment and development problems which face the entire planet and to develop appropriate partnerships for this purpose.

Because of the emphasis on synthesis and inter-disciplinarity, the Conference dynamics required that the sixteen working groups each include not only those whose primary expertise was in that particular subject but also others whose expertise lay elsewhere but whose special perspective could enhance the discussions and ensure that the interlinkages were taken into full account. Accordingly, participants were asked to commit themselves to a full week of attendance. Again, it was the Conference dynamics that demanded this: if the aim of the building process at the heart of ASCEND 21's design was to succeed, all participants needed to be party to the development of the recommendations as they proceeded from theme chapter and working group through section synthesis to the overall recommendations at the close. Each participant was assigned to a working group in each of the three sections, to assist in the group objective of reviewing the chapters presented, adjusting them if necessary, and formulating recommendations for future research needs and action. After each set of working groups a Section plenary was held with the purpose of synthesizing the conclusions and recommendations of the Section as a whole.

OUTPUTS OF ASCEND 21

The ASCEND 21 Conference provided a unique platform for the scientific community to reflect on the issues of environment and development and on the role of science in finding solutions to these. It underlined the classical responsibility of all scholars to undertake research, to publish the results and to explain the implications for society. It asserted that it is especially important to draw attention to the implications of what is not known: indeed, the precautionary principle advocates that complex, inadequately understood systems should be protected against disturbances until further investigation has provided a scientific understanding of their vulnerability. The Conference also provided a model for a future *modus operandi* in which assessment of scientific results serves as a basis for predicting impacts and formulating policy options, leading to authoritative advice on the basis of international consensus among scientists.

Direct outputs of ASCEND 21 were:

(a) contributions to the finalization of the chapter on Science for Sustainable Development for the UNCED "Agenda 21";

(b) a Conference Statement dealing with the issues of environment and development and proposing recommendations for action, and

(c) recommendations for the revision of the Conference chapters for publication in this volume.

All of these provide important inputs to the formulation of a program of action in environment and development by the scientific community leading up to and reaching beyond the 1992 UNCED Conference.

The Conference was characterized by a positive spirit of culturally neutral scholarly collaboration aimed at achieving analysis and recommendations for research and other action. In its final statement, participants recognized a number of essential problems, identified certain courses of action in which science could contribute, and issued a set of recommendations which were at the same time destined for consideration by the Rio Conference and intended to be discussed by the scientific community and shaped into its own action program for the coming years.

The text of the Conference Statement was finalized in the days after the closing session of ASCEND 21, taking into account the written and oral comments made by participants there. It was immediately made available to the UNCED Secretariat for incorporation as appropriate in its preparations for the Rio Conference. The text has also been circulated to the entire ICSU membership, and National Members in particular have been urged to ensure that the Statement is brought to the attention of the national delegations to the Rio Conference so that where relevant the conclusions and recommendations of ASCEND 21 can help in the shaping of national positions regarding the scientific issues which should have a central position in the discussions and negotiations to take place in June 1992 in Brazil.

In his closing remarks at the Conference, the President of ICSU, M.G.K. Menon, announced ICSU's intention to consolidate the co-operation and coherence of ICSU's major international research programs and to promote the collaboration with relevant programs outside ICSU; strengthen ICSU's role in the evolving partnership among science, government, intergovernmental organizations, business and industry; strengthen ICSU's capacity to prepare objective scientific assessments; report on scientific issues to the general public and decision-makers; strengthen its own activities in capacity-building, and help review the implementation of UNCED's "Agenda 21" following the Rio Conference.

Conference Statement

EXECUTIVE SUMMARY

The International Conference on an Agenda of Science for Environment and Development into the 21st Century (ASCEND 21) was convened by ICSU in Vienna during the last week of November 1991. The results of ASCEND 21, the first international conference of its kind, will serve to make a major contribution to the formulation of the future directions of world science, as well as to the preparation of the 1992 UNCED.

ASCEND 21 stressed a **new commitment** on the part of the international scientific community as a whole to work together so that improved and expanded scientific research, and the systematic assessment of scientific results, combined with a prediction of impacts, would enable policy options in environment and development to be evaluated on the basis of sound scientific facts.

Furthermore, it forcefully asserted the **responsibility of science** (encompassing the natural, social, engineering and health sciences), to provide independent explanations of its findings to individuals, organizations and governments. In this context, ASCEND underlined the central importance of the precautionary principle, according to which any disturbances of an inadequately understood system as complex as the Earth System should be avoided.

Members of the scientific community participating in ASCEND agreed on the nature of the major **problems** that affect the environment and hinder sustainable development, and identified a number of specific areas through which the scientific community could begin to tackle those problems considered by ASCEND as being of the highest scientific priority: population and *per capita* resource consumption; depletion of agricultural/land resources; inequity and poverty; climate change; loss of biological diversity; industrialization and waste; water scarcity; energy consumption.

ASCEND **recommended**:

(a) intensified research into natural and anthropogenic forces and their inter-relationships, including the carrying capacity of the Earth and ways to slow population growth and reduce over-consumption;

(b) strengthened support for international global environmental research and observation of the total Earth System;

(c) research and studies at the local and regional scale on: the hydrologic cycle, impacts of climate change; coastal zones; loss of biodiversity; vulnerability of fragile ecosystems; impacts of changing land use, of waste and of human attitudes and behavior;

(d) research on transition to a more efficient energy supply and use of materials and natural resources;

(e) special efforts in education and in building up of scientific institutions as well as involvement of a wide segment of the population in environment and development problem-solving;

(f) regular appraisals of the most urgent problems of environment and development and communication with policy-makers, the media and the public;

(g) establishment of a forum to link scientists and development agencies along with a strengthened partnership with organizations charged with addressing problems of environment and development;

(h) a wide review of environmental ethics.

In his closing speech at the ASCEND Conference, the President of ICSU announced ICSU's intention to: consolidate the co-operation between, and coherence of, ICSU's major international research programs; strengthen ICSU's role in the evolving partnership among science, government, IGOs, business and industry; strengthen ICSU's capacity to prepare objective scientific assessments; report on scientific issues to the general public and decision-makers; strengthen its own activities in capacity building and help review the performance of Agenda 21 after the UNCED.

THE NATURE OF ASCEND 21

1. The International Conference on an Agenda of Science for Environment and Development into the 21st Century (ASCEND 21) was held in Vienna, Austria, from 25 to 29 November 1991. The objective was to define the agenda of science (a term used as encompassing the natural, social, engineering and health sciences) for the next 10 to 20 years relating to environment and development. The Conference was convened by the International Council of Scientific Unions (ICSU) in co-operation with several other organizations, and hosted by the Government of Austria.

2. The Conference recommendations provide:
    (a) a perspective for the future direction of international science in the field of environment and development, and in related areas;
    (b) a contribution by the scientific community to the United Nations Conference on Environment and Development (Rio de Janeiro, June 1992).

3. Some 250 invited persons from nearly 70 countries attended the Conference. Participants included natural scientists, health scientists, social scientists and engineers. Approximately half of the participants came from developing countries in Africa, the Arab region, Asia and Central and South America. The Conference was organized around three Sections, each meeting in parallel working sessions and followed by a plenary session to synthesize and draw up recommendations. The three Sections dealt with:
    (a) problems of environment and development;
    (b) scientific understanding of the Earth System;
    (c) contributions of science to environment and development strategies.

4. The background chapters, together with the summaries of the discussions at the parallel working groups and the presentations and discussions at the plenary sessions, will be published in April 1992 by Cambridge University Press on behalf of ICSU.

5. Stimulated by the call for ICSU to contribute to the United Nations Conference on Environment and Development (UNCED), and by the precedent set by the 1990 Bergen Conference, ASCEND 21 was an important landmark in the history of international science.

6. ASCEND 21 provided a unique platform to start a long-term partnership between the natural and the social sciences. It underlined the classical responsibility of all scholars to undertake research, to publish the results and to explain the implications for society. It asserted that it is especially important to draw attention to the implications of what is not known: the precautionary principle rests on the need to avoid disturbing a system that is inadequately understood. The Conference provided a model for a future *modus operandi* in which assessment of scientific results serves as a basis for predicting impacts and formulating policy options, leading to authoritative advice on the basis of international consensus among scientists. The Conference was characterized by a positive spirit of scholarly collaboration aimed at achieving culturally neutral analysis and recommendations for research and other action.

THE PROBLEMS

7. Achievement of sustainable societies requires an improved understanding of the complex forces which generate global environmental problems, and hinder social and economic development. The impacts of population pressure, over-consumption, land degradation, deforestation, climate change, loss of biological diversity, industrialization, waste accumulation, as well as water and energy use, on human society and the environment create major problems for which the skills of the scientific community will be critical in clarifying remedial policies and options for action. Scientists and technologists cannot by themselves solve the problems, but they can supply knowledge and informed opinion, as well as technical know-how, for consideration by governments and society. They can also assist in devising solutions.

8. The world faces serious inequities – economic, political and environmental – between countries and within countries; the causes of these are complex. Along with high *per capita* consumption rates there is a continuous and unacceptable process of impoverishment and deprivation aggravated by high population growth rates. The landless and the homeless are extreme examples. To survive, they almost inevitably compromise the sustainability of the natural resources available to them.

9. The precautionary principle was endorsed by ASCEND 21 as providing a scientifically sound basis for policies relating to complex systems that are so poorly understood that the consequences of disturbances cannot yet be predicted. According to this principle, highest priority should be given to reducing two greatest disturbances to Planet Earth: the growth of human population and the increase of

resource use. Unless these disturbances are minimized, science will become powerless to assist in responding to the challenges of global change, and there can be no guarantee of sustainable development.

10. It is urgent to address the problems of growth in world population. Much can be done on the basis of existing knowledge: improving living standards and education, particularly of women; reducing infant mortality; providing acceptable means of contraception and developing other appropriate methods. Further scientific research is also needed to provide additional means of reducing population growth.

11. Policies for the management of *per capita* resource consumption in the industrialized countries will have to be focussed on energy and materials efficiency, recycling and economic regulatory mechanisms. There is also urgent need for behavioral research on ways to encourage alternative consumption patterns more compatible with global resource changes.

12. Recent decades are remarkable for the increasing rate and scale of transformation of the environment, with effects both locally and in distant regions of the globe; e.g. the stratospheric ozone hole in Antarctica resulting from CFCs released mainly in the Northern Hemisphere. But these are only early warnings. Scientists are persuaded that if humankind does not alter its behavior and priorities, unprecedented crises may ensue within the lifetime of a half of the world's present population, arising from such changes as:

    (a) world population doubling to 10 billion in only 35 years;

    (b) migration and urbanization, assuming dramatic proportions, with notable consequences on coastal zones;

    (c) continuing rise of energy consumption exerting increasing pressures on the global ecosystem;

    (d) climate change, sea level rise and associated impacts on the biosphere;

    (e) irreversible loss of a substantial part of the total number of living species;

    (f) continued reduction and deterioration (including chemical pollution) of quality of the natural resource base including the exhaustion, degradation, salinization and loss of a major proportion of the world's soils;

    (g) growing and widespread water scarcity.

13. Some countries are using up the stores of their natural environmental capital. This includes the excessive depletion of agricultural and forest resources, such as the transformation of prime agricultural land for urban and industrial development, the burning of fossil fuels at high rates and the pollution of water, land and air through toxic and other wastes. Trade in raw materials and natural resource products has added an international dimension to the depletion of the environmental endowment of humanity.

14. The factors conditioning land use and degradation are closely linked to population and resource use generally. Although there is no reliable picture of the extent or costs of land degradation or enough knowledge of the institutional arrangements needed to improve land use management, it is presently doubtful whether available land is sufficient for meeting growth in agricultural and non-agricultural demand world-wide projected for the next 35 years.

15. There is a disparity between energy consumption of the poor and of the rich. The production and use of fossil fuel-based energy in the industrial countries causes trans-boundary pollution in the form of acidification and the build-up world-wide of atmospheric carbon dioxide, which has dangerous consequences for global warming. In view of the rapidly increasing demand for the heating, cooling, lighting and power services provided by energy, it will be essential to plan a new, more sustainable, more efficient energy supply system which delivers these services using less primary energy, generating less pollution and narrowing the rich–poor energy gap.

16. Although industrial activities have hitherto been associated with environmental pollution, modern industry is becoming waste conscious. In some sectors the current trend towards energy and raw materials efficiency and "clean" production may be the precursor of a new era of industry, operating in a much more sustainable mode. The importance of clear and stable regulatory and financial signals from governments as a critical factor during such a transition phase needs to be emphasized. However, a crucial question for further research is the amount of time available for restructuring before environmental stress becomes widespread.

17. A significant portion of humankind, especially in developing countries, suffers from poor health owing to a variety of factors associated with poverty. Morbidity, especially in children, is closely linked to malnutrition, and environmental contamination with significant incidence of communicable diseases and avoidable disabilities. Preventive primary health care, adequate nutrition and an improved environment are central for long-term solutions.

18. In the framework of these issues, problems of environmental ethics are increasingly being discussed by a wide range of citizen, religious and scientific groups. Environmental ethics embrace many

questions, including intergenerational equity and the ethical responsibilities of citizens and scientists in respect of the environment.

WHAT CAN SCIENCE DO?

19. Earth is the only celestial body where life is known to occur and the only known habitable planet for man (the term "man" being used generically to cover all human beings). This will only endure as long as the planet maintains its unique life support function. All the three compartments of the land, oceans and atmosphere are connected with one another through the web of the interlocking biogeochemical cycles of water and nutrients. Together they form the Earth System that has evolved towards its present state of high complexity in continuous interaction with the biota that inhabit it, including man.

20. Systematic investigation on a global scale has only recently become feasible. But it has already been shown that our species has brought about alterations in the planetary life-support system which put the sustainable development of present and future generations at considerable risk. Enough is also known to identify – at least qualitatively – the main mechanisms (population, resource use, agriculture, industry) of human intervention, and to consider some of the impacts, and response strategies, including precautionary action needed to avoid disturbing an inadequately understood system as complex as the Earth System. This development marks the transition of man's status from that of a passive product of evolution towards a position in which humankind attempts to use its intelligence in the evolutionary process that includes itself.

21. Research results indicate the vast complexity of the system with its many dynamic interconnections and control mechanisms which display great variations in sensitivity: the relatively small increments made by man to the CO_2 flux into the atmosphere have a global climate impact, whereas man's doubling of the fluxes of sulfur and nitrogen compounds into the atmosphere has resulted in more regionalized impacts such as acidic deposition. We know that the collective, integrated mass of biota on Earth are the prime pumps in the major biogeochemical cycles which are a vital part of the planet's life-support system, and we must gain more understanding of these interactions if we are to be able to manage the systems sustainably.

22. The lack of knowledge of control mechanisms and their sensitivity to disturbance must prepare society for surprises, which means that methodologies for impact assessment must closely follow research as it advances. For example, the impacts of natural disasters can be considerably reduced by focussed Earth System research and technological solutions (e.g. through IDNDR). Such disasters, however, have effects directly related to high population densities and poverty.

23. Disturbances of the Earth System are most often the result of a sum of local actions while changes on the global scale, in turn, have results that may be local as well as global. Thus global change studies require iterative focussing on the global, regional and especially local scales at which changes are experienced most acutely.

24. Many of the institutional arrangements for embarking on systematic interdisciplinary investigation of the Earth System are now in place (e.g. GCOS, GOOS, HDGECP, IGBP, IHP, MAB, SCOPE, STEP, WCRP, etc.) and they are capable of delivering new data and more detailed model-based simulations, and hence new understanding and useful foresights to the extent that support is given to these programs.

25. One bottleneck appears to be the lack of scientists and engineers, and a supportive research context. This is particularly severe in certain disciplines and regions. In view of the lesser number of scientists in developing countries, focussed capacity building is necessary to permit world-wide balanced participation in research as well as to enable developing countries to strengthen their scientific capabilities to solve local problems and to participate on the international scene, including in the negotiations of relevant international conventions. Existing regional networks for research and training should be supported and new ones established. A recent example is the initiation of the Global Change System for Analysis, Research and Training (START).

26. The process of understanding the Earth System is now beginning and, if intensified, will produce valuable results for sustainable development in the first decades of the 21st Century. It will be of great importance to take stock periodically of advances made and of new priorities that emerge both from results obtained and surprises encountered and to share these with policy-makers and the public at large.

27. This is not the only role of science in sustainable development. Since it is now already abundantly clear that "more has to be done with less", science and technology have a crucial role in raising productivity so as to make this possible. The challenge of shifting technological growth onto an ecologically sound path includes applying the precautionary principle without endangering the fulfilment of basic needs so as to achieve improved response strategies.

28. Science plays an essential role in the search for pathways to produce goods and services without

degradation of the natural environment through:

(a) better understanding of the Earth System, of its carrying capacity and of environmental change;

(b) scientific support for determination of priorities between competing development endeavors and assessment of economic and ecological tradeoffs;

(c) improved ways to cope with natural and man-made disasters and their effects, such as storms, floods, drought and war;

(d) economic analyses that embrace ecological factors and internalize environmental costs and benefits in the long term;

(e) engineering studies for technological innovations that raise the efficiency of resource use and limit discharges;

(f) identification of development–environment interactions and emerging problems in specific areas.

29. Science can help resolve development issues on local, regional and global scales. The separate institutions charged with addressing these must work together because the different scales interact. The larger-scale environmental picture must be considered as an essential overlay when addressing local and sectoral development problems. The Brundtland challenge to consider the welfare of the next generation calls for a lengthening of the time scale over which prediction is required and an intensification of the efforts to reduce uncertainties still further.

30. One of the contributions of science lies in its power to predict how the environment will change both through natural fluctuations and in response to man's activities. Such predictions will include not only global average conditions but, more importantly, regional variation due to both global and local pollution of the atmosphere and ocean and their natural variability. Operational forecasting systems derived from research are needed to produce environmental predictions.

31. The role of the sciences in effectively linking environment and development is vital if the political and economic transformation of development patterns world wide is to be put on a sustainable basis. This may call for science itself to become a more active partner in guiding development. A recognition of the link between science, environment and development would also be an opportunity for scientists in developing countries to raise their standing in society. Indeed, many people and governments, especially in developing countries, do not yet see any "practical" use of science and therefore tend to under-use it. A visible application of science to environment and development should certainly help in this context.

32. Science must seek to improve public awareness of scientific principles and encourage greater

participation to ensure that public concerns and indigenous understanding are taken into account. By the 21st Century science should be more actively involved in helping to shape development in ways that guarantee environmental security for present and future generations.

33. The Conference confirmed that the scientific community is willing and able to enter into partnerships with organizations charged with addressing problems of environment and development, in such areas as:

(a) Understanding the natural resource system

(b) Availability of relevant data

(c) Informing society

(d) Understanding human behavior in response to global and local change

(e) Laws relating to resources, environment and property (including patents)

(f) Environmental ethics.

RECOMMENDATIONS

Research and Monitoring of the Earth System

34. Research attention is needed to discriminate between environmental changes due to **natural fluctuations and those induced by anthropogenic forcing**, as a pre-requisite for early warning of change.

35. The international **global environmental research programs**, WCRP and IGBP, provide a clear strategy for research aimed at improving understanding of the total Earth System. The goals of these programs should be vigorously pursued through related national projects.

36. There is a need for research directed towards improved understanding at the local and regional scales, of the **hydrologic cycle**, including its interaction with soil and vegetation, and its response to changing land use and pollution, environmental degradation and rising industrial, agricultural and domestic demands, particularly in semi-arid regions. There is also a need to develop technology designed to raise efficiency of water use.

37. In order to establish more clearly the likely **impact of climate change** on natural and managed ecosystems and on society it is necessary to improve predictions of the regional characteristics of environmental change.

38. It is necessary to intensify interdisciplinary research on **coastal zones** where 75% of the world's population will live by the year 2000, and which will be receiving strongly increasing loads of nutrients and pollutants from the land, endangering its major role in marine food production through traditional fisheries and mariculture.

39. There is a need for research on the rapid and irreversible loss of **biodiversity** as a basis for developing appropriate management tools, with attention to strengthening human and institutional capabilities in critical areas. Because biodiversity depends on habitat, research should concentrate on biosphere dynamics, including anthropogenic influences, one of the goals being to promote restoration of degraded ecosystems.

40. Attention should be paid to integrated regional studies of **vulnerability in fragile ecosystems** (e.g. in mountains) or where environmental degradation threatens human well-being and capacity to respond. These studies should emphasize the need to determine ways to promote ecosystem integrity.

41. There is a need to define and assess the **carrying capacity of the Earth** at all scales; to find acceptable ways to slow population growth; to reduce overconsumption; and to examine alternative consumption patterns and lifestyles.

42. There is a need to establish **operational systems for environmental forecasting** (based on research models and supplied by new kinds of data) to predict how the environment will change globally, regionally and locally in the atmosphere, oceans, coastal seas and the biosphere.

43. All components of the Earth System (including atmosphere, ocean, land surface, cryosphere, hydrosphere and biosphere) must be monitored on a global basis and in sufficient detail as a basis for detecting and predicting environmental change. This requires support for substantially enhanced **programs of Earth System observation** planned under the WWW, GAW, GEMS, GCOS and GOOS. Appropriate means need to be developed for the management of data obtained from those observing systems, including, storage, analyzis, quality control, dissemination, free exchange between researchers and assimilation into numerical models.

Research on Impacts and Responses

44. Research is needed to predict the **impact of changing land use** on the environment and on options for land development and increasing productivity. Account needs to be taken of the widest possible range of factors, such as urbanization, land allocation, loss of soil fertility, agriculture and forestry and the ability to sustain biodiversity. Attention should be given to classification of landscapes in terms of sustainability.

45. Further research is required on the **impact of waste and pollution on public health and on ecosystems**, in particular on alternative modes of avoiding and controlling solid, liquid and gaseous pollution,

including the use of economic mechanisms, through increasing process efficiency and waste recycling.

46. Research and development is needed on the **epidemiology** associated with dietary patterns and nutritional states, the production of new vaccines and the social factors controlling their distribution and use. Improved disease and health indicators need to be designed and implemented, and the factors affecting the spread and control of old and new diseases should be researched.

47. An interdisciplinary research program is needed on the technical, economic and socio-political implications of timed and costed transitions to a more efficient **energy** supply system and lower energy-intensive pathways in industry, agriculture, and transportation.

48. There is a need for improved yardsticks to measure quality of life and to define alternative development objectives which contribute to sustainable use of resources. Improved **monitoring is needed of social and economic processes**, in particular as they relate to population growth and modes of consumption, settlement patterns, health, epidemiology and related factors.

49. Attention is needed in future research on **human attitudes and behavior** as driving forces central to an understanding of the causes and consequences of environmental change and of human responses to them.

Science and Policy

50. There is a need for increased **integration of the natural and social sciences** in order to address the issues of environment and development.

51. In devising and implementing development strategies, existing scientific knowledge often goes unused. Natural, social and engineering scientists, working through their professional organizations, should regularly produce **appraisals** of the more urgent problems in environment and development.

52. An international forum should be established to facilitate improved **links between scientists and development agencies**.

53. **Current scientific capacities** are inadequate to respond to the challenges posed by the crises of development, environmental degradation and rapid global change. Special efforts are required in education and in building up scientific institutions in particular in developing countries in order to allow them to join in international environmental programs. New methods and approaches must be developed to facilitate the participation of all sectors of society in environmental matters in the transition to sustainable

development. The role of women in relation to the environment should be recognized in this context.

54. The **organization of science** should be adapted in order to to cope with the explosive growth in scientific literature, meetings and institutions, to encourage creativity and to reduce the inequalities in access to and distribution of scientific literature and information.

55. Increased efforts should be made by scientists to **communicate** with policy-makers, the media and the general public about the implications for society of the results of their research.

56. Continued efforts should be made to build bridges between leaders of **science, business and industry** and consumers.

57. A mechanism needs to be established to examine the effectiveness of the **institutional arrangements** of science for the furtherance of the environment and development process discussed during the ASCEND 21 Conference.

58. Further development of institutional arrangements for the **stewardship of the World's Commons** including the ocean beyond exclusive economic zones, should be based on the best available scientific knowledge.

59. The international scientific community should undertake a wide review of **environmental ethics** related to such issues as: the intrinsic value of nature; environmental rights of citizens; intergenerational rights; communal rights for common properties (atmosphere, oceans, etc.); environmental codes of conduct; and ethical responsibilities of scientists and of the world scientific community.

Opening Remarks by Franz Vranitzky
Federal Chancellor of Austria

Ladies and Gentlemen,

The Austrian Government is deeply conscious of the fundamental importance of the United Nations Conference for Environment and Development (UNCED 92), and deeply committed to make this conference a success. On this premise, we are equally committed to its thorough preparation and try to be as actively involved in it as possible. It is thus a great pleasure for us to host ASCEND 21, the preparatory conference organized by the International Council of Scientific Unions, and I welcome all of you, Ladies and Gentlemen, most warmly to this event here in Vienna.

It is 7 months till UNCED 92 will convene in Rio de Janeiro. Some will argue that this is ample time for a thorough preparation of the Conference. True enough. On the other side, however, it is not so much if you measure it on the great – even if sometimes very divergent – expectations which this Conference has provoked in the international community.

It would certainly be premature to venture a forecast of the results of UNCED at this point. One thing nonetheless is certain: it will leave a strong imprint on this decade – either as an instrumental event and a starting point for a new global co-operation, or as a memento for a great chance that was lost.

Ecological damage in the industrialized world is very often the result of an over-expansion of industrial and technological growth. Ecological damage in the countries of the Third World very often is the result, but also the root cause of poverty.

We cannot ignore that developing countries very often are forced into a damaging exploitation of their resources, just in order to survive. Nevertheless, they have to maintain their place in a world in which the gap in wealth that separates them from the North is continuously growing, and in which the industrialized states are using up an ever-increasing portion of our global ecological capital.

It will be the task of the preparatory work as well as of the UNCED Conference itself to forge this verbal consciousness into a concrete agenda acceptable to all. The chance for a success is great. After all, the end of the cold war and of super-power rivalry, the diminishing threat of nuclear war, the substantial successes in disarmament negotiations, they all have changed the political environment, in which UNCED 92 will take up its work, for the better. Nonetheless, we also have to be aware that the road before us will be steeper and more difficult than the one behind. It will be our task to embark on this road with great determination and seriousness.

Not a blue-eyed, uncritical and unchallenging belief in technology is called for, but a well-balanced co-operation between economy and ecology, between science and technology is the order of the day. And this order is addressed to scientists and politicians alike – many of the problems which confront us today have their roots not in scientific, but in political or economic decisions.

An enormous scientific and technological potential has been placed in our hands. If used wisely and with circumspection it will open great perspectives for the future of mankind. If we manage the transition to intelligent production, to a production of more with less, of more goods with less resources, less energy, less waste, than we will be able to meet the demands of a solid and durable development.

The high level of science and technology almost naturally gives rise to the belief that each and every problem in our natural or in our man-made environment can be solved by proper analysis and suitable corrective measures. A number of unexpected setbacks, however, have taught us that this is not always, it is not necessarily the case.

But this is not enough. Sheer acceptance of new values does not promote change. It does not change the basis of international economic relations, it does not break the acceleration of the global environmental crisis, it does not create the necessary new institutional framework.

UNCED 92 will be the test-case whether, and to what extent, the international community will be able to convert its words into concrete action. All our joint efforts must aim at creating a new model of wealth and well-being, way beyond the old paradigms.

During the last centuries science has proven its extraordinary capability to change and to shape the world. Thus, it has also chartered its prominent role in coping with the new challenges for our time, for our generation and for future generations.

ASCEND 21 will be a significant stage in this process, as – and I said once before – UNCED 92 will be only as successful as the quality of its preparation permits. I wish all of you an interesting, stimulating and challenging conference, as well as a very pleasant stay here in Vienna.

Message from Gro Harlem Brundtland

Prime Minister of Norway

Throughout human history, responsibility for the future of our own children and grandchildren has always been inherent in human nature. It has been the pride of the present generation to leave a heritage on which the next one can bring society forward. This was possible in earlier generations when the future appeared stable and predictable.

Today, the future appears neither stable nor predictable. We all know the signs of the global crises approaching – signs that represent threats to vital support systems for life on Earth. These trends must be reversed to ensure that there will be a future world worth living in.

Science and technology play an integral part on the interaction between mankind and environment. On the one hand, it has been part of the problems facing us. It is our scientific discoveries and our technological knowledge which has given us the power to modify nature and the capability to destroy life on Earth.

Science and technology must now become part of the solution. Science can give us warning and time to act in fields which are invisible to the human eye, such as the threat to the stratospheric ozone layer and to long-term climate change. Science can give us new insights in how we can turn around present processes, and it can give us new tools to repair damage already done. Rapid advances in the development of cleaner environmental technology give us hope that we can reverse present negative trends if we put our hearts and best minds to the task.

The World Commission on Environment and Development concluded on sustainable development as the basic strategy for global change. Sustainable development is more than a concept. It is more than a policy. It is above all a process of change, in which the exploitation of resources, the direction of investments, the orientation of technological development and institutional change are made consistent with future as well as present needs.

We did not claim that such a process would be easy or straightforward. We are still on a learning curve on how we can achieve a sustainable world society. We will have to adjust our goals and our methods along the way, as we acquire more scientific knowledge about the state of the environment and more experience in our attempts to integrate ecological and economic concerns.

The dramatic changes in international politics over the last few years have created an increasing opportunity. Indeed a historical opportunity, to change the way the increasing interdependences are met. This opportunity must not be lost.

We need to build new global coalitions in the efforts to save our common future. We need a new partnership between North and South to stage a committed attack against world poverty, and to enable the countries of the South to join the global agreements now needed. We need a new partnership between governments, the private sector and the scientific community to achieve the transition to a sustainable world society.

The United Nations Conference on Environment and Development in Brazil next June must become an expression of a shared global vision. It must also become an instrument for translating this vision into concrete actions on the issue before us.

Science has a central and legitimate role to play in the UNCED process. I am encouraged by the initiatives taken at the Bergen Conference in 1990 and the Vienna Conference in 1991 to mobilize the scientific community to join the partnerships now needed. The uniqueness of the Bergen and Vienna Conferences is that they bring in interdisciplinary scientific groups to identify the research needed to grapple with genuinely sustainable development on a world scale.

I feel certain that both the preparations and the conduct of ASCEND 21 will have a positive effect. We need the assistance of the scientific community to help us address the vital questions of our time such as: why is the Earth being degraded? What are its resiliences and vulnerabilities? What is the role of human behavior and how does it need to change?

We face challenges of new dimensions. Climate change, which goes to the core of growth pattern and lifestyles in the industrialized world and to the heart of the North/South-gap, is probably the most difficult issue we have dealt with. It is a fundamental part of the economic life and industrial level of states. To meet such challenges, we must develop a new generation of environmental agreements and international regimes which seek maximum environmental benefit at a minimum cost.

Never have our knowledge and capacity to address vital challenges been greater. For science to make maximum impact on the societies it must interact with politics, with democratic debate, and it must be geared towards defined needs. It is the responsibility of scientists to take an active part in the shaping and directing of our common future. To make full use of human knowledge, we need better interplay between science, politics and public opinion. If we succeed in forging this alliance, we can make the necessary changes and offer concrete solutions.

I very much welcome your involvement in the UNCED process. I believe ASCEND 21 testifies to the genuine willingness of the international science community to take increased responsibility for our common future, and I wish you a successful Conference.

Address by M. G. K. Menon
President, International Council of Scientific Unions

We meet at a time which must be regarded as an important turning point in history: to redefine the direction of further human development. The question is whether we will recognize, in time, the challenges and the opportunities that lie before us – for we need to chart new pathways that can lead to sustainable progress, on an equitable basis, for all of humanity. We meet here together, as natural and social scientists, because we know that we have a responsibility, and a role to play, to bring to the notice of the governments of the world, when they meet next year at the UN Conference on Environment and Development in Rio de Janeiro, the possibilities that can lead to sustainable development.

Until recently, humankind was a relatively small entity on what appeared to be a very large home – the Earth with its seemingly infinite life-support systems. We are now becoming aware that this may not be the case.

The progress of the human race since its earliest beginnings has been marked by the use of the brain and its great powers of logical reasoning. Human creative genius, marked by tremendous innovation, has led to the great achievements, cultures and civilizations of which we can be truly proud.

In particular, we have witnessed over the past few hundred years the onward revolutionary march of science. While the development of techniques during earlier periods was empirical, in recent times, scientific advances have given rise to wholly new technologies and revolutionized old ones. Technology, in turn, has made the advance of science more rapid. There is now a symbiotic and synergistic relationship between science and technology. These developments have made possible the modern industrial societies, characterized by high levels of affluence and material standards of living. We see all round the manifestations of these great powers of science and technology: in the generation and use of energy; in transport and communication systems which have made the world a small place, now frequently referred to as a global village; in the small numbers of the populations that are engaged in developed countries to produce food needed by all, and much more; in the advances of modern medicine; in our greatly increased understanding in many basic areas of knowledge relating to the microstructure of matter and reaching out to the far reaches of the universe characterized by astrophysics and cosmology; in our understanding of the structure, functioning and development of living, self-organizing systems that have given use to the new biotechnologies; in the spectacular growth of electronics and informatics; in the new materials technologies; and much more.

Apart from this continuing advance of science and technology, and all that it has given rise to, we have also witnessed in recent decades another growth – that of population. From a population of 3 billions in 1960, we will shortly be crossing double that figure, namely 6 billions by the turn of the century; this has happened in less than four decades. This increase in population has taken place in the developing countries; and has been due to various factors. The advances of modern medicine have led to the conquest of many diseases, resulting in greatly reduced death rates and increase in life expectancy. At the same time, primarily because of poverty and lack of human development, birth rates have not fallen correspondingly. The developing countries are faced with a vicious circle: lack of development leads to a situation of high population growth rates, while the latter acts as a brake on achieving greater development. Clearly a major effort or input is called for to cut this vicious circle. In the absence of this, continuing population growth will occur, which is an irreversible process.

The progress of science and technology has thus led to modern industrial societies, characterized by affluence and high consumption patterns; and to high population growth rates in developing countries. These are the two driving forces of environmental stress. Many developing countries, particularly the most populous, have had, for several decades, policies and plans for stabilization of population. Their leaders have spoken eloquently about the gravity and dangers of the situation. In the Report of the South Commission chaired by Julius Nyerere it is stated: " Rapid population growth presents a formidable challenge for most developing countries..." A strong commitment to slowing down population growth through integrated population and human resource planning can bring large personal and social benefits in most developing countries..." The present trends in population, if not moderated, have frightening implications for the ability of the South to meet the twin challenges of development and environmental security in the 21st Century...." In the long run, the problem of over-population of the countries of the South can be fully resolved only through their development. But action to contain the rise of population cannot be postponed... The task is indeed formidable, but the consequences of inaction can be disastrous. The South must summon sufficient political will to overcome the various obstacles to the pursuit of a sensible policy on population." The question is not, therefore, one of awareness – but what to do about it. Development and the motivation that that gives rise to far smaller families, as well as methods that are easy and suitable, are needed.

I would be happy if I could be assured that there is similar concern among the policy-makers and societies of the North about the over-use and wastage of resources that have come with affluence.

We have been conscious for the past several decades of the steady deterioration that has been taking place in the environment. This is mostly related to deterioration at a local and to some extent regional level, in the form of air, water and industrial pollution, as well as through deforestation. This led to the convening of the UN Conference on the Human Environment in 1972 in Stockholm. There has been progress on various issues since then. But over the last two decades, there has been increasing awareness of human activities producing changes more insidious, more global and more long term, with what could be extremely serious consequences for human functioning on Earth. This includes: deforestation and irreversible loss of biodiversity; increase in greenhouse gas concentrations leading to global warming causing climate changes and sea-level rise; ozone depletion; large-scale marine and coastal zone pollution, etc. As a result of the increase in public awareness concerning these, peoples and governments, for the first time in history, are devoting a large fraction of their time to the global environment and the impact of human activities on it, which could produce on a historical time-scale, what may have happened naturally on geological time scales. There have thus been many conferences on these questions over the past few years.

Twenty years after Stockholm, there will be in 1992, in Rio de Janeiro, the UN Conference on Environment and Development (UNCED). This Conference will deal with issues of the Environment – but not by themselves; these will be considered in conjunction with and as part of issues relating to Development. We thus see a recognition of the close relationship between Environment and Development.

It is interesting to recall at this stage that the UN had set up three Commissions in recent years. These were: the Palme Commission to consider matters of disarmament and security – and fortunately we are much better placed now with the Cold War and Super-power confrontation being a matter of history; the Brandt Commission to consider economic issues, particularly the North–South relationships on which there has been little progress; and most recently, the Brundtland Commission on the area of Environment and Development – which brought to world attention the concept of *sustainable development*, that only scientists had been concerned about earlier. It is this issue which is before us today, at this Conference: namely, what is the pathway to be followed which will ensure development for all – for the so-called developed countries, the North, and for the developing countries, the South – and also ensure that this is sustainable in the long term; we need to achieve both intra- and inter-generational equity which is the key to a just and long term social order.

It is equally interesting to recall that, initially, the development of science was motivated by human curiosity to understand more about the surroundings and the Laws of Nature, this will always be the case. But then science began to have the now-seen impact on technology, which has resulted in increasingly rapid applications of great value to society; this is manifested in the world which we see around us today; thus science started to interface more closely with the sectors of production and services. But now, as we look ahead, we see the role that science has to play in bringing about development in a form in which it is sustainable. This represents changing responsibilities in a world of rapid change. This is the challenge before us – and I am confident that we will rise to this. However, for this, as natural scientists, we will need to relate strongly to our colleagues in the social sciences as also with governments, and with intergovernmental and non-governmental organizations. It is this interfacing and way of working together that we have to learn to establish.

For this reason, Mr Maurice Strong, Secretary General of UNCED, requested ICSU to act as principal scientific adviser, and to organize this International Conference, at which we are all present today: An Agenda of Science for Environment and Development into the 21st Century (ASCEND 21). This Conference brings together natural and social scientists to consider: aspects that constitute the driving forces of environmental change, and the manner in which these are changing with time; the present status of our environment and the extent to which these driving forces are altering it; and what we need to do about it in terms of attitudes, creation of awareness and understanding about these issues, building of scientific capacity to deal with these, development of suitable technologies and of needed institutions. It is these issues which are discussed under the sixteen themes of this Conference. We hope to provide an objective independent input of value to UNCED. As we look into these problems on a global scale – which arise as a build-up of activities at a local level – and try to find solutions to these, there will also be the need to work out action plans that can be agreed to internationally, and to act upon these locally, taking note of the various levels of development and of human needs, varied cultures and the like. On these issues we need to think globally; and we need to act globally and locally.

The international scientific community has been involved in efforts to study the environment on a global scale since the first International Polar Year in 1982–83. Specifically, under ICSU there has been the International Geophysical Year (1957–58) followed by the International Quiet Sun Year (IQSY); the International Biological Programme (1964–74), which was the direct predecessor of UNESCO's Man and the Biosphere Programmes; the Global Atmospheric Research Programme (GARP) (1967–80) jointly with the World Meteorological Organization, which was followed by the World Climate Research Programme (WCRP) since 1980; and the various environment-related activities of its discipline-based participating, Scientific Unions and the Scientific Committees it has set up on Problems of the Environment (SCOPE), on Oceanic Research (SCOR), Antarctic Research (SCAR), Solar Terrestrial Physics (SCOSTEP) and Space (COSPAR). ICSU, therefore, has had a long and continuing program on the Earth's environment. It carried out, under SCOPE, a scientific study of the Environmental

Consequences of Nuclear War (ENUWAR). It is today engaged on the International Geosphere–Biosphere Programme (IGBP) on Global Change, the most ambitious wide-ranging international program that it has ever undertaken. With this background, ICSU is naturally concerned about what the scientific community should be doing, as we move into the next century, on questions of Environment and Sustainable Development; a Conference of this nature is something that ICSU would have organized at some time – but it has been arranged now so that we can ensure that outputs from it are available to UNCED.

The dimensions of global change and the driving forces responsible for it are such as to have far-reaching consequences for the capacity of the Earth to accommodate life and to support future generations. We are entering a new stage in the existence of human society that calls for an extraordinary response. These environmental issues are linked to economic development from which they cannot be separated. The issue of technologies needed for economic and human development and the issue of equity within and between nations needs immediate attention. This calls for a spirit of international co-operation – between the North and the South. We have to bend the great powers of science and technology to chalk out new pathways of development which will be environmentally sound and sustainable, rather than subject ourselves to an acceleration of the present trends that can spell disaster. For this there is need as much for moral resources as for physical resources. The international conscience of science must help in the marshalling of such resources through an effort such as we start on today at ASCEND 21.

Address by Umberto Colombo

President, European Science Foundation and Chairman of ENEA, Rome, Italy

THE NEED FOR A WIDELY INTERDISCIPLINARY APPROACH IN COPING WITH GLOBAL PROBLEMS

Throughout the early history of man, the world of "scientia" encompassed in a seamless fabric all human understanding. Renaissance man was a supreme generalist, his mind untrammelled by demarcation between compartments of knowledge. The 17th and 18th Centuries, the time of Galileo and Newton, saw the rise of modern science with the partition of knowledge into broad disciplines, which since have split and subdivided into the ramified system we have today. There was a parallel in everyday life, where the artisan – quintessentially a Jack of All Trades – gave way with the onset of the Industrial Revolution to the specialized worker, just one element fulfilling one role in a much larger productive system.

This retreat from horizontal knowledge probably reached its apex in the decades immediately following World War II. Now, we are more aware of the limitations of a social system founded on specialization and the mass production based upon it. A series of incidents dented what was previously an almost blind popular enthusiasm for technology as the panacea to all man's problems. Changes in the economic paradigm governing Western advanced societies as a result of technological innovation have called for a much more sophisticated system, demanding pluridisciplinary approaches to complex and largely interdependent issues. Thus, mass production has given way to a more personalized approach, quality has become a greater source of value added than quantity, man increasingly focusses on the performance of functions, rather than being satisfied with the possession of goods.

Moreover, modern society has become an intricate mass of highly complex systems and, in the effort to dominate them, man must rely ever more on capabilities deriving from the whole spectrum of his knowledge. In societal organization too, social change both determines and reflects technological and scientific advance. These relationships are best studied in terms of disequilibrium, the dynamics of non-linear phenomena, theories of chaos and catastrophe. The interchange between the social sciences and humanities on the one hand, and the natural and life sciences on the other, is thus expanding exponentially. It becomes most intense and fruitful in the approach to the host of global problems which are now affecting the planet.

Global problems, *par excellence*, demand the harnessing of all of human knowledge. Interdisciplinarity is the only way by which man can first comprehend their significance, and then seek solutions. Of these global problems, special relevance is assumed by the environment–development nexus. The relationship between the protection of global climate and development, a central issue for our Conference, is an outstanding example of the importance of a wide interdisciplinary approach. I shall consider in particular its energy aspects, my own main field of expertise and a problem area at the crossroads of environment, climate and development.

Ever since the discovery of fire, energy has been a major factor in development. Each of the great transformations of civilization has been accompanied by new ways to produce and utilize energy. Conversely, each new source of energy has made possible, or has required, transformations in social organization. By permitting conservation of food and protection from cold, fire created the bases for a more stable and complex society. Draught animal energy was a major component in the advent of agriculture. Refinement of techniques to harness the wind in ships and mills expanded cultural and commercial horizons and made a real contribution to the Renaissance. The Industrial Revolution was grounded on coal, and started in countries with large reserves of this fuel (such as Great Britain and Germany). Electricity stimulated new forms of industry and changed the urban environment, while oil has fuelled the great transformations of industrial countries after the Second World War.

The importance of energy for development is well recognized. In fact, 25% to 30% of the finance made available to the Third World in aid to development programs is targeted to energy projects.

Energy from fossil fuels and from the unsustainable use of biomass is also the main source of anthropogenic accumulation of greenhouse gases (in particular carbon dioxide) in the atmosphere, and therefore of the threat to the stability of global climate. Technologies to reduce the production of CO_2 and its release to the atmosphere do exist, and others could be developed in the short to medium term. They include technologies that raise the efficiency of energy production, transformation and use; non-fossil energy sources like renewables and nuclear; and shifts between fossil fuels, favouring those emitting less CO_2 (in particular, natural gas). There is also some promise of technologies, largely still at the development phase, to capture CO_2 and dispose of it permanently on the sea bed or underground.

We are increasingly aware, however, that the problem of greenhouse warming is not going to be solved by technological fixes alone. Economic conditions limit the range of acceptable solutions. Financial difficulties, especially shortage of capital, hamper the diffusion of technologies with lower running costs but requiring higher initial investment.

Even where economic and financial obstacles are not present, many factors of a non-technical nature have to be overcome in order to modify the energy scenario. Let me mention a few. The most obvious, immediate and economical way to attain a more efficient use of energy is to avoid its wastage. Lighting an empty room, keeping the heating or cooling system on when windows are open, letting the engine of a car idle for a long time in a traffic jam, and so forth, account for a non-negligible part of energy consumption, without adding anything to our comfort. These are habits that have grown up in an epoch of unlimited low-cost energy, which are difficult to eradicate. A major effort in education is needed. A social conscience must evolve, and this brings us into the sphere of ethics. Changes such as these could actually be more effective and realistic than any artificial increase in the price of energy.

Energy-efficient technologies and new sources of energy have been developed mostly in industrialized countries, and respond to the contextual conditions of these countries. Energy technologies, in fact, are often highly specific. They depend on the availability and quality of energy sources, on the specific applications for which they are required, on environmental and climate conditions, on the presence of industrial and other infrastructures, on specialized skills and, last but not least, on the organization of society. Developing countries do not meet most of the above requirements, and a process of adaptation of a technology to their particular needs and conditions is necessary, even if not in itself sufficient. The process must be carried out in large, part by the developing country itself, since it is there that the conditions in which the technology is actually going to be implemented are best understood. It has to be noted, however, that countries importing new technologies often do not have the capability in technology assessment required to evaluate all the social, economic and environmental consequences of the diffusion of a given technology. In fact, many failures in technology transfer (especially in the energy sector) are due to the prevalence of supply-side, over market-driven, considerations.

In particular, the social effects of energy choices, which I have already referred to, should not be underestimated. The introduction of renewable energy technologies, for instance, does not require investments in heavy centralized infrastructure, such as large interconnected electricity grids. Obstacles to the diffusion in developing countries of renewable energy systems may derive, however, from resistance to change in local social structures or from the reluctance of potential users to pay for a commodity (energy), that, when consisting in self-gathered firewood and wastes, is considered as free, even if the quality of the new service offered is far superior. Sociological studies can help identify such obstacles in advance and point to strategies to overcome them.

At the other extreme of technological complexity, obstacles to a further diffusion of nuclear energy (a source that generates virtually no greenhouse gas, and that has relatively few adverse environmental effects under normal operating conditions) are often again of a non-technical nature. One key point in the public acceptability of nuclear energy (which is often the limiting factor) is the perception of risk. Probabilistic evaluation of risk has been rejected by the public. The concept of extremely low probability of a severe accident has little or no meaning for a man ruled by common sense, while its gravity – however unlikely – grips his imagination. Another important problem of nuclear power derives from radioactive waste, especially long-life alpha emitters. Technologically acceptable solutions to cope with this issue have been devised and are actively being pursued. The real issue here is, again, educating the public. The whole problem of the acceptability of nuclear power is not so much technical; much more important is to convince the layman of the significance of the modifications for his perception of risk.

The possibilities for the great majority of the Third World to undergo economic development in the presence of constraints represented by the limited availability of resources and by the exigency to preserve climate stability and environmental quality would be increased by adequate economic and social policies – aiming at accelerating the stabilization of population and supported by interdisciplinary research – were set in place. For the Third World it is imperative to undertake less energy- and materials-intensive paths of development than those of the past. In other words, to "leap-frog" directly to solutions involving technologies which are currently available only in the most advanced countries, and which can be grafted on to traditional technologies, leading to appropriate technological blends. Thus, the concept of appropriate technologies proposed by Ernst Schumacher in the 1960s as an improvement over his previous concept of intermediate technologies can now be better understood if we refer appropriateness to the whole system rather than to the individual technologies.

This radically different pattern of development implies profound social changes and a different perception of values and priorities. In the conventional idea of the industrial society, increasing consumption of manufactured goods is, in practice, a prerequisite for sound growth of the industrial sector and is regarded almost as if it were a moral imperative. This model of a "consumer society" is still the aspiration of many citizens in less developed countries. Communicating the message that the paradigm has changed, that development must be sustainable and therefore sparing, efficient and frugal, is no simple matter. It challenges both human and social sciences. If we in the North intend to continue to preach to the Third World the adoption of a new pattern of development inspired by the concept of sustainability, we have to set an example, but this implies a radical change in our own culture and lifestyle, and a new spirit of solidarity as citizens of the world.

Global problems of the type of anthropogenic greenhouse warming are relatively new. They require new approaches and new tools in international relations. The present start of negotiations for a world-wide Convention including protocols aimed at curbing greenhouse gas emissions, which will culminate in the UN Conference on Environment and Development in Rio de Janeiro next June, is unprecedented, except for the much smaller-scale, though very important, endeavor which led to the Montreal Protocol achieving a gradual ban on CFCs and halons to protect the stratospheric ozone layer. In the case of greenhouse warming, an approach based on a balance of power among different countries, or groups of countries, is clearly inapplicable. Consideration of respective rights and duties is extremely difficult in a situation which sees the costs and benefits of a preventive policy to protect climate not only very unevenly distributed, but also extremely difficult to assess.

Another problem area which may seem less complex, yet is certainly fundamental, is that of water, and I am happy to note that a specific section in ASCEND 21 is devoted to it. I shall limit myself to underlining that this problem too requires a multi-disciplinary approach, making massive use of the natural sciences and engineering skills, but also of the social sciences and decision-making capacities in the political arena. The European Science Foundation has an important project on water resources, drawing on its competence in many different fields. The ESF project studies the possibilities for efficiently sharing scarce freshwater resources among countries and regions of the Mediterranean basin. Special attention is given to exploring the potential for using market mechanisms, such as tradeable water rights, in furthering this goal. Economists from Bulgaria, Greece, Israel, Italy and Turkey are involved in the project, with linkage to the World Bank and the United Nations Environment Programme (UNEP), both of which have already identified scarcity of freshwater (due to physical exhaustion or contamination) as probably the major environmental problem facing the region in the next century.

ESF undertakes considerable work in environmental fields. There is an on-going research project into eutrophication and waste management in the North Sea, and other programs cover such areas as paleoclimatology, arctic geosciences and ecotoxicology. There is ample evidence of how fundamental questions about human behavior can be built into the framework of scientific programs in the environmental sciences. The ESF program "Environment, Science and Society" addresses these concerns in a pragmatic way, focussing on selected economic and institutional issues. It seeks to enhance environmental economics as an established area of scientific enquiry on the international level. Theoretical work is being promoted which involves assigning economic value to environmental resources. All this is quite challenging, as it means recognizing different sorts of values – scientific values, rarity values, bequest and option values, in addition to the more conventional economic ones. Environmental economics addresses the essential task of integrating the relative importance of these different sorts of value in resource allocation and environmental protection.

Institutional adaptation is a much less widely researched field, deserving priority development. In dealing with major environmental problems and related policy-making, even the most robust scientific recommendations can founder because institutions are not always able properly to implement them. We as scientists have a responsibility here too. Our recommendations are frequently prescriptive, without paying sufficient attention to what is practical and feasible in actual situations. We must foster interdisciplinary co-operation, interacting not only with engineers, which is now common practice, but also with social scientists, juridical, economical and political scientists in research on institution building and re-structuring that can effectively tackle regulatory needs and establish the broader frame within which the environment must be viewed.

Continued encouragement must be given to interdisciplinary perspectives in the environmental sciences, including the social sciences and the humanities, coupling this with greater international co-operation between the world's scientific communities. We must recognize the essential linkage between our scientific findings and policy-making, and try to widen the channels of communication through which scientific knowledge is transmitted to the political arena. Moreover, the essentially international dimension of modern science, which knows no man-made frontiers, may help a new international approach to global problems to evolve, based on a broad interdisciplinary contribution. In fact, I think we can say that this is what is already gradually emerging, if not without great difficulties, as we approach the 21st Century.

It is now over thirty years since C.P. Snow's famous lecture entitled "The Two Cultures" in which he attempted to define scientific culture. Snow was right to affirm that scientists share "common attitudes, common approaches and assumptions . . . in shorthand, I should say that naturally they had the future in their bones". The other culture, pre-eminently dominated by the literary-historical tradition, he saw as, on the contrary, dominated by the past. Things have changed in the intervening years. Now it is clear that, in order to provide convincing and acceptable solutions to man's problems, science and technology, like the social science and the humanities, must be profoundly rooted in the past, so that they can respond with strength and fortitude to the needs of Mankind and build together

the future. It is clear that, with an increasingly important contribution from all disciplines, science can indeed drive technological innovation permitting the advancement of peoples. But this advancement, this social progress, is in turn demanding more of science and technology than before, and more of the social sciences too. Snow's two cultures have ceased to stare out at one another: modern man looks to a global understanding and appreciation of the fruits of human knowledge. As Francis Bacon put it: "Knowledge is not to be sought for the pleasures of the mind, or for contention, or for superiority to others, or for profit, or fame, or power, or any of these inferior things, but for the benefit and use of life."

In surveying the global problems besetting us on this Earth, we now see that neither science nor technology alone has the answer; neither has politics nor the "dismal science" of economics. Each is imperfect; each requires reinforcement from the others. Humility is needed to recognize that no single expertise is sufficient to provide a truly satisfactory answer. As scientists, we must embrace the contribution that other disciplines can make to our knowledge base, and then together approach policy-makers in a constructive manner. It is through partnerships such as these that today's key issues will best be resolved. ASCEND 21 is testimony to the wisdom of the approach. I am pleased that ESF is a co-sponsor of this ICSU conference which seeks to take this process further.

Address by Abdus Salam

Director, International Centre for Theoretical Physics
President, Third World Academy of Sciences

I have long maintained that this globe of ours is inhabited by two distinct species of humans, the developed and the developing, the rich and the poor. What distinguishes one type of human from the other is the ambition, the power, the elan which basically stems from their differing mastery and utilization of present-day Science and Technology.

I was delighted with Professor Menon's invitation to the ASCEND 21 Conference because the present concern about environment in the North presents one of the few but potent ways for bringing the two species of humans together and increasing their collaboration. Environment is certainly one of the few areas where multilateral co-operation can make real headway.

Let me first start by giving you the idea of why we in the South are so backward. Here I present statistics to show that, by and large, we are just not serious about Science and Technology, and that is shown by the present numbers of people trained in Science and Technology (Table 1) plus the amount of money which is spent among us on Science and Technology (Table 2). These Tables are self-explanatory.

I shall be speaking mainly about two distinct problems. The first one concerns an environment-related subject, specifically the idea that the tropical climate is easier to predict than the temperate climate. The important point about this example is that it should be done in a centre of excellence which should be run by the United Nations family. This, and twenty other centres of Science and High Technology, are proposed to be created world-wide under the auspices of the UN family of nations.

This brings me to the second point which I wish to discuss in more detail. That is the reforms we should make to the UN family of nations, which will make it possible to sponsor such creations of a first-class collections of talent in one place. This represents a real challenge to the authorities of the UN, and should be treated as such in order to exploit the major recent scientific advances which are at present shared between the rich and poor countries. These are the two items on which I wish to speak.

Regarding the first idea, an international institution could be created for training and research on prediction of climates which impacts the social and economic development of tropical countries. The Second World Climate Conference (Geneva, 1991) strongly recommended that the scientific basis for tropical climate predictions be well established, and the time has now come for nations of the world to take concrete steps to establish prediction facilities.

Scientific research by the world community of scientists during the past 10-20 years has clearly established, and demonstrated, that tropical climate fluctuations, especially at seasonal and inter-annual time-scales, are indeed predictable. It is quite extraordinary that now, for the first time in the history of the world, we have the opportunity to ameliorate the effects of these climate catastrophes in the tropics by predicting them, using advanced mathematical models and very high-speed computers.

Tropical climate forecasts already have some impact on tropical societies. For example, government agencies in Peru, Ecuador, Brazil, India and China and Ethiopia regularly monitor the evolution of tropical Pacific ocean temperature. In Peru, an assessment of the likelihood of a warm or cold phase of tropical Pacific ocean guides a decision as to whether to plant cotton or a more responsive crop, rice. However, these forecasts are rudimentary. They are carried out by small research groups which lack the resources to make use of all available observational data.

Therefore, a centralized facility for training and research on prediction of tropical climate variations, with the resources and the sense of mission to be responsive to the need of tropical societies, is crucial to the quality of life for people from these countries.

The computing power needed to process and prepare the large volumes of data and to make the forecasts, is available in developing countries. The people with the knowledge and will to make these investigations and train the future generations are ready. What is lacking are the resources needed to focus these elements into a centralized training and research institute for the benefit of the tropical societies.

Table 1: Scientists and Engineers engaged in R&D and expenditures to R&D

Region, area	Year	R&D Scientists and engineers		R&D expenditure	
		Estimated number	Estimated number per million population	Estimated amount in millions of US$	% of GNP
WORLD TOTAL	**1980**	**3,920,754**	**894**	**208,370**	**1.85**
	1990	**5,223,614**	**1,000**	**452,590**	**2.55**
Industrial countries	1980	3,452,128	3,038	195,798	2.22
	1990	4,463,798	3,694	434,265	2.92
North America	1980	688,020	2,734	66,796	2.23
	1990	930,722	3,359	193,721	3.16
Europe	1980	893,482	1,859	70,712	1.81
	1990	1,091,003	2,206	104,956	2.21
USSR	1980	1,373,300	5,172	32,273	4.69
	1990	1,694,430	5,892	55,712	5.66
Developing countries	1990	759,816	189	18,325	0.64
Africa (excluding Arab States)	1990	34,963	74	746	0.29
Asia (excluding Arab States)	1990	1,190,369	396	88,533	2.08
Arab States	1990	77,261	363	3,078	0.76
Latin America	1990	162,930	364	2,360	0.40

Source: Table reprinted from the 1991 Statistical Yearbook.

It is ironic that the tropical countries, which will be the largest beneficiaries of these climate predictions, do not have, at this time, the required resources of trained scientific personnel and advanced computation-communication facilities which are needed to process large volumes of data, make predictions and disseminate the results.

It is proposed that this conference endorse the idea of establishing an international institution with a centralized supercomputer facility to train the scientists from tropical countries in climate modeling, data analysis and climate prediction. Such an institution will also greatly enhance the national capabilities in climate modeling and prediction of the participating countries. We would like to suggest that a formal announcement for the establishment of an international institution for training and research on prediction of tropical climate variations be made at the United Nations Conference on Environment and

Table 2: *Comparative expenditure on Science and Technology vs. proposed funding for Science and Technology by country*

	Population (million) 1988	GNP per capita (US $) 1987	Military expenditure (% of GNP) 1986	Health expenditure (% of GNP) 1986	Education expenditure (% of GNP) 1986	Science and Technology Actual expend. (% of GNP)	Scient. & engineers (million pop.)	Visits at ICTP 1970-91*
	1	2	3	4	5	6	7	8
Developed Market Economy countries								
1 New Zealand	3.3	7,750	2.20	4.60	4.80	0.90	..	13
2 Ireland	3.7	6,120	1.90	7.80	6.90	1.00	1076	50
3 Norway	4.2	17,190	3.20	5.50	6.80	2.10	2882	55
4 Israel	4.4	6,800	19.20	2.10	7.30	3.70	4836	297
5 Finland	4.9	14,470	1.70	6.00	5.90	1.80	..	51
6 Denmark	5.1	14,930	2.10	5.30	7.50	1.70	1907	166
7 Switzerland	6.5	21,330	1.90	6.80	4.80	2.80	2299	508
8 Austria	7.5	11,980	1.30	5.30	6.00	1.30	1013	541
9 Sweden	8.3	15,550	2.90	8.00	7.60	3.00	2724	526
10 Belgium	9.9	11,480	3.10	5.60	5.60	1.70	793	451
11 Greece	10	4,020	5.70	3.50	2.50	0.30	54	428
12 Portugal	10	2,830	3.30	5.70	4.30	0.50	439	308
13 Netherlands	15	11,860	3.10	7.50	6.60	2.30	2518	415
14 Australia	16	11,100	2.70	5.10	5.10	1.30	2174	102
15 Canada	26	15,160	2.20	6.60	7.40	1.40	2243	281
16 South Africa	34	1,890	3.90	0.60	4.60	..	629	..
17 Spain	39	6,010	2.30	4.30	3.20	0.60	436	519
18 France	56	12,790	3.90	6.60	5.90	2.30	1973	1,571
19 Italy	57	10,350	2.30	4.50	4.00	1.20	1232	5,711
20 United Kingdom	57	10,420	5.00	5.30	5.30	2.30	..	1,827
21 Germany FR	61	14,400	3.10	6.30	4.50	2.80	2724	2,037
22 Japan	122	15,760	1.00	4.90	5.00	2.80	5029	468
23 USA	245	18,530	6.70	4.50	5.30	2.60	3317	3,451

	Population (million) 1988	GNP per capita (US $) 1987	Military expenditure (% of GNP) 1986	Health expenditure (% of GNP) 1986	Education expenditure (% of GNP) 1986	Science and Technology Actual expend. (% of GNP)	1 % of GNP (million US$)	Scient. & engineers (million pop.)	Visits at ICTP 1970-91*
	1	2	3	4	5	6	7	8	9
Eastern Europe and Asian Socialist countries									
1 Mongolia	2.1	..	10.50	1.30	..	..	..	..	4
2 Albania	3.1	..	4.00	2.30	..	..	..	..	17
3 Bulgaria	9	4,150	3.60	3.20	4.40	3.30	..	5641	442
4 Hungary	11	2,240	2.40	3.20	3.80	2.40	238	2028	422
5 Czechoslovakia	16	5,820	4.10	4.20	3.60	4.50	..	4295	410
6 German DR	17	7,180	4.90	2.60	3.80	4.70	..	7774	110
7 Korea DR	22	..	10.00	1.00	..	..	..	1325	5
8 Romania	23	2,560	1.60	1.90	1.80	..	..	..	304
9 Yugoslavia	24	2,480	4.00	4.30	3.80	1.20	58	1452	1,551
10 Poland	38	2,070	3.30	4.00	4.50	1.50	724	908	1,304
11 USSR	284	4,550	11.50	3.20	5.20	6.20	..	5387	755
12 China	1105	290	6.00	1.40	2.70	0.97	3,198	..	1,991
Developing countries - population up to nearly 3 millions									
1 Gabon	1.1	2,700	3.80	2.00	4.80	..	29	190	5
2 Mauritius	1.1	1,490	0.20	1.80	3.30	0.20	15	214	15
3 Botswana	1.2	1,050	2.30	2.90	9.10	0.20	12	..	6
4 Trinidad & Tobago	1.2	4,210	1.00	3.00	5.80	0.80	51	235	11
5 Oman	1.4	5,810	27.60	3.30	6.60	..	78	..	3
6 Bhutan	1.5	150	..	..	..	..	2	..	-
7 United Arab E.	1.5	15,830	8.80	1.00	2.20	..	228	..	3
8 Lesotho	1.7	370	2.40	1.60	3.50	..	6	..	13

Table 2: (cont.) Comparative expenditure on Science and Technology vs. proposed funding for Science and Technology by country

	Population (million) 1988	GNP per capita (US $) 1987	Military expenditure (% of GNP) 1986	Health expenditure (% of GNP) 1986	Education expenditure (% of GNP) 1986	Science and Technology			
						Actual expend. (% of GNP)	1 % of GNP (million US$)	Scient. & engineers (million pop.)	Visits at ICTP 1970-91*
	1	2	3	4	5	6	7	8	9
Developing countries – population up to nearly 3 millions (contd.)									
9 Namibia	1.8	..	..	..	..	..	..	..	-
10 Mauritania	1.9	440	4.90	1.90	6.00	..	8	..	21
11 Congo	1.9	870	4.60	2.00	5.00	0.00	18	509	22
12 Kuwait	1.9	14,610	5.80	2.90	4.60	0.90	273	929	128
13 Yemen PDR	2.3	420	22.00	2.00	6.00	..	10	..	4
14 Panama	2.3	2,240	2.00	5.40	5.50	0.00	51	..	8
15 Libera	2.4	450	2.20	1.50	5.00	..	10	..	21
16 Jamaica	2.4	940	1.50	2.80	5.60	0.00	23	8	19
17 Singapore	2.6	7,940	5.50	1.20	5.20	0.90	207	1287	59
18 Central African R.	2.8	330	1.70	1.20	5.30	0.20	9	78	3
19 Lebanon	2.8	..	..	..	..	..	..	67	133
20 Costa Rica	2.9	1,610	0.00	5.40	4.70	0.30	43	..	70
Developing countries - population from 3 to 10 millions									
21 Uruguay	3.1	2,190	1.60	2.70	6.60	0.20	66	684	29
22 Togo	3.3	290	3.20	1.60	5.50	1.40	10	..	41
23 Nicaragua	3.6	830	16.00	6.60	6.10	0.00	30	207	1
24 Papua New Guinea	3.8	700	1.40	3.30	5.60	..	26	..	12
25 Lao PDR	3.9	170	..	..	1.20	..	6	..	-
26 Sierra Leone	4	300	1.20	0.70	3.00	..	12	..	72
27 Jordan	4	1,560	13.80	1.90	5.10	0.40	44	115	174
28 Paraguay	4	990	1.00	0.40	1.40	0.20	39	248	2
29 Libya	4.2	5,460	12.00	3.00	10.10	0.20	223	361	173
30 Benin	4.5	310	2.30	0.80	3.50	..	13	..	51
31 Honduras	4.8	810	5.90	2.60	5.00	..	36	..	16
32 El Salvador	5.1	860	3.70	1.00	2.30	1.60	42	1034	6
33 Burundi	5.2	250	3.50	0.70	2.80	0.30	12	48	47
34 Chad	5.4	150	6.00	0.60	2.00	0.30	8	..	2
35 Hong Kong	5.7	8,070	..	..	..	..	453	..	26
36 Haiti	6.3	360	1.50	0.90	1.20	..	22	..	-
37 Guinea	6.6	..	3.00	1.00	3.00	..	..	216	50
38 Niger	6.7	260	0.70	0.80	4.00	0.10	19	..	13
39 Rwanda	6.8	300	1.90	0.60	3.20	0.50	20	12	33
40 Bolivia	6.9	580	2.40	0.40	2.40	..	42	..	42
41 Dominican R.	6.9	730	1.40	1.40	1.60	..	49	..	11
42 Senegal	7	520	2.30	1.10	4.60	1.00	` 35	335	84
43 Somalia	7.1	290	4.40	0.20	6.00	..	17	..	34
44 Yemen AR	7.6	590	6.80	1.10	5.10	0.30	49	..	57
45 Tunisia	7.8	1,180	6.20	2.70	5.00	..	90	..	137
46 Malawi	7.9	160	2.30	2.40	3.70	0.20	12	..	23
47 Kampuchea	7.9	..	..	..	..	..	..	..	-
48 Zambia	7.9	250	3.20	2.10	4.40	0.50	17	..	42
49 Burkina Faso	8.6	190	3.00	0.90	2.50	0.50	14	..	15
50 Guatemala	8.7	950	1.30	0.70	1.80	0.90	68	236	6
51 Mali	8.9	210	2.50	0.70	3.20	..	16	..	61
52 Zimbabwe	9.2	580	5.00	2.30	7.90	..	53	..	28
53 Angola	9.5	470	12.00	1.00	3.40	..	..	..	5
Developing countries - Population from 10 to 20 millions									
54 Ecuador	10	1,040	1.60	1.10	3.60	..	103	..	21
55 Cuba	10	..	7.40	3.20	6.20	0.30	..	1108	111
56 Madagascar	11	210	2.40	1.80	3.50	0.20	22	23	55

Table 2: (cont.) Comparative expenditure on Science and Technology vs. proposed funding for Science and Technology by country

	Population (million) 1988	GNP per capita (US $) 1987	Military expenditure (% of GNP) 1986	Health expenditure (% of GNP) 1986	Education expenditure (% of GNP) 1986	Science and Technology Actual expend. (% of GNP)	1 % of GNP (million US$)	Scient. & engineers (million pop.)	Visits at ICTP 1970-91*
	1	2	3	4	5	6	7	8	9
57 Cameroon	11	970	1.70	0.70	2.80	0.90	104	..	39
58 Côte d'Ivoire	12	740	1.20	1.10	5.00	0.30	83	..	40
59 Syrian Arab R.	12	1,640	14.70	0.80	5.70	..	204	..	179
60 Saudi Arabia	13	6,200	22.70	4.00	10.60	0.50	905	..	92
61 Chile	13	1,310	3.60	2.10	5.20	0.50	165	363	178
62 Ghana	14	390	0.90	0.30	3.50	0.90	53	..	277
63 Mozambique	15	170	7.00	1.80	..	..	21	..	5
64 Afghanistan	16	..	..	..	..	..	..	..	11
65 Uganda	17	260	4.20	0.20	1.10	0.20	41	..	74
66 Sri Lanka	17	400	5.70	1.30	3.60	..	66	173	233
67 Malaysia	17	1,810	6.10	1.80	7.90	..	296	183	253
68 Nepal	18	160	1.50	0.90	2.80	0.30	28	22	135
69 Iraq	18	3,020	32.00	0.80	3.70	..	..	..	207
70 Venezuela	19	3,230	1.60	2.70	6.60	..	482	279	222

Developing countries - population from 20 to 50 millions

	Population (million) 1988	GNP per capita (US $) 1987	Military expenditure (% of GNP) 1986	Health expenditure (% of GNP) 1986	Education expenditure (% of GNP) 1986	Science and Technology Actual expend. (% of GNP)	1 % of GNP (million US$)	Scient. & engineers (million pop.)	Visits at ICTP 1970-91*
71 Peru	21	1,470	6.50	1.00	1.60	0.20	297	273	240
72 Kenya	23	330	1.20	1.10	5.00	0.80	75	..	124
73 Sudan	24	330	5.90	0.20	4.00	0.20	76	..	277
74 Morocco	24	610	5.10	1.00	5.90	..	142	..	276
75 Algeria	24	2,680	1.90	2.20	6.10	0.30	636	..	272
76 Tanzania	25	180	3.30	1.20	4.20	..	52	..	140
77 Colombia	31	1,240	1.00	0.80	2.80	0.10	360	40	248
78 Argentina	32	2,390	1.50	1.60	3.30	0.50	745	352	866
79 Zaire	34	150	3.00	0.80	0.40	..	53	..	61
80 Myanmar	40	200	3.10	1.00	2.10	..	..	..	1
81 Korea R.	43	2,690	5.20	0.30	4.90	1.90	1,129	..	225
82 Ethiopia	45	130	8.60	1.00	4.20	..	55	..	139

Developing countries - population from 50 to over 100 million

	Population (million) 1988	GNP per capita (US $) 1987	Military expenditure (% of GNP) 1986	Health expenditure (% of GNP) 1986	Education expenditure (% of GNP) 1986	Science and Technology Actual expend. (% of GNP)	1 % of GNP (million US$)	Scient. & engineers (million pop.)	Visits at ICTP 1970-91*
83 Egypt	51	680	8.90	1.00	4.80	0.20	360	428	1,202
84 Iran	53	..	20.00	1.80	3.50	..	..	67	552
85 Turkey	54	1,210	4.90	0.50	2.10	0.70	636	224	892
86 Thailand	54	850	4.00	1.30	4.10	0.20	448	104	266
87 Philippines	59	590	1.70	0.70	1.70	0.10	346	90	144
88 Viet Nam	64	..	..	..	..	0.63	..	333	203
89 Mexico	85	1,830	0.60	1.70	2.80	0.60	1,494	215	559
90 Nigeria	106	370	1.00	0.40	1.40	0.30	395	..	768
91 Bangladesh	110	160	1.50	0.60	2.20	0.20	174	..	393
92 Pakistan	115	350	6.70	0.20	2.20	1.00	36	58	814
93 Brazil	144	2,020	0.90	1.30	3.40	0.40	3,146	390	1,047
94 Indonesia	175	450	2.50	0.70	3.50	0.20	768	183	193
95 India	820	300	3.50	0.90	3.40	0.90	2,413	109	2,760

.. Data not available.

Source: from column 1 to column 5: Human Development Report, 1990.

Column 6: Statistics on science and technology, Extracts from "Unesco Statistical Yearbook", 1990.

Column 7: Desirable expenditures on classical "low technology" are taken into account in the UNESCO proposed expenditure of 1% of GNP. GNP in billion US$ has been taken from Human Development Report, 1990.

Column 8: Statistics on science and technology, Extracts from "Unesco Statistical Yearbook", 1990.

Figures in column 6 and 8 relate to different years ranging from 1975 to 1988.

Column 9: *ICTP statistics as of 31 August 1991.

Table 3a: Donor Contributions to CGIAR Center Programs, 1972–86 (in US$ million)

Donor	CORE PROGRAMS				TOTAL (CORE + NON-CORE)
	1972-76	1982	1984	1986	1986
Australia	4.00	3.77	4.00	4.52	4.85
Austria				1.00	1.01
Belgium	3.48	1.85	1.71	1.77	2.48
Brazil			1.00		0.01
Canada	17.37	8.29	10.03	10.66	14.26
China			0.50	0.48	0.48
Denmark	1.71	0.96	1.24	1.65	1.67
Finland			0.50	0.99	0.99
France	1.05	0.90	0.88	2.07	2.15
Germany Fed. Rep.	13.27	7.84	6.67	8.03	8.90
India		0.50	0.50	0.50	0.50
Iran	1.98				
Ireland		0.21	0.41	0.58	0.58
Italy	0.10	1.59	6.62	8.33	9.73
Japan	2.49	8.85	9.72	15.89	18.92
Mexico		0.10	1.22	0.20	0.25
Netherlands	4.11	3.24	3.28	6.65	7.88
New Zealand	0.11	0.02	0.02	0.01	0.01
Nigeria	1.30	1.13	1.00	0.19	0.38
Norway	3.33	1.87	1.92	3.12	3.40
Philippines		0.45	0.32	0.27	0.27
Saudi Arabia	1.00		1.50		
Spain		0.46	0.52	0.50	0.50
Sweden	7.19	3.18	3.07	4.20	4.21
Switzerland	1.87	2.76	6.70	7.11	9.08
United Kingdom	9.02	6.34	5.66	8.40	8.55
United States	41.60	40.79	45.25	46.25	60.22
Country Subtotal	114.98	95.11	114.23	133.36	161.30

Development (UNCED). This would become a major contribution to the cause of development of tropical countries.

Where does the UN come into this? Clearly the World Meteorological Organization (WMO) or other such organizations would like to operate in this area. They need a mandate at the present time for their work. Such mandates are hard to come by because this is not specified as part of the Agency work.

As an example one may take the Consultative Group on International Agricultural Research (CGIAR), founded in 1978, which has since been viewed with suspicion by trans-national agroindustrial corporations. The centers of CGIAR, located mostly in developing countries in different regions of the world, are creating seeds which are highly resistant to disease and frost, and are yielding maximum productivity.

The "Group of 13" includes three centers devoted to research on tropical agriculture in Colombia, India and Nigeria, one in Syria concentrating on agriculture in arid zones, a fifth in the Philippines on cross-breeding of rice, and three on genetic improvement of cattle in Ethiopia, Kenya and the Côte Ivoire.

Perfection of strains of corn resistant to the weevil, wheat resistant to blight, potatoes capable of reproduction by seed instead of by the usual tuber, and rice which grows easily near salt-water shores, are among the projects scientists have undertaken in the 13 CGIAR centers.

On the staff of CIMMYT (the International Center for the Improvement of Corn and Wheat, in Mexico) which belongs to CGIAR, is Norman E. Borlaug, who oversees research on a highly resistant and highly productive strain of dwarf wheat, and whose work earned him the Nobel Peace Prize in 1970.

Varieties of wheat obtained at the CIMMYT are now cultivated on 50 million hectares of land in 70 countries, while corn hybrids originating from Mexican breeds have been sown on 3 million hectares.

The CGIAR was created (in 1978) with the aims of establishing other such centres, guiding their management, and guaranteeing them long-range economic backing. Starting out with an annual budget of 9 million dollars, the CGIAR – an organization of simple structure and non-bureaucratic functioning – has aroused such an interest that it can presently dispose of 250 million dollars a year. The annual 250 million dollars are really not enough to achieve the aims of these thirteen centers, whose ultimate goal is the feeding of 800 million undernourished people throughout the world.

Donor contributions to CGIAR programs are illustrated by the Tables 3a and 3b.

The thirteen centers were set up with the help of the

Table 3b: *Donor Contributions to CGIAR Center Programs, 1972–86 (in US$ million)*

Donor	CORE PROGRAMS				TOTAL (CORE + NON-CORE)
	1972-76	1982	1984	1986	1986
Ford	16.79	0.81	0.99	0.90	1.73
Kellogg	1.32		0.34		
Kresge	0.75				
Leverhulme		0.65	0.81	0.62	0.62
Rockefeller	17.10	0.80	0.50	0.93	1.22
Foundation Subtotal	*35.96*	*2.26*	*2.64*	*2.45*	*3.57*
ADB	0.30				0.71
ASDB		0.02		0.59	0.59
AFESD		0.24	0.23	0.34	0.34
EC		4.72	4.72	7.14	8.47
IDB	11.15	8.10	8.73	9.39	9.44
IDRC	3.95	1.20	1.01	1.18	3.51
IFAD		5.94	7.02	0.45	1.22
OPEC		3.58	2.19	0.47	0.87
UNDP	7.42	6.19	8.06	8.42	8.87
UNEP	0.94	0.18	0.03		0.03
World Bank (IBRD)	16.15	16.30	24.30	28.40	29.61
Int'l Donor Subtotal	*39.91*	*46.47*	*56.29*	*56.39*	*63.66*
Other Donors					7.01
TOTAL	**190.85**	**143.84**	**173.16**	**192.20**	**235.54**

ADB	African Development Bank
ASDB	Asian Development Bank
AFESD	Arab Fund for Economic and Social Development
EC	European Community
IDB	Inter-American Development Bank
IDRC	International Development Research Centre
IFAD	International Fund for Agricultural Development
OPEC	Organization of Petroleum Exploring Countries
UNDP	United Nations Development Programme
UNEP	United Nations Environmental Programme

World Bank. I am told this cannot be repeated for the twenty centers of Science and Technology. The Third World Academy of Sciences has been reduced to requesting that the countries concerned should be asked to apply individually to the World Bank for their own country uses.

I would like to see the centers grow up with the help of the World Bank and other such organizations.

At the present time we have one center located in Trieste, the International Center for Theoretical Physics (ICTP). This Center, although ostensibly for theoretical physics, is essentially representing all scientific disciplines. It has now reached a stage of maturity after nearly thirty years of operation. It is sponsored mainly by the government of the host country where it is located, to the tune of 17 million dollars, while it gets 1 million dollars from the International Atomic Energy Agency in Vienna, and one-third of a million dollars from UNESCO every year.

There is a need for stable funding from UN sources, independent of the Board of Governors which should be more academically inclined. Likewise, the 3 new centers proposed to be set up in Trieste, comprising the International Center for Science, are the International Institute for Science and High Technology, the International Institute for Earth, Environmental and Marine Sciences, and the International Institute for Pure and Applied Chemistry. These have made requests to the Italian government. These requests should be examined and, if possible, the UN bodies like UNIDO should be given the mandate to fund these organizations. These will then serve as the technological arm of ICTP.

This is the plea by which I wish to end my talk: that the UN should wake up to the organization of such centers which will then supplement the contribution to Science and Technology made by the developed countries. This is the only way in which we can hope to have plans for action turned into real action.

Address by Nay Htun

Special Advisor/Director of Programmes
U.N. Conference on Environment and Development (UNCED)
Geneva, Switzerland

THE UNCED OPPORTUNITIES

On behalf of Mr Maurice Strong, the Secretary-General of UNCED, who will be joining us at the closing session of this important Conference and Mr Nitin Desai, the Deputy Secretary General, who regrets very much for being unable to attend, may I say at the very outset how grateful we are to the organizers for the outstanding work that has been undertaken in arranging this Conference. Clearly, the conclusions and recommendations will be important inputs into the preparatory process for the 1992 Conference. The ASCEND 21 Report will be made available to the fourth and last session of the UNCED Preparatory Committee.

The national and global issues relating to environment and development needs no elaboration at this Conference. Also, the urgent need to integrate environment and development to increase standards of living and improve the quality of life of all the people is now recognized. The challenge is how to achieve this.

The 1992 Conference provides very important and unique opportunities. Some are describing it as *the* opportunity of our generation to launch a process for integrating environment and development. This process could include seven major areas, and scientists have opportunities to make major contributions to all of them.

These are:

(a) *Principles*
UNCED provides the opportunity to develop and elaborate principles on general rights and obligations of States in the field of environment and development for incorporation into an appropriate charter/declaration. The Preparatory Committee is discussing and negotiating a list of elements, based on a check-list compiled from existing texts. The preliminary list, which will be negotiated and elaborated further at the fourth and final session in March–April 1992, include the following potential principles:

- integration of environment an development;
- common but differentiated responsibility;
- individual and group rights;
- responsibility to present and future generations;
- sovereignty and responsibility to others;
- precautionary principle and prior assessment;
- co-operation;
- special needs of developing countries;
- environment and world trade;
- peace and security;
- changing production and consumption patterns;
- information and education;
- cost internalization;
- liability participation and democracy;
- public participation and democracy;
- eradicating poverty;
- resource transfer.

(b) *Programs*
There is an imperative need to identify and describe the momentous changes that are taking place and intensified by the growing scale and speed of the interactions among biophysical, economics, social and demographic trends. While change is continuous, the speed of the changes taking place now is unprecedented. Many of these changes have global implications with far-reaching consequences for human civilization and planet Earth. Better information and data collection systems to indicate more accurately the rate and scale of the changes are needed in order to narrow the uncertainties so that appropriate response options can be taken. However, the lack of scientific information and absolute certainty should not, and must not, be an excuse for inaction. The UNCED provides the opportunity to formulate programs that would provide transitions towards a more environmentally sound and sustainable future. Programs and activities to help conserve natural resources, alter consumption and production patterns, produce less wastes, lessen or eliminate industrial impact on human health and the environment, promote equity, abolish poverty and improve the overall welfare and quality of life

of the individual, need to be formulated and implemented.

A tentative set of about 120 programs is being negotiated to address the wide range of environmental, social and economic issues identified by the UN General Assembly. These programs, while addressing specific environmental issues such as atmosphere, biodiversity and toxic chemicals, have a very prominent and inherent development orientation with focus on the social and economic transitions that are needed. The programs not only identify the scientific issues involved but also help facilitate the integration of environment and development, since the means for implementing them require a multidisciplinary approach that could operationalize the principles and concepts of sustainability.

One of the most significant products of the 1992 UNCED promises to be AGENDA 21, a comprehensive and far-ranging program of action plans for the 21st Century. The program areas include, for example:

- eradicating poverty and meeting basic needs;
- international and domestic policies to accelerate sustainable growth;
- dealing with links between trade and the environment;
- changing consumption patterns;
- reducing demographic pressures;
- protection of human health conditions;
- promoting a sustainable pattern of human settlements;
- integration of environment and development in decision-making;
- protecting the atmosphere;
- planning and management of land resources;
- combating deforestation, desertification and drought;
- conserving biological diversity;
- environmentally sound management of biotechnology;
- protection of the oceans and all kinds of seas, and the rational use and development of their living resources;
- protection of the quality and supply of freshwater resources;
- environmentally sound management of toxic chemicals and prevention of illegal international traffic in toxic and dangerous products and wastes;
- environmentally sound management of solid and hazardous wastes, as well as radio-active wastes;
- strengthening the role of major groups, e.g. women, youth, indigenous people, NGOs, cities and local authorities, trade unions, business and industry, and the scientific community.

The means of implementing these substantive programs are elaborated and include financial resources and mechanisms, transfer of environmentally sound

technology, science for sustainable development, promoting education, training and awareness, and capacity building in developing countries.

The international institutional arrangements need to help facilitate implementation of Agenda 21 will be proposed later.

These are but some of the program areas that make up Agenda 21, and the ASCEND 21 Conference could help provide more scientific precision.

(c) *Participation*

One of the major features of the 1992 Conference is that it is catalyzing increasing interest and participation, and not only at the preparatory committee meetings. During the past 18 months an increasing range of activities at the national, regional and global level has been and is taking place to help provide inputs either to the national strategies and reports for UNCED or directly to the work of the secretariat, of which this Conference is an excellent example. Similarly, many other professions and independent sectors such as business and industry, women, youth, indigenous people, trade unions and religious groups are participating in the process. An average of 700 to 800 participants representing 120–130 countries, numerous intergovernmental organizations and over 400 non-governmental organizations are attending the Preparatory Committee meetings. As the Conference is a United Nations-wide Conference, all the UN agencies are also closely involved and taking an active part in the preparations.

(d) *Process*

This broad participation has opened up the involvement of governments and major groups of society in the discussing, planning and preparing of policy and strategy options for integrating environment and development at the national and international levels. Clearly, this democratic feature is a very important aspect of the preparatory process and will have important future consequences.

(e) *Policies*

A major recurring theme articulated by the scientific community is the weakness and lack of integration between science and policy. While neither the UNCED process nor the Conference will solve this shortcoming, they nevertheless provide the opportunity to increase the use of scientific knowledge in policy formulation. The preparatory process at the national level and the discussions and negotiations at the preparatory meetings have already provided opportunities for the application of scientific knowledge. In the Preparatory Committee, working groups which include delegates with scientific expertise, are formally established to address the substantive aspects of the issues; their deliberations are then fed into plenary session where further negotiations and decisions occur. It is critically important that good scientific information and advice continue to aid the preparatory process and that scientists remain active as well. Hopefully, the benefits of integrating science with policy to promote sustainable development, as exemplified by the UNCED process, will become more pronounced, and help strengthen the participation of scientists in decision-making. The institutional interface between scientific opinion and political decision-making should come together increasingly.

(f) *Political support*

The very large and widespread participation of governments in the preparatory process and at the four Preparatory Committee meetings, as well as the public awareness and interest generated by the 1992 Conference should hopefully provide the necessary impetus to increase political awareness and hence support for the principles, programs and policies which they the politicians themselves have taken the main part in negotiating and agreeing. Furthermore, the decision taken at the last session of the General Assembly inviting governments to attend the 1992 Conference at the Head of State or Government level underscores the unprecedented opportunity for political support.

(g) *Partnership*

In the overall analysis, the 1992 UNCED Conference is about partnership. It is an opportunity to foster greater partnership between environment and development at concepts, principles, policies and program levels.

It provides the opportunity for stronger partnerships between and amongst the scientific disciplines.

It provides the opportunity for developing more co-operative partnerships between government, industry, the scientific community, the independent sectors and the non-governmental organizations.

It provides the opportunity to build more harmonious partnerships between countries.

It provides the opportunity for a more compassionate and understanding partnership between human being and with nature.

The 1992 Conference, the opportunity of a generation could provide the catalyst to set us on path towards a more secure, safe, prosperous and equitable millenium.

Can we seize this opportunity now ?

Address by Margaret J. Anstee

Director-General of the United Nations Office at Vienna (UNOV) and Head of the Center for Social Development and Humanitarian Affairs

Your Excellencies, Ladies and Gentlemen

On behalf of the United Nations and the United Nations Industrial Development Organization (UNIDO) and the International Atomic Energy Agency (IAEA), our co-habitors in Vienna International Center (VIC), may I welcome you most warmly to the Center. I should like, first of all, to pay a special tribute to all the organizations whose initiative and hard work have made this important event possible.

May I also thank our host country, and particularly the Chancellor of the Federal Government of Austria, whose address we heard earlier this morning.

The agencies located here at UNIDO and IAEA, as well as UNOV itself, have a direct concern with the issues before you – from a scientific perspective, UNIDO and IAEA especially. UNIDO has substantially augmented its efforts to provide advisory services to industry in regards to environmental protection.

I should also mention the work done to promote the definition and adoption of:

- a Voluntary Code of Conduct for the Release of Genetically Modified Organisms into the Environment; this is of major significance to the development of biotechnology-based industries.

For IAEA, environmental concerns are clearly at the forefront of its work, and about which you which be informed in some detail by colleagues from the Agency who will be participating in the various sessions and workshop.

In UNOV, itself, environmental issues have received increasing attention.

As part of our crime prevention and criminal justice activities, we are exploring the potential of using the criminal justice system in the battle against pollution and other environmentally harmful practices.

In our programs for the advancement of women, we are exploring the particular roles of women in the safeguarding of the environment, and the special burdens on women of environmental degradation in developing countries.

These are but two specific areas of interest to us. In addressing this distinguished scientific audience, I am very conscious that, in our professional lives, your and my paths do not frequently cross. Not frequently, but *not never*.

The most recent occasion on which these paths crossed was when the Secretary-General entrusted me with the responsibility of co-ordinating the international effort to tackle the environmental consequence of the Gulf War. In that role, I have been able to witness at first hand the damage done to the environment. I have a vivid impression of my last visit to the scene.

I also retain a vivid impression of the ingenuity shown by the scientific and technological community in tackling the enormous task of clean-up and damage control. I thus keenly await the results of your deliberations and wish to hear more about the thinking on environmental issues that is taking shape in the scientific community.

I currently also am responsible for co-ordinating the international response to the Chernobyl accident. In this case, my attention is focussed on the longer-term issues stemming from this environmental tragedy. In this task I am on more familiar ground, since my principal responsibility as Director-General of UNOV is with social issue; I might mention that UNOV is the designated nucleus for social policy and development in the UN Secretariat.

And it is thus from *the social perspective* that I would like to address a few words to you on this central contemporary issue: environment and development. When viewed from a social perspective, the issue of environment leads naturally to the concept of sustainable development. This concept has now gained a certain standing – an *intellectual* acceptance, at least.

Most simply put, sustainable development means – to me at least – a form of development – a development path – that allows us, the present generation, to leave to the next generation an environment that is at least no worse – and preferably better – than the one we inherited. I think it is now beginning to be appreciated – indeed very much appreciated – that we cannot continue with the patterns and means of development that have brought us to the present stage of global development.

In the countries that are in the early stages of industrialization, where poverty and deprivation are common and in many places becoming more acute, we cannot conceive of *their* quest to higher living standards following the path already trodden by the highly industrialized countries – neither the environment of developing countries, nor that of the planet, will bear this.

And, clearly, for the highly industrialized countries, we cannot condone any attitude of "business as usual". The price cannot, most certainly, be levied only on those countries which have come later into the development race.

Science and technology will have a role to play in enabling all of us to change course, by devising new and cost-effective ways of cleaning-up past damage, in finding safer and cleaner ways of doing things, and in conserving and restoring.

This, as I intimated earlier, will not come without its price, however, cost-effectiveness notwithstanding, – at least in the medium-term. And this price will not fall equally on all members of society or on all countries. And some – individuals or countries – will be in a better position than others to pay this price.

And, much as science and technology can do to put right past abuses, and to blunt the edge of the social choices that need to be made, science and technology cannot provide that complete answer. Society will be faced with difficult choices; regarding goals and means of pursuing our various ways of life. To be specific, many of us will have to desist from wanting some of the things we think of now as contributing to the "good life".

In changing direction from the development path towards the path of sustainable development, the short-term gains to some and losses to others have to be reconciled. What burdens have to be borne by certain groups or sectors to promote the long-term common interest, have to be determined and balanced.

In most cases, governments, in this balancing act, are not making decision between the "good" and the "bad". After all, activities that degrade, despoil or pollute the environment are but the means to an end: a better life, as *particular* people see it.

This balancing act is, quite essentially, a social issue.

Social policy will thus have a central role to play in bringing about the needed changes – in leading us toward the path of sustainable development.

Let me illustrate this with a few specific examples. Sustainable development may well require drastic conservation of energy, and a higher price for energy may be the instrument of choice. If the price of energy has to increase, there are many possibilities in the industrialized countries to shift its costs from one sector or to another, from one social group to another. Those particularly hard hit can be compensated in various ways. Social benefits to the poor can be increased; tax regimes can be adjusted; heating and transport allowances can be introduced in favour of specific target groups: the range of options is wide indeed.

If jobs are lost, mechanisms exist to shift workers to other sectors to sustain them while unemployed. Many such examples can be provided. In developing countries, in sharp contrast, these possibilities are few, or in some cases non-existent. Poverty, as we all know, is also a major cause, itself, of environmental degradation. It is not easy to ask people to desist from development activities harmful to the environment when their livelihood is at stake and there are not easy alternatives.

I think you can see what I am leading up to: that we should take seriously the idea of *international solidarity*, and that the rich North provide greater support to the developing countries in managing the hard environmental choices that we all face – but that are hardest of all for them.

Those in the North also need to convince those in the South that "environmental concerns" are not yet another non-tariff barrier imposed by the strong on the trade of the weak.

And let us frankly acknowledge what is involved: an element of transfer of resources from North to South. This is not now a popular notion in the North – at least not among Governments of the North. But, movement towards recognizing that help from North to South is necessary if we, the global community, are to take seriously what is so often said: that we are all in the same boat, and ultimately sail or sink together.

The word "price" has figured large in what I have just said. The overwhelming question, surely is: *What price global survival?*

Address by Peter E. de Janosi
Director, International Institute for Applied Systems Analysis (IIASA)

It is a pleasure and privilege to be here and to have IIASA as a co-sponsor of this major scientific event. We are grateful to ICSU for taking the initiative, and for inviting IIASA's participation.

IIASA is not only present as an institutional co-sponsor, but also as an active participant in the preparation of the scientific content of our deliberations. It is a source of pride and satisfaction that numerous speakers and attendees are IIASA staff members and alumni. And that is precisely the way it should be in view of IIASA's fine history of studying complex problems, including the environmental field and in view of our ambitious plans for the future.

ICSU's leadership in bringing to bear the best science has to offer on one of the critical sets of issues confronting our societies must be applauded, and those of us who are supporting this effort happily follow ICSU's lead. The connections between the environmental and social, economic and political development provide ample opportunity for scientific studies. Such studies could indeed provide a better understanding of these complex interrelationships and to point to ways to reconcile some of the conflicts necessarily embedded in them. And it is particularly heartening to see that this conference will place special emphasis on the links between topics that traditionally have been examined primarily from disciplinary perspectives. While we clearly must reply on the deep insights offered by disciplines and the understandings they offer, the world and its problems rarely coincide with disciplinary boundaries.

It is right for us to be proud of the achievements of science and technology in allowing much of our world to improve its standards of living and the quality of life. Yet these improvements have been uneven, and frequently have come at the expense of our environment. It is now realistic to look to science and technology to grapple with these problems, and to make a critically important contribution to their amelioration.

Too frequently the impacts of international conferences are short-lived, and are restricted to those attending the gatherings. I am confident, however, that our present undertaking will not only help set the course of scientific action for years to come, but will also influence the course of national and international policy action.

Our responsibilities as scientific organizations and networks are to contribute to the knowledge base for ways to satisfy a world-wide aspiration for a better life for all, and to fulfil them in a responsible and sensitive way, especially regarding our planetary environment. Our gathering here this week is an important and most appropriate step toward meeting those responsibilities, and I can assure you that IIASA intends to continue to play an active role in these matters.

Address by Jostein Mykletun
on behalf of the Norwegian Research Council for Science and the Humanities (NAVF) and the Norwegian Academy of Science and Letters

Mr President,

Ladies and Gentlemen,

In May last year the Norwegian Research Council for Science and the Humanities organized, together with the European Science Foundation, the Bergen Conference on "Sustainable Development, Science and Policy".

Several of the participants here today also attended the Bergen Conference, which, in several ways, was a forerunner to ASCEND 21.

The Bergen Science Conference concluded that:

In order to ensure that the science community is actively and broadly involved in the process leading up to the Brazil Conference, it is strongly recommended that a major science conference be held 6 months in advance.

This Conference should take the form of a Global Science Summit. It should underscore the need for a comprehensive, yet focussed, scientific approach which is required to face up to the challenges put to the UN Conference in 1992. Beyond Rio '92, the aim is to produce an enduring global research and action agenda in the field of sustainable development.

Clearly, ASCEND 21 here in Vienna is very much the sort of necessary and desirable meeting ground envisaged by the Bergen Conference last year.

The Norwegian Research Council and the Norwegian Academy of Science and Letters are proud and pleased to be among the co-sponsors of ASCEND 21. Our hope is that ASCEND 21 will indeed take us several steps forward in identifying the research required to grapple with genuinely sustained development on a world scale.

May I take this opportunity, Mr President, to congratulate you Sir, and your highly professional ICSU Secretariat, for having prepared so efficiently for this important Conference. As co-sponsors, we will do our utmost to promote an enduring and action-oriented outcome of this Conference.

I should like to conclude with a line of urgency, quoting from the Executive Summary of the Bergen Science Conference, and addressed to the international science community:

It is better to be roughly right now than to be precisely right too late.

The line of urgency relates directly to the question of how to move forward. It seems to me that in this crucial context, we as members of the international scientific community have a broad societal and ethical commitment to provide guidance for appropriate policy action, both before and after Rio '92.

Thank you for your attention.

Address by Michael J. Chadwick
Director, Stockholm Environment Institute (SEI)

I took great heart from the words of the Chancellor this morning because here at a meeting of mainly scientists and in a country that has given a home to IIASA – the International Institute of Applied Systems Analysis – he told us that we must think in a manner that interlinks and connects the various components in the environment–development nexus. Nearly every speaker this morning has underlined this message. It is not good enough to focus individually on the various components. As any scientist who has made use of a technique like the Analysis of Variance knows, the interactions are invariably of more interest and importance than the main effects. At the SEI we have been striving in our work on natural resources, technology, environmental impacts and on environmental management policies within an economic framework, to indicate the necessity to see this equation as a whole – not to focus for too long on one aspect, such as the environmental impacts or the demand on natural resources, and think that the problem may be resolved by intensive work and intervention on one of these main effects. This

approach has usually failed to offer solutions to our global problems – it has even at times added to the number and scale of the problem areas.

SEI tries to follow through to economic considerations within a policy and management framework from the use (and misuse) of natural resources, from the appropriate (and inappropriate) technologies employed to do this and from the range of environmental impacts that result, but need not only to be recognized but be brought to manageable proportions. No single aspects hold the key to the improvement of the present situation in the area of development needs and environmental impacts. This is why the global situation is so difficult to keep in balance and influence for the good. There are so many linkages, so many feedbacks that must be considered. A new mode of thinking about problem-solving as well as new lifestyles are needed. This is why SEI, through many of the Institute staff members, has been pleased to be involved in the UNCED process and contribute to supporting ASCEND 21 as a part of this process.

Address by Maurice F. Strong

Secretary-General, United Nations Conference on Environment and Development (UNCED)

I am pleased to have the opportunity and privilege of joining you at this important Conference. There could be no better or more appropriate place for it than Vienna. You are to be congratulated on the impressive quality and range of the work that has gone with the preparations. The UNCED Secretariat has benefited enormously from its association with ICSU which has acted as its principal source of scientific support and advice and we are deeply grateful for the co-operation and support we have received from President Professor Menon, Secretary-General Professor la Rivière and their colleagues.

I am also extremely grateful for the inspiration and guidance of my good friend Gordon Goodman, who is one of my most valued special advisers.

We regard this as an extremely important meeting, which has provided a valuable input to the UNCED process, for science is intimately bound up with the main challenge for the 1992 Conference: to translate sustainable development from concept to action.

I am impressed with the ambitious programs you have set yourselves, and appreciate in particular the structure of the meeting, which reflects the critical stages in formulating an agenda for science for sustainable development. The increasing demand upon natural resources by a growing population, the sustainable productive capacity of the Earth's natural systems to supply these goods, and its capacity to assimilate our wastes, are the parameters within which humankind must manage the global risks it has brought upon itself.

The unprecedented increase in human numbers and activities since the industrial revolution, and particularly in this century, has given rise to a deterioration of the environment and depletion of natural resources that threaten the future of the planet. It is ironic that these impacts have occurred largely as a result of the same processes that have produced such unparalleled levels of wealth and prosperity in the industrialized world.

As the human community moves towards the important milestone of the third millenium, it is coming to realize that it confronts an unparalleled confluence of new challenges. Spectacular advances in our mastery of science and technology have created standards of life and opportunity for the privileged minority which lives largely in the more industrialized countries and created the possibility, and the expectation, of a better life for the underprivileged majority which lives primarily in the developing countries. At the same time human activities have produced ominous new risks to all life on our planet that rich and poor must share.

The fundamental challenge that faces the human community at this stage in history is to control its impacts on conditions that make life on Earth possible. For the environmental impacts of human behavior, in the minute fraction of the history of the world that has passed since the industrial revolution, have reached a level that is altering the planet's capacity to support life. Science has provided us with the levers that have made this possible.

Accordingly, how science is developed and applied will be a major factor determining the future of mankind, and of other life on the planet.

Science has taught us that environment and development issues are global in scale and systemic in nature, and, if we are to remain within safe limits we must manage the changes that are shaping our future. To do this, we must adopt an integrated, systems approach. UNCED is striving to provide the basis for this in the proposals it is developing, particularly Agenda 21, and it is equally important for the scientific community. This approach calls for a combination of specialties, accompanied by excellence in specialization, as well as the interdisciplinary expertise to analyze systemic issues. An important example is the need to put more emphasis on a closer interaction between the physical and biological sciences on the one hand, and the economic and social sciences on the other. At the same time, scientists must pay much greater attention to the economic and social consequences of their work.

Each action must now make its contribution both to the evaluation of action needs and to their implementation within the framework of an integrated and collaborative process through which its contributions can be made as an integral part of the total cause and effect system.

Pure science has been outstandingly successful in clarifying the nature of matter and energy systems in the universe. But the traditional subdivision of science into specialist disciplinary elements is less well equipped to deal with the biosphere. The highly complex and interactive problems posed by the structure and functioning of biosphere systems are made even more complex by their intimate relationships to the complicated socio-economic and political systems governing the development of human societies world-wide. Understanding these problems is at the heart of understanding environment and development. And in this sense pure science, together with the social and engineering sciences, faces a powerful and new challenge on a global scale.

Humanity needs the help of analytical methods and technical skills evolved by these sciences (and for all of you scientists as their practitioners to survive and develop in the next century!) So this powerful new challenge for science is also a new and sobering responsibility for its practitioners.

The sciences must develop more trans-disciplinary skills, interactively connecting formerly disparate fields of knowledge.

But there is a far more subtle and, for many scientists a more difficult, imperative than this. When scientists of all specialties work on environment and development issues, they must dedicate themselves not to building cathedrals for the sciences, but to fashioning the households of humanity. The sciences must see themselves in an important but supportive role to our common human endeavour to live healthier and more fulfilling lives on a healthy planet. And so the scientist should be concerned less with "what does science need for its furtherance?", and more with "what does humanity need for its development and how can the sciences help to provide it?".

While science and technology are intrinsically universal, their practitioners are not necessarily so. I have noted with concern that, despite the growing international dimensions of virtually all of the activities that bear on sustainable development, there is still a great deal of parochialism amongst some of the important actors in respect of these issues in the scientific, technological and educational communities. This tendency is undoubtedly due, at least in part, to the increasing degree to which knowledge and technology are regarded as primary sources of competitive advantage. Indeed, at the very time when the world is moving towards freer trade in goods and services, the commercialisation of basic research and accompanying attempts to transform knowledge into intellectual property are threatening to impose new barriers to the free-flow of knowledge which has been one of the most important features of science.

In this respect, I regard this as a most significant meeting. For this landmark conference brings together a wealth of knowledge and experience from the natural, engineering and social sciences, and is a very encouraging signal.

What is to be the output from a meeting with such an agenda and interdisciplinary participation? In collaboration with UNCED you have prepared for this conference an impressive range of expert background chapters and developed a detailed agenda for action highlighting the priorities in our search for the knowledge and capacity to develop sustainably. This is an appropriate and practical response that goes a long way towards helping us isolate programs to translate sustainable development from concept to action. It will enable the scientific community to (1) re-examine its activities and adjust them to focus on the priorities that sustainable development presents, and (2) develop its relationship with other sectors of society: particularly decision-makers and the public.

An examination of the figures illustrating growth in population, *per capita* consumption of resources, energy use, and gross world product reveals that this growth is not simply incremental or arithmetic but exponential, so that it has a doubling time. The doubling time is of the order of a little more than a generation: that is, approximately 35 years. We can reasonably predict, therefore, a strong trend towards a further doubling by around 2020. There is a good deal of evidence to suggest that we have already reached a point on the logarithmic growth curve when the available planetary resources are being used to a point where it is doubtful that the planet can sustain such a doubling beyond the next generation. Even at today's levels of population and consumption, the effects of human intervention on environment and development are already serious. Driven by poverty in developing and over-consumption in industrialized countries, the symptoms of over-use of the resource base are manifold from the degradation of land and shortage of freshwater resources, to the local and global effects of pollution. And all the problems we have experienced to date have occurred at levels of population and human activity that are bound to increase.

The spiral of environmental degradation threatens not only individuals' lives, but peace and environmental security. In a typical chain of events, land degradation leads to hunger and poverty, ill health and child mortality. This leads to migration into increasingly overcrowded cities, which, by a series of positive feedbacks, will generate further poverty. The growing problem of environmental refugees and conflicts over land show how these problems can escalate into civil and even international strife.

Renewable natural resources are being consumed faster than they are able to replenish themselves. Having used up their yielding capacity, we are now eating into their stocks.

In short, we are living off our natural capital as well as its annual interest. More and more of the Earth's surface must be processed for scarce non-renewable resources, and to keep up the supply, increasing large-scale mechanization, more and more dependent on still inexpensive fossil fuels, has to be deployed to meet our escalating demands. The natural chemical and biological clean-up mechanisms in the atmosphere, and in water and soil systems are being damaged and swamped by the explosive growth of all these resource-use activities, and widespread pollution builds up.

In short, Mankind is in the process of overwhelming nature. And there is a penalty we must all pay for our domination of the natural world upon which we have always depended absolutely – we are forced more and more to take up the burden of being responsible for its survival forever. As a species we have not yet begun to accept the apocalyptic implications of this let alone apply ourselves to learning how to manage environments for the purpose of sustained development. So, instead of being a minor guest at the Earth's table, the human species is on the way to consuming the whole feast and we are in danger of being forced into a partnership with nature to provide the feast forever. We must not turn the gift of nature into a Faustian bargain.

It is clear that human activities are now the principal determinant of the future of the Earth as a hospitable home for our and other species. Our future is literally in our hands. This means that we must learn to manage that future much more carefully and effectively. Particular responsibility rests on those who shape economic policy and carry it out, as well as those who provide the scientific and technical guidance to policy- and decision-makers. What we need is far greater partnership between these various sectors. They have functioned for too long in isolation from each other, and must be brought together. The forging of such partnerships is one of the prime tasks for UNCED. And in no sector is this more vital than the scientific community. Thus we attach particular significance to this meeting.

The evidence of history suggests that many past civilizations have foundered upon unsustainable use of the resource base. Despite the imperfections in the understanding of the biogeochemical cycles which govern the Earth's carrying capacity, this generation nonetheless has a unique understanding of humankind's effects upon the Earth. Might we, nonetheless, allow our global civilization to founder on a planetary scale, notwithstanding? It is unthinkable.

The gross imbalances that have been created by the concentration of economic growth in the industrialized countries and population growth in the developing countries, are at the center of the current dilemma. Redressing these imbalances will be the key to the future security of our planet in environmental and economic as well as in traditional security terms. This will require fundamental changes in both our economic behavior and our international relations. Effecting these changes peacefully and co-operatively is, without doubt, the principal challenge of our time.

In this important undertaking, co-operation can only be based on common interests. While there is widespread acknowledgment, at the level of principle, of the need to achieve a sustainable balance between environment and development, it should be no surprise that the perspectives of developing countries on the issues differ substantially from those of industrialized countries.

The lifestyles of the rich are the source of the primary risks to our common future. They are simply not sustainable. We who enjoy these lifestyles are all "security risks". Our patterns of production and consumption have brought the whole human community to the threshold of risks to our survival and well-being which rich and poor alike must share. We must reduce our impacts on these thresholds and make room for the growth of developing countries. This requires a transition to lifestyles that are less wasteful and indulgent, more modest in their use of resources and the pressures they exert on the environment. The shifts in culture and values which must provide the basis for this transition will impose challenging responsibilities on our educational system, and on the religious leaders, communicators and public figures who are the primary agents of change.

Developing countries point out that the industrialized countries are largely responsible for these risks, and have been the main beneficiaries of the wealth that has accumulated through the processes of economic growth that have produced them. They insist that they cannot divert resources required to meet the most immediate and fundamental needs of the people, to pay the additional costs of incorporating into their development policies and practices the measures needed to reduce major global risks. At the same time, they ask that the industrialized nations take seriously the compelling case made by the Brundtland Commission that the common need for global environmental security requires a substantial and sustained increase in the flow of financial resource to support the broad development needs of developing countries.

Developing countries have, in many respects, become the victims, rather than the beneficiaries, of the recent globalization of the world economy. Interdependence has made their fragile economies highly vulnerable to changes in the world economic conditions over which they have no control. It compels them to compete in an international market-place in which the principal sources of added value and comparative advantage are technology, capital, management and marketing skills and scientific knowledge. In all of these areas the developing countries are seriously handicapped. They are often compelled to over-exploit the natural resources on which their future development depends.

The primary responsibility for the future of developing countries rests, of course, with them, and their success will depend largely on their own efforts. But they deserve and require an international system that lends strong support to these efforts. This includes substantially increased financial assistance, and much better access to markets, private investment and technology to enable them to build stronger and more diversified economies, to effect the transition to sustainable development and to reduce their vulnerability to changes in the international economy. They must be helped to break the vicious circle of ecological deterioration and economic decline and to adopt environmentally sound patterns of natural resource use and agricultural production. They should be especially encouraged and supported in their effort to upgrade their capacity to add value to the natural resources and commodities on which they depend primarily for their export earnings. Traditional programs of technical co-operation and educating and training people from developing countries in our institutions have made only limited progress towards meeting this need. Indeed, the situation is becoming more acute. To some extent these well-meaning programs have compounded the problem by educating developing country people without supporting the institutional development that would enable them to pursue their professional careers in their own countries. Thus, while some of their brightest and best work outside of their countries, foreign experts fill the gap at much greater cost than it would take to retain indigenous expertise. In some countries, like Tanzania, the cost of supporting foreign experts exceeds the entire budget of the national public service.

The South Commission recognized the pressing need for developing countries rapidly to build their own human skills and institutional capabilities so that the pace of absorption and diffusion of science and technology can be substantially accelerated. The key to self-reliance is to foster a pool of indigenous talent that can adapt and innovate, in a world where knowledge is the primary basis of competition.

A new thrust for capacity building would mark a major contribution towards mobilizing the South towards greater co-operation among developing countries, as the South Commission advocated. The UNDP's Sustainable Development Initiative offers a promising framework for co-operative arrangements of this nature at the regional level. There is a strong regional component to the UNCED preparatory process, and the UN Regional commissions are collaborating with the regional development banks, the UNDP, UNEP and other regional and international organizations to design a sustainable development and

technology support system that I hope will bring a major new impetus and support for the strengthening of institutions and human development programs in all the regions of the developing world. There must be a special emphasis on the least developed countries, and particularly on sub-Saharan Africa.

The answer to the global predicament lies beyond our actions in the realm of the values which motivate them. This is especially true of science. The scientific community, playing as it does a seminal role in the informing of decision-makers upon these issues, should take greater responsibility for the direction of its development and for the applications of its discoveries. It should, for example, develop a code of environment and development conduct, readily accessible to the public, so that its guiding principles are transparent and the scientific community itself accountable to the other sectors of society. The scientific community could then base its operations on a societally acceptable set of ethical principles, much as the medical profession has done for over a thousand years. Codes of practice or ethical rules to protect the biosphere and future generations need to be codified, agreed and accepted.

UNCED itself is addressing the issue of the conduct of nations and peoples in respect of environment and development in its "Earth Charter" or Declaration of basic principles to ensure the future viability and integrity of the Earth as a hospitable home for human and other forms of life. A contribution from the scientific community of elements for inclusion in the Charter would be most valuable, as would the adoption by the scientific community of its own Charter for Science.

Science must develop a higher visibility within other sectors of society. In particular, the scientific community must be more accessible, and be seen to be responsible in its relations to policy-makers and the general public.

For this reason I want to touch for a moment on a particularly important partnership: between scientists and policy-makers. The importance of new research to expand our understanding of the Earth's natural systems cannot be underestimated, but another kind of response is called for from the scientific community, at the interface of science and policy-making. Clearly, the scope of the science input to the policy-making process should be commensurate with complexity of the problem at hand, and the relevance and need for interpretation of scientific evidence surrounding it. Too often in the past, however, scientists have limited their involvement with policy-makers to merely presenting them with their research findings, and have often been far from satisfied with the manner in which their discoveries have been employed. The forte of scientists has been in pure research, and has, ironically, been weakest in just that area where it might best inform decision-makers upon the options and implications of their decisions. Scientists have a vital role in advising upon the possibilities and limitations of social change and governance that lie at the heart of the transition to sustainable development.

The relationship between scientists and decision-makers should not end with the results of research being made public, but should be continuous, for expert advice, interpretation and prediction is required at all stages of the decision-making process, and thereafter through implementation and monitoring.

The integrated management approach of IPCC, with its three working groups focussing on assessment, potential impacts and response strategies, is an important development, but it may be that its intergovernmental character would limit its usefulness as a model in some instances.

The science/society relationship also needs an overhaul. In the modern world, science, and its application in technology, have dramatically changed the ways in which a large percentage of the world's population lives. But the impact of science on people's social attitudes and our institutions is lagging far behind.

In the absence of good communications between science and society, a high level of distrust of scientists and their work is engendered in the public mind. This is visible in public concern about sensitive issues such as genetic manipulation, and is inevitable in a world where a large element of the scientific community has been employed in research and development for the military and nuclear industries.

Scientists bear a heavy responsibility to convey their knowledge in a manner that gives a fair and timely presentation of the facts, uncertainties and risks surrounding it. Of great value to public and decision-makers alike, would be a mechanism by which the best scientific views on major issues, even where there was as yet no consensus, can be made available. The state of knowledge on these issues, the consensus or varied opinions within the scientific community as to the interpretation of their findings and their significance, could be conveyed by means of intermittent reports, or even public hearings. This is an area where ICSU could take the lead and this conference could provide guidance.

Scientists have traditionally been reluctant to provide guidance before all the evidence is available, because they are, quite rightly, mindful of the risks of action upon incomplete evidence. But policy-makers and the public need continuous information and guidance from the scientific sector, so that action must be guided by the best available evidence. Uncertainty is a feature of life, and if we are to rise to the challenge of managing the change brought upon ourselves, then we must learn to respond appropriately to uncertainty.

Equally, decision-makers should not be allowed to hide behind science in the name of seeking for scientific certainty before acting. Certainty in science is an impossibility, and decision-makers are used to making policy decisions under conditions of considerable uncertainty – that is what they are good at – weighing the balance of risks and, where appropriate, taking out "insurance" policies. This is why the "principle of precaution" is so important to be understood clearly.

A Canadian geologist W.S. Fyfe, summarizing the proceedings of a symposium on global change and the IGBP, concluded: Many of the problems with which the IGBP is concerned are too complex to be solved by the traditional disciplines. Members of our very traditional universities always have a little fear of moving out from the secure havens of specialization but, when the best minds are involved, I think the rewards which follow from such motion are great. We are involved in new science." There is a growing movement within the scientific community that endorses the need for a new science. While this raises some difficult and controversial questions that have yet to be adequately addressed within the scientific community, I believe the new approach has great promise.

Fortunately we do not start from scratch. There are a number of excellent programs underway that offer an embryonic glimpse of the new science.

The Man and the Biosphere Programme encourages interdisciplinary research, demonstration and training in natural resource management. MAB thus contributes not only to better understanding of the many factors that affect environment–development interactions but to a greater involvement of scientists in deciding how resources can be used more wisely.

ICSU's own Scientific Committee on Problems of the Environment, SCOPE, focusses on major international environmental problems, selecting those that are both interdisciplinary and tractable. Examples are the landmark SCOPE Study on the Environmental Consequences of Nuclear War or the promising work on the "Greenhouse Effect". The work of SCOPE has also drawn attention to the importance of understanding the major biogeochemical cycles in order to understand the Earth's ability to produce resources sustainably, and to assure stability for ecosystems and indeed the conditions that make life on Earth possible.

Despite some excellent work, scientists are the first to admit that what is needed is a more root and branch approach to understanding global cycles. The IGBP (International Geosphere–Biosphere Programme) initiative, attempting to describe and understand the interacting physical, chemical and biological processes that regulate the total Earth System, is an important step in the right direction as is the focus on the atmosphere, ocean, terrestrial biosphere, ice and the hydrologic cycle in the World Climate Research Programme. HDGEC (Standing

Committee on Human Dimensions of Global Environmental Change) drawing as it does on expertise from the physical and social sciences, will, I am sure, provide very useful and timely work as it considers how human activities contribute to, and are affected by, global environmental change.

A heavy responsibility for the transition to sustainable development falls upon science, and from the results of this conference I can see that this responsibility is taken very seriously by the scientific community and is in good hands.

Closing address by M.G.K. Menon,
President, International Council of Scientific Unions

This Conference came into being through a very elaborate preparatory process. In September last year, in Paris, during the Scientific Advisory Council meeting of the International Geosphere–Biosphere Programme, a Global Change Forum was also organized under the auspices of ICSU's Advisory Committee on Environment. This Forum was organized because ICSU's responsibility to harmonize its many major activities in the environment and to make sure that information about each of them is circulated fully.

We requested Mr Maurice Strong, Secretary General of the UN Conference on Environment and Development, to come to the Global Change Forum and to address us. It was during a preliminary discussion that he suggested that he would very much like ICSU to be the principal scientific advisor and to provide scientific inputs which would be of value to UNCED. Mr Strong reiterated this invitation at the Global Change Forum. We accepted that responsibility and proceeded to organize what has come to be called ASCEND 21 as well as a number of other activities.

As you are also aware, the Norwegian Research Council had organized the Bergen Conference in May of 1990 and initiated what is referred to as the "Bergen process". This is why ICSU invited the Norwegian Research Council and the Norwegian Academy of Sciences to join with the other organizations co-sponsoring this Conference.

The approach adopted, bringing together natural and social scientists to look at these global problems on a highly interdisciplinary basis, was in a number of ways unique. Normally ICSU, and natural scientists in general, tend to discuss these questions among themselves; and to achieve interdisciplinarity within this framework is itself usually difficult. Hence, extraordinary preparation was called for to ensure the necessary interaction among natural and social scientists.

In this audience you can see some of the very best experts in specific areas relating to the atmosphere, the oceans, the terrestrial system, biodiversity, population, and indeed all the other issues covered by the sixteen ASCEND Conference themes. There have been many scientific seminars, discussions and conferences which relate to these specific themes, but what has been attempted at the ASCEND 21 Conference is to bring them all together in an integrated manner.

The physiology of the Earth's environment, what happens when various driving forces act upon it, what our responses should be, how swiftly we can initiate such responses, what science can do about these issues – such are the questions that have been before us the whole of this week. We have not discussed them purely as technical questions in a confined, narrow framework. Instead, we have felt it more important to consider them within an overall interactive approach.

In organizing this Conference we had to have appropriate facilities and support, and I would like to thank most warmly and express a deep sense of gratitude on the part of ICSU and its partners to the Government of Austria. Thanks to our hosts, not only were all of the excellent facilities provided here in Vienna which have made this Conference possible, but also generous support was given to enable a very large number of specialists, particularly from developing countries, to attend. The inaugural address by the Chancellor of Austria, Dr Vranitzky, is a clear indication of the importance that the Government of Austria attaches to this Conference.

We are also grateful to the McArthur Foundation, to UNDP, and to each of the co-sponsoring organizations for all the support they gave: financial, intellectual and material. The Conference was organized under the auspices of our Advisory Committee, which set up a Bureau responsible to a great extent for the detailed structuring and arrangements for the Conference. The Bureau was chaired by Professor J. Dooge to whom we are most grateful, as also to the other members of the Bureau.

ICSU organized this Conference through its Secretariat, headed by Julia Marton-Lefèvre, Executive Director, who was supported by all the individual members of the Secretariat staff with whom you are all well familiar by now particularly Tish Bahmani Fard, Maureen Brennan, Bruno Durand, Elisabeth Merle and Mike Millward. They are part of the small structure that keeps ICSU's activities going all through the year; and they all enthusiastically assumed the additional burden of organizing ASCEND.

I believe that we are at an extraordinary moment. Often in such a situation, as in the eye of a storm, or a hurricane, there might be a sense of calm: all the same there is a storm around though one may not realize it. Scientists normally regard their lives as wholly built around their research, motivated by curiosity. And this they will always continue to do. However, we have seen how the rapid development of science results in technology, which can then result in applications which are transforming society and its functioning. One has evidence of the manner in which scientists become technologists, or technologists divert or relate to science. The tempo of this symbiosis and the rate of development of capabilities in science and technology have given rise to new global situations. As we move into what are totally new domains, we are asking basic questions about the direction and the manner in which we are being led by the rapidly increasing knowledge and the powers that science is conferring on humanity. Francis Bacon had said "Knowledge and power are co-extensive". We have taken a certain pathway up to now and been brought to the present situation. The question we now ask is: what is the nature of the situation we encounter today, and where do we go from here?

Some aspects of the predicament are described in the Conference Statement, based on the extensive interactions over the last week. Many of you have raised your hands to say the situation is of great concern and we must ring the alarm bells. One need not be alarmist, but at the same time one must have a sense of urgency in dealing with the problems with which we are are confronted. There is no room for complacency. That has come out very clearly here this week. We are indeed in an extraordinary situation at this Conference for, while we take stock of the situation, we are asking ourselves: what can science begin to do about this as of now? As scientists we understandably wish to do good science, but we wish to do more, too-we wish to ensure that our future is not endangered. It has also emerged most distinctly that, with our present knowledge and capabilities, there is a great deal that can be done immediately to treat some of the problems confronting us. We need not wait for more research or more studies. We should not make up alibis for waiting much longer. We have to deal now with those problems which can be tackled, and we must do it at the lowest cost, with the greatest returns, and benefits for society as a whole.

On the other hand, we also realize that we need to know more. Although many uncertainties prevail and are often reflected in the scientific data that are published, society tends to regard such information as absolutely correct and reliable because it has come from science. We have to beware of this. There has to be still more research, which is

why major efforts such as the World Climate Research Programme or the International Geosphere–Biosphere Programme need to be supported. Furthermore we have to realize that global change is not just climate change. The great problems of loss of biodiversity, of destruction of precious soil cover, of population increase and the like are immediate problems which we face and on which we must work. All of this must be supported by society.

We scientists have to demonstrate to the world our concern about immediate problems of environment and development. Many at this Conference have come from the developing countries, and are aware of the poverty, the deprivation and the severe degree of underdevelopment which prevail there, with all the consequences which we would like to see disappear from this Earth. Therefore we have a very significant responsibility in respect of stimulating development. We must tackle this in a spirit of international solidarity. Science is international and that is what ICSU stands for.

I should like to point out that ICSU, in its present form, was founded in 1931. So this year, 1991, marks the fulfilment of 60 years of existence of our organisation. Many wonder what is the significance of 60 years. In these parts of the world it is regarded as a diamond anniversary. In mine – India – it is also regarded as a very important milestone in human life. When a person reaches the age of 60, he or she moves out of the pathways of materialism into a new world of contemplation, of meditation, of the pursuit of truth and the searching for the content and meaning of life. This is when the human spirit is given pre-eminence. We in ICSU, at this age of 60, are looking toward this new life, toward new pathways and new developments which will be sustainable – where science can play the extraordinarily important role of which it is capable: that of contributing to human welfare. However, it cannot play that role on its own, by remaining confined to laboratories and narrow disciplines. It has to ensure interdisciplinarity within the natural sciences and then relate to all of its partners in the social sciences, industry and business, governments, intergovernmental organizations, non-governmental organizations and, last but not least, various elements in society not all of which are institution-alized. As awareness of all these areas and domains is raised, it must be pointed out that we do not know everything precisely. Yet the approach has to be the method of science, seeking to look at problems on a rational and objective basis, and demonstrating that such an approach can solve many problems. We have demonstrated that in the past and I think we can do so again on issues of far greater importance to humanity, namely, the achievement of sustainable development with social justice, equity and solidarity. That is the challenge before us.

This particular Conference is not just an event whereby we came together, talked about various important issues, dispersed and then produced a book. We also have a Statement which has been delivered to UNCED. But most importantly, we have to start a new and continuing process – an Agenda of Science – as we move forward in these areas of vital human concern.

In this regard I would like to make a formal statement which will outline ICSU's plans.

The contributions of science to environment and development have been rapidly growing over the past few years and so have the expectations of those responsible with respect to the assistance science can provide in their policy making and programs. Consequently, and in accordance with its responsibility in the scientific community ICSU has decided to take the following steps as a follow-up to both ASCEND 21 and UNCED:

(a)   to further consolidate the co-operation and coherence of ICSU's major international research programs on the global environment and to promote their collaboration with relevant programs outside ICSU with special attention to be given to co-operation with the social sciences;

(b)   to strengthen further its capacity to play its role in the evolving partnership among science, governments, intergovernmental organizations and business and industry;

(c)   to strengthen the capacity of ICSU to prepare objective scientific assessments of issues of environment and development within its competence;

(d)   to enhance the efforts of ICSU to report scientific issues to the general public as well as to decision-makers in a readily understandable manner;

(e)   to strengthen its program for capacity building in science in the developing countries;

(f)   to declare its willingness to play a role in helping whatever mechanism is set up to review Agenda 21 performance, e.g. by evaluating adequacy of scientific inputs or by helping define indicators by which sustainability and progress towards it can be measured.

At this Conference old partnerships have been reaffirmed and new ones have been forged. These partnerships will be of great importance in carrying forward actions that are on the agenda of ASCEND 21 and UNCED.

And with this determination on the part of ICSU to press ahead into the future, I hope that we can contribute meaningfully, through your efforts and those of all our colleagues who are not here today, towards sustainable development.

Thank you all very much for the contribution you have made during the past week.

I. Problems of Environment and Development

Introduction

G. T. Goodman

The papers in this Section deal with population growth and *per capita* resource demand; energy demand; food demand (from the point of view of "agriculture, land use and degradation"); industry and waste production; and human health. During the discussion at ASCEND 21 it became clear that the closely interlinked, twin topics of population growth and *per capita* resource demand were widely regarded by the participants as the crucial driving forces at the center of the whole relationship between environment and development and this should be recognized in formulating the Agenda of Science for Environment and Development in the decades ahead.

Chapters 1 to 5 which follow treat the above topics on the basis of the discussions at ASCEND 21. Some of the more important issues and conclusions from these chapters are outlined below. Many recommendations emerged; these are reflected in the revised papers and the Conference Statement. Many others were concerned with accelerating the application of existing knowledge or the implementation of policies already known to be effective. Although these cannot form part of a new science agenda and are not reported in this volume, the frequency with which such items came up was a revealing testimony to the enormous amount of work that could be done immediately given appropriate resources and political will.

Population and Natural Resource Use

The present rapid rate of population growth is predicted to deliver a global total of at least 8 billion, and more probably, of 10–14 billion before stabilizing around the closing decades of the next century. Population growth, particularly rapid in many developing countries, often places demands on resources which exceed the sustainable yielding capacity of local ecosystems.

Present inequities in living standards world-wide indicate that industrial countries representing about 20% of the global population consume almost 80% of global resources. The patterns of high and often wasteful resource consumption in industrialized countries cause serious environmental damage from waste emissions not only in such countries themselves but also on a global scale.

There is widespread agreement that two complementary policy goals must be addressed:

1. Population stabilization at as low a level and as rapidly as possible, by reducing fertility in developing countries, coupled with improvement in land management to reduce soil and land degradation.

2. Reduction of excessive *per capita* resource use and waste production to a lower level as rapidly as possible in industrialized countries, coupled with the repair of pollutant damage.

In view of the complementarity of population and resource use in determining the total impact of human populations on the planet, it is only meaningful to use the product of population and resource use *per capita* as a measure of human impact. This makes it easier to select an appropriate balance between human numbers and individual consumption for the purposes of policy development.

There remains an important question of timing. Can societies respond to these imperatives of population and resource use management faster than the time scales of the expected environmental and developmental deterioration? This must form the subject of a major systems research effort. The work could be carried out by some amplified form of co-ordinated monitoring and evaluation of development and health indicators linked to economic and resource variables.

Chapter 1 discusses the nature of these problems and emphasizes the need for agreement on human goals and research priorities. The authors stress the need for a hierarchy of goals for planning and management; the need to adjust incentives to reflect long-term global costs including the cost of uncertainty, and the necessity of arresting the decline in natural capital. They outline a priority research agenda in relation to the key areas of resource use, technology population and health, culture and education. In the discussion on this theme at the ASCEND 21 Conference, it was repeatedly stressed that success in this general area depended on the promotion in all countries of a full social role for women.

Agriculture, Land Use and Degradation

The physical, socio-economic and cultural factors conditioning land use and degradation are closely linked to population and *per capita* resource use generally. Any solutions will require multidisciplinary analysis and policy formulation by the natural, social and engineering sciences and are long overdue. Current statistics appear to show that available land is insufficient for meeting growth in agricultural and non-agricultural demand world-wide projected for the next 35 years.

Clearly defined property and user rights of rural communities would give local farmers a vested interest in preventing degradation. New institutional arrangements are needed for this and for fostering closer links between farmers and scientists responsible to government so that research is sensitized to farmer needs, and good local practices can be incorporated into official policy.

Most of the priority research and development (R&D) needs could be captured by a system of monitoring and evaluation of land degradation, coupled with the development of methods for the more realistic economic valuation of land as a natural asset. Last but not least, the need for integrated monitoring and research will require changes in management organization and funding and a major investment in capacity building world-wide which assures adequate career structures and financial rewards in many countries where the absence of these causes loss of trained expert personnel.

During the working group discussion on this theme at ASCEND 21, it was stressed that the topic should be considered as including the whole of rural economy and development. Chapter 2 reviews the resource base issues, assesses present and projected land use requirements, and examines the key problem areas for ecologically sound and sustainable development. The authors recommend three research areas as having top priority: (1) filling the serious gaps in the basic data base on actual land use, rates of land degradation, and the physical results and economic costs of land degradation; (2) research on institutional structures required at all levels from local to national to improve land use management; (3) continued research to improve land productivity.

Industry and Waste

Industry is the main agent for converting primary resources and other materials into a wide range of products useful to human society. Its manufactured goods are indispensable to the modern, densely populated,

urbanized world of affluent consumption. These products are made using great quantities of raw materials and energy, and generating large amounts wastes, often toxic in nature.

Industrial activities have traditionally been associated with environmental pollution, but modern industry is becoming very waste conscious and so consumer disposal is assuming an increasingly larger percentage of total waste. This industrial trend towards energy and raw materials efficiency and "clean" production is likely to be the precursor of a new era of industry, operating in a much more sustainable mode.

But how much time is available to human institutions to restructure before running into widespread environmental stress? There is a wide spectrum of views: at one extreme there are those whose experience leads them to believe that the Earth System is relatively robust and resilient. At the other extreme, there are some who see evidence that the biosphere is already more stressed than is healthy, with the response times of human institutions too slow to adapt to avoid serious trouble.

This crucial question has already been raised in relation to population and resource use generally and requires a major systems research effort to clarify the likely outcomes. Connected with this are important institutional questions which are developed in the paper on Industry and Waste.

Because of the individual views of the three authors, Chapter 3 reflects two conflicting viewpoints on whether a gradual approach based on speeding up present tendencies towards developing clean technologies would be adequate to meet potential global risks or whether a more radical approach is required. The chapter thus encompasses the elements of a debate that is central to the whole subject of the relationship between environment and development.

During the discussion of this theme at ASCEND 21, it was pointed out that the topic of industry and waste had close linkages with other themes of the Conference. Consequently, the material in Chapter 3 should be considered in conjunction with Chapter 12 which deals with the quality of life and its effect on consumer demand, with Chapter 13 on public awareness which can play a key role in influencing lifestyles, with Chapter 15 on policies for technology which should include the tools for transition to a new form of industry, and Chapter 16 on institutional arrangements which deals with the structural reforms required for effective action.

Energy

As a result of intense study and research, energy is one of the best known and understood of our global resources. Yet the current system of energy supply is totally inadequate to meet the needs of the majority of the world's population, particularly the poor, a situation which frequently causes land degradation from over-harvesting of woody biomass for fuel in developing countries. By contrast the generous and often wasteful use of fossil energy in the industrialized countries causes trans-boundary pollution in the form of acidification and the build-up of atmospheric carbon dioxide with potentially dangerous consequences from global warming. Expansion of the present supply system would further damage the environment world-wide to unacceptable levels, is unlikely to solve the shortages of the poor, and will be immensely expensive.

A large transdisciplinary research program will be needed on the technical, economic and socio-political implications of a timed and costed transition to a new energy supply system. The economic and institutional requirements to catalyse this transition and to transfer the new technologies to developing countries also need detailed elaboration. Once again there is an important time-frame question surrounding this transition. The pricing policies adopted throughout should reflect environmental and health costs and could be considered as a mechanism for guiding the nature of the supply mix.

The debate on the energy future at ASCEND 21 has been characterized by conflicting claims and statements. The authors of Chapter 4 in their survey of the present position and future outlook reduce this degree of conflict to a variation in emphasis. The working group discussion which reviewed the earlier draft by the authors reflected a general acceptance of their approach modified by an occasional variation in degree of emphasis.

Chapter 4 reviews the current global energy situation and its impact both on the environment and on living standards. Future options are explored in the perspective of more or less "business as usual" scenarios. The ingredients of an alternative energy policy are also outlined. The chapter describes the research priorities and financial costs required for the implementation of such an alternative strategy.

Health

Health, especially in children, is closely linked to nutrition (particularly consumption patterns associated with lifestyle), environmental contamination and poverty. "Health for All" is now regarded as a basic human right. It is the responsibility of governments to ensure that their populations receive basic hygiene and health education, are given opportunities to lead healthy lives and are protected from the spread of communicable diseases.

Although a great deal of work is actively being done in the fields of nutrition, public health, disease prevention and immunology, as well as environmental effects, a great deal more could be implemented without any new R and D. But R and D is still needed on the epidemiology associated with various dietary patterns and nutritional states, the production of new vaccines and the social factors controlling the improved monitoring and surveillance of diseases and health indicators, as well as the factors affecting the spread and control of old and new diseases. Like energy supply, many of these issues are related to the rich–poor gap which needs to be narrowed for health matters among others.

Chapter 5 sets forth the reasons why health should constitute a major subject on the agenda of UNCED and in the actions undertaken following that Conference. During the discussion of the original draft at ASCEND 21, it was emphasized that the diagnosis of the world health situation in that draft paper implied the necessity of including a number of topics as priority areas in the agenda of science for health. The working group identified the following as high priority research areas: human reproduction, human feeding patterns and nutrition, new vaccines and therapy based on recent advances in science and technology, new methods of diagnosis, surveillance and prevention of communicable diseases, systematic studies of environmental epidemiology, local effects on human health of ecosystem degradation and inappropriate technology, and socio-behavioral studies of human attitudes to health and sickness.

Chapter 1: Population and Natural Resource Use

L. Arizpe, R. Costanza and W. Lutz

EXECUTIVE SUMMARY

Human use of natural resources has continued to increase dramatically in recent years, as a result of complex linkages between increasing population and increasing *per capita* consumption. This has begun to have significant effects on the planet's climate and long-term life-support functions. Future projections of human population range over an order of magnitude by the year 2100, depending on assumptions about rates of fertility and mortality, and indicate a significant aging and urbanizing of the population.

On a finite planet, growth in resource use and population is not sustainable indefinitely. To achieve a sustainable pattern of resource use and population we must understand and control the interactions of population and *per capita* resource consumption as mediated by technology, culture, and values. We must also be careful to differentiate between economic growth (increase in quantity of production and consumption) and economic development (increase of human well-being without a necessary increase in consumption). Technological optimists assume that new technology will be developed to eliminate any resource constraints to continued economic growth. Technological skeptics argue that while this *might* happen it is irrational to count on it. Because of the large uncertainty about the long-term impacts of population and resource use growth on ecological sustainability, we should *at least provisionally assume the worst* and plan accordingly. We need to focus research on the understanding and control of linked population, consumption, and distribution patterns and not count on a technical fix and continued economic growth to solve the problem.

1. INTERACTIONS BETWEEN HUMAN POPULATIONS, TECHNOLOGY, VALUES, AND NATURAL CAPITAL

The crux of the problem of assuring a sustainable world is understanding the full range of possible interactions between, and among, humans and their natural environment, and choosing from this spectrum forms of interaction that are sustainable. While acknowledging that the sustainability concept requires much additional research, we employ the following working definition: *Sustainability* is a relationship between dynamic human cultural/economic systems and larger dynamic, but normally slower-changing ecological systems, in which a) human life can continue indefinitely, b) human individuals can flourish, and c) human cultures can develop but in which effects of human activities remain within bounds, so as not to destroy the diversity, complexity, and function of the ecological life-support system (Costanza *et al.*, 1991).

In this context the way human populations interact with natural capital (renewable and non-renewable natural resources) is critical. We need to develop a much deeper understanding of the relationships between human populations, their technologies, cultures, and values, and the natural capital they depend on for life support if we are to achieve sustainability. Science and technology alone, however, will not achieve this: it is essential that governments and local peoples are mobilized in a major collective effort to ensure the survival of humankind.

1.1 Trends in human population and environmental transformations

The one common agreement of all human societies is the wish to overcome death; hence the unprecedented alliance of science, government, economic and cultural agents, and communities which effectively brought down mortality all over the globe. Can such agreements be built on fertility, *per capita* resource use, competition for resources and the environment? Such issues touch needs and values which are at the heart of different societies so it is no wonder that ". . . in the late 1980s, despite the long history of intellectual efforts invested in the population and resources issue, the field appears to be in disarray, with little prospect for the emergence of scholarly consensus" (Demeny, 1988: 237).

While scientific research is becoming more specialized

and precise, the debate about the global relationship between population and the environment remains polarized between two positions, each of which has a partial scientific understanding of the problem. One position holds that an increasing population is the principal threat to the environment because of the planet's finite resources[1]; the other that human creativity will continue to find technological solutions to expand the planet's carrying capacity. Neither position represents the state of the art of scientific understanding (Johnson and Lee, 1987; Repetto, 1987; Keyfitz, 1991).

These positions, which have moved from the scientific to the public area in popular renditions of the Ehrlich-Simon debate, focus on uni-causal linear explanations and on macro-level analysis using world-wide historical statistics or aggregate simulations. The debate has not taken into account the wealth of data generated at the micro-level, data which provide information on the determinants of fertility and child mortality, and on patterns of production, consumption and distribution. These complex patterns of human relationships overlay, alter and modify the relation of people to land and natural resources. Science must move away from single factor explanations about numbers[2] of people and resource limits and be directed to the analysis of the increase in *per capita* use of renewable and nonrenewable resources rather than to numbers of people alone (Demeny, 1988: 217; Harrison, 1990; Durning, 1991).

A scientific framework must be built overcoming the division between contending camps of North and South, rich and poor, since environmental problems are global and regional in causation and in their effects, and they require a concerted effort to arrive at global solutions. In addition, such a division oversimplifies the complex issues that are really a continuation of over-population and underpopulation, overconsumption and underconsumption, and sustainable and unsustainable livelihoods. Adopting a different framework will serve as a corrective in itself: much of the present debate in the North is about population policies to implement in the South rather than about curbing overconsumption in the North (Worldwatch Institute, 1988).

No simple correlations can be established between population and environmental transformations. In a recent study it was found that the time scales of population variability are asynchronous with given environmental transformations and recovery (Whitmore *et al.*, 1990). The authors stress "the need for caution in using population as a simple surrogate for environmental transformation" (Ibid.).

A major heuristic problem is the great regional heterogeneity of population growth. Whitmore *et al.* (1990) conclude that "if the experience of past regional population changes and their accompanying environmental

transformations has relevance for the future, the projected global scale population "leveling out" need not diminish the scale and profundity of global environmental change. This is particularly true on the regional or local scale, where global zero population growth (of population or transformation) need not be accompanied by local or regional equilibrium" (Whitmore, 1990).

If one shifts the analysis to the *per capita use* of resources, then priority research areas are the perceptions and assessments of people as to their needs, incomes and desires, as well as their comparative perceptions of the needs, income and desires of other groups. Research on the latter must be extended to analyzing the acceptability of different negotiation options as to who is going to bear the costs of adjustments of *per capita* resource use. We need to know much more about the institutional, and social capacity for adjustment and adaptation (Jacobson, 1990).

1.1.1   Population growth up to 1990

With all the qualifications given above, there is little doubt that the recent extremely rapid increase and the projected further increase of the species *Homo sapiens* on our planet is a major factor affecting global resource use which, together with technology and consumption patterns, contributes to the stress put on the natural environment.

Figure 1 gives a picture of this extreme acceleration of the growth of the world population. While little is known about prehistorical times, estimates put the total world population of 7000–6000 BC at about 5–10 million people (UN, 1973). For millenia there seem to have been fluctuations in population size with a fundamental influence exerted by climatic conditions. Fossil remains show that life was short and frequently ended in violent death. By the year 1 AD the world had an estimated population of 200–400 million. This increase had been made possible by improved agricultural techniques and the development of human civilization. But it was not until the 17th Century that population growth started to really take off. The first billion people living on the Earth was reached somewhere around 1830. The second billion was reached in the 1930s, the third in 1960, the fourth in 1975, the fifth in 1987, and the sixth will most likely be reached around 1997. Hence the time it requires to add an additional billion to the world population decreased from more than 100 years for the second billion to about 10 years for the sixth one. Under current growth rates the world population would double again in 40 years.

This unprecedented growth in the number of humans has been brought about by a decrease in death rates while birth rates remained at their traditional high levels[3]. In Europe and North America significant mortality declines were already occurring during the second half of the 19th Century, mostly due to improved sanitary conditions and food security. Since then life expectancy has steadily

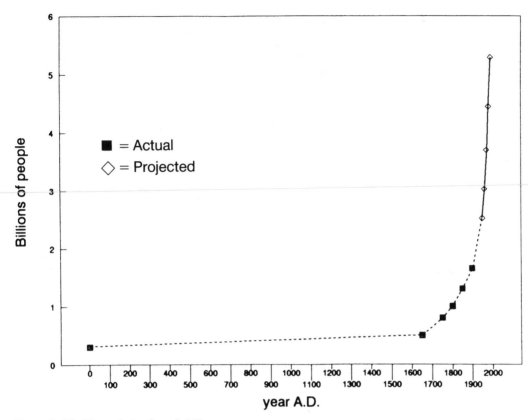

Figure 1: World population from 0 A.D.

increased in the more developed world. On a global scale the most significant change in mortality came in today's less developed countries (LDC) after World War II with a generally very successful fight against infectious diseases. More recently rates of mortality improvement have declined because the easy measures had already been taken and those requiring infrastructural or behavioral changes proved to be much more difficult.

On the fertility side, birth rates entered a secular decline in Europe and North America during the first decades of this century. This led to a reduction in the growth rate of the population. The phenomenon that birth rates follow the decline of death rates with some lag has been described by the notion of demographic transition. While in the more developed regions this transition is essentially completed today, in most LDCs it is still under way; in some regions – especially Africa – the fertility transition has hardly started.

1.1.2 Future population growth

Changes in population size and age structure have great inertia. Unless wars, famines, or epidemics kill significant proportions of the population, or massive migratory streams empty some regions and fill others, the future population of a certain region can be projected with high certainty in the short run. Projections for the next 20–30 years are rather reliable and insensitive to minor changes in mortality, migration, and fertility. In the following scenarios, we estimate that even widely differing

assumptions will yield almost identical results up to the year 2010–2020. But thereafter the range of possible futures opens widely (Lutz and Prinz, 1991).

The inertia of population changes also implies that there can be a great impact in the long run of only minor differences in assumed fertility trends over the next few decades. To take an extreme example from the scenarios: Whether a certain low fertility level in Africa will be reached by 2025 or 2050 will make a difference in total population size by the year 2100 of more than 1.5 billion, which is almost three times Africa's present population size. This incredible inertia that makes it such a difficult and long-term issue to stop population growth is also called the momentum of population growth, the fact that the age structure of a fast growing population is so young that even if fertility per woman declined to a very low level, the increasing number of young women entering reproductive age will cause the population to grow further for quite some time.

The scenario approach chosen here calculates the implications of several alternative possible future paths of fertility and mortality that need not necessarily reflect the present "mainstream" thinking. The main value of such a set of alternative scenario projections (that are based on controversial but informed guesses about the future) lies in giving a picture of the possible range of future population sizes and structures. This will help to distinguish almost inevitable trends from changes that are very sensitive to slight modifications in the assumption.

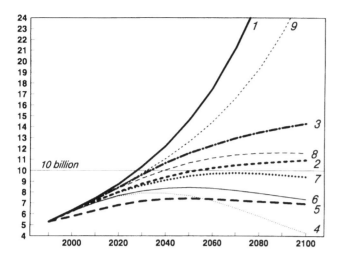

Scenario 1: **Constant Rates**; constant 1985-1990 fertility and mortality rates
Scenario 2: **UN Medium Variant**; strong fertility and mortality decline until 2025, then constant
Scenario 3: **Slow Fertility Decline**; UN fertility decline 25 years delayed, UN medium mortality
Scenario 4: **Rapid Fertility Decline**; TFR = 1.4 all over the world in 2025, UN medium mortality
Scenario 5: **Immediate Replacement Fertility**; assumed TFR = 2.1 in 1990, UN medium mortality
Scenario 6: **Constant Mortality**; TFR = 2.1 all over the world in 2025, constant mortality
Scenario 7: **Slow Mortality Decline**; UN mortality decline 25 years delayed, TFR = 2.1 in 2025
Scenario 8: **Rapid Mortality Decline**; life expectancy of 80/85 years and TFR = 2.1 in 2025
Scenario 9: **Third World Crisis**; constant fertility and 10% increase in mortality in Africa and
    Southern Asia; TFR = 2.1 in 2025 and UN mortality for the rest of the world

Note: "TFR" is the Total Fertility Rate (= average number of children per woman)

Figure 2: Total projected world population 1990–2100 according to scenario.

Figure 2 plots the total population sizes resulting from the nine scenarios considered here with the projections performed separately for six major regions of the world. In 1990, our planet accommodated 5.3 billion people. Under all scenarios considered over the next 30–40 years, the world population will increase to a size of at least 8 billion. Even immediate replacement fertility in all parts of the world would result in an additional two billion or more people, due only to the momentum of population growth. Under the unlikely rapid fertility decline scenario, assuming only two-thirds of replacement fertility by 2025, the total population size would peak in 2040 at around 8 billion and only decline thereafter.

We may conclude that, on the lower side of the spectrum, unless unexpected major threats to life kill great proportions of the world population over the coming 30–40 years, the world will have to accommodate an extra number of people that is at least as large as half of the world population today. Under the assumption of sustained subreplacement fertility in all regions of the world, the population might then decline again in the very long run, and possibly by the year 2100 reach a size that is lower than that of today. But still in the transition period the 8 billion mark will be touched.

At the higher end of the spectrum, we have to distinguish between the scenarios that look at the case of continued growth and those assuming a levelling off. Obviously exponential growth cannot continue forever and therefore in the longer run is not only unrealistic but also

impossible. Nevertheless, it is instructive to study the results especially in the short- to medium-term future, and compare them to other scenarios. Furthermore, an assumed continuation of currently observed levels is in almost every scientific discipline a standard for comparison unless there is certainty that this level will change in one specific direction. In the case of population growth, we have good reason to assume a change in rates, but we are far from any certainty about the extent and the timing of this decline. In any case, because of the inertia, over the coming three decades the Constant Rates Scenario (1) will not result in very different total population sizes than most other scenarios. Around 2015 the 8 billion mark would be reached and 10 billion only after 2025. Under this scenario a world population of 15 billion would first appear after 2050. For the second half of the next century continued exponential growth would lead to ever-increasing absolute increments resulting, under Scenario 1, in about 40 billion in the year 2100. Further continuation of this growth would then soon result in a "standing room only" situation. By definition, these exponential scenarios do not assume a feedback from population size to fertility or mortality.

For the scenarios assuming a decline of fertility to replacement level at some point in the future, it appears that even relatively small differences in assumptions concerning the timing of fertility decline have a major impact on population size. Projecting UN Medium Variant assumptions up to the year 2100 gives a levelling off in population growth at around 11 billion, the population size in 2050 already being 10 billion (Scenario 2). Delaying the fertility decline by 25 years, Scenario 3 gives a population size of more than 14 billion in 2100 and population growth does not seem to stop before having reached 15–16 billion in the 22nd Century. Likewise, a rapid linear fertility decline to a TFR of only 1.4 children per women in the year 2025 for every region (Scenario 4) gives a totally different picture: After an increase to 8 billion in 2030–2040, population size may decline to a figure below 5 billion in the very long run (2100) as long as no major wars, famines or epidemics occur.

Population size is, to a lesser extent, influenced by assumptions about mortality. In the medium term, a constant mortality level in conjunction with replacement fertility by 2025, Scenario 6, would delay the growth in population size by some 10 years as compared to the UN Medium assumptions on mortality improvements. In the very long run, population size in the absence of mortality improvements tends to level off at around 7.5 billion, which is two billion below that in a corresponding scenario with increasing life expectancy. Delaying the assumed improvement in life expectancy by 25 years, Scenario 7 has virtually no impact on total population size. Assuming a rapid increase in life expectancy to 85 years for women and 80 years for men in all the regions in the year 2025,

Scenario 9 increases population size by one billion in 2050 and two billion in 2100.

1.1.3. Age distributions

The age structure of the population is significant not only because of the above mentioned momentum of population growth, but because it has implications for economic structure and dependency burdens ranging from educational expenses for the young to health care and other support for the elderly. A society's age distribution also has significant impact on consumption patterns, changes in the value system and even culture.

Aging of the population is a universal trend even in less developed countries. While in the more developed countries (MDCs) the proportion of the population above age 60 increased from 11.4% in 1950 to 17.1% in 1990 (with some countries such as Germany already above 20%), in the LDCs this proportion also increased slightly from 6.3% in 1950 to 6.9% in 1990 despite the large numbers of newborns added at the base of the age pyramid.

In the future, further aging seems to be inevitable. Even the Constant Rates Scenario, which is very unlikely in the long run, gives an increase in the mean ages of all regions over the next 50 years. Scenario 2, based on the UN Medium Variant, will result in much more significant aging in all regions of the world. While in today's industrialized countries the mean age is expected to increase from the present 35 years to more than 43 years by 2070, the extent and pace of aging will be even stronger in Eastern Asia. There, the UN Medium Variant expects an increase in the mean age from the current 29 years to 44 years by 2070. Even in Africa the mean age is expected to increase by more than 12 years to about the same level we find in Europe and North America today.

As could be expected, the Rapid Fertility Decline Scenario (4), which is the only one that would ultimately bring down the world population below its present size, results in the most extreme aging of the world population. Under this scenario, in almost every continent the mean age of the population would reach about 50 years by the end of next century. What this means in terms of changes of the social and economic structure is hard to imagine, not to mention medical expenses and retirement benefits.

In summary, this comparison of the consequences of various scenarios on total population size and the age structure of the population makes clear the fundamental dilemma of future population trends under low mortality conditions. All scenarios that limit population growth even at a level two to three times of today's world population, will result in extreme aging of the population. Only further exponential growth of the population will keep the populations young.

Put in simple words: either the population explodes in size or it ages to an unprecedented extent. The explosion

Table 1: *Population sizes and distributions, 1950 and 2000*

	1950		2000	
World	2516	(100%)	6122	(100%)
Developed countries	832	(33%)	1277	(21%)
Developing countries	1684	(67%)	4846	(79%)
Africa	224	(9%)	872	(14%)
Latin America	165	(7%)	546	(9%)
North America	166	(7%)	297	(4%)
Asia	1376	(55%)	3549	(58%)
China	555	(22%)	1256	(21%)
India	358	(14%)	964	(16%)
Europe	392	(16%)	512	(8%)
Oceania	13	(5%)	30	(5%)
USSR	180	(7%)	315	(5%)

The estimates given for 2000 result from Scenario 2 as described above.

will sooner or later result in higher mortality levels because it cannot go on forever, the aging makes painful social adjustment processes necessary and a complete remodeling of both family and state support systems for the elderly.

1.1.4 Regional distributions

The distribution of the population of the major regions of the world has changed significantly over the past years and is likely to change even more in the future. As Table 1 indicates, in 1950 one-third of the world population lived in the developed world. In 2000 this fraction will have declined to about one-fifth despite the fact that the population of the developed countries grew by about 50%.

By the end of our century the population of Africa will have quadrupled from 224 million to 872 million and increased its share of the world population from 9% to 14%. On the losing end are Europe with a halving of its proportion from 16% to 8% and North America with a decline from 7% in 1950 to only 4% in 2000. Latin America and Asia will have slightly increased their share. Due to its successful population stabilization policy, the share of the Chinese population in the world has even declined slightly.

For the year 2050 the above described scenario calculations yield the following pattern. The proportion of the world population living in Africa will inevitably further increase to as much as 25% under Scenarios 1 and 9. Even under the rapid fertility decline scenario the proportion will further increase to 17%. At the other extreme the proportion of the world population living in Europe, North America, and the USSR together would decline to as little as 7% by the year 2050 under the constant rate scenario. Even under the unlikely rapid fertility decline scenario the proportion in those more developed regions would decline from the present 20% to around 12%. This also demonstrates a strong momentum in the growth and shrinking of relative population sizes in the major world regions.

1.1.5   Migration and urbanization

The human capacity for adaptation (the highest among natural species) allowed *Homo habilis* to roam the Earth for 3 million years, dispersing all the way from East Africa to Tierra del Fuego in the tip of South America. But never has the magnitude of such geographical shifts been as high as those caused by the demise of agrarian societies since the 17th Century. The four major agricultural crises in Europe – mostly Central and Baltic – between 1844 and 1913 sent out most of the 52 million emigrants that went overseas to settle in thinly populated areas (Brinley, 1961:11)[4]. With the spread of the market economy in the second half of this century, the rural exodus has become pervasive, though with different outcomes, in Latin America, Africa, and Southern Asia[5].

In the last few decades, rural outmigration in the latter three regions has changed: push factors have become stronger than pull factors; internal migrations now overflow into the international arena; gender imbalances in migratory cohorts tend to decrease as economic opportunity costs and cultural determinants begin to change the gender division of labor and of geographical mobility, though, at the same time poverty tends to push more women and older people out of rural villages into the cities; and, most recently, outmigration from cities towards other locations is also on the rise.

Projections tend to indicate that migrations will continue to increase in all developing regions, fostered by a combination of factors including the spiral of population growth and poverty; land or wealth concentration; economic polarization in agricultural production; and inefficient, corrupt or mistaken government policies. All of these lead to rural outmigration: loss of livelihoods, land degradation and desertification, and food and land scarcity[6].

Another important factor, which will very likely increase the importance of the "pull" factors in rural outmigration, is the spread of a cultural urban bias through education and the mass media (Swaminathan, 1988). Additionally, if competition for control and access of scarce resources (land, capital, technology, water) increases (especially in rural areas of less developed countries and in Eastern European countries, most probably running along traditional lines of ethnicity and nationalism) political refugees will also most likely increase in numbers.

Migration has been a survival strategy for the poor, be they a Punjabi family having several sons working in the

Gulf region, or a Peruvian couple who live off the remittances sent by their daughter from the United States. Regional studies which analyze the relationship between economic and migration flows are necessary in order to map out the intricate web of migratory movements that may have global effects.

Future research priorities related to migration might seem rather simple; people will continue to flow to where wealth, meaning possibilities of livelihoods, amenities and the picture of the "good life" are concentrated. Studies that disaggregate migrant cohorts by gender, age, ethnic group, etc. may provide valuable information to "color" the maps of migration flows, but will yield few possibilities for prediction. In contrast, studies on the selectivity of migrants, that is who migrates within the local community and *why* they settle at a given destination may provide useful insights for policy and planning options.

At the global and regional level, migration studies should focus on probabilistic and simulation models of major outflows which may be expected with increasing climatic events and with cumulative environmental changes that may destroy local people's livelihoods. First in line would be outflows in Africa due to famines and desertification; and from regions most vulnerable to possible climatic changes, especially coastlines, deltas and islands. Most migration will tend to flow from South to North, although also from East to West and from South to South. A great uncertainty in attempts to forecast international migratory streams are also short term changes in the immigration policies of the MDCs and the question how effectively such policies can be enforced.

Presently 45% of the world population lives in urban areas (37% in LDCs, 73% in MDCs). This percentage is likely to increase to 51% by 2000 and 65% by 2025 (UN, 1991). While urbanization at different levels is a universal trend in less developed regions the rapid growth of some mega-cities is most visible. In 1960 seven of the world's ten largest urban agglomerations were in North America, Japan and Europe, with New York, Tokyo and London at the top. Presently Mexico City is the largest one with more than 20 million, and seven of the top ten are in LDCs.

1.1.6   Population pressure and per capita resource use

A primary question is: are there limits to the carrying capacity of the Earth System for human populations? Kenneth Blaxter gives an unequivocal yes, but cautions that "where doubt sets in is on the precise definition of the number of people that can be supported, about the way in which population will increase, about the way in which food production will reach the limit imposed by the carrying capacity and about the availability of resources to push back this limit" (Blaxter, 1986). These issues must be the priority research topics for the next decades. Various estimates of global carrying capacity of the Earth for people have appeared in the literature, ranging from 7.5 billion (Gilliand, 1983) (cited in Demeny, 1988) to 12 billion (Clark, 1958), 40 billion (Revelle, 1976) and 50 billion (Brown, 1954). However, many authors are skeptical about the criteria – amount of food, or kilocalories – used as a basis for these estimates. "For humans, a physical definition of needs may be irrelevant. Human needs and aspirations are culturally determined: they can and do grow so as to encompass an increasing amount of "goods", well beyond what is necessary for mere survival" (Demeny, 1988).

Cultural evolution has a profound effect on human impacts on the environment. By changing the learned behavior of humans and incorporating tools and artifacts, it allows individual human resource requirements and their impacts on their resident ecosystems to vary over several orders of magnitude. Thus it does not make sense to talk about the "carrying capacity" of humans in the same way as the "carrying capacity" of other species (Sanchez *et al*, 1989; Blaikie and Brookfield, 1987) since, in terms of their carrying capacity, humans are many subspecies . Each subspecies would have to be culturally defined to determine levels of resource use and carrying capacity. For example, the global carrying capacity for *Homo americanus* would be much lower than the carrying capacity for *Homo indus*, because each American consumes much more than each Indian does. And the speed of cultural adaptation makes thinking of species (which are inherently slow changing) misleading anyway. *Homo americanus* could change its resource consumption patterns drastically in only a few years, while *Homo sapiens* remains relatively unchanged. We think it best to follow the lead of Herman Daly[7] in this and speak of the product of population and *per capita* resource use as the *total impact* of the human population. It is this total impact that the Earth has a capacity to carry, and it is up to us to decide how to divide it between numbers of people and *per capita* resource use. This complicates population policy enormously, since one cannot simply state a maximum *population*, but rather must state a maximum number of *impact units*. How many impact units the Earth can sustain and how to distribute these impact units over the population is a very dicey problem indeed, but one that must be the focus of research in this area.

Many case studies indicate that "there is no linear relation between growing population and density, and such pressures (towards land degradation and desertification)" (Caldwell, 1984). In fact, one study found that land degradation can occur under rising pressure of population on resources (PPR) under declining PPR and without PPR (Blaikie and Brookfield, 1987). Therefore, the scientific

agenda must look towards more complex, systemic models where the effect of population pressures can be analyzed in its relationships with other factors (Garcia, 1990). This would allow us to differentiate population as a "proximate" cause of environmental degradation from the concatenation of effects of population with other factors as the "ultimate" cause of such degradation (Asian Development Bank, 1990).

Research can begin by exploring methods for more precisely estimating the total impact of population. For example, William Clark (1991) suggests that the "Ehrlich identity" (Pollution/Area = People/Area × Economic Production/People × Pollution/Economic Production) can be operationalized as (CO_2 Emissions/km^2 = Population/km^2 × GNP/Population × CO_2 Emissions/GNP). Clark and his colleagues examined data for twelve countries from 1925 to 1985 and concluded that the same loading of pollution on the environment can come from radically different combinations of population size, consumption, and production. Thus no single factor dominates the changing patterns of total impact across time. This points to the need for local studies of causal relations among specific combinations of populations, consumption and production, noting that these local studies need to aim for a general theory that will account for the great variety of local experience (see also Lutz, 1991).

Along this line of research, Barbara Torrey and Gretchen Kolsrud (1991) examined population growth and energy efficiency in several countries and concluded that the very small population growth forecast for developed countries over the next forty years will add a burden of CO_2 emissions that will be equal to that added by the much larger population growth forecast for the less developed countries. Decreasing energy consumption in developed countries could dramatically decrease CO_2 emissions globally. It is only under a scenario of severe constraints on emissions in the developed countries that population growth in less developed ones plays a major global role in emissions growth. If energy efficiency could be improved in the latter as well as the former, then population increase would play a much smaller role. José Goldemberg has suggested that enabling developing countries to "leap-frog" in adopting new energy efficient technologies could accomplish this goal.

Research priority should also look at situations where demand (either subsistence or commercial) becomes large relative to the maximum sustainable yield of the resource, or where the regenerative capacity of the resource is relatively low, or where the incentives and restraints facing the exploiters of the resource are such as to induce them to value present gains much more highly than future gains (Repetto and Holmes, 1983).

1.1.7 Curbing population growth and per capita resource use

Some authors single out a high rate of population growth as a primary root cause of environmental degradation and overload of the planet's carrying capacity, and therefore point to population control as the appropriate policy instrument. Ehrlich and his colleagues maintain "There is no time to be lost in moving toward population shrinkage as rapidly as is humanly possible" (Ehrlich *et al.*, 1989). Certainly, one step we should take immediately is to provide universal access to simple birth control measures to allow potential parents the full range of options in planning their families. Concerning population policies the same standards, such as voluntary choice of family size and family planning methods, should be applied in both more and less developed countries. Policies focussing solely on the control of population size are known to be insufficient. It has repeatedly been shown that the level of fertility in a population cannot be simply regulated but depends crucially on social and economic changes such as improved educational status, especially of women, improvements in social infrastructure and the reduction of poverty. There are also examples showing that curbing population growth does not necessarily result in an improvement of welfare and of environmental conditions, although one could make the argument that conditions might be even worse if rapid population growth had continued.

The opposite position is taken by those who see high rates of population growth as stimulating economic development through inducing technological and organizational changes (Hirschman, 1958; Boserup, 1965), or as a phenomenon which can be solved through technological change (Simon, 1990). Aside from the important question whether there is enough time for such reactions to be effective, such positions ignore the dangers of environmental depletion implicit in unchecked economic growth: consumption increases and rapidly growing populations that can put a very real burden upon the resources of the Earth, and bring about social and political strife for control of such resources. This position also assumes that technological creativity will have the same outcomes in the future as in the past, and in the South as in the North, a questionable assumption. Finally, it heavily discounts the importance of the loss of biodiversity – a loss which is irreversible and whose human consequences are as yet unknown.

A different approach is taken by those authors for whom, "population can only be expected to fall when livelihoods (of the poor) are secure, for only then does it become rational for poor people to limit family size" (Chambers, 1988; DAWN, 1988). According to a World Bank study of 64 countries, when the income of the poor rises by 1%, general fertility rates drop by 3% (Lappe and

Schurman, 1988). When making such statements, however, one must be aware of the great social and cultural heterogeneity among the poor in different parts of the world which is highly relevant for the way in which their fertility reacts to improving living conditions. In contrast to the previous positions, some authors state that "population is not a relevant variable" in terms of resource depletion and stress that *resource consumption*, particularly overconsumption by the affluent, is the key factor (Hardoy and Satterthwaite, 1991; Sanchez, 1990; Harrison, 1990; Durning, 1991). OECD countries represent only 16% of the world's population and 24% of land areas; but their economies account for about 72% of the world gross product, 78% of road vehicles, 50% of global energy use. They generate about 76% of world trade, 73% of chemical products exports, and 73% of forest product imports (OECD, 1991). The main policy instrument, in this case, in the short term is then, reducing consumption and this can be most easily achieved in those areas where consumption *per capita* is highest.

Thus a new framework should expand the definitions of issues: focus not only on population size, density, rate of increase, age distribution, sex ratios, but also on access to resources, livelihoods, social dimensions of gender, and structures of power. New models have to be explored in which population control is not simply a question of family planning but of economic, ecological, social and political planning (U.N., 1990; Jacobson, 1988); in which the wasteful use of resources is not simply a question of finding new substitutes but of reshaping affluent lifestyles and in which sustainability is seen not only as a global aggregate process but also as one having to do with sustainable livelihoods for a majority of local peoples.

1.2 Trends in resource use

What are the biophysical limits to economic and population growth and economic development and how should we deal with uncertainty about these limits? These questions are critical for designing and implementing policies to assure a sustainable ecological economic system. To answer them we must broaden our perspective on the problem to include the entire ecological economic system. First, it is important to differentiate between growth and development, and to investigate the linkages between resource use, population, growth, and development.

1.2.1 Growth vs. development

Human welfare can improve by either pushing more matter-energy through the economy (more production and consumption), or by squeezing more human want satisfaction out of each unit of matter-energy that passes through (Daly and Cobb, 1989). These two processes are so different in their effect on the environment and on the prospects for sustainability that we must stop confusing them. It is better to refer to energy and matter throughput increase as *growth*, and efficiency increase as *development* (Daly and Cobb, 1989). Economic *growth* is destructive of natural capital and beyond some point will cost us more than it is worth – i.e. sacrificed natural capital will be worth more than the extra human-made capital whose production necessitated the sacrifice. At this point growth becomes anti-economic, impoverishing rather than enriching humankind. *Development*, or qualitative improvement, is not at the expense of natural capital.

There are clear economic limits to growth – but not to development – although there are some cases in which the differences between them may not be so marked. This does not mean there are no limits to development, only that they are not so clear as the limits to growth, and consequently there is room for a wide range of opinion on how far we can go in increasing human welfare without increasing resource throughput. How far can development substitute for growth – this is the relevant question, not how far can human-made capital substitute for natural resources (the answer to which is hardly at all).

Some believe that there are truly enormous possibilities for development without growth. Energy efficiency, they argue, can be vastly increased (Lovins, 1977; Lovins and Lovins, 1987; Holdren, this volume) as can the efficiency of water use and the use of other natural resources. Others (Cleveland *et al.*, 1984; Hall *et al.*, 1986; Gever *et al.*, 1986) believe that the coupling between growth, development, and energy use is not so loose. This issue arises in the Brundtland Commission's Report (WCED, 1987) where on the one hand the authors recognize that the scale of the human economy is already unsustainable in the sense that it requires the consumption of natural capital, and yet on the other hand they call for further economic expansion by a factor of five to ten in order to improve the lot of the poor without having to appeal too much to the "politically impossible" alternatives of serious population control and redistribution of wealth. The big question is, how much of this called-for expansion can come from development, and how much must come from growth? This question is not addressed by the Commission. But statements from the leader of the WCED, Jim MacNeil (MacNeil, 1990) that "The link between growth and its impact on the environment has also been severed" (p.13), and "the maxim for sustainable development is not limits to growth; it is the growth of limits", indicate that WCED expects the lion's share of that factor of five to ten to come from development, not growth. They confusingly use the word "growth" to refer to both cases, saying that future growth must be qualitatively very different from past growth. When things are qualitatively different it is best to call them by different names. Hence the importance of the distinction between growth and development. Our view is

that WCED is too optimistic – that a factor of five to ten increase cannot come from development alone, and that if it comes mainly from growth it will be devastatingly unsustainable. Therefore the welfare of the poor, and indeed of the rich as well, depends much more on population control, consumption control, and redistribution, than on the technical fix of a five to ten-fold increase in total factor productivity.

It is most important, then, to explore new methods to measure development. A recent, highly improved method is the Human Development Index put forth by UNDP (1990).

1.2.2 Energy, economic performance, and population

What is the evidence on the linkages between energy and resource consumption and economic growth? One popular way to look at the relationship between energy and economic performance has been the ratio of fuel use to GNP (or GDP) over time. This ratio has declined by about 25% in the past decade in most OECD countries, and this has been interpreted as indicating an increase in the energy efficiency of the economic systems. This interpretation has been extended to imply that energy efficiency can be increased much further still, requiring no basic changes in lifestyle, population, or wealth distribution to achieve a sustainable energy supply, only continuing efficiency improvements.

While this may be the correct interpretation, it is certainly not the only possible one. There are severe conceptual and accounting problems in both the numerator and denominator of this ratio. Let us start with GNP.

Gross National Product, as well as other related measures of national economic performance, have come to be extremely important as policy objectives, political issues and benchmarks of the general welfare. Yet GNP as presently defined ignores the contribution of nature to production, often leading to peculiar results.

For example, a standing forest provides real economic services for people by conserving soil, cleaning air and water, providing habitat for wildlife, and supporting recreational activities. But as GNP is currently figured, only the value of harvested timber is incorporated in the total. On the other hand, the billions of dollars that Exxon spent on the Valdez cleanup – and the billions spent by Exxon and others on the more than 100 other oil spills in the last 16 months – all actually improved our apparent economic performance. Why? Because cleaning up oil spills creates jobs and consumes resources, all of which add to GNP. Of course, these expenses would not have been necessary if the oil had not been spilled, so they shouldn't be considered "benefits". But GNP adds up all production without differentiating between costs and benefits, and is therefore not a very good measure of

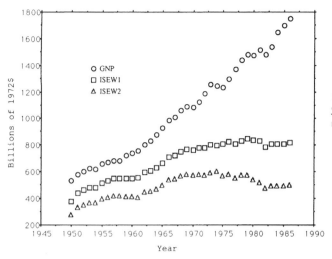

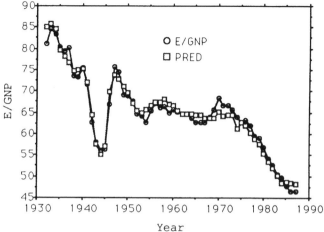

Figure 3: US GNP compared to the Index of Sustainable Economic Welfare (ISEW, from Daly and Cobb 1989) for the interval 1950 to 1986. ISEW2 includes corrections for depletion of non-renewable resources and long term environmental damage, ISEW1 does not.

Figure 4: The Energy/GNP ratio for the US economy from 1932 to 1987 and the predicted ratio (PRED) based on a regression model with percent of primary energy from petroleum (%PET) from electricity (%ELEC) and percent of Personal Comsumption Expenditures spent on direct fuel (%PCE) as the independent variable (R2 = .96)

economic performance or health.

In fact, when resource depletion and degradation are factored into economic trends, what emerges is a radically different picture from that depicted by conventional methods. For example, Herman Daly and John Cobb (1989) have attempted to adjust GNP to account mainly for depletions of natural capital, pollution effects, and income distribution effects by producing an "index of sustainable economic welfare" (ISEW, Fig. 3). They conclude that, while GNP in the US rose over the 1956–86 interval, ISEW has remained relatively unchanged since about 1970. When factors such as loss of farms and wetlands, costs of mitigating acid rain effects, and health costs caused by increased pollution, are accounted for, the US economy has not improved at all. If we continue to ignore natural ecosystems we may drive the economy down while we think we are building it up. By consuming our natural capital, we endanger our ability to sustain income.

Estimating total energy consumption for an economy is also not a straightforward matter, because not all fuels are of the same quality (ability to do work). Electricity, for example, is more versatile and cleaner in end use than petroleum, and it also costs more energy to produce. In an oil-fired power plant it takes from 3–5 kcal of oil to produce each kcal of electricity. Thus, adding up the raw heat equivalents of the various forms of fuel consumed by an economy without accounting for fuel quality can radically distort the picture, especially if the mix of fuel types is changing over time.

Studies that have tried to adjust for fuel quality and distortions in GNP have painted a much less rosy picture of past energy efficiency gains and the potential for future gains. Cleveland *et al.* (1984) and more recently Kaufmann

(1990) have shown that almost all of the changes in E/GNP (or E/GDP) ratios in the US and OECD countries can be explained by changes in fuel quality and the percent of personal consumption expenditures (PCE) spent directly on fuel. This latter effect is due to the fact that PCE is a component of GNP and spending more on fuel directly will raise GNP without changing real economic output. Figure 4 is an example of the explanatory power of this relationship for the US economy from 1932 to 1987 (from Kaufmann, 1990). Much of the apparent gain in energy efficiency (decreasing E/GNP ratio) is due to shifts to higher quality fuels (like primary electricity) from lower quality ones (like coal). Renewable energy sources (like biomass and direct solar) are generally of lower quality and major shifts to them may therefore cause significant increases in the E/GNP ratio.

1.3. The limits and uncertainty of technological change

One key element that has frustrated population and resource policy is the enormous degree of uncertainty about long-term human impacts on the biosphere. The argument can be summarized as differing opinions about the degree to which technological progress can eliminate resource constraints. Current economic world views (capitalist, socialist, and the various mixtures) are all based on the underlying assumption of continuing and unlimited economic growth. This assumption allows a whole host of very sticky problems, including population growth, equity, and sustainability to be ignored (or at least postponed), since they are seen to be most easily solved by additional economic growth. Indeed, most conventional economists define "health" in an economy as a stable and high *rate of growth*. Energy and resource limits to growth, according to these world views, will be eliminated as they arise by clever development and deployment of new technology. This line of thinking is often called "technological optimism."

An opposing line of thought (often called "technological skepticism") at least provisionally assumes that technology will *not* be able to circumvent fundamental energy and resource constraints and that eventually economic growth will stop. It has usually been ecologists or other life scientists that take this point of view[8] largely because they study natural systems that *invariably* do stop growing when they reach fundamental resource constraints. A healthy ecosystem is one that maintains a stable level. Unlimited growth is cancerous, not healthy, under this view. The technological optimists argue that human systems are fundamentally different from other natural systems because of human intelligence. History has shown that resource constraints can be circumvented by new ideas. Technological optimists claim that Malthus' dire predictions about population pressures have not come to pass and the "energy crisis" of the late 1970s is behind us.

The technological skeptics argue that many natural systems also have "intelligence" in that they can evolve new behaviors and organisms (including humans themselves). Humans are therefore a part of nature, not apart from it. Just because we have circumvented local and artificial resource constraints in the past does not mean we can circumvent the fundamental ones that we will eventually face. Malthus' predictions have not come to pass yet for the entire world, the pessimists would argue, but many parts of the world are in a Malthusian trap now, and other parts may well fall into it.

This debate has gone on for several decades now. It began with Barnett and Morse's *Scarcity and Growth* (1963) and really got into high gear with the publication of *The Limits to Growth* by Meadows *et al.*[9] and the Arab oil embargo in 1973. There have been thousands of studies over the last 15 years on various aspects of our energy and resource future and different points of view have waxed and waned. But the bottom line is that there is still an enormous amount of uncertainty about the impacts of energy and resource constraints, and we doubt that the argument will ever be decided on purely scientific grounds.

In the next 20–30 years we may begin to hit real fossil fuel supply limits as well as constraints on production due to global warming. Will fusion energy or solar energy or conservation or some as yet unthought of energy source step in to save the day and keep economies growing? The technological optimists say yes and the technological skeptics say no. Ultimately, no one knows. Both sides argue as if they were certain, but the worst form of ignorance is misplaced certainty.

The optimists argue that, unless we *believe* that the optimistic future is possible and behave accordingly, it will never come to pass. The skeptics argue that the optimists

	Real State of the World	
	Optimists Right	Skeptics Right
Technological Optimist Policy	High	Disaster
Technological Skeptic Policy	Moderate	Tolerable

Figure 5: Payoff matrix for techologically optimistic vs. skeptical policies.

will bring on the inevitable leveling and decline sooner by consuming resources faster and that to sustain our system we should begin to conserve resources immediately. How do we proceed in the face of this overwhelming uncertainty?

We can cast this optimist/skeptic choice in the form of the "payoff matrix" shown in Figure 5. Here the alternative policies that we can pursue today (technologically optimistic or skeptical) are listed on the left and the real states of the world are listed on the top. The intersections are labeled with the results of the combinations of policies and states of the world. For example, if we pursue the optimistic policy and the world really does turn out to conform to the optimistic assumptions then the payoffs would be high. This high potential payoff is very tempting and this strategy has paid off in the past. It is not surprising that so many would like to believe that the world conforms to the optimist's assumptions. If, however, we pursue the optimistic policy and the world turns out to conform more closely to the technological assumptions of the skeptics, then the result would be "Disaster". The disaster would come because irreversible damage to ecosystems would have occurred and technological fixes would no longer be possible.

If we pursue the skeptical policy and the optimists are right then the results are only "Moderate." But if the skeptics are right and we have pursued the skeptical policy then the results are "Tolerable".

This simplified choice matrix has a fairly simple "optimal" strategy. If we *really* do not know the state of the world then we analyze each policy in turn, look for the worst thing (minimum) that could happen if we pursue that policy, and pick the policy with the largest (maximum) minimum. In the case stated above we should pursue the skeptical policy because the worst possible result under that policy ("Tolerable") is a preferable outcome to the worst outcome under the optimist policy ("Disaster"). Given this analysis, what can one recommend for population and resource policy? Because of the large uncertainty about the long-term impacts of population and resource use growth on ecological sustainability, and the enormous costs of guessing wrong, we should *at least provisionally assume the worst*. If the dire predictions about population impacts are correct, we will still survive. If they are wrong, we will be pleasantly surprised. This is a much different scenario than the consequences of provisionally assuming the best about the impacts of population growth. If we assume that the predictions of no impacts of population and resource use growth are right and they are not, we will have irreversibly degraded the planet's capacity to support life. We cannot rationally take that risk.

The problem is even more severe if we consider expectations about social as well as technological change. Most of the problems we are considering in this chapter are not soluble through purely technical means. This implies that the chances of a combined social and technological "fix" are even worse than the chances of a purely technical fix, and precaution is called for to an even higher degree.

1.4 Adjusting incentives and disincentives

Current systems of regulation are not very efficient at managing environmental resources for sustainability, particularly in the face of uncertainty about long-term values and impacts. They are inherently reactive rather than proactive. They induce legal confrontation, obfuscation, and government intrusion into business. Rather than encouraging long-range technical and social innovation, they tend to suppress it. They do not mesh well with the market signals that firms and individuals use to make decisions and do not effectively translate long-term global goals into short-term local incentives.

We need to explore promising alternatives to our current command and control environmental management systems, and to modify existing government agencies and other institutions accordingly. The enormous uncertainty about local and transnational environmental impacts needs to be incorporated into decision-making. We also need to better understand the sociological, cultural, and political criteria for acceptance or rejection of policy instruments.

One example of an innovative policy instrument currently being studied is a flexible environmental assurance bonding system designed to incorporate environmental criteria and uncertainty into the market system, and to induce positive environmental technological innovation (Perrings, 1989; Costanza and Perrings, 1990; Perrings, 1991).

In addition to direct charges for known environmental damages, a company would be required to post an assurance bond equal to the current best estimate of the largest potential future environmental damages; the money would be kept in interest-bearing escrow accounts. The bond (plus a portion of the interest) would be returned if the firm could show that the suspected damages had not occurred or would not occur. If they did, the bond would be used to rehabilitate or repair the environment and to

compensate injured parties. Thus, the burden of proof would be shifted from the public to the resource-user and a strong economic incentive would be provided to research the true costs of environmentally damaging activities and to develop cost-effective pollution control technologies. This is an extension of the "polluter pays" principle to "the polluter pays for uncertainty as well". Other innovative policy instruments include tradeable pollution and depletion quotas at both national and international levels. Also worthy of mention is the newly emerging Global Environmental Facility of the World Bank that will provide concessionary funds for investments that reduce global externalities.

We should implement these fees, taxes, quotas, and bonds on the destructive use of natural capital to promote more efficient use, and ease up on income taxes, especially on low incomes in the interest of equity. Fees, taxes and subsidies should be used to change the prices of activities that interfere with sustainability versus those that are compatible with it. This can be accomplished by using the funds generated to support an alternative to undesirable activities that are being taxed. For example, a tax on all greenhouse gases, with the size of the tax linked to the impact of each gas, could be linked to development of alternatives to fossil fuel. Gasoline tax revenues could be used to support mass transit and bike lanes. Current policies that subsidize environmentally harmful activities should be stopped. For example, subsidies on virgin material extraction should be stopped. This will also allow recycling options to effectively compete. Crop subsidies that dramatically increase pesticide and fertilizer use should be eliminated, and forms of positive incentives should also be used. For example, debt for nature swaps should be supported and should receive much more funding. We should also offer prestigious prizes for work that increases awareness of, or contributes to, sustainability issues, such as changes in behavior that develop a culture of maintenance (i.e. cars that last for 50 years) or promotes capital and resource saving improvements (i.e. affordable, efficient housing and water supplies).

Economic incentives or disincentives to change behavior are the first step towards achieving sustainability, but must be encompassed within broader, more profound change in the fundamental values that shape and stimulate behavior.

2. RESEARCH NEEDS

2.1 Increasing Scientific Capacities

It is essential that research be carried out locally and regionally by scientists trained at the highest professional level. A global understanding of the impacts of population and resource use can not be based simply on the analysis of global averages since the results in a heterogeneous world depend crucially on the level of aggregation at which the interactions are studied. In depth, methodologically sound research must be carried out focussing on local and regional differences, so that global models can adequately assess and account for such differences. Such a goal cannot be achieved, however, if 95% of scientific research is carried out in developed countries. The highest priority should be given, then, to increasing the scientific capacities of developing nations to do research on the links between environmental transformations and the dynamics of population trends.

2.2 New Modeling Approaches

Since ecosystems are being threatened by a host of human activities, protecting and preserving them requires the ability to understand the direct and indirect effects of human activities over long periods of time and over large areas. Computer simulations are becoming important tools to investigate these interactions and in all other areas of science as well. Without the sophisticated global atmospheric simulations now being done, our understanding of the potential impacts of increasing CO_2 concentrations in the atmosphere due to fossil fuel burning would be much more primitive. Computer simulations can be used to understand not only human impacts on ecosystems, but also our economic dependence on natural ecosystem services and capital, and the interdependence between ecological and economic components of the system (see, for example, Braat, 1991; Costanza *et al.*, 1990).

Several recent developments make such computer simulation modeling feasible, including the accessibility of extensive spatial and temporal data bases and advances in computer power and convenience. Computer simulation models are potentially one of our best tools to help understand the complex functions of integrated ecological economic systems. But even with the best conceivable modeling capabilities, we will always be confronted with large amounts of uncertainty about the response of the environment to human actions (Funtowicz and Ravetz, 1991). Learning how to effectively manage the environment in the face of this uncertainty is critical (Costanza and Perrings, 1991; Perrings, 1991).

The research program for sustainable natural resource use should pursue an integrated, multiscale, trans-disciplinary, and pluralistic, approach to quantitative ecological economic modeling, while acknowledging the large remaining uncertainty inherent in modeling these systems and developing new ways to effectively deal with this uncertainty.

Such models should not only be conducted at a global level but the complex interactions between the relevant factors can often be better understood at the level of case studies. For a specific place at a specific level of its development, consumption pattern and use of technology, one can more directly measure the impact of population and resource use on the environment. How to generalize the lessons learned from case studies is an interesting research question in itself which deserves more attention.

2.3 Social and behavioral factors

In the realm of behavior and the study of decision-making, we need to broaden the research agenda to include all aspects of human behavioral responses, at all scales of operation, from the individual to the group, community, nation and world. The field of social psychology and anthropology, and the new fields of experimental and evolutionary economics are based on this idea and have begun to bear some fruit. A key aspect of this research is developing enough understanding of the way incentives of various kinds influence behavior to be able to modify incentive systems to achieve sustainability. The concept of "social traps"[10] is one way to generalize dysfunctional incentive systems with an eye toward fixing them. Research in this general direction should be expanded and strengthened.

There is ample evidence that human behavior is non-linear and that people react many times in an "irrational" way. In many cases, vested interests and bureaucratic dealings change the course of policies intended to protect the environment or to achieve sustainability in economic activities. Understanding such processes is essential to overcome obstacles to policy implementation that will benefit local communities. We know that in human actions related to the environment there can be no purely "technical solutions" nor are there likely to be purely "political or education solutions." We know for certain that solutions must be multi-dimensional covering technical expertise, managerial skills, political negotiation, lifestyle changes and new ethical and philosophical perspectives. In spite of this, a narrow definition of the role of science in environmental matters is sometimes taken, where it is asked to provide only the measuring, monitoring and observation of populations and resources, knowledge which would then be channeled through appropriate educational institutions to the general public.

Instead what is needed is for science to become deeply involved as well in the observation and monitoring of values, perceptions and assessments of populations about resources and environmental change, and in offering scholarly views of a more imaginative and creative kind to evolve a new vision of the relationship of humans to the geophysical environment. At present, this most important field of philosophical and ethical discussion is being pulled in many directions.

We know for certain that the philosophical underpinnings of science provide an implicit frame of mind which channels human actions. Such a framework must also be examined, since its foundations were set in tune with a 19th Century society which no longer exists. Empirical observation, then, must be complemented with a rethinking of our scientific world-view, touching on the following issues: science as an instrument for mastering the biophysical universe rather than for co-inhabiting it; the deep cleavage between the natural and social sciences following the Judeo-Christian splicing of body and soul; disciplinary boundaries, especially in the social sciences, which are based on arbitrary divisions.

As to the role of education in the population/resources question, we know it plays an ambiguous role: on the one hand, literacy and higher levels of education are needed to enable individuals and societies to better understand and manage the technological and administrative processes of the modern global economy, and hence, the process of rationalizing environmental transformation[11] (UN, 1989). On the other hand, higher levels of Western education also encourage higher levels of resource consumption among the elites and, through social prestige, among low income groups as well[12] (Durning, 1991).

2.4 Developing institutions to achieve sustainable resource use

Institutions with the flexibility necessary to deal with ecologically sustainable development are lacking. Indeed, many financial institutions are built on the assumption of continuous exponential growth and will face major restructuring in a sustainable economy. Many existing institutions have fragmented mandates and policies, and often have not optimally used market and non-market forces to resolve environmental problems. They also have conducted inadequate benefit/cost analyzes by not incorporating ecological costs; used short-term planning horizons; inappropriately assigned property rights (public and private) to resources; and made inappropriate use of incentives.

There is a lack of awareness and education about sustainability, the environment, and causes of environmental degradation. In addition, much environmental knowledge held by indigenous peoples is being lost, as is knowledge of species, particularly in the tropics. Institutions have been slow to respond to new information and shifts in values, for example, concerns about threats to biodiversity or the effects of rapid changes in communications technologies. Finally, many institutions do not freely share or disseminate information, do not provide public access to decision-making, and do not devote serious attention to determining and representing the wishes of their constituencies.

Many of these problems are a result of the inflexible bureaucratic structure of many modern institutions.

Experience (i.e. Japanese industry) has shown that less bureaucratic, more flexible, more peer-to-peer institutional structures can be much more efficient and effective. We need to debureaucratize institutions so that they can effectively respond to the coming challenges of achieving sustainability.

3. HUMAN GOALS AND RESEARCH PRIORITIES

To summarize, our conclusion as to what is needed to achieve a sustainable, just, and equitable pattern of natural resource use and population is first, to establish a hierarchy of goals for national and global ecological economic and social planning and management. Sustainability, and human development should be the primary long-term goals, replacing the current GNP growth mania. Issues of justice, equity, and population are intimately tied in with sustainability as preconditions. Only sustainable levels of human population are desirable. Economic growth in this hierarchy is a valid goal only when it is consistent with sustainability. The goals can be put into operation by having them accepted as part of the political debate, and implemented in the decision-making structure of institutions that affect the global economy and ecology (like the World Bank).

It is necessary to adjust current incentives and disincentives to reflect long run, global costs, *including uncertainty*. To paraphrase the popular slogan, we should: model globally, adjust local incentives accordingly. In addition to traditional education, regulation, and user fee approaches, a flexible assurance bonding system has been proposed to deal specifically with uncertainty (Costanza and Perrings, 1990).

Further decline in the stock of *natural capital* must not be allowed, preferably by taxing natural capital consumption (Costanza and Daly, 1992). This policy will encourage the technological and social innovation that optimists are counting on while conserving resources in case the optimists are wrong. Revenues can be used to mitigate problems and to ease up on income taxes especially for the lower end of the income scale.

The issues of consumption in developed countries and population in developing countries more and more are seen to be linked: research should be directed at identifying the mechanisms whereby loops connecting such phenomena are perpetuated. This means crossing both socio-political borders in the real world and crossing disciplinary borders in the sciences. It is imperative that research be oriented in this direction. Time is at a premium and thus, a new sense of urgency, given current unsustainable situations, must be given to research on the issues raised in this book.

3.1 Research priorities

3.1.1. Resource use, technology and employment

(a) Data and modeling on effective educational and cultural policies to change lifestyles towards sustainable population patterns, consumption control, and redistribution.

(b) Develop better global ecological and socio-economic modeling capabilities to allow us to see the range of possible outcomes of our current activities, especially the interrelated impacts of population, *per capita* resource use, and wealth distribution.

(c) Employment is an area requiring special research attention, as the aging of the population and technological trends require new strategies to create sustainable livelihoods and jobs.

(d) The elasticity of consumption is such that even with a decrease in population growth, consumption will increase, so research should focus on technological and organizational strategies whereby more outputs can be produced with less inputs. Blending of local knowledge and technology with new technologies should be especially important.

(e) Monitoring and dynamic spatial representation of zones of urgency is needed through mapping of the carrying capacity of different ecosystems, especially in relation to their state of fragility, noting that polar and subtropical zones generally are less resilient to ecological change. Digital maps using remote sensing and computer technology are important tools for this task. The goal here is to identify "red" zones, while being able to better assess the potentialities of given socio-ecosystemic contexts.

(f) Local institutional forms of resource management must also be given priority in research, including the establishing of new institutions such as an Ombudsman for the Environment.

(g) Methodologically, science should give attention to developing:
   (i) more precise methods for identifying and measuring natural resources, while at the same time advancing in creating national inventories of such resources.
   (ii) new forms of data collection that will allow for regional and national differences in the management of resources and population regulation to be better understood and monitored.

3.1.2 Population and health

(a) Research focussing on local factors affecting the interactions between basic health care, family planning and women's education programs.

(b) Study the influence of cultural norms and the role of political and religious leaders in transformation of family structures and implementation of family planning.

(c) Put greater emphasis on the development of better contraceptives for women and men with low cost, easy use, high use-effectiveness and high cultural acceptability.

(d) Investigations into the danger of a new increase in infectious disease as a consequence of increasing population density with special attention given to rapid urbanization.

(e) Research focussing on individual and local factors affecting the implementation of mother and child health care, family planning and women's education programs. Including the links of population growth leading to a new increase in infectious diseases which will require more health services, especially in developing countries.

3.1.3 Culture and education [13]

(a) Local perceptions and assessments on environmental transformations especially in relation to the costs and benefits which people believe they will have to bear in changes of livelihood or lifestyle. Also, their views on the relative advantages that other local, ethnic, national or international groups have in losing actual or potential benefits.

(b) The visions of major world religious and philosophical systems as to the relation between human beings and the Earth System. Three topics are central here:
   (i) beliefs as to the natural or cultural right or ability of human beings to alter the Earth System.
   (ii) the role of religious or philosophical leaders in controlling and directing their followers in matters related to environmental transformations and which may or may not derive from doctrinaire assumptions.
   (iii) messianic or apocalyptic beliefs which lead to *laissez faire* or indifference to environment and development.

(c) The way in which environmental or developmental proposals are filtered through cognitive, philosophical, theological or other values and which may be entirely different in diverse social and national contexts.

(d) Priorities in the area of education include:
   (i) examining perceptions and attitudes transmitted in school curricula about the Earth System and desirable consumption patterns and lifestyles.
   (ii) programs that have successfully incorporated environmental and population questions into school and university curricula.
   (iii) analyzing gender differences in schooling which may lead to differential perceptions and attitudes between men and women towards population and resource use.

    (iv) understanding the links between educational systems and social, ethnic, political and governmental groups with different perceptions and attitudes towards development and environment policies.

    (v) recovering useful traditional knowledge and technologies transmitted through informal education, especially in the fields of agronomy, botany, zoology and related subjects, and mechanisms to blend it with new technologies and give it widespread use.

NOTE

Much of this chapter is taken from the previous work of the authors, which benefited from the collaboration of many individuals, among them: Herman Daly, Joy Bartholomew, Richard Rockwell, Charles Perrings, Cutler Cleveland, John Cumberland, Margarita Velazquez, Bruce Hannon, Bob Ulanowicz, Christine Halvorson, Veronica Behn and Alan Scholefield.

END NOTES

1. A more extreme position likens human populations to a cancerous growth bound to kill its hospitable planet.

2. We refer to population numbers in the sense that the physical presence of human bodies on the surface of the Earth becomes significant because of what these bodies do: they breathe, eat and use the resources of the Earth. A body which does none of these things has no impact on the Earth system.

3. In some instances birth rates even increased slightly due to better maternal health conditions.

4. Other kinds of outflows have involved fewer numbers; the slave trade across the Atlantic, six to twelve million; political refugees in all countries of the world, 14 million.

5. "Perhaps nine-tenths of the population increase in Northern America and Oceania and two-thirds of that in Latin America could be directly attributed to European migrant populations within Europe's demographic outshoots in the vast, formerly thinly populated land of the Americas, Oceania and Northern Asia. Altogether, the areas of European settlements that comprised 20% of the world's population in 1700 claimed 36% of that total by the middle of the 20th Century" (Demeny, 1990).

6. A typical combination of such factors at a regional scale in Latin America is given by Stonich (1989): "Agricultural development in the region has been highly uneven not only in terms of the spatial distribution of people. Political–economic factors related to the expansion of export-oriented agriculture constrain access to the most fertile lands of the region (in South Honduras). This has resulted in a highly uneven distribution of population in which the greatest population densities occur in the highlands, the areas that are the most marginal for agriculture. The growing population in the highlands with inadequate opportunities to earn a living has led to a parceling of land among more and more people, with agricultural production expanding into even more marginal areas. Growing rural poverty stimulated outmigration from the more densely patched South into other parts of the country, thereby decreasing population pressure in the region while simultaneously augmenting urban populations and escalating pressure on tropical forest areas in the remainder of the country."

7. Daly, H.E. 1977. *Steady State Economics*. W.H. Freeman, San Francisco, 185 pp.

8. Notable examples are Paul Ehrlich and Garrett Hardin. Herman Daly is a rare economist who shares this view.

9. H.J. Barnett and C. Morse. 1963. *Scarcity and Growth: The economics of natural resource availability*. Johns Hopkins, Baltimore.

    D.H. Meadows, D.L. Meadows, J. Randers, and W.W. Behrens. 1972. *The Limits to Growth*. Universe, New York.

10. Including, but not limited to, the "tragedy of the commons" (cf. Hardin, G. 1968. The tragedy of the commons. *Science* **162**:1243-1248), and the "prisoner's dilemma" (cf. Axelrod, R. 1984. The Evolution of Cooperation. Basic Books, New York), Platt, J. 1973. Social traps. *American Psychologist* 28:642-651; Cross, J.G. and M.J. Guyer.1980. *Social Traps*. University of Michigan Press, Ann Habor; Teger, A.I. 1980. *Too Much Invested to Quit*. Pergamon, New York; Brockner, J. and J.Z. Rubin, 1985. *Entrapment in Escalating Conflicts: A social psychological analysis*. Springer-Verlag, New York. 275 pp; Costanza, R. 1987. Social traps and environmental policy. *BioScience* **37**:407-412.

11. The success of education, primary, secondary or third level, is strongly associated to the level of development of a region, as recent UN study showed (UN, 1989). In particular, "the greatest impact on rural development can be made where education is part of a package of measures (degree of access to credit, extension services, new needs, other inputs)" (Ibid).

12. The idea that environmental education automatically decreases environmental depletion is too optimistic: for example, several case studies show that even if cultivators are educated and convinced of the need to preserve the rain-forest, they are unable to do so because they have no other alternatives for livelihood (Schmink, 1989; Arizpe and Aranda, 1990).

13. See Jacobson, 1988 and Clark, 1991.

REFERENCES

Arizpe, L. and Aranda, J. 1990. The comparative advantages of women's disadvantages: women workers in the strawberry agroindustry in Mexico. In: *On Work: Historical, Comparative and Theoretical Approaches*. Basil Blackwell, Oxford.

Asian Development Bank. 1990. Population Pressure and Natural Resource Management: Key Issues and Possible Actions. Paper No. 6.

Barnett, H. and Morse, C. 1963. *Scarcity and Growth: The Economics of Natural Resources Availability*. Baltimore: Johns Hopkins University Press.

Blaikie, M. and Brookfield, B. 1987. *Land Degradation and Society*. Methuen, London.

Blaxter, K., 1986. *People, Food and Resources*. Cambridge University Press. Cambridge.

Boserup, E. 1965. *The Conditions of Agricultural Growth: The Economics of Agrarian Change under Population Pressure*. Aldin, Chicago.

Braat, L.C. and Steetskamp, I. 1991. Ecological economic analysis for regional sustainable development. pp. 269-288 In: R. Costanza (ed.). *Ecological Economics: The Science and Management of Sustainability*, Columbia University Press, New York, 525 pp.

Brinley, T. 1961. *International Migration and Economic Development*. UNESCO, Paris.

Brown, H. 1954. *The Challenge of Man's Future*. The Viking Press, New York.

Caldwell, J. 1984. Desertification: Demographic Evidence, 1973-83. Australian National University. Occasional Paper No. 37.

Chambers, R. 1988. Sustainable Livelihoods, Environment and Development: Putting Poor People First. Brighton, U.K.: Institute of Development Studies, University of Sussex.

Clark, C. 1958. Population growth and living standards. In: A.M. Agarwala and S.P. Singh (eds). pp. 32-53. *The Economics of Underdevelopment*, Oxford University Press, London.

Clark, W. 1991. Paper presented at the Annual Meeting of the American Association for the Advancement of Science, Washington, D.C.

Cleveland, C.J., Costanza, R., Hall, C.A.S. and Kaufmann, R. 1984. Energy and the United States economy: A biophysical perspective. *Science*, **225**, 890-897.

Costanza, R. (ed.). 1991. *Ecological Economics: The Science and Management of Sustainability*. Columbia University Press, New York. 525 pp.

Costanza, R. and Daly, H.E. 1987. Toward an ecological economics. *Ecological Modeling*, **38**, 1-7.

Costanza, R. and Daly, H.E. 1992. Natural capital and sustainable development. *Conservation Biology* (in press).

Costanza, R., Daly, H.E. and Bartholomew, J.A. 1991. Goals, agenda, and policy recommendations for ecological economics. pp. 1-20 In: Costanza, R. (ed.). *Ecological Economics: The Science and Management of Sustainability*, Columbia University Press, New York. 525 pp.

Costanza, R. and Hannon, B.M. 1989. Dealing with the "mixed units" problem in ecosystem network analysis. Pages 90-115 In: F. Wulff, J.G. Field, and K.H. Mann (eds). *Network Analysis of Marine Ecosystems: Methods and Applications*. Coastal and Estuarine Studies Series. pps. 90-115, Springer-Verlag, Heidelberg.

Costanza, R. and Perrings, C.H. 1990. A flexible assurance bonding system for improved environmental management. *Ecological Economics*, **2**, 57-76.

Costanza, R., Sklar, F.H. and White, M.L. 1990. Modeling coastal landscape dynamics. *BioScience*, **40**, 91-107.

Daly, H.E. and Cobb, J.B. Jr. 1989. *For The Common Good: Redirecting the Economy Toward Community, the Environment, and a Sustainable Future*. Beacon, Boston.

DAWN (*Development Alternatives for Women in a New Era*), 1988, G. Sen, K. Gowen (eds.).

Demeny, P. 1988. Demography and the limits of growth. *Population and Development Review Supplement*, **14**, 213-244.

Demeny, P. 1990. Population. In: B.L. Turner II, *et al.* The Earth Transformed by Human Action. Global and Regional Changes in the Biosphere over the Past 300 Years, pp. 41-54. Cambridge University Press with Clark University, New York.

Durning, A. 1991. Asking how much is enough. In: L.R. Brown, *et al. State of the World 1991. A Worldwatch Institute Report on Progress Toward a Sustainable Development*. pp. 153-169.

Ehrlich, P, et al. 1989. Global change and carrying capacity: Implications for life on Earth. In: Ruth DeFries and Thomas Malone (eds). *Global Change and Our Common Future: Papers From a Forum*, pp. 19-27. National Academy Press, Washington, DC.

Funtowicz, S.O. and Ravetz, J.R. 1991. A new scientific methodology for global environmental issues. In: R. Costanza (ed.). *Ecological Economics: The Science and Management of Sustainability*. pp.137-152, Columbia University Press, New York. pp.525.

Garcia, R. 1990. Metodologia para el estudio del cambio global. In: *ISSC Workshop: Issues in Global Environmental Change*. Mexico City.

Gever, J., Kaufmann, R., Skole, D. and Vörösmarty, C. 1986. *Beyond Oil: The Threat to Food and Fuel in the Coming Decades*. Ballinger, Cambridge, MA, 304 pp.

Gilliand, B. 1983. Considerations on world population and food supply. *Population and Development Review*. **9**(2), 203-211.

Hall, C.A.S., Cleveland, C.J. and Kaufmann, R. 1986. *Energy and Resource Quality: The Ecology of the Economic Process*. Wiley, New York, pp.577

Hardoy, J. and Satterthwaite, D. 1991. Environmental problems of the Third World cities: A global issue ignored? In: *Public Administration and Development*, Vol. II.

Harrison, P. 1990. Too much life on Earth? *New Scientist*, **126**, no.1717, 28.

Hirschman, A 1958. *The Strategy of Economic Development*. Yale University Press, New Haven, Connecticut.

Jacobson, J. 1988. Planning the global family. In: L.R. Brown, *et al. State of the World 1988: A Worldwatch Institute Report on Progress Toward a Sustainable Development*, pp.151-169.

Jacobson, H. with Price, M. 1990. *A Framework for Research in the Social Sciences for Global Environmental Change*. Paris, ISSC.

Johnson, D. Lee, Gale and D. R. (eds.). 1987. *Population Growth and Economic Development: Issues and Evidence*. University of Wisconsin Press, Madison. A Publication of the National Research Council Committee on Population.

Kaufmann, R. 1990. An analysis of the energy/GDP ratio in the big five: Confirming biophysical principles, rejecting neoclassical principles. Paper at the 1st Annual Conference of the International Society for Ecological Economics, Washington, DC., May 21-23, 1990. Abstract published in: Costanza, R., Haskell, B., Cornwell, L., Daly, H., and Johnson, T. (eds.). The ecological economics of sustainability: making local and short-term goals consistent with global and long-term goals. World Bank Environment Working Paper No. 32. Washington, DC.

Keyfitz, N. 1991. Population and development within the ecosphere: one view of the literature. *Population Index*, **57**(1), 5-22.

Lappe and Schurman. 1988. *Taking Population Seriously.* Earthscan.

Lovins, A.B. 1977. *Soft Energy Paths: Toward a Durable Peace.* Ballinger, Cambridge, MA.

Lovins, A.B., and Lovins, L. H. 1987. Energy: the avoidable oil crisis. *The Atlantic*, pp. 22-30, December.

Lutz, W. 1991. Towards the holistic understanding of a microcosmos: a case study on Mauritius. In: *Population, Economy and Environment in Mauritius*, CP-91-01. International Institute for Applied Systems Analysis, pp.3-8. Laxenburg, Austria.

Lutz, W. and Prinz, C. 1991. Scenarios for the World Population in the Next Century: Excessive Growth or Extreme Aging. WP-91-22. International Institute for Applied Systems Analysis, Laxenburg, Austria.

MacNeil, J. 1990. Sustainable Development, Economics, and the Growth Imperative. Workshop on the Economics of Sustainable Development, Background Paper No. 3, Washington DC, September.

Meadows, D. 1988. "Quality of Life". *In: Earth '88: Changing Geographic Perspectives.* National Geographic Society. Washington DC., pp.332-349.

OECD. 1991. *The State of the Environment.* OECD, Paris.

Pearce, D. 1987. Foundations of an ecological economics. *Ecological Modeling*, **38**, 9-18.

Perrings, C. 1989. Environmental bonds and the incentive to research in activities involving uncertain future effects. *Ecological Economics*, **1**, 95-110.

Perrings, C. 1991. Reserved rationality and the precautionary principle: Technological change, time and uncertainty in environmental decision-making. In: R. Costanza (ed.). *Ecological Economics: The science and management of sustainability*, pp. 331-343. Columbia University Press, New York. pp.525.

Repetto, R and Holmes, T. 1983. The role of population in resource depletion in developing countries. *Population and Development Review*, **9**(4), December.

Repetto, R. 1987. *Population, Resources, Environment: An Uncertain Future.* Population Reference Bureau, Washington, DC.

Revelle, R. 1976. The resources available for agriculture. *Scientific American*, 165-178, September.

Sanchez, V. 1990. Poblacion y medio ambiente, Mexico, Fundacion Nuevo Mundo.

Sanchez, V., Castillejos, M. and Bracho, R.R. 1989. Poblacion, recursos y medio ambient en Mexico. Fundacion Universo Veintiuno, A.C. Mexico.

Schmink, M. 1989. The Rationality of Tropical Forest Destruction – Paper presented at the USAID/USDA Conf. on Management of the Forests of Tropical America – Prospects and Technologies, San Juan, Puerto Rico, 1986.

Simon, J. Population Matters: 1990. *People, Resource, Environment and Immigration.* Transaction Publishers, New Brunswick.

Stonich, S. 1989. The Dynamics of Social Processes and Environment Destruction: A Central American Case Study. In: *Population and Development Review*, **15**, No.2.

Swaminathan, M.S. 1988. Global agriculture at the crossroads. In: *Earth '88: Changing Geographic Perspectives.* pp.316-331. National Geographic Society. Washington, DC.

Torrey, B. and Kilsrud, G. 1991. Paper presented at the Annual Meeting of the American Association for the Advancement of Science, Washington, DC.

United Nations. 1973. The determinants and consequences of population trends. *United Nations Population Studies*, No. 50, New York.

United Nations. 1989. Department of International Economic and Social Affairs. *World Population Prospects 1988.* New York.

United Nations. 1990. Global Outlook 2000. An Economic, Social and Environmental Perspectives. United Nations Publications.

UNDP (United Nations Development Programme) 1990. *Human Development Index*, New York.

United Nations Population Fund. 1991. *The State of the World Population.*

Victor, P.A. 1972. *Pollution, Economy and Environment.* University of Toronto Press.

Whitmore, T.M. *et al.* 1990. Long-term population change. In: B.L. Turner II, *et al. The Earth Transformed by Human Action: Global and Regional Changes in the Biosphere over the Past 300 Years.* pp.25-39. Cambridge University Press with Clark University, New York.

World Commission on Environment and Development. 1987. *Our Common Future.* Oxford University Press.

Worldwatch Institute. 1988. State of the World 1988. A Worldwatch Institute Report on Progress Toward a Sustainable Society.

Worldwatch Institute. 1991. State of the World 1991. A Worldwatch Institute Report on Progress Toward a Sustainable Society.

Chapter 2: Agriculture, Land Use and Degradation

D. Norse, C. James, B.J. Skinner and Q. Zhao

EXECUTIVE SUMMARY

Available land resources are probably insufficient to meet the anticipated growth in demand for agricultural and non-agricultural land projected for the next 35 years, although mineral needs can probably be met. Consequently, unless there is urgent action on a number of fronts, economic and social development in many countries will be held back up to and beyond UNCED's year 2025 horizon. On the one hand, the physical status and agricultural productivity of the land presently in use are being seriously undermined by mismanagement. On the other hand, the unutilized land resources are generally too small or of insufficient quality to compensate. Both aspects set challenges for the research community.

The issues and options can be visualized as a tetrahedron, the four faces of which represent, respectively the driving forces for land use and degradation; the resource problems the research solutions and the policy decisions. No face has an existence without the others. Take away one of them and the tetrahedron collapses, and with it the global land system.

The edges of the base represent the three major driving forces of land use and degradation. Rapid population, which outstrips the ability of countries to invest adequately in education and other aspects of economic and social development. Poverty, which forces many people to adopt unsustainable land use practices. And its reciprocal, income growth, which drives the demand for goods produced from the land, and presently tends to encourage consumption preferences that over-burden the environment and under-value research.

Similarly, the first face encapsulates the three major resource problems confronting mankind. Competition for shrinking land resources between crop production, livestock rearing and forestry and sometimes losing to a fourth demand, namely urban and industrial development. Land degradation, most importantly in the form of soil fertility decline, as well as through soil erosion, salinization, and other mechanisms. Finally, the other biotic and abiotic stresses, such as weeds and acid soils, that constrain the intensification of production needed to compensate for the lack of land.

The second face represents the three priorities for the research sector, if it is to help overcome the above resource problems. The first addresses gaps in our information base, particularly on actual land use and the physical and economic costs of land degradation. The second is concerned with institutional research to improve land use management at local, regional and national levels, and to strengthen farmer–researcher links so that scientific research is appropriate to farmer needs. The third is scientific research to raise productivity and thereby reduce pressure on the land. The opportunities range from basic research to widen the options for technology development, to mission-oriented applied research into such priorities as the restoration of degraded lands, development of sustainable alternatives to shifting agriculture and some forms of high input agriculture, and the application of biotechnology in the limitation of biotic and abiotic stresses.

The policy decisions encompassed by the final face are prerequisites for widespread and successful implementation of research. Once again there are three major thrusts. The establishment of clear property or user rights to land, so that farmers and rural communities have a vested interest in preventing or reversing degradation. Introduction of new institutional mechanisms for land use planning so that they are more decentralized to the district and user level. And last, but not least, revision of the incentives and regulations guiding resource use so that commodity prices reflect the full environmental cost of land use and appropriate environmental standards are set.

1. INTRODUCTION

Sustained development is impossible without sustainable agricultural land use. Although there are technological alternatives to land-based agriculture (e.g. soil-less hydroponics), they are either dependent on non-renewable resources or so costly that the half to one billion people who are currently malnourished, plus the other one to two billion with low incomes, would be unable to afford them.

Current land use systems and practices are not sustainable in many parts of the globe. Land degradation, in particular is undermining social and economic development by lowering the productivity of the soil (Lal, 1987). In extreme cases, land has been degraded beyond economic recovery. Soil erosion and other forms of land degradation in some developing countries, for example, have been estimated to be lowering Gross National Product (GNP) by as much as 5–10% per year (Bishop, 1990; Pearce, 1991).

Future land use will be even more difficult to sustain given present trends. Most countries have no, or limited, land reserves suitable for crop production. Therefore population growth will inevitably lower *per capita* land resources. Unless there are corresponding improvements in technology to provide higher yields without increased stress on the soil, land degradation will inevitably increase. Moreover, growth in population and in income-led consumer demand will increase the competition for land both within and outside the agricultural sector. Within the sector there will be increased competition between crop, livestock and forestry production. Outside the sector there will be the competition between agriculture, mineral production, and urban/industrial development for both land and water.

The conjunction of forces governing land use places the underlying science, and particularly land husbandry, high on the long-term research agenda for environmentally sound and sustainable development. Mankind's failure over much of the globe to reverse the speed of soil degradation has highlighted major gaps in our knowledge, and notably of:

(a) the spatial and temporal dimensions of land resource utilization

(b) how the resource base is likely to respond to different forms of use and management of soil processes and

(c) the causes of land degradation and the consequent implications for agricultural productivity.

This chapter therefore reviews resource base issues, assesses present and projected land use requirements, examines the key problem areas for ecologically sound and sustainable development, and finally outlines the main response options and research needs.

2. THE RESOURCE BASE

Underlying all issues of land use and resource production is the immutable fact that the world's land area is fixed in size but the population is not. With a growing population, each individual must rely on an ever smaller fraction of

that land for their supplies of food and natural resources.

The world's land area is approximately 15 billion hectares. Excluding inland water bodies and areas covered by ice and regions above 5000 m in altitude, there are some 11.5 billion hectares of continental land on which humans can conceivably live, on which they can produce food, and from which they can recover natural resources. In reality, of course, the amount of land suitable for agriculture is much less because of a variety of soil and agro-ecological constraints. Of the total land area only some 25%, i.e. 3.2 billion hectares are potentially suitable for arable agriculture. Of this 3.2 billion hectares, only about 10–15% has neither inherent physical nor chemical soil constraints. The remainder suffers from one or more constraints, particularly steep slopes, shallow soil, poor drainage, macro- and micro-element deficiency, low nutrient retention, aluminium toxicity, acidity, phosphorus fixation, excessive cracking, low potassium reserves, free calcium carbonate, salinity, excess sodium, gravel, and amorphous materials. These constraints are a major factor behind the fact that only 1.5 of the 3.2 billion hectares are currently used for arable agriculture.

These land and soil constraints are commonly compounded by, or are in part a consequence of, agro-climatic constraints, such as high rainfall or high evapo-transpiration and hence they tend to vary considerably across agro-ecological zones. In the humid, and, to some extent, the seasonally dry, warm tropics, aluminium toxicity and acidity are widespread constraints. Low potassium reserves and phosphorus fixation also occur in these zones. Steep slopes, hence potential erosion, are a major constraint in the cool tropics of Africa and Latin America, but in Asia, the humid, warm tropics is the zone most affected. The agro-climatic constraints carry their own problems for land use and sustainable development in that they restrict the range of crops which can be grown, e.g. because of temperature or soil moisture limitations.

2.1 Land: the widening mismatch between supply and demand

The demand for land and the minerals that can be extracted from it or the agricultural commodities which can be grown on it, are driven primarily by population and income growth. In the past the land resources have been sufficient to meet the demands although the visible environmental costs have been high and the invisible ones even higher.

The visible costs are exemplified by the 5–7 million ha (0.3–0.5%) of arable land lost every year through soil degradation (FAO, 1991a). The invisible costs, on the other hand, are only now being understood and estimated, albeit in a partial manner because of severe weaknesses in our monitoring data on the physical losses, and in our inadequate understanding of the biological and economic consequences of such losses. Recent estimates range from

Table 1: Past changes in land use (billion ha)

	Actual		Percent change
Category	1882	1991	1882 – 1991
Arable and permanent crops	0.86	1.5	+ 74
Permanent pastures	1.5	3.2	+ 113
Forest and woodland	5.2	4.1	– 21

7% GNP for the direct costs of land degradation in Ghana to 17% GNP for the costs of soil erosion and other forms of pollution in Nigeria (Pearce, 1991). Such costs are clearly unsustainable.

2.2 Present land use

Although 1.5 billion of the 13.3 billion ha of land (Table 1) are currently in use for arable and permanent crops, the area actually cultivated each year is less because of fallows. A further 3.2 billion hectares is used currently as permanent pastures or rangeland and 4.1 billion hectares are in forest and woodland with some overlap between these two categories.

The quantitative picture for other land uses – primarily mineral extraction and urban/industrial development – is not known with precision, and estimates could be wrong by a factor of 2 or 3. The area devoted to mining (including sand, gravel, cement, phosphates as well as fossil fuel) is difficult to estimate because, as well as the actual area of working mines, quarries, etc., there is the infrastructural area, and the area abandoned, or waiting reclamation. The total land area affected is of the order of 50–100 million hectares. Hence it is small compared with the agricultural area, and with the area currently used for urban/industrial/transport development which is of the order of 550 million hectares.

The mineral resource implications for sustainable development, however, go beyond the question of the area used for mining. The world population consumes about 9 tonnes of mineral resources *per capita* per year; a total of some 50 billion tonnes/year. The estimate includes fuel, metals, fertilizers, building materials, and all other resources extracted from the ground. It does not include all of the soil and rock moved to facilitate building and construction. What makes the mass of material consumed such an impressive figure is that all of the sediment and dissolved matter transported annually to the sea by all the rivers of the world is estimated to be only 16.5 billion tonnes/years. The human population is obviously a geological entity to be reckoned with (Skinner, 1989).

Two key questions arise from this resource use. What are the supply limits, if any, to the continued extraction

and use of mineral extraction resources? And what are the possible consequences for global chemistry through the use of mineral resources on this scale? The first question is dealt with in the next section, and the second in the section on biotic and abiotic stresses.

2.3 Future land requirements

Leaving aside the question of fuels because they are treated elsewhere in this Volume, we focus on non-fuel resources extracted from the land. The economy is engaged, in essence, in a continual struggle between two forces: advances in technology which enhance the productivity of labour and the exhaustion of resources that diminishes the productivity of other factors. Over the past two centuries technology has been the clear victor but it is not clear that the same pattern will hold for the century ahead.

No country today is completely self-sufficient with respect to mineral resources. The more industrialized a country the wider is the range of materials required and the less likely it is that the country can achieve self-sufficiency. Indeed, it is not clear that any country today could achieve self-sufficiency and also develop a full technological and industrial capacity. To some extent therefore, all countries must participate in the global market in order to meet their resource needs.

The global magnitude of all mineral resources appears to be adequate for the next 35 years (McLaren and Skinner, 1987), but heterogeneities in the geographic distribution of raw materials will continue to present major challenges. There is no compelling reason to believe, however, that in the short run of 35 years a stable world trading system cannot cope with the supply problems. Lesser assurances are forthcoming concerning the consequences of extracting and using mineral resources at the rate we do.

Looking ahead 110 years to the end of the 21st Century when world population is projected to be more or less stable, or less ambitiously to the year 2025 adopted by UNCED for its long-term assessments, it is clear that the historic growth pattern of land use cannot be sustained. Between 1882 and 1991 the area of arable land in use expanded by 74% (Table 1) through forest clearance and the ploughing up of rangeland. A similar rate of expansion over the next 110 years would exceed the area of land potentially suitable for ecologically sound and sustainable agriculture. Even for the next 35 years the expansion of land use could not be sustained because :

(a) a high proportion of the present reserves are only marginally suited for crop production. They need to be managed very carefully to avoid environmental damage and in many instances are best left under forest;

(b) most reserves are currently under forest or permanent pastures or rangeland, and the demand for both forests and pastures is growing.

Table 2: Population growth – UN medium variant population projection (in billions)

	1990	2020	2050	2100
Developed countries	1.1	1.2	1.1	1.0
Developing countries	4.2	6.9	8.8	10.0
Africa	0.6	1.4	2.2	3.5
Latin America	0.5	0.7	1.0	1.2
Asia	3.1	4.8	5.6	5.3
World	5.3	8.1	9.9	11.0

Table 3 : Current and projected average annual growth of gross domestic product per capita

	1981–1990	1991–2000	2001–2010	2011–2030
Developed market economies	2.4	2.4	2.0	1.5
Sub-Saharan Africa	-1.3	0.1	2.1	2.9
Latin America	-1.3	1.1	2.3	3.0
S. and S.E. Asia	4.1	3.6	3.0	2.5

It is also clear that the pressure for additional land is largely an issue for the developing countries and not the developed ones.

In most developed countries population growth is virtually static (Table 2) and incomes have reached the level at which demand for agricultural commodities is close to saturation. They have achieved high productivity growth resulting in some relatively marginal lands being taken out of cultivation. Their problems are commonly ones of over-production rather than under-production. Consequently agricultural land requirements in developed countries are unlikely to grow appreciably, and may even contract further, although pressures to reduce some of the negative environmental impacts of intensive production systems could result in some area expansion.

The developing country situation is the reverse. Land requirements will continue to grow even though the rate of expansion has been declining since the 1960s and 1970s. Population will be growing at some 900 million per decade (Table 2). Population and income growth (Table 3) will rapidly drive the demand for food, fibre and other agricultural commodities. Agricultural productivity growth, however, may, remain slow or stagnant because of institutional or infrastructural weaknesses which cannot be overcome in the short- to medium-term, or because of the lack of technologies appropriate to the physical and

Table 4: Projected decline in potentially cultivable land resource per capita in the developing countries. (ha per capita)

	1990	2025
Sub-Saharan Africa	1.6	0.63
West Asia and North Africa	0.22	0.11
Asia, excluding China	0.20	0.12
Central and South America	2.0	1.17

economic situation and to farmers' perceptions of their needs. Hence there will be insufficient capacity to substitute modern production inputs for land.

However, most developing countries have little or no potentially arable land left to develop, although a few countries like Brazil and Zaire still have substantial reserves (Norse, 1988). Consequently the area of arable land *per capita* will decline, in some instances to a tenth of a hectare. This will intensify the competition for land between crop production, the rearing of livestock and forestry. Crop production generally wins the competition, so the grazing area per animal will decline even more rapidly, with the probability of rising environmental degradation and declining sustainability.

3. PROBLEM AREAS

This lack of land resources to meet future demands exposes three main problem areas to be addressed by the science agenda:
(a)   competition for land;
(b)   land degradation and land management;
(c)   other biotic and abiotic stresses.

3.1 Competition for land

There has been increasing competition for land (and water) in recent decades, and as indicated above, it will intensify in the future. In parts of the USA irrigated agriculture is no longer possible because of a combination of over-extraction of ground-water resources and growing urban/industrial demands. Similar problems are arising in the Near East and other developing regions. Consequently the greatest challenge for some areas will be competition for water rather than for land with serious impacts on freshwater availability and quality, as explored in Chapter 10 of this Volume.

Agricultural expansion is the major cause of deforestation particularly in the Tropics (FAO, 1991; World Resources Institute, 1990). The broad orders of magnitude are that tropical deforestation is currently running at about 17 million hectares a year compared with about 10–11 million ha in 1980 (note that these estimates include land under shifting cultivation and hence cleared for short periods before being abandoned to natural cover). Shifting agriculture has been responsible for about a half and permanent agriculture about one-quarter to one-third of the recent deforestation. Expressed as a proportion of total forest cover in the tropical regions, annual rates of deforestation in the period 1981–1990 are 1.7% in Africa, 0.9% in Latin America, and 1.4% in Asia (FAO, 1991).

The consequences of deforestation and forest mismanagement and their negative impacts on stream flow, soil erosion, agricultural productivity, availability of fuelwood, loss of biodiversity and on both local and global climate have been well documented (IPCC, 1990). It does not automatically follow, however, that reforestation is always the answer.

The competition between crop and livestock production is a major cause of desertification, particularly in Africa and the Near East. The ploughing up of rangeland and pastures for crop production has forced pastoralists onto drier and more marginal land. This pressure, together with the increase in livestock numbers as a result of human population growth, has led to overgrazing and eventual desertification.

It seems likely that this competition will continue. The arable land area is projected to expand by some 80 million ha between 1980 and 2000 (Norse, 1988) and by a similar area over the following 25 years. Much of this land will come from existing forests, permanent pastures, and reclaimable waste lands.

Finally, there is the competition between agriculture and urban/industrial development. Some 0.1 – 0.25 ha of land *per capita* is used currently for housing development plus its urban and industrial infrastructure. This area increases with income growth so *per capita* requirements may rise in the future. Thus it seems probable that an additional 300 million ha of agricultural land will be lost in this way by 2025. Moreover, much of this loss is likely to be of high-quality land because the major urban centers are commonly located in river valleys and coastal plains.

3.2 Land degradation

Degradation of the resource base, and particularly of agricultural land through soil erosion, is the fundamental constraint to sustainable development. Soil is produced from underlying bedrock through a complex series of chemical and physical interactions with the hydrosphere, atmosphere, and biosphere. The rate at which new soil is produced varies as a function of such variables as climate, rock type, and steepness of slope, but rates are everywhere slow – of the order of a millimetre/year. Because soil is the medium within which the world's crops, pastures and forests areas grow, a sustainable global economy would be one in which soil erosion and degradation do not exceed

Table 5: Areas of different developing country regions moderately, severely or extremely affected by various types of human-induced land degradation – in million ha (Oldeman et al., 1990)

Type	Land area		
	Africa	Asia	South and Central America
Water erosion	170	315	77
Wind erosion	98	90	16
Nutrient loss	25	104	3
Salinization	10	26	–

soil formation. At present the world is far from that ideal balance and soil is a wasting asset.

Over the past 1000 years, the land area degraded by man's activities has amounted to two billion hectares (Qiguo Zhao, 1990) although only a small proportion of this area is too degraded to remain in agricultural use. More recent estimates suggest that the situation may be even worse, with two billion hectares being degraded over the past 50 years alone (Oldeman *et al.* 1990), suggesting that about 25% of present arable and grazing land is affected by human-induced soil degradation (Table 5). Rapid population growth and consequent expansion of agriculture on to marginal lands plus mismanagement of good land has accelerated the annual loss to 5–6 million ha (FAO, 1991a).

In many countries, soil loss through erosion by wind and water represents one of the major consequences of agro-ecosystem mismanagement, especially in semi-arid areas and parts of the humid tropics. The underlying cause of the erosion is inappropriate land use, such as the clear cultivation of steep slopes, loss of vegetative cover due to overgrazing or clearing and the extension of continuous cultivation into marginal arid areas.

These data provide only an approximate indication of the scale of the different major soil degradation problems, and their full physical and economic implications for future agricultural production are not well established, and hence research priorities or requirements are not easy to define and will not be sufficient in themselves to overcome the problem.

At the global level the figures are very disturbing. One of the first comprehensive attempts to estimate the rate of global soil erosion was by Judson in 1968. Judson used geological data to estimate that the pre-agriculture rate of erosion of soil was 9.3×10^9 tonnes/year. This figure is a base from which departures can be estimated and it is also an estimate of the rate at which new soil is created around the world because in pre-agriculture days, soil loss and soil production were in balance.

When Judson estimated the global rate of erosion in 1968 he arrived at a figure of 24×10^9 tonnes/year. The difference between 24×10^9 and 9.3×10^9, that is 14.7×10^9 tonnes/year, is an estimate for the wastage of soil above production in 1968. Subsequently, Brown (1984) made a similar and somewhat more detailed estimate, namely that the erosion in excess of new soil produced was 23×10^9 tonnes/year, suggesting a jump of more than 50% over an 18-year period. The increased rate is alarming enough, but if we consider that the global population in 1968 was about 3.7×10^9 and by 1984 it had arisen to 4.8×10^9, the *per capita loss of soil had increased from 4.0 to 4.8 tonnes/year.*

Neither of the data sets used by Judson and Brown is entirely satisfactory so there are large uncertainties in the estimates they made. The trends indicated by the numbers are not in doubt, however. Both magnitude of the global rate of erosion, and the *per capita* rate of erosion are increasing.

Simple arithmetic emphasizes the possible future dangers posed by soil erosion. The total reservoir of soil has been estimated by Brown (1988) to be 3.5×10^{12} tonnes. Assuming that the current rate of erosion above replacement is 23×10^9 tonnes/year (a figure that is probably conservative because the global rate of erosion has surely increased since 1984), the soil reservoir could be declining at a rate of 0.7%/year. By the year 2025, and even presuming humankind is successful in holding the present annual rate of erosion to 23×10^9 tonnes/year, between 20 and 25% of the total soil reservoir could be lost.

The losses are not restricted to soil particles. Cultivation of soils and soil degradation through various practices has resulted in the loss of billions of tonnes of organic carbon during the last three centuries, as well as stores of nitrogen.

Unfortunately these apparently horrifying soil erosion and organic matter loss figures are relatively meaningless in terms of their implications for long-term agricultural production and sustainability. Physical and social scientists have yet to determine adequately the relationship between soil erosion and productivity loss on different soils (Stocking, 1984). For example, we do not know with any certainty the yield loss or economic costs of losing 50 or 100 tonnes of soil per hectare per year. Even 50 would be disastrous for a strongly weathered, poor soil with thin topsoil, while 100 might have small costs and effects on a deep soil in loess.

There is, however, increasing evidence for declining base yields of crops of the order of 2–10% per year (Twyford, 1988). Part of this decline will be the result of physical damage to the soil through erosion. Part will be from the loss of organic matter and the subsequent reduction in soil moisture holding capacity. The latter is an under-researched area, but one of vital importance to the stability of crop and livestock production, particularly in areas of low and uncertain rainfall.

Much of the remainder of the decline appears to be related to chemical damage of soil through the loss of plant nutrients by erosion, from soil acidification because of mismanagement of fertilizer practices, and as a result of nutrient mining. The latter is an increasingly important problem in many areas of the developing world. It arises most commonly where soil fertility is lowered through the removal of nutrients in harvested plant and animal products without adequate counter-balancing inputs of fertilizer. This process can be exacerbated by the loss of nutrients in the topsoil due to erosion. For example, in some East African countries, where soil erosion is serious, and only small amounts of fertilizer are used, net country-wide average rates of nutrient loss are as high as 80 kg $(N + P_2O_5 + K_2O)$ per hectare per annum (Stoorvogel and Smaling, 1990).

Another vitally important dimension of land degradation is that in irrigated areas. Some 76 million hectares are suffering from human-induced salinization. Of this total area, about 40 million hectares of land world-wide, mainly but not exclusively irrigated, are moderately or strongly salinated. It is particularly serious in the developing countries (Table 5) where irrigated agriculture has contributed enormously to recent gains in food production, and in the future is likely to play an even more critical role. Salinization, caused by design faults, poor water management and inadequate drainage systems, threatens the sustainability of large areas of irrigated land. According to the World Commission on Environment and Development (1987), millions of hectares of productive irrigated land are abandoned annually. FAO estimates, however, suggest that actual area losses are perhaps only 2 million hectares per year, but the productivity losses on the remainder of the affected land are equally if not more important. These problems are particularly disturbing because the areas affected are major food bowls in the countries concerned.

Water resources needed for irrigation are also threatened indirectly by land degradation. In particular, siltation from upland watersheds in some areas is reducing the life of water storage reservoirs and distribution canals to as little as 25% of their designed life. These problems point to the need to manage and utilize available water resources and irrigation systems more efficiently, and to give greater attention to soil erosion control in major watersheds.

3.3 Biotic and abiotic stresses

There are no systematic global quantitative studies that assess the relative importance of the various constraints that impede the intensification of crop and livestock production to compensate for the lack of land or to limit its degradation. Nonetheless many studies have generated data which provide an indication of the importance of specific factors. These constraints can be conveniently categorized into abiotic and biotic stresses.

3.3.1 Abiotic stresses

Drought or insufficient water is probably the most pervasive abiotic stress. It is estimated to affect one-third of all the area under crop production, and is intensified by poor land management. Mineral toxicities and deficiencies are also major stresses. For example, the acid soils that affect large areas of the Cerrado in Latin America suffer from aluminium toxicity and phosphate unavailability. Lack of an adequate supply of nitrogen is a stress that precludes optimal production in many crops, particularly the major cereal staples which unlike leguminous crops do not fix their own nitrogen.

As the food need increases in developing countries, the traditional ecological limits of crops, such as wheat, are being challenged and the limited heat tolerance of wheat in the tropical lowlands represents an increasingly important constraint. Cold tolerance has also been a constraint in both industrial and developing countries, and there have been a number of successful attempts to adapt temperate or tropical crops to higher altitudes and latitudes. Cold tolerance will, however, become less of a constraint in the future, if global warming follows its projected pattern.

Man is adding to natural abiotic stresses. The annual anthropogenic redistribution of 50 billion tonnes of mineral resources is affecting the Earth in several obvious ways – increased CO_2 in the atmosphere and acid rain are two. There are probably many more subtle but still unappreciated changes. No place on Earth is, or can be, free from anthropogenic environmental changes. As pointed out by Professor W.S. Fyfe in an address to the 28th International Geological Congress in 1989, "If we change the local geochemistry there will be a related change in the local species distribution. A stable population of a given species requires a high degree of geochemical stability. It seems certain that we must now be changing the metabolism and distribution of all living species on the planet. One of our greatest challenges is to describe and quantify the change in the microbiota occurring in all our major systems containing liquid water, including soils" (Fyfe, 1989). What is not yet clear, however, is whether these changes are of a magnitude to affect sustainable development in the very long term. Although they are not yet a major issue for the first half of the 21st Century, the changes must be monitored if we are to fully appreciate the extent to which humans are changing global chemistry.

3.3.2 Biotic stresses

Biotic stresses due to pests are a major constraint to increased crop production. Huge losses occur from insects, diseases, weeds and nematodes at both pre- and post-harvest stages. Pre-harvest losses due to each of these groups of pests have been estimated at up to 1 billion tonnes of food, feed and fibre, equivalent to US$100 billion per annum. This significant loss occurs despite the fact that $23 billion is spent on pesticides annually, a cost that must be added to the $400 billion lost to pests overall. The original methodology expresses losses as percentage of potential yield in the absence of various pests (Cramer, 1967); lacking information about the interactions of losses from different pests the Cramer-derived estimates assume that losses from different pests are additive, whereas in practice the collective effect of pests will almost certainly be over-estimated. In addition post-harvest losses resulting from pests, many of which are carried over from pre-harvest infections or infestations, are estimated to decrease availability of food by up to 40%. About a half of the losses caused by biotic stresses occur in developing countries which can least afford the loss.

Infectious and non-infectious diseases of animals are stresses that significantly reduce livestock productivity in both industrial and developing countries. Infectious diseases have been classified into three categories. The first category is represented by epidemic diseases, such as foot and mouth disease, Newcastle disease, rinderpest, swine fever, theileriosis and trypanosomiasis. These epidemic diseases often result in high mortality rates and severely affect the productivity of survivors. The second category is zoonotic diseases like rabies, brucellosis and salmonellosis, which affect both humans and livestock, and the third is diseases that are present in both extensive and intensive livestock farming systems but which exert severe effects in intensive-rearing systems: examples include mastitis in cows and coccidiosis in poultry. The other diseases affecting livestock are non-infectious and include metabolic diseases such as milk-fever in cows and various genetic disorders.

4. RESPONSE OPTIONS

Only general options can be presented here because land use and land degradation is governed by a range of socio-economic and biological forces that vary with the institutional and agro-ecological situation. They will have to be combined and modified at the national level to match local needs. Nonetheless, three sets of options can be distinguished which either have a global validity or appear applicable to a significant number of countries: improving the information base; changing institutional mechanisms; promoting better husbandry together with land saving or land protecting technologies. All three sets need complementary action on human resource development and institutional capacity building as argued in Chapter 14. Particular attention needs to be given to the training of biologically oriented agro-scientists and the fostering of interdisciplinary research between natural and social scientists.

4.1 Improving the information base
It should be clear from the foregoing that there are major gaps in our knowledge concerning:
(a) the socio-economic costs of competing land-use demands on a declining resource;
(b) the physical, biological and economic costs of land degradation;
(c) the subtle effects of the anthropogenic distribution of minerals and chemicals on the environment, and especially on soil microbiota and soil water.

These gaps need to be filled by national and international efforts to improve the methodologies for impact assessment, and to establish monitoring stations at suitable sites as part of a global network. Specific recommendations are given in the next section.

4.2 Changing institutional mechanisms
Land use planning is still in its infancy. Until the 1960s and 1970s few of the developed countries had effective land use policies to protect good agricultural land from urban/industrial development (OECD, 1976). Some of them still have weaknesses in the implementation of such policies.

In the developing countries, most land use decisions remain in the hand of individual farmers and communities. Governments can, however, strongly influence these decisions through price signals and other income-related incentives; through zoning and certain other regulatory measures; or, in the few countries where state farms exist, they can stipulate the cropping pattern to be adopted, and other land management decisions.

There have been successful attempts at land use planning, but in the main efforts have failed because they have been too insensitive to local needs and perceptions, too bureaucratic and too centralized. One option, therefore, is to give greater stress to resource management through local-level husbandry and on-site development "controlled" by local community or user-based institutions, e.g. grazing associations and water-user groups for integrated watershed development, rather than on sophisticated macro- or meso-level institutional structures, e.g. for national or regional land use planning or investment allocation. This is not to deny that such structures have a role to play and high priority should be given to the formulation of integrated land-use plans as a framework for sustainable development planning. But the institutional structures must be economically sustainable by the nations concerned and provide incentives rather

than top-down directives for resource development and management. The issue is not top-down or bottom-up; both approaches are needed, but they must be consistent with each other, and must promote convergence between national goals and local priorities.

However, this increase in the decentralization of land use planning can only lead to a socially desirable and sustainable allocation and use of land resources if three other requirements are met. First, the responsibilities and entitlements of farmers and rural communities to land must also be decentralized through the clear allocation of property rights. Secondly, commodity prices must reflect the full environmental costs of land use, as well as giving adequate incentives to producers. Thirdly, once governments have set appropriate environmental standards and taxes on emissions, etc., to protect the environment, their interventions should be kept to a minimum so that markets can function efficiently.

4.3 Promoting land saving or land protecting technologies
Three options deserve particular attention: wider application of ecologically sound conventional technologies; introduction of the new opportunities for sustainable agriculture arising from advances in biotechnology; and finally, new approaches to soil conservation.

The developed countries have been able to take marginal land out of production through the use of more intensive land saving technologies and external inputs, notably mineral fertilizers, improved seeds, higher yielding pastures and forage legumes and feed concentrates. In some instances they have resulted in a new set of environmental problems, e.g. soil compaction, ground-water and surface water pollution, etc., but in general these problems are not inherent in the technologies themselves but in the way they are managed. The developing countries are well able to learn from this experience, but it is more than a scientific issue. Institutional changes are required, for example, to ensure that farmers have access to the technological inputs required, and research systems develop technologies that are appropriate to farmer needs and perceptions, rather than scientists' perceptions of those needs as is commonly the case now. This mismatch of perceptions and the failure to build on indigenous knowledge are major factors underlying the low uptake of technologies by farmers. Thus it is vitally important that farmers become more directly involved in the research process.

Biotechnology offers a range of applications for plant and animal production than can be expected to have an impact in the near, mid and long term. A summary of the applications are listed in Table 6, with an assessment of the likely time frame for adoption.

Table 6: *Summary of biotechnology applications in agriculture*

CROP APPLICATIONS	
Applications	*Time Frame*[1]
Tissue culture	
Rapid propagation	Near
Production of disease-free stock	Near
Anther culture	Near
Embryo rescue	Near
Somaclonal variation	Near–Mid
Genetic mapping	
RFLPs	Near
Diagnostics	
Monoclonals/Polyclonals	Near
Nucleic acid probes	Near–Mid
Use of microorganisms	
Nitrogen Fixation	
legumes (selection of better N fixers)	Near–Long
cereals	Long
Plant growth promoter microbes (mycorrhiza)	Mid
Biological control	Near–Long
Post harvest food processing	Near–Mid
Transgenic crops	
Horticulture/root crops	
potato	Near
tomato	Near
cassava	Near
Cereals	
rice	Near
maize	Mid
wheat	Long
Fibre	
cotton	Long
Oil seed	
rapeseed	Near
soyabean	Mid
Tree crops	
walnut	Near
poplar	Near
coconut	Long
cocoa	Long

ANIMAL APPLICATIONS	
Applications	*Time Frame*[1]
In-vitro fertilization	Near
Embryo transfer	Near
New diagnostics	Near
Novel vaccines	Near
Novel hormones	
BST	Near
PST	Near
Modification of rumen microflora	Near
Use of RFLP for genetic mapping	Near
Transgenics	Mid–Long

[1]Estimated time-frame for commercialization of products:

Near-term	0–5 years
Mid-term	5–10 years
Long-term	over 10 years

Most of the applications listed in Table 6 will, to a lesser or greater extent, make a contribution to more sustainable land use by raising yields and therefore reducing land requirements for a given level of production. The new technologies to decrease the dependency on toxic pesticides applied to feed, food and fibre crops deserve special mention. Following several years of field testing, advanced technology is now available involving transgenic crops which include tomato, potato, rapeseed and cotton, maize and rice that have genes that confer resistance to selected insect pests (e.g. *Bacillus thurengiensis* (Bt) genes). Similarly field tested transgenic crops of tomato, potato, sugar beet, cucumber, alfalfa and squash with genes that confer non-conventional resistance (e.g. coat protein) to various viruses are at an advanced stage of development. Considering that approximately 25% of the total global pesticides, with a market value of $23 billion annually, is applied to vegetables and fruit, and that an additional 11% of total pesticides is applied to cotton, the potential for substituting for a portion of these insecticides applied to selected pests and virus diseases with transgenics in the near to mid-term is significant.

However, before society can realize the global benefits of the biotechnology option and particularly of these new field tested transgenic crops, regulatory standards must be satisfied that will allow commercialization of these products in the industrial countries and also the sharing of these technologies with the Third World using responsible and effective technology transfer mechanism.

Finally, turning to the soil conservation option, there is the need for a major shift in strategy. Until recently most soil conservation and restoration efforts were centerd on physical or engineering approaches involving the use of heavy machinery, and erecting barriers to soil and water movement. These actions were commonly expensive, and imposed on farmers or farming communities by central authorities. Failure rates were high, both in economic terms and physical terms, because farmers had little or no motivation to maintain conservation structures.

The present option differs in two important ways. It is centered on biological approaches to soil erosion prevention, restoration and conservation aimed at maintaining ground cover to protect the soil surface and using living fences of bushes or grasses to build up terraces during periods of heavy rain. Secondly, it aims to involve farmers and local communities in the design, implementation and maintenance of conservation schemes, and to provide economic benefits to them in the short term through higher crop yields.

5. RESEARCH AND DATA NEEDS

Actions are needed at two levels. First, the creation of baseline data sets on key indicators of degradation to clarify long-term trends and impacts, and as a preliminary step in selecting and designing appropriate actions of a more specific nature. Secondly, physical and economic research on degradation processes, impacts, and control, and on minimum standards for ecologically sound and sustainable production and land use.

5.1 Monitoring and data collection

A prime candidate is estimation and monitoring of actual land use. With few exceptions, we do not have geographically referenced data for which land is being used for what purpose. This will require improved Geographical Information Systems (GIS) that can bring together land use data with socio-economic information.

A second candidate is the establishment of a global monitoring network for soil carbon, both as a mechanism for determining changes in the soils capacity to act as a sink or a source of carbon dioxide, in the context of greenhouse warming and climate change, and as an indicator of soil fertility and agricultural sustainability. This could involve protocols for assessing soil degradation and the monitoring of other soil properties affected by man, e.g. buffering capacity as a result of acid rain or heavy fertilizer applications (Paces, 1985) to eventually develop a global digital data base of land degradation.

A third candidate is geochemical mapping to determine today's situation, and a series of geochemical monitoring stations to know how soils, rivers, lakes and ice fields are changing.

Both the existing and the new monitoring networks proposed here need to be fully integrated with those for freshwater.

5.2 Degradation processes and impacts
Priorities include research on:
(a) The genetic conditions, processes and mechanisms, especially the rehabilitation and restoration of different types of global soil degradation, and the theory and practice of controlling them;
(b) Technology for monitoring the global changes of land use and the dynamics of soil degradation as well as the methodology of soil degradation evaluation;
(c) The function of agro-ecosystem and its relationship with land use changes and soil degradation. Various simulation methods and modeling may be employed to predict the existing and new management measures and the long-term effect of varying climates on soil quality and agricultural productivity, so as to build the optimized artificial ecosystems of agricultural land use;
(d) The rate and consequences of deforestation. This should involve the monitoring of the dynamics of deforestation rate and its effect on rivers, soil erosion, agricultural productivity, fuel wood supply and the loss of biodiversity, as well as the local or global weather;
(e) The monitoring and evaluation of the global dynamics of soil erosion quantity and rate to identify enviromentally critical areas;
(f) Evaluation of the real value of natural resources;
(g) Fire management in forest and savanna areas;
(h) Minimum standards for ecologically sound production so that the land degradation costs of traded commodities can be reflected in export prices.

5.3 Degradation control, land restoration and land substitution
Prime candidates here are:
(a) Research on soil biology and on biological methods of soil conservation and restoration, such as the selection or development of live leguminous mulches and organic matter management;
(b) Development of sustainable production systems and their component technologies, notably to maintain or increase in soil fertility through integrated plant nutrition systems that avoid soil acidification;
(c) Promotion, by traditional breeding and biotechnological methods, of improved crops, including minor ones, for mixed cropping systems;
(d) Adoption of biotechnology derived crops that confer resistance to pests, that will provide alternatives to traditional toxic pesticides that currently degrade land, environment and ground water through accumulation of pesticide residues;
(e) Use of biotechnology to develop higher yielding crops and to process agricultural wastes, such as bagasse, oil cake and straw, to upgrade their nutrient content for livestock feed or to promote their use as fuel substitutes;
(f) Development of improved forest and agro-forestry practices for land stabilization and sustainable multi-purpose land use.

REFERENCES

Bishop, J. 1990. The cost of soil erosion in Malawi. Draft report for World Bank, Washington, November 1990.

Brown, L.R. 1984. Conserving soils. In: Brown, L.R., Chandler, W., Flavin, C., Postel, S,. Starke, L., and Wolfe, E. [eds]. *State of the World*, pp.53-73. 1984, W.W. Norton and Co.

Brown, L.R. 1988. The changing world food prospect: The nineties and beyond, Worldwatch Paper 85, 58p.

Cramer, H.H. 1967. Plant protection and world crop production, *Pflanzenschutznachr*, Bayer, **20**.

FAO, 1991a. Sustainable development and management of land and water resources. Background document no.1 for FAO/ Netherlands Conference on Agriculture and the Environment, 's-Hertogenbosch, The Netherlands, 15-19 April 1991.

FAO, 1991b. Global Overview of Status and Trends of World's Forest presented to Technical Workshop to explore the feasibility of forest options, Bangkok, 24-30 April 1991.

Fyfe, W.S. 1989. Personal communication.

IPCC (Intergovernmental Panel on Climate Change). 1990. *Climate Change, the IPCC Scientific Assessment*, New York, Cambridge University Press.

Judson, S. 1968. Erosion of the land, or what's happening to our continents. *American Scientist*, **56**, 356-374.

Lal, R. 1987. Effects of soil erosion on crop productivity, *Critical Reviews in Plant Sciences*. No.4.

McLaren, D.J., and Skinner, B.J. (eds) 1987. *Resources and World Development*. Dahlem workshop reports: New York, John Wiley and Sons.

Norse, D. 1988. *Sustainable Growth in Production in World Agriculture: toward 2000*, Alexandratos, N. [ed.] FAO, Rome/Belhaven Press, London.

OECD, 1976. *Land Use Policies and Agriculture*, OECD, Paris.

Oldeman, L.R., Hakkeling, R.T.A. and Sombroek, W.G. 1990. *Global Assessment of Soil Degradation*, ISRIC/UNEP study.

Paces, T. 1985. Sources of acidification in central Europe estimated from elemental budgets in small basins, *Nature*, **315**, 31-36.

Pearce, D. 1991. Report of the British Association Meeting, New Scientists, London.

Skinner, B. 1989. Resources in the 21st Century: Can supplies meet needs?, *Episodes*, **12**, 627-275.

Stoorvogel, J.J. and Smaling, E.M.A. 1990. Assessment of Soil Nutrient Depletion in Sub-Saharan Africa: 1983-2000, Winand Staring Center, Wageningen/ FAO, Rome.

Stocking, M. 1984. *Erosion and Soil Productivity: a Review*, FAO, Rome.

Twyford, I.T. 1988. Development of smallholder fertilizer use in Malawi, Paper to FAO/FIAC meeting 26-29 April 1988, Rome.

World Commission on Environment and Development, 1987. *Our Common Future*, Oxford University Press, Cambridge.

World Resources Institute (WRI),1990. *World Resources 1990-91*, Oxford University Press, New York, pp.363

Zhao, Qiguo. 1990. Land degradation, *Acta pedologica Sinica*, **25**, No. 4.

Chapter 3: Industry and Waste

R.U. Ayres, H.L. Beckers and R.Y. Qassim

EXECUTIVE SUMMARY

In the larger context of human interaction with the environment, "industry" is shorthand for productive economic activities, especially those involving materials. Industrial activities include raw materials extraction, physical and chemical separation and refining, energy conversion, shaping and forming, assembly, construction and distribution. Manufacturing *per se* is only a small part of the spectrum activities, and not necessarily the most important from an environmental perspective. For example, both extractive industries and final consumption are likely to contribute more to the generation of wastes and residuals. It is important to bear in mind that all materials extracted from the Earth's surface (plus atmospheric oxygen) eventually return to the environment in some degraded form.

Many industrial leaders are convinced that industry has already made progress in reducing waste emissions (and environmental impacts thereof). As society becomes increasingly concerned with preserving environmental amenities, and as economic incentives are adjusted, industry will continue to respond positively. The adoption of "greener" products and low waste, energy-conserving production technologies will surely spread. In fact, adopting a more holistic "systems approach" to the problem of production has already revealed unexpected profit opportunities in a number of industries. Nevertheless, much R and D needs to be done in this field.

It is questionable, however, whether the adoption of a "waste minimization" strategy at the plant level (or even the sector level) will suffice to achieve long-run sustainability. The problem is that continued reliance on virgin raw materials and fossil fuels may still impose unacceptable burdens on the environment, regardless of how efficiently intermediate processes are carried out. The systems approach must ultimately be applied at the national and global levels.

1. INTRODUCTION

The subject of industry and waste can be viewed within the framework of two archetypal world views. These two worldviews differ in respect to implicit risk perceptions. Persons who believe that the Earth System, including human institutions, is relatively robust and that it can adjust to anthropogenic perturbations without near-terms danger of collapse are probably in the majority at present. Two of the authors of this paper belong to this group.

On the other hand, there are people who are worried that:

(a) human institutions may be too slow to adopt, or
(b) the natural system may be more fragile than it currently appears to be. The corresponding worldview, shared by one of the authors (R.A.), can be characterized as "risk averse".

Another point of considerable importance is worth emphasizing at the outset. It is tempting to think of industrial structure as "fixed" or, at least, very slow to change. In the short run this is true. Nevertheless, roughly 15% of gross world product (GWP) is invested each year in capital goods (plant and equipment). The entire world industrial base is effectively rebuilt or replaced roughly once each decade. There is ample opportunity for introducing new, more environmentally "friendly" technologies over the next 30 years.

In this paper, after giving an overview of industry, we present both worldviews, discuss their challenges and implications. We conclude by summarizing the principal facts, needs, and actions.

2. OVERVIEW OF INDUSTRY

The first question to be answered is: what do we mean by industry? A schematic chart that will help to clarify this question is presented as Figure 1, below:

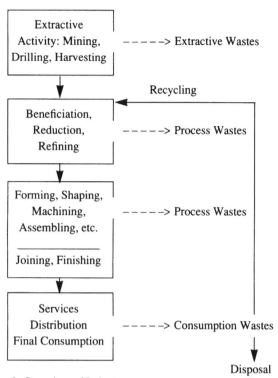

Figure 1: Overview of Industry

In brief, industry can be described from the physical stand point as the sum total of materials/energy transformational activities, converting raw (or recycled) materials into products. Products, in turn, provide useful services for consumers (i.e. all of us).

In the physical context, we note that large quantities of natural resources, both renewable and non-renewable, are extracted each year from the environment. They enter the transformational process stream in relatively crude form. Typical raw materials range from timber and pulpwood to crude oil, coal, gas, metal ores and other minerals such as phosphate rock and limestone.

The first step in materials processing normally involves both physical and chemical separation and removal of undesired components of the raw material. This unwanted fraction includes the branches and bark of trees, the hulls and stalks of plants, the ash in coal, the overburden and gangue associated with metal ores, etc. For instance, in the case of copper, more than 550 tonnes of materials must be displaced to obtain 1 ton of pure copper; nearly 200 tonnes of this material is intimately associated with the copper-bearing minerals, and has to be removed by a complex series of process such as flotation, roasting, smelting and electrolysis.[1] It is important to realize that all of these unwanted materials become wastes and must be disposed of. Most of them have no economic value, actual or potential. Even the "cleanest" production technology cannot eliminate these waste streams.

Subsequent process steps, such as shaping, forming, machining, joining, finishing and assembly are more amenable to "waste minimization". We discuss this topic later.

From the organizational point of view, industry consists of firms. There are small and very large companies. There are state-controlled and private industries. Industrial companies range from very low-tech to very high-tech oriented, from companies which are not much more than trading outfits to research intensive and technically oriented outfits. On top of all these differences, one has to make a distinction between the various industrial sectors. They vary from food, energy (coal, gas, nuclear, oil, etc.), agricultural, shipping, railways, cars, textile, and chemical companies to computer, medical, pharmaceutical, and micro-electronic multinationals. Last but not least, one has to keep in mind that the majority of industrial activities is not carried out by large industrial firms but by small and medium-sized companies. This is important, because small and medium-sized firms may lack the financial and/or technical resources to adopt the newest most advanced production technologies. Alternatively, they may lack the advantages of large-scale production which permit the more efficient recovery and re-use of certain waste materials (as in the paper industry).

Then we come to the next question: what is the objective of an industrial company ?

Typically the answer might be something along the lines of "providing goods and/or services to their customers in an effective and efficient way while having regard for the interests of their shareholders and employees". Profit is considered to be a means of measuring the efficiency of a company, while competition in a free market forces companies to chase their objectives vigorously. By choosing the most efficient and effective methods, companies will try to do their job better than their competitors to ensure their market share and the highest return allowed under the various societal legal conditions and constraints. Companies will try to produce products and services which are wanted by the customers and they will use those raw materials, other resources and manufacturing processes which are most efficient.

It is up to Society (the customers) to set the conditions by which products and services will have to be produced, to lay down rules with respect to the resources which can be used and to set other conditions and constraints but it is important to note that the variety of products and services and the total quantity will ultimately be determined by the customers. As long as the number of these customers is growing and needs for products and services are increasing, industrial output (and therefore waste) will tend to increase too. Therefore, population growth and required living standards will be the overriding factors in the analysis of industrial activities and their total impact on the environment.

Industry reactions to changes in waste rules and regulations, and to the introduction of new environmental legislation can vary from negative to highly positive depending on the size and type of company and the sector it belongs to. Consider the case of one common type of regulation: the banning of a product for reasons of toxicity or hazard. A producer of one single product or service that is banned will react quite differently from the company that produces the product or service that is an acceptable alternative to the banned product or service. When sunk investment in the manufacture of an existing product is high, resistance to banning that product will be strong and the logic and factual basis for banning the product will be challenged thoroughly.

On the other hand, a competitor with less investment in the production of the banned product will see the ban as a new opportunity to penetrate the market with a possible alternative product. In short, when introducing new rules, there will always be misery and pain due to capital destruction, job losses and uncertainty about the necessity of the measures introduced. Therefore, general statements for industry as a whole are very difficult to make. Small operators who will find it difficult to go for a new alternative when their product is banned, could face bankruptcy; bigger ones, who could have a heavy impact on their resources, will not be easily convinced of the need to enforce that particular rule and will vigorously scrutinize it.

Of course there is a wide variety of other types of regulations, ranging from "Best Available Technology" (BAT) standards to emission standards, emission taxes, exchangeable emission permits, and so on.

It requires a great deal of expertise and professionalism to balance the pros and cons of such new rules or legislation and to judge whether they are based on solid technical information. In particular, the human element and drive in analysing proposed new constraints will have to be purely objective and be scrutinized all the time. Often the objective of such a rule is so overwhelming that the negative side, such as job and capital destruction, but also possible costs of Research and Development for alternative products and services, is not highlighted enough. There is definitely a need for more prenormative research to investigate the consequences and the pros and cons of proposed new rules.

As had also always been the case with prenormative research on standards, the vast majority of this research will have to be carried out in industry. In most cases, resources needed will also be provided by industry.

3. GRADUAL CHANGE SCENARIO

This scenario is based on the assumption that the risk of environmental disaster is perceived to be negligible, at least in the short run. In this case, one might suppose that new proposals for improving the environment are not too

drastic, but might be viewed as business opportunities. Consequently, one can expect that industry in general will take up the challenge and grasp the new opportunities. On the 10–12 April 1991, the International Chamber of Commerce, in co-operation with the United National Environment Programme (UNEP) and the United Nations Conference on Environment and Development (UNCED) organized the Second World Conference on Environmental Management in Rotterdam and in the Proceeedings of this Conference one can find support for the statement made.

Changing circumstances and introduction of new rules always produce new opportunities for innovative companies. Often their R and D departments will in their hearts welcome these new challenges and hasten to search for alternative solutions. They are used to working year after year to adapt company products and processes to changing circumstances. Market requirements change all the time, feedstock prices are certainly not constant over long periods and also the continuous change of capital and operating costs will ask for continuous review and adaptation of existing products and processes to new circumstances or the introduction of new ones. Introduction of increasingly restricted environmental constraints will not change the name of the game. From the point of view of an R and D manager, one could even say that the required additional effort is more of the same. Most bigger companies are therefore used to adapting existing products and process to changing requirements or to introducing alternative processes and products when existing ones are no longer competitive. In many cases the effort needed does not always require the introduction of new technology. More often, it requires clever application of existing technology in such a way that the economic constraints forced upon a company by such new rules will be overcome. One could almost say: "tell industry what one wants and given the right economic conditions it will technically almost always be able to provide it".

This does not mean that industry does not need time to adapt. Apart from the economic adaptation, the time needed for further development and introduction of new products and processes is often underestimated. It would, of course, be impossible from an economic as well as from a technical point of view to replace, as an example, all petrol driven cars in the near future by electric ones. But to create the circumstances whereby industry can adapt over time will definitely lead to results. The way the State of California forces the car industry to proceed with the development of the electric car is a good example of how, on the one hand, to stimulate the car industry to consider the introduction of the electric car, but on the other hand also to give that industry time to adapt. Still it would take years to introduce electric cars on a large scale and spend a vast amount of money to set up a new infrastructure. Many examples of the introduction of new concepts in the energy field point to time scales in the order of 10 to 20 years to arrive from an original idea to a prototype. Examples are the introduction of new clean coal gasification plants, the building of a new nuclear reactor or the setting up of a plant to convert natural gas in fluid hydrocarbons. These are projects which often require cash outlays in the order of more than one billion dollars for the development of an original idea in research to, and including, the first demonstration plant.

Increase of complex rules and constraints will often be welcomed by those companies who can afford the required investment in Research and Development, followed by the necessary capital investment. These companies will use their knowhow and technology assets to deal with changing rules and constraints to keep or increase their competitive edge. After all, complex rules and regulations will make it difficult for newcomers into that industry to compete. But, on the other hand, customers will have to provide the incentive for companies to invest in the necessary Research and Development and subsequent capital investment. Companies will not invest if they cannot see a possible benefit for their efforts. Obviously, governments can assist to provide the necessary incentives.

Finally, it is very important that, in this game of competition, the introduction of new rules and constraints is non-discriminative for all competitors in a particular sector of industry. If not, incentives disappear quickly. This is a great challenge for governments. Many markets in the various industrial sectors are growing towards a global size. Having different rules in different regions will not be very encouraging for a company to take up such a challenge on a global basis.

Given the vast R and D resources of industry, it is vital that appropriate incentives and conditions be created so that this reservoir of knowhow will be mobilized to provide new environment-friendly processes, products and services. But, in order to be able to cope with new environmental conditions and rules, technical and R and D departments of industrial companies will need a continuous refreshment of their skill pools. They need new skills and professionals who know the new environmental requirements and their significance intimately. It will be a great challenge for universities and other schools to motivate young people to go and work in these important fields. New courses will have to be set up to provide the new knowledge. As examples, how does one approach issues in waste management, new developments in optimizing low-waste manufacturing processes, how can one use biotechnology to avoid poisonous chemicals in agriculture, what role will recycling play in existing manufacturing processes etc.?

Going back to Figure 1, one can say that research on energy-efficient processes and products has already been getting attention by industry for a number of years.

Challenged by the expected future high energy prices, industrial research laboratories took up energy-saving projects very early. This is also the case for replacement of relevant raw material used.

The potential for improving industrial processing and manufacturing so as to fulfil stricter environmental requirements, are vast. When challenged by the right incentives, the industrial innovation engine, helped by the enormous industrial R and D resources, will in time produce new processes and manufacturing concepts and reduce waste created industry itself. This is even more so in the field of designing alternative environment-friendly products to fulfil services required by customers, whereby the opinion and choice of the customer will have a great influence on the speed of progress.

Industrial R and D outfits are only beginning to explore the possibilities for large-scale introduction of alternative environment-friendly products. Dealing with the waste of used products and the possible reuse of them is a more difficult issue, not only from a technical but also from a socio-political point of view. A lot of research will still have to be done in this field to come up with larger-scale solutions. One concrete step has already been taken: waste minimization. This is dealt with in the next Section.

4. WASTE MINIMIZATION

As mentioned earlier, the "waste minimization" strategy is not a total solution to the environment problem. It is applicable, primarily, to the intermediate stages of materials processing. Nevertheless, since the late 1970s, there has been an increasing trend toward deliberate minimization of the amount of waste generated. Essentially this approach is based on the premise that it is necessary to look at the causes of waste generation before considering its effects. In addition to its environmental attractiveness, there is a substantial economic incentive in that the reduction of waste generation implies lesser cost of waste treatment and disposal.

In waste minimization, a systems approach is employed. In particular, the following hierarchy is adopted:
(a) Source reduction;
(b) Recycling;
(c) Recovery;
(d) Treatment and incineration;
(e) Storage and disposal.

The basic idea is that, in a given industrial plant, every effort is made continuously to reduce the generation of waste, then to recycle it, recover it, treat and incinerate it, and finally store the waste and dispose of it (Freeman, 1990; Hunt and Schauhter, 1989).

Source reduction may be achieved in a variety of ways:

raw material substitution, product alteration, change of process operating conditions, and modification of operational procedures. In all of these methods, the industrial process under consideration is analyzed, simulations are made, etc. In many cases in addition to reduction of the waste generated, other benefits are incurred due to the fact that process analysis uncovers certain hidden inefficiencies in plant operation.

Recycling also involves process analysis, with a view to discovering possibilities of the substitution of fresh reagents and solvents by the waste generated. Recovery requires an investigation into possibilities of transforming the waste from one industrial process into useful raw materials for other industrial processes.

Having exhausted the potential to source reduction, recycling, and recovery, one proceeds to consider waste treatment and incineration. Finally, the storage and disposal of waste has to be envisaged. Although a fairly recent idea, the systems approach of waste minimization is already being employed in industry. There have been two incentives to the development and application of waste minimization procedures in industry. On the one hand, environmental governmental and societal agencies and organizations have been encouraging industry to conduct waste minimization audit procedures. On the other hand, a number of industrial firms have taken significant waste minimization initiatives which have been yielding substantial economic benefits. To cite but two examples of the first type of action, mention may be made of the Ontario Waste Management Corporation (Richmond, 1989) and the International Chamber of Commerce (ICC Guide, 1991).

Industry has not lagged behind. A pair of initiatives may be mentioned. One is by CEFIC (European Council of Chemical Industry Federations), which has put waste minimization as the principal component in its industrial waste management approach (CEFIC, 1989). The other initiative is that by Dow Chemicals, with its WRAP (Waste Reduction Always Pays) Programme. Dow Chemicals has achieved substantial economic benefits in a number of its plants (Nelson, 1990). There is an increasing volume of literature reporting case studies of waste minimization projects in various industrial sectors and in R and D Centers. These case studies demonstrate not only the environmental attractiveness of waste minimization projects, but that substantial economic benefits are usually involved. A wide spectrum of industries and wastes have been investigated with a view to obtaining waste minimization results. To cite but a few examples, mention can be made of the aerospace industry (Evanoff, 1990), chromium-contaminated wastewater at plating facilities (Meltzer, 1990), used oil recycling (Courtright, 1990), recovery and recycle of valuable constituents in spent pickling acids (Thornburg, 1990), the pharmaceutical

industry (Venkataramari, 1990), electric arc furnace steel plants (Drabkin and Rissmann, 1990), and R and D research laboratories (Boortz, 1990).

The study of waste minimization as a systematic project activity is little more than a decade old. It is then hardly surprising to find that a lot of research remains to be done. The research required lies on three fronts: scientific, economic, and legislative. However, an integrated point of view must be maintained: it is not of much use obtaining scientific waste minimization techniques which are economically unfeasible. What are sought are research results which are scientifically proven with adequate considerations of economic and global industrial competitiveness. Having recognized the interdependence between the scientific, economic, and legislative aspects of waste minimization, one can then proceed to raise some of the research necessities and future perspectives pertaining to each of these facets of the question of waste minimization.

Taking the scientific aspect first, the most urgent necessity lies in source reduction. This is the area which needs most research and which promises most benefits. Research in this area goes to the root of the question of waste generation: waste should be minimized if not eliminated.

A waste audit and reduction study usually involves seven phases:
(a) Understanding the process in the plant;
(b) Defining process inputs;
(c) Defining process outputs;
(d) Material balance calculations;
(e) Identification of waste reduction alternatives;
(f) Economic evaluation;
(g) Implementation plan.

The first three phases involve work in the plant itself and should provide no difficult research problems. The seventh phase is one of execution which depends on the conditions and is plant-specific. It is the fourth, fifth, and sixth phases, which require the bulk of research needed.

In so far as material balance calculations are concerned, the necessary research aspects are:
(a) Precise process simulation, almost certainly with the aid of computers. Computer-aided process simulation has taken large strides since the late 1950s. However, more sophisticated calculations are required when dealing with hazardous wastes, particularly at low concentrations. There is room for a substantial amount of research in this direction. An example of such research with case study is given in Vasconcellos and Qassim (1991).
(b) Accurate data for the process simulation is essential

for reliable material balance calculations. This means that good measurements must be made in the plant. The quality of the data obtained must be as high as possible. This would entail research into adequate techniques of data reconciliation and parameter estimation.

Turning to the identification of waste reduction alternatives, research is definitely necessary in the development of low-waste and clean technologies. Two recent seminars have been dedicated to this question (Stockholm Environment Institute, April and Sept. 1991). A start has been made as shown in the literature. However, a more systematic approach is required. Waste reduction opportunities must become an essential criterion in the selection and development of process technology. Furthermore, the search for waste reduction possibilities must be pursued in process design (Berglund and Snyder, 1990).

In addition to the search for clean or rather cleaner production technologies, waste minimization may be enhanced by a strategy of long-life goods and product-life extension. This is an extremely important area of research with a large number of unanswered questions. A significant contribution to this area may be found in (Stahel, 1991).

Last but by no means least, there is the question of the economic evaluation of waste minimization projects. Little work has been done here (Karam *et al*, 1988). This is not surprising in view of the large uncertainties involved. In fact, the true cost of hazardous wastes is still almost unknown. Data management systems are an essential requirement. The most suitable technique to be used in economic evaluation is an open question and needs an urgent answer. Should the technique be purely deterministic ? Is it necessary to include risk considerations? Should new techniques of project evaluation be developed for waste minimization? These and other questions will have to be answered.

To conclude, waste minimization is the cornerstone of current waste management systems. Future lines of research must be performed along three fronts in unison: scientific, economic, and legislative. The basic premise is to attack the causes of waste generation, rather than to reply on "end of pipe" treatment of wastes.

5.   RISK-AVERSE INDUSTRIAL ECO-STRATEGY

The uncertainties in our current scientific understanding of the Earth System are very great. This is not necessarily a good reason to delay action pending the outcome of research. On the contrary, one can make a cogent argument that more radical changes in the industrial system must be considered precisely because we cannot be sure that the Earth System (including the biosphere) is "robust" enough

to withstand ever-increasing environmental pressures without danger of catastrophic collapse. The argument is summarized in the following paragraph.

The basic conditions of life-support on the Earth include a stable climate (temperature, rainfall, etc), a secure and stable food supply (the "food chain"), biological waste assimilation capacity and efficient recycling of those nutrient elements that are inherently scarce (i.e. concentrated in living organisms), reactive or difficult to mobilize. The list includes carbon, nitrogen, sulfur and phosphorus. It is noteworthy that climate, the global cycle of carbon, oxygen, nitrogen and sulfur, the ozone layer in the stratosphere, the salinity and pH of the soils and oceans, mature forests, soil fertility and bio-diversity are not technologically replaceable (or reparable) to any meaningful degree, at least for the planet as a whole. Thus, we have no real choice but to preserve them from harm.

Long-term ecological sustainability is clearly incompatible with major anthropogenic changes in the radiation balance of the atmosphere, human interference with the carbon cycle, anthropogenically induced desertification or deforestation of the tropics, accumulation of toxic heavy metals and non-biodegradable halogenated organics in soils and sediments, or sharp reductions in biodiversity, for instance.

It also follows that human activities leading to such degradation are not sustainable. It is assumed hereafter that ecological sustainability in the long run also implies that the *carrying capacity, renewal capacity* or *assimilative capacity* of environmental systems should not be exceeded consistently, either regionally or globally. Carrying, renewal or assimilative capacity *per se* are not easy to determine precisely, in many cases, except (as with fisheries) by empirical observation after the fact. However, all of these concepts are related to environmental resilience or the ability to absorb stress and (if necessary) recover from damage. Thus, an approximately equivalent condition is that renewable resources should not be harvested faster than they are replaced, and irreversible damage should not be done to environmental support systems, regardless of apparent short-term economic benefit. Examples of environmentally non-sustainable human activities in the long run include the extraction and use of fossil fuels, dissipative use of halogenated organic chemicals and plastics, irrigation with "fossil" ground-water, and dissipative uses of heavy metals.

It can be argued that several specific environmental conditions must be met to assure sustainability (Ayres, 1991). These include:

(a) No further (anthropogenic) change in the determinants of climate beyond some unknown limit. This means no further accumulation of greenhouse gases in the atmosphere and consequently a sharp reduction in fossil fuel use and general use of synthetic nitrogenous fertilizers and chloro-fluorocarbons (CFCs).

(b) No further net increase in the acidity of the environment, especially the freshwater lakes and rivers and forest soils. This means imposing strict limits on levels of emissions of sulfur oxides (SO_x) and nitrogen oxides (NO_x), which tend to oxidize to the corresponding acids in the atmosphere. This is a regional problem, especially in regions where soils have little or no buffering capacity.

(c) No further net accumulation of toxic heavy metals, radioactive isotopes or long-lived halogenated chemicals in soils or sediments. This implies an end to virtually all dissipative uses of these – indeed, most – metals and many other extractive resources i.e., *closing the materials cycle*. It also implies a restriction on the use of synthetic chemical pesticides in agriculture.

(d) No further net loss of topsoil.

(e) No further net withdrawal of fossil ground-water.

(f) No further net loss of ocean fisheries, estuaries, coral reefs, wetlands, old-growth forest or biological diversity.

In addition to the above environmental tests of ecological sustainable industrial development, at least two other operational requirements for sustainability must be added. These are economic *efficiency* and *equity*.

Efficiency can be considered both from an economic and an industrial perspective. We discuss the latter perspective below in the context of "clean production". Regarding the question of economic efficiency, the major point that has to be emphasized, as often it is forgotten, is the following: both theory and experience strongly suggest that "command and control" type approaches are invariably inefficient. The economic system itself is extremely complex, and the interface between economics and environmental concerns does not introduce any simplifications. Economic efficiency is only achievable, in practice, by the "invisible hand" of competitive markets. This means that, to achieve economic efficiency along with sustainability in the long run requires that market mechanisms must be allowed to operate, or even created, whenever possible.

Environmental problems are, in a fundamental sense, due to economic inefficiency, as reflected in *market failures*. It is a fact of life that certain critical environmental resources are so-called "public goods", due to inherent attributes such as indivisibility or the impossibility of capture or enforcement of ownership rights. Such goods (or bads) – examples include fresh air, air pollution, scenery, sunshine, rainfall, or wild animals – cannot be individually owned or exchanged in markets.

They may have positive or negative value, but they cannot be priced, at least by conventional market mechanisms. This type of market failure results in price distortions and, consequently, wrong price signals. To be concrete, market prices of certain commodities, namely those whose extraction and use is environmentally damaging (e.g. coal), are currently much too low. This is because the environmental damages are not borne by either the sellers or users of such commodities. To achieve efficient resource allocation in the long run, therefore, the missing link between production costs of environmentally damaging products and damage costs must be created. This link has been termed the *polluter pays principle* (PPP). It is this principle that lies behind current proposals to impose emissions charges or pollutions taxes. However, up to now, it is more a slogan than a reality.

The question of equity arises in most discussions of economic development. Here again, there are two important ethical perspectives. One is the contemporary ("horizontal") perspective, which emphasizes reducing current income differentials between North and South, urban and rural and so on. In particular, it is argued that those who have benefited least from the over-exploitation of extractive resources and environmental assimilative capacity, up to now, should not be expected to pay a price, in terms of reduced growth prospects, for the ecological protection and economic restructuring that is now required. It is also commonly argued that any excess costs of meeting stricter environmental standards should be met by those whose present wealth is (partly) due to having benefitted from unlimited use of natural resources, including environmental waste assimilative capacity. If past abuses must be corrected and limits must now be imposed, it would seem that the wealthy countries should pay the costs.

From a practical viewpoint, it is also clear that people who are very poor, or starving, cannot and will not concern themselves with long-term environmental considerations. The first priority is survival. Meanwhile, however, cumulative environmental damage – especially from deforestation and soil erosion – is now becoming a serious impediment to development itself. Moreover, damage done by ill-considered short-term development projects – especially timber cutting and clearing of tropical rain forest to encourage cattle raising – often reduces the long-term productivity of areas that were formerly self-sufficient. Given that extreme poverty can create the conditions for major natural disasters (as in Bangladesh) which, in turn, generate waves of poverty-stricken and desperate refugees, even the wealthiest countries cannot insulate themselves totally from the ripple effects. Development assistance, on an increased scale (but allocated to more environmentally friendly projects) can easily be justified by these considerations.

The second ("vertical") perspective on equity is intergenerational. It recognizes that we are now paying a price for past sins such as the desertification of North Africa and much of the Middle East, due to overgrazing and deforestation. Future generations may have to pay a much higher price for our own failings. As a matter of equity, it is argued, we who are alive today should not leave the unpaid bill to them. Yet, in the way we evaluate future costs and benefits of present actions, we do not (as a rule) give future generations any significant "vote" in the use of resources that are exhaustible (as in the case of high quality mineral resources, "fossil" ground water or mature forest trees) or vulnerable to irreversible change (as in the case of desertification, soil erosion, and the build up of greenhouse gases or toxics).

Since the environmental movement first became a significant part of the public policy agenda, following "Earth Day" in 1970, the creation of the Environmental Protection Agency (EPA) in the United States, and the UN Stockholm Conference of 1972, the primary concern has been with controlling pollution of the air and water. Regulation by setting emission limits, or standards, has been the major tool of public policy, and industry has been the major target of regulation. It is therefore appropriate to consider how far the regulatory strategy can be used to solve the problem.

Several increasingly restrictive and increasingly comprehensive types of environmental standards may be envisioned. They correspond roughly to increasing effectiveness, as well as increasing transitional adjustment cost, increasing length of time likely to be needed for implementation, and increasing level of controversy with respect to enforcement and allocation of costs. Beginning with the weakest and lowest in immediate cost, they can be characterized as a hierarchy, as follows:

(a) To set resource input and emission standards at the establishment level, based on the present average performance for the industry (presumably achievable with ordinary "good housekeeping" practice), supplemented by locational restrictions to preserve particularly sensitive areas, such as densely populated areas, cultural treasures, virgin forests and wetlands.

(b) To set technology standards at the establishment level, based on "best industry practice" or "best available technology" (BAT), with similar locational restrictions to protect the most sensitive areas.

(c) To set emissions or efficiency targets at the product or sectoral level (e.g. housing, autos), either fixed or variable according to some schedule. Possible measures of compliance might include (i) continually increasing energy use efficiency in each product category or sector, (ii) continually increasing materials productivity in each sector and (iii) continually reducing dissipative uses of inherently

toxic non-biodegradable materials, by sector.

(d) To set general ambient emissions standards at the local or regional level, aimed at preserving the quality of specific airsheds or watersheds. Standards can be fixed or varying according to a specified schedule.

(e) To set emissions standards at the continental or global level, based on the criteria stated at the beginning of this section for assuring long-term environmental sustainability. At the regional level, such standards would also have to assure that emissions from both production and consumption do not exceed regional renewable resource regeneration rates and waste assimilative capacity, as appropriate. (The latter would be applicable, in principle, to biological wastes, but not necessarily to other toxic wastes.) For example, heavy metal emissions (including consumption wastes) would be limited, based on not exceeding "natural background" exposures, similar to the approach that has generally been adopted for radioactive wastes. Possible measures of compliance include:

(i)   declining absolute use of non-renewable energy in the economy as a whole;

(ii) declining use of non-renewable sources of energy by the economy as a whole;

(iii) increasing share of secondary and recycled materials in the economy as a whole;

(iv) declining absolute use of materials per $ of output. Of course, renewable resource regeneration capacity and waste assimilative capacity are regarded as fixed (local or regional) natural capital assets. The same philosophy could be applied to the use of non-renewable natural resources, viz "rights" to extraction. Rights to emit, or to extract, not having a market at present, must somehow be allocated among all potential claimants. The allocation might be done on the basis of "tradeable emissions (or extraction) permits" or some variant.

Given the local nature of product/sectoral Levels a-c norms, it is easy to see that even with very tight restrictions on emissions or performance, such regulations do not even assure a decrease in emissions at the regional, national or global levels. Nor can uncoordinated regional standards (Level d) assure that global needs will be met. It is necessary to face an unpalatable truth: *even the cleanest conceivable production technologies, applied at the establishment, product or sectoral levels, cannot guarantee long-run sustainability. The latter goal requires fundamental economic restructuring together with corresponding changes in consumption patterns.* The necessary restructuring includes two elements that can be stated simply: (1) phasing out fossil fuels, especially coal, and (2) closing the materials cycle.

The logic is straightforward. Any use of fossil fuel – for electric power production or transportation – results in the production of combustion products, including carbon dioxide and oxides of nitrogen. Considering "greenhouse gases", for instance, fossil fuel combustion accounts for most (95%) anthropogenic emissions of carbon dioxide (the biggest single contributor), and some of the nitrous oxide. Agriculture (including animal husbandry) probably accounts for most of the anthropogenic methane and nitrous oxide. Leaks from refrigeration and air-conditioning units, and foamed plastics are the source of much of the chlorofluorocarbon (CFC) emissions. (These, in turn, are also mainly responsible for the depletion of ozone in the stratosphere.) Similarly, we have already noted that acid rain is mostly due to fuel combustion, either of sulfur-containing fuels or of any fuels burned at very high temperatures. The industrial contribution to SO_x and NO_x is minor.

Technology has been enormously effective in creating new resources to replace scarce ones. Coal, itself, was a substitute for increasingly scarce and expensive charcoal, as the forests of Britain were largely cut down by the end of the 17th Century. Kerosine, derived from petroleum, was introduced as a substitute for whale oil, for lighting purposes. Whales were becoming extremely scarce by the middle of the 19th Century, as demand for "illuminating oil" skyrocketed. Other examples can be cited, *ad infinitum*.

However, technology cannot be relied on to solve all the emerging problems. In the first place, technology is not autonomous: technological innovation responds to market forces and profit opportunities. Environmental problems, being in the public sector, do not necessary create such opportunities. Waste residuals and environmental assimilative capacity are not a part of the market system (they are unpriced) and market incentives have not been applicable in the past to stimulate technological substitutes for environmental services, or "technological fixes" for environmental damages. In the second place, even if the relevant market mechanisms were operating and the "shadow prices" of all environmental assets were known in principle, capital constraints may inhibit the implementation of projects that are inherently very large in scale. And, thirdly, some environmental problems may be simply beyond the reach of any conceivable technology that could be available in the next half century. There is no technology to re-create an extinct species, for instance.

To be sure, it may be possible with modern technology to protect some of the world's vulnerable coastline by building dikes, albeit at great expense. Similarly, various geo-engineering schemes have been proposed; e.g. to counter greenhouse warming by injecting condensation nuclei into the stratosphere, fertilizing the oceans or capturing carbon dioxide and pumping it into deep wells. Scientists and engineers have proposed massive projects; e.g. to use nuclear power – perhaps from "second generation" nuclear fission or even fusion reactors – to convert coal into hydrogen, while disposing of the resulting carbon dioxide in old gas wells. Another interesting proposal is to cover the sunward face of the moon with photovoltaic "farms", transmitting electric power to Earth by way of coherent beams of microwaves. But there is no assurance so far that such hypothetical large-scale technologies could be implemented safely, or that they would be environmentally benign and cost-effective. The history of large projects is not especially encouraging in this regard.

Nevertheless, technological progress along appropriate lines does appear to offer real hope. For instance, economic development strategies dependent on use of renewable resources, solar energy, biotechnology, and (above all) information technology and human services are generally consistent with the long-term eco-sustainability constraint. A critical element of any sustainable strategy must be enhanced re-use and recycling of manufactured products. This issue can be approached from two main directions. One – with an obvious analogy in nature – is to focus on exploiting the energy available from the sun and "closing the materials cycle". The other is to focus on "closing the product cycle".

Other technologies are more attractive in the near term. Biomass gasification together with high efficiency aircraft-type gas turbines appear to be an extremely promising technology for utilizing crop wastes such as biogas. New "high tech" wind turbines are declining rapidly in cost. Similarly, photovoltaic systems should be commercially competitive with diesel-powered electric power plants well before the end of the present decade, and should be competitive with coal or nuclear power by 2010 or 2020 at the latest.

Methane from controlled anaerobic fermentation of biomass (e.g. "biogas") is an ideal domestic fuel for tropical countries, wherever a reliable supply of waste cellulose can be assured. Methane is a much more potent greenhouse gas than carbon dioxide, and natural decay process produce large amounts of it in swamps, rice paddy fields, and the guts of cellulose-digesting animals such as termites, cattle and sheep. Composting (and land-fills) also generate methane. It is desirable from an environmental perspective to capture and burn as much of this methane as possible, not only to make use of its energy value, but to prevent it from contributing to climate warming. Since the technology is essentially quite simple, many rural villages, and even some cities, may be able to supply domestic needs (e.g. for cooking) from this source.

Near-term opportunities for converting organic wastes to biogas on an industrial scale exist already in some countries, particularly in juxtaposition with sugar-cane, cotton, coconut, coffee, cocoa, pulp and paper, rice and other grain harvesting, animal feeding and food processing. Tropical Africa, Southeast Asia, Latin America and Central America could all benefit enormously from accelerated introduction of large-scale biogas plants. These would, among other benefits, sharply reduce the need for more conventional "end-of-pipe" treatment for organic wastes. Even in the industrialized countries, there may be a strong argument for converting some existing aerobic sewage-treatment facilities into anaerobic biogas plants.

Where biogas is not a feasible source of energy due to scarcity of cellulosic wastes, natural gas must be the fuel of preference in the near term for all domestic heating requirements and stationary boilers (and even some mobile uses) except where direct coal combustion can be adequately controlled, as noted above. Natural gas is widely available to consumers in the industrialized world. There is a large current surplus supply in the Middle East and some other locations, where much of it is flared for lack of markets. Where coal is locally available and natural gas is not, syngas plants and/or cooking plants would be a third-best option. Unfortunately, gas distribution systems do not exist as yet in most parts of the world. The building of urban gas distribution systems (and biogas or syngas plants where appropriate) should be given high priority by international development assistance agencies.

In the longer run (toward the middle of the next century) it is likely that electricity can be generated economically on a large scale by photovoltaic (PV) cells, either in large PV "farms" located on wastelands or on Earth-orbiting satellites (or even on the moon itself). Of course, other less exotic possibilities will also be explored, including safer nuclear reactors, wind turbines, geothermal turbines, and so on. But the most promising "source" of new energy, over the intermediate term is undoubtedly increased energy-use efficiency.

There are very large opportunities for increased efficiency in the transportation sector and the housing sector, especially. Public transportation is much more energy efficient than private cars, for instance, and even private cars can be made several times as efficient as they are today. The average private car in the US today obtains only 19 miles per gallon (5.4 km per litre) of fuel. Yet there are now cars on the market that get 55 miles per gallon (15.6 km per litre), and prototype vehicles have been demonstrated that can do more than twice that well. In the field of household heating, cooling, water heating and so on, the opportunities are actually even more dramatic. Today, about a third of all energy is used for this purpose in the US, and most of it is certainly wasted. Proper design and orientation with respect to the sun, insulation, and intelligent use of waste heat can reduce overall energy use in buildings to a small fraction of the

present level. This has been repeatedly demonstrated. The major issue is how much extra efficiency would cost and what kinds of regulation would be most effective.

With regard to energy conservation in general, it is increasingly clear that major opportunities exist to save energy while simultaneously saving money. The reason these opportunities are not fully exploited is mainly due to lack of information, subsidies and institutional barriers favoring established energy industries, and lack of economic incentive to innovate on the part of middle managers in bureaucratic hierarchies. Apart from this, however, it seems clear that the present price of energy is much too low to provide effective economic incentives for people to design and use more efficient cars and buildings, or to use energy efficiently in manufacturing processes.

Closing the materials cycle is particularly important for the toxic heavy non-ferrous metals;, such as arsenic, bismuth, cadmium, chromium, copper, lead, mercury, plutonium, silver, thallium, uranium and zinc. Of these, the artificial element plutonium is by far the most dangerous. It is carcinogenic and toxic to an extreme degree, as well as being fissionable and hence usable for nuclear explosives. Many people now believe that the various risks associated with use of plutonium (or even uranium, from which plutonium is manufactured) are too great to justify its use in nuclear power production. In any case, the closing of the global uranium–plutonium cycle is extremely urgent. However, this subject cannot be realistically discussed without also discussing the entire question of nuclear armaments and their disposition.

Following uranium–plutonium on the list would be lead, mercury and those toxic metals (arsenic, bismuth, cadmium, thallium) that are by-products of copper, lead or zinc mining and refining, and which are widely used in chemical products mainly because there is a steady low-cost supply and no constraint on such use. Finally, copper, chromium, manganese, nickel, silver, tin and zinc constitute lesser but significant problems worthy of consideration.

6. CONCLUSION

Main facts about industry and waste:

(a) The link between economic growth and materials/energy use is not absolute (it has been loosened in recent years) but it remains quite close. Large quantities of raw materials are extracted and processed to make products. Large amounts of the resulting wastes are guaranteed (by the materials-balance principle) as long as the economy depends on non-renewable energy sources and fails to recycle materials.

(b) Optimization with respect to waste must be carried out on the whole system (global) level. Micro-optimization at the level of individual plants and firms will not suffice, because major material substitutions, lifestyle changes and new sources of energy (for instance) are not considered. There are many opportunities for improvement. The industrial system is far from optimal now.

(c) The problem of Third World industrial development cannot be neglected. Yet the current development trajectory is excessively (and unsustainably) energy/ materials intensive, and driven by rapid population growth.

The immediate needs are:

(a) Research on raw materials substitution and process alternatives to facilitate resource recovery and recycling.

(b) Objective assessment of "no regrets" and "low regrets" opportunities for redeployment of investments.

(c) Continued research by industry on new technologies such as biotechnology for medical, agricultural and industrial waste applications.

The necessary actions are:

(a) Institute policies – both regulatory and economic – that will motivate industry to adapt to environmentally sound technologies.

(b) Development of institutional mechanisms and frameworks for facilitating the sharing of new technological applications on a global basis particularly with the Third World.

(c) Human resources training for implementation of technologies which will lead to sustainable development.

END NOTE

[1] US Bureau of Mines 1990, Minerals Yearbook, US Government Printing Office, Washington, DC.

REFERENCES

Ayres, R.U. 1991. Eco-instructuring: Managing the transition to an ecologically sustainable economy. *Humankind in Global Change Symp.*, AAAS Annual Meeting, Washington DC.

Ayres, R.U. 1989. Energy inefficiency in US Economy. Res. Rep. WP-RR-12, IIASA, Laxemburg, Austria.

Berglund, R.L. and Snyder, G.E. 1990. Minimize waste during design. *Hydro. Proc.*, 39-42.

Boortz, M.J. 1990. Hazardous waste minimization at an R and D Laboratory. *Environm. Prog.*, **9**, 30.

CEFIC. 1989. *Industrial Waste Management*. CEFIC. Brussels .

Courtright, M.L. 1990. Used oil: don't dump it, recycling it. *Mach. Des.*, 81-86.

Drabkin, M. and Rissmann, E. 1990. Waste minimization opportunities at an electric arc furnace steel plant producing speciality steels. *Environm. Prog.*, **8**, 88.

Evanoff, S.P. 1990. Hazardous waste reduction in the aerospace industry, *Chem. Eng. Prog.*, 51-61.

Freeman, H.M. 1990. *Hazardous Waste Minimization*, McGraw-Hill, New York.

Hunt, G.E. and Schauhter, R.N. 1989. Waste minimization and recycling. In: Freeman, H.M. (ed.), *Standard Handbook of Hazardous Waste Treatment and Disposal*, McGraw-Hill, New York.

Karam, J.G., St. Cin, C., and Tilly, J. 1988. Economic evaluation of waste minimization options. *Environm. Prog.*, **7**, 192.

ICC. 1991. Guide to Effective Environmental Auditing. International Chamber of Commerce, Paris.

Meltzer, M. 1990. Trivalent chromium electroplating systems offer cost savings as well as environmental protection. *Environm. Prog.* **9**, M10.

Nelson, K.E., 1990. Use these ideas to cut waste. *Hydro.Proc.*, 93-98.

Richmond, J. (ed.). 1989. *Industrial Waste Audit and Reduction Manual*. Ontario Waste Management Corporation, Toronto.

Stahel, W.R. 1991. Product innovation in the service economy. *Stockholm Environment Institute Seminar on Clean Production: Challenges and Opportunities*, Prague.

Stockholm Environment Institute. 1991 (April). *The Principles of Clean Production*. Stockholm.

Stockholm Environment Institute. 1991 (September). *Clean Production: Challenges and Opportunities*. Prague.

Thornburg, G. 1990. Recovery and recycling of valuable constituents in spent pickling acids. *Environm. Prog.* **9**, N 10.

Vasconcellos, R.V. and Qassim, R.Y. 1991. The use of computer-aided simulation in waste minimization, *Seventh Int. Conf. on Solid Waste Management and Secondary Materials*, Philadelphia.

Venkataramari, E.S. Backer, S., and Olsen, W., 1990. waste minimization in a leading ethical pharmaceutical company. *Environm. Prog.* **9**, A 10.

Chapter 4: Energy

J. P. Holdren and R. K. Pachauri

EXECUTIVE SUMMARY

Civilization's use of energy in 1990 amounted to the equivalent of 13 billion tonnes of coal, four times more than in 1950 and 20 times more than in 1850. The 1990 energy supply came 77% from fossil fuels, 18% from "renewable" energy forms (primarily hydropower, wood, crop wastes, dung and wind), and 5% from nuclear energy. More than two thirds of the supply went to the 1200 million people living in industrialized countries, less than one third to the 4100 million living in developing countries.

This pattern of energy supply and use is unsatisfactory not because of imminent depletion of fossil fuels (which with heavier reliance on coal could last for centuries) but rather for environmental, political, and economic reasons. Environmentally, we face oil spills, acute urban and rural air pollution from fuel combustion, degradation of forests and soils by biomass energy use, nuclear-accident risks, and the threat of intolerable climate change by fossil-fuel-produced carbon dioxide and other greenhouse gases. Politically, there is the potential for conflict over oil supplies, the contribution of nuclear energy to the spread of nuclear-weapons capabilities, and resentments and tensions over the inequitable distribution of energy's benefits and risks. Economically, the poor who need more energy the most cannot afford it. These problems can only be exacerbated by the practically unavoidable addition of 4000 million more people to the world's population over the next four decades.

Expansion of today's energy-supply system along business-as-usual lines is not the solution; that would only make the biggest problems bigger. Required instead is a worldwide effort consisting of five main elements: rapid increase in the efficiency of energy end-use; reductions in the environmental impacts of current energy technologies; development of advanced (low impact) energy sources; increased East-West and North-South technological cooperation and financial assistance for environmentally sustainable energy supply; and a global commitment to stabilize world population below 10,000 million.

1. THE NATURE OF THE ENERGY PROBLEM

Energy is both an indispensable ingredient of material prosperity and a source of some of the most damaging impacts of human beings on their environment. Where and when energy is scarce or expensive, people suffer from lack of direct energy services (such as cooking, heating, refrigeration, lighting, and transport) and from unemployment, inflation, and reduced economic output. But where and when energy supply is expanded with insufficient attention to the environmental and sociopolitical costs of doing so, there arises the danger of excessive damages to the environmental and social fabric, undermining current well-being and the prospects for sustainable prosperity.

The world energy situation today combines these syndromes. Much of the world's population has insufficient access to affordable energy services to meet basic human needs or to fuel the economic and social development on which hopes for higher living standards depend. Yet the environmental costs of the energy supplies available to this poorer population – forest degradation from over-harvesting of fuelwood, reduced soil fertility because crop wastes and dung are being consumed as fuel, and acute local air pollution from burning dirty fuels with no pollution control – are already excessive, and the high monetary costs of less disruptive energy sources are a major barrier to environmentally sound development. In the richest industrial nations, on the other hand, the monetary costs of energy supply are tolerable for all but the poorest citizens, but the environmental impacts of the huge quantities of energy being supplied are becoming intolerable locally, regionally, and globally; the political risks of excessive dependence on imported oil are substantial; and cleaner, safer sources are resisted because they are costlier in monetary terms. These problems of the present will be made even more difficult to solve by the practically unavoidable addition of four more billions to the world's population over the next four decades.

Table 1: World energy supply in 1970 and 1990

| | 1970 | | 1990 | |
	terawatts	share %	terawatts	share %
"Industrial" Energy Forms	7.4	88.2	11.7	88.7
Petroleum	3.4	40.4	4.6	34.9
Coal	2.2	26.1	3.2	24.2
Natural gas	1.4	16.6	2.4	18.2
Hydropower	0.4	4.7	0.8	6.1
Nuclear fission	0.03	0.4	0.7	5.3
"Traditional" Energy Forms	1.0	11.8	1.5	11.3
Fuelwood and charcoal	0.6	7.1	0.9	6.8
Crop wastes and dung	0.4	4.7	0.6	4.5
TOTAL	8.4	100.0	13.2	100.0

Notes: One terawatt (TW) = 10^{12} watts = 31.5×10^{18} joules/year = 1 terawatt-year (TWy) per year 1 TWy is about 1 billion tonnes of coal or 5 billion barrels of oil. Hydropower contribution calculated as quantity of fossil fuel energy required to generate same amount of electricity. Industrial energy data from British Petroleum (1989, 1991), converted from lower to higher heating values. Data for traditional energy are authors' estimates based on various sources, e.g.: Hughart (1979), Hall *et al.* (1982), World Bank (1983), Goldemberg *et al.*, (1987). Excluded here are animate energy flows in the form of human metabolic energy use (5.3 billion persons x 2400 kcal/person-day = 0.6 TW) and that of domestic animals (circa 1 TW in 1990). [Percentages may not add up due to rounding.]

2. CURRENT ENERGY PATTERNS AND THEIR IMPLICATIONS

The primary sources of the inanimate energy used by humankind can be divided into two categories: the "industrial" energy forms, consisting today mainly of oil, coal, natural gas, hydropower, and nuclear energy; and the "traditional" energy forms, consisting of fuelwood and charcoal, crop wastes, and dung. Table 1 shows the global rates of use of these energy forms in 1990 and, for comparison, in 1970. The 1990 total of some 13 terawatts is about four times larger than the corresponding figure for 1950 and 20 times larger than the world total in 1850 (Holdren, 1991). Particularly striking in Table I is the continuing preponderance of oil in particular (39% of all "industrial" energy and 35% of total energy in 1990) and of fossil fuels in general (87% of "industrial" energy and 77% of total energy in 1990). "Renewables" (biomass and hydropower) made up only 17% of the 1990 supply, and nuclear energy barely more than 5%.

The global energy system is characterized, further, by an elaborate set of operations and facilities in which primary energy forms are converted into secondary ones of greater usefulness and value, and by extensive energy-distribution networks by which energy commodities are delivered to their users. By far the most important secondary energy forms worldwide are refined petroleum products and electricity, the former accounting for nearly 40% and the latter about 35% of all uses of "industrial" energy forms (British Petroleum, 1991; EIA, 1991).

World production of refined petroleum products was about 60 million barrels per day in 1990, from a refinery capacity of about 75 million barrels per day (average utilization factor about 80%). The disposition of crude petroleum in 1990 was, approximately: 27% to gasoline; 34% to middle distillates, e.g. jet fuel, kerosine, diesel fuel; 20% to heavier fuel oils for, e.g. industrial boilers and electricity generation; and 19% to other uses, including lubricants, solvents, and refinery fuel and losses (British Petroleum, 1991).

World electricity generation in 1990 was 11.3 trillion kW-hours, of which some 63% was generated from fossil fuels (accounting for a quarter of all fossil fuel use worldwide); 19% came from hydropower, 16.5% from nuclear energy, and perhaps 1% from geothermal, solar, wind, and biomass energy combined. Electric generating capacity worldwide totalled about 2700 GW in 1990, with an average utilization or capacity factor of about 50% (EIA, 1991).

In the "traditional" fuels sector, the principal secondary energy form is charcoal, produced from wood at an efficiency (charcoal energy/wood energy) typically around 25% (Hughart, 1979). In addition, alcohol fuels are produced from biomass in quantities that are regionally significant (e.g. in Brazil) but globally modest (in the range of 0.01-0.02 TW worldwide). The global pattern of end uses of both industrial and traditional energy forms is shown in Table 2.

Another dimension of the current world energy picture is the striking disparities in energy availability among countries, mirroring the enormous differences in *per capita* economic product. These patterns are displayed in Table 3, which for simplicity divides the countries of the world into just three groups – "poorest", "intermediate", and "richest" – based on GNP per person. The industrialized countries, consisting mainly of the OECD nations and the former members of the Soviet bloc, contain less than 25% of the world's population but account for 85% of global economic activity and more than 75% of the use of "industrial" energy forms. Even when the "traditional" energy forms – proportionately much more important in the less developed countries than in the industrialized ones – are included, the average energy use per person among the poorest 3 billion people on the planet is about ten times smaller than among the richest billion. Even more

Table 2: *Pattern of world energy uses*

	Percent to Electricity	Residential and Commercial		Industrial		Transportation	Non-energy Uses
		Percent to End-Use Sectors					
Oil	3.9	5.2	(+1.8)	4.5	(+2.1)	16.8	5.2
Coal	12.0	3.6	(+5.6)	7.8	(+6.4)	0.6	0.6
Natural gas	4.9	5.6	(+2.3)	5.6	(+2.6)	–	1.4
Hydropower	5.4		(+2.5)		(+2.9)	–	–
Nuclear	3.6		(+1.7)		(+1.9)	–	–
Traditional		9.0		4.2		–	–
TOTAL	29.8	23.4	(+13.9)	22.1	(+15.9)	17.4	7.2

Notes: All figures are percentages of 1985 world supply of industrial and traditional energy forms (11.6 TW). Figures in parentheses are electrical contributions, measured as thermal equivalent. Source: Davis (1990).

Table 3: *Distribution of energy and economic activity by national income*

	POOREST (<$1000)		INTERMEDIATE ($1000-4000)		RICHEST (>$4000)	
	Country grouping by 1988 GNP/person					
Population, billions	3.1	(61%)	0.8	(16%)	1.2	(24%)
GNP, billion 1988 US dollars	1100	(6%)	1500	(8%)	16400	(86%)
"Industrial" energy use, TW	1.6	(14%)	1.1	(10%)	8.5	(76%)
"Traditional" energy use, TW	1.1	(73%)	0.2	(13%)	0.2	(13%)
Total energy use, TW	2.7	(21%)	1.3	(10%)	8.7	(69%)
Electricity use, trillion kWh/yr	1.1	(10%)	1.1	(10%)	8.4	(80%)
Electric generating capacity, GW	240	(9%)	280	(11%)	2030	(80%)
Refinery capacity, million bbl/day	6	(8%)	1.3	(18%)	55	(74%)
Average GNP/person, 1988 US dollars	350		1900		13700	
"Industrial" energy use/persons, watts	500		1400		7100	
"Traditional" energy use/persons, watts	350		250		200	
Electricity use/persons, kWh/yr	350		1400		7000	
Refinery capacity/persons, bbl/yr	0.7		5.9		16.7	

Notes: All figures are for 1988. Parenthetical figures are percentages of category; totals may not add due to rounding. Sources are same as for Table 1, plus World Bank (1990), Population Reference Bureau (1990), and author's calculations.

disturbing is the enormous disparity in *per capita* use of industrial energy forms between some of the poorest countries and the richest, e.g. Bangladesh with less than 100 kilograms of oil equivalent per person per year (140 watts) against 9000 kilograms of oil equivalent per person per year (12,700 watts) in Canada (Pachauri, 1991).

2.1 Implications for resource availability

Civilization's heavy reliance on fossil fuels has sometimes led to concern about the adequacy of the geologic stocks of these fuels to sustain such demands into the medium-term and long-term future. Table 4 summarizes the situation in terms of reserves (resources that have already been found

Table 4: Reserves and resources of conventional fossil fuels (TWy)

	Petroleum + NGL		Dry Natural Gas		Coal	
	rsrvs	RURR	rsrvs	RURR	rsrvs	RURR
North America	12.1	30	12.7	50	200	2000
Europe + USSR	24.3	50	58.3	120	290	3500
Middle East	114.0	140	43.0	80	0	0
Africa	11.6	20	7.1	20	60	200
Asia + Oceania	8.8	30	9.9	35	290	1000
Latin America	14.1	30	8.4	13	16	20
World Total	184.9	300	139.4	318	856	6720

Notes: 1 terawatt-year (TWy) = 5.25 billion barrels of crude petroleum = 7.08 billion barrels of natural gas liquids (NGL) = 29 trillion cubic feet of natural gas = 1 billion tonnes bituminous coal. RURR = remaining ultimately recoverable resources, of which reserves are a subcategory. The uncertainty range on the RURR figures is from about a third smaller to a third larger than the indicated values. Sources: World Bank (1980), WEC (1983), Masters *et al.* (1990); British Petroleum (1991).

and that are extractable with current technology under current economic conditions) and remaining ultimately recoverable resources (RURR, which comprise the reserves plus material inferred to exist but not yet found, and material that will only become exploitable if technology improves and/or prices rise).

Tables 1 and 4 indicate that, worldwide, the RURR for petroleum could sustain the current rate of use of oil for about 70 years. In the case of natural gas, the ratio of the RURR to the current use rate is about twice as long. Higher use rates would of course shorten these periods. Such numbers make clear that although current rates of use of oil and gas could be sustained at least into the middle of the next century, much higher use rates – as would be required for these energy sources to maintain their current roles in the industrialized countries and fuel significant increases in material well-being for the majority of the world's population who are now poor – are not realistic.

If substantial increases in the use of fossil fuels are to be contemplated, they will have to be based on coal and/or on unconventional oil and gas resources (e.g. oil shale, methane clathrates): the RURR of coal are about 10 times larger than those of conventional oil and gas combined, and the potential resources of oil shales and unconventional gas are five to ten times larger than those of coal (Hubbert, 1969; MacDonald, 1990). The rates at which these more abundant fossil fuels are used will be

limited, then, not by the quantities of the resources that exist but by the monetary and (probably above all) the environmental costs of exploiting them, compared to the costs of other energy options available in the same time frame.

As for uranium reserves and resources, world reserves of uranium – about 2.5 million tonnes (Harris, 1979) – are sufficient to fuel for 30 years the 400 electrical GW of nuclear capacity now in operation or under construction. Resources likely to be recoverable up to $130/kg, however, probably are in the range of 20–50 times greater than current reserves, hence could fuel 4–10 times more reactors than today for 5 times as long, or 150 years. Use of such relatively costly uranium would increase the cost of electricity generated by today's (non-breeder) reactor types by less than 10% (Holdren, 1975). Heavy reliance on nuclear fission for periods much longer than a century would eventually dictate a transition to breeder reactors. For many decades to come, however, the extent and form of civilization's reliance on nuclear energy will be governed not by resource constraints but by a combination of comparative monetary costs and public acceptance (or lack of it) of nuclear risks.

The main "renewable" energy forms in use today are, as we have seen, bio-mass and hydropower. Despite the "renewable" appellation, the current biomass use rate of about 1.5 TW is not sustainable under current practices (which are responsible for considerable forest degradation, erosion, and loss of soil quality). With wiser practices, a somewhat higher use rate probably would be sustainable, and scientific developments to increase biomass yields without costly inputs might raise the ceiling still further nonetheless, it seems likely that the energy uses of biomass will be constrained within 3–10 times the current use rate (hence, 5–15 TW) by competing uses for the resource (food, fodder, fiber, fertilizer, chemical feedstock, and ecosystem function). As for hydropower, estimates of the ultimate harnessable potential (Hubbert, 1969; WEC, 1983) range from 4 times to 8 times the current installed capacity (some 600 electrical GWs, peak, worldwide). The current level and even somewhat larger ones are surely sustainable, but attempts to increase the electricity derived from hydropower by more than about two-fold will have to overcome rising monetary, environmental, and social costs (discussed below) and corresponding public opposition.

2.2 Environmental impacts

Between 1890 and 1990, global use of fossil fuels increased 20-fold and that of biomass fuels increased threefold. These increases were major factors in the transformation of civilization, over this 100-year period, from a mainly localized disrupter of environmental

Table 5: Energy's role in some global environment impacts

Indicator of Impact	Natural Baseline	Human Disruption Index	Share of Human Disruption Caused By:			
			Industrial Energy	Traditional Energy	Agriculture	Industry & other
SO_x to atm	50 Mt S/yr	1.4	85%	1%	1%	13%
Pb to atm	25 kt/yr	15	63%	negl	negl	37%
oil to ocean	500 kt/yr	10	60%	negl	negl	40%
PM to atm	500 Mt/yr	0.25	35%	10%	40%	15%
NMHC to atm	800 Mt/yr	0.13	35%	5%	35%	25%
N fixation	200 Mt N/yr	0.5	30%	2%	67%	1%
Hg to atm	25 kt/yr	0.7	20%	1%	2%	77%

Notes: atm = atmosphere, PM = particulate matter, NMHC = nonmethane hydrocarbons, Mt = megatonnes, kt = kilotonnes, negl = negligible. The Human Disruption Index is the ratio of the anthropogenic contribution to the Natural Baseline. Estimates are based on a variety of sources and are very approximate. See, e.g. Holdren (1987, 1991); Lashof and Tirpak (1989; Graedel and Crutzen (1989), and IPCC (1990).

Table 6: Indoor and outdoor population exposure to particulate matter

	Industrialized Countries (1.2 billion people)				Developing Countries (3.8 billion people)			
	Urban		Rural		Urban		Rural	
	Out-door	In-door	Out-door	In-door	Out-door	In-door	Out-door	In-door
average PM concentration ug/m^3	75	100	25	75	200	250	50	300
Distribution of total person-hours	7%	68%	4%	21%	3%	30%	27%	40%
Total population exposure, 10^9 EU	6	82	1	19	23	285	51	456

Notes: PM = particulate matter, ug = micrograms, EU = exposure unit = 1 "person-yr-ug/m3". The estimates given are very approximate. Source: Smith (1988)

conditions and processes to a global ecological and geochemical force. Today, human activities rival or exceed natural processes worldwide as mobilizers of sulfur oxides, nitrogen oxides, hydrocarbons, lead, mercury, cadmium, and suspended particulate matter; the actions of humans have increased the global atmospheric burden of carbon dioxide by more than 25% and that of methane by more than 100%, compared to preindustrial levels; and, among all human activities, the technologies of energy supply – above all, fossil fuel technologies – are the dominant sources of most of these global pollutants and significant sources of all of them (see Table 5).

The effluents from fossil fuel combustion are the principal contributors to the problem of acid precipitation (Harte,1988) and to the potential for global climate change (Schneider, 1989), and they share with the effluents from biomass fuels the main responsibility for the world's pervasive problems with air pollution. (Total population exposure to suspended particulates – perhaps the most dangerous of air pollutants in terms of public health – is even more a matter of indoor than outdoor exposures; and the largest contributor to the indoor exposures is the combustion of biomass fuels for heating, cooking, and boiling water, above all in the rural sector of the South

Table 7: Energy and nonenergy sources of anthropogenic greenhouse gases

	Share of Warming	Main Contributors to Emissions of this Gas	
CO_2	66%	fossil fuel	(76%)
		fuelwood	(3%)
		land clearing	(19%)
		cement manufacture	(2%)
CH_4	17%	fossil fuel	(20%)
		biofuels	(3%)
		agriculture	(47%)
		land clearing	(15%)
		landfills	(15%)
CFCs	12%	refrigeration	
		foams and solvents	
		aerosol cans	
N_2O	5%	fossil fuel	(12%)
		biofuels	(4%)
		agriculture	(64%)
		land clearing	(20%)

Notes: Based on 100-year warming commitment. Sources: IPCC (1990). Lashof and Tirpak (1989); British Petroleum (1991) and authors' calculations.

(See Table 6). Use of biomass fuels is also an important contributor, in many circumstances, to deforestation and impoverishment or erosion of soils (Smith, 1987).

Global warming, to which carbon dioxide release from fossil fuel burning is the largest single contributor, is arguably the most dangerous of all the environmental impacts of human activity (IPCC, 1990; Mintzer, 1990). This is so because climate effects – and climate change can drastically disrupt – most of the other environmental conditions and processes on which the well-being of the human population critically depends: magnitude and timing of runoff, frequency and severity of storms, sea-level and ocean currents, soil conditions, vegetation patterns, and distribution of pests and pathogens, among others. Altogether some 54% of the global warming potential of contemporary human activities comes from fossil fuel use, and another 3% or so from the use of fuelwood (see Table 7).

The operation of hydroelectric energy facilities does not produce air pollution or contribute to global warming; but the construction of the facilities contributes at least modestly to these problems; and hydro dams may flood fertile agricultural lands or forests, reduce the fertility of land downstream, block the migration of commercially important fish species, facilitate the spread of parasitic diseases, and pose a risk of catastrophic loss of life through failure of the dam. Nuclear energy, like hydropower, does not produce conventional air pollutants or contribute to global warming except through construction of the facilities; but it poses radiological hazards that, although generally modest in routine operation, can become severe in the event of accidents at reactors, fuel reprocessing plants, or waste repositories, or in shipment of spent fuel and radioactive wastes (NAS, 1979).

2.3 Sociopolitical impacts

Prominent among the sociopolitical impacts of energy supply are the risks of dependence on energy supplies imported from regions whose volatile politics may threaten unpalatable conditions for access; the stakes in this connection include, of course, the possibility of going to war over access or price (Pachauri, 1985). This is understandably a particular concern in the Middle East, which is the source of nearly half of the oil that moves in world trade and contains some 60% of the world's oil reserves.

Another energy-conflict connection that is arguably as dangerous as the potential for fighting over oil is the link between nuclear energy and nuclear weapons. The essence of this problem is that the spread of nuclear energy technology spreads access to nuclear-explosive materials and related capabilities in ways that make it easier for additional countries to acquire nuclear weapons (Sweet, 1984; Holdren, 1989). In the future, as subnational criminal groups become more sophisticated, the danger may increase that these groups, too, could acquire nuclear bombs or radiological weapons by misusing nuclear energy technologies.

A third energy-related impact on international relations – and one likely to grow in the years ahead – is the aggravation of international tensions by trans-boundary pollution (such as from acid precipitation or radiological impacts from nuclear accidents) or disputes about responsibility and compensation for impacts of global environmental change. Particularly important in the latter category may be impacts of climate change on water availability in rivers shared by more than one nation, and changes in regional carrying capacity that produce large flows of environmental refugees (Gleick, 1989, 1990).

Deserving of mention, finally, are the domestic sociopolitical impacts associated with disputes and resentments about the inadequacies of energy supplies to particular regions and socioeconomic classes, or over the siting of energy facilities and activities, including, e.g. oil drilling in offshore and wilderness environments, dislocation of native peoples by hydropower development, and the siting of nuclear facilities from uranium mines to radioactive-waste repositories. Resolution of such problems by imposition of the will of the central governmental authorities may increasingly be seen as a significant political cost of particular kinds of energy expansion.

2.4 Economic aspects

At 1990 energy prices, expenditures on energy by individuals, firms, and governments typically amounted to about 10% of GNP in market economies; similarly, valuation of the 1990 quantities of industrial energy forms world-wide at market prices gives a figure of about $2 trillion, compared to a world economic product around $21 trillion. In 1990, 44% of world oil production was traded internationally, as was 14% of natural gas and 10% of coal; the value of this energy trade was about $250 billion, or nearly 10% of all international trade.

Poor people pay a substantially larger fraction of their income for energy than do rich people, and energy imports and exports are typically a larger fraction of all foreign trade for developing countries than for industrialized ones. The high costs of oil imports were a major contributor to the trade deficits and debt problems experienced by many developing countries in the 1970s and 1980s.

The total investment in world electric generating capacity, calculated at replacement value, is about $2.5 trillion, and the total investment in electricity transmission and distribution facilities is comparable. The replacement cost of facilities in the non-electric part of the energy supply system is perhaps another $3–4 trillion, so the total value of the global system supplying industrial energy forms is in the range of $8–9 trillion. The total annual investment in energy facilities, perhaps $350 billion, is only 8–9% of gross domestic investment worldwide; but this fraction tends to be higher in developing countries. The large investments in energy facilities and the long lifetimes of such plants (typically 30 to 40 years) mean that the composition of world energy supply can change only slowly. The world could not afford, for example, to throw away its multi-trillion-dollar investment in fossil fuel-based energy supply and replace it quickly with something else.

It is often assumed that decisions about what energy resources to develop and what kinds of technologies to deploy are made mainly on economic grounds, that is, based on systematic and consistent comparisons of expected monetary costs of the different options. But that assumption is rarely correct. Political considerations – such as concern with balance of payments, import vulnerability, and energy independence – have often played major roles in energy choices (such as Brazil's fuel alcohol program and France's nuclear program); it is to be expected that this will continue, and that the political considerations increasingly will be joined by environmental ones in altering the choices that would be dictated by monetary considerations alone.

Even where countries or companies *believe* they are choosing on monetary grounds, moreover, the monetary comparisons used for these purposes are often distorted by inconsistent assumptions (about, e.g. time frames, discount rates, and the treatment of "indirect" costs) and by failure to account for subsidies enjoyed unequally by different energy sources (through, e.g. preferential tax treatment, government assumption of liability for major accidents, and subventions for technology development and facility construction). To the extent that monetary costs are supposed to be important in energy choices, they ought to be calculated with much more careful attention to these potential distortions than has been the rule in most countries until now.

3. WHAT IF THE ENERGY FUTURE IS LIKE THE PAST, EXCEPT BIGGER?

The simplest (although almost invariably wrong) depictions of the energy future are obtained by extrapolating from the recent past. From 1970 to 1980, world use of "industrial" energy forms grew at an average rate of 2.9% per year, and from 1980 to 1990 it grew at about 1.9% per year. If, after 1990, the growth rate were maintained at 2.0% per year, the world's use of industrial energy forms would reach 17.3 terawatts in 2010 and 25.7 terawatts in 2030.

These results are not very different from those obtained, in a slightly more disaggregated way, by combining World Bank population projections (World Bank, 1990) with extrapolations of the 1980–90 growth rates in *per capita* energy use in industrialized and developing countries. This version of a "business as usual" energy future is shown in Table 8, along with a more sophisticated "energy efficient demand scenario" developed by one of the authors of the World Bank's 1992 World Development Report (Anderson, 1991). The latter scenario is based on the assumption that energy intensities (ratio of energy to economic output) in developing countries will follow the same rising and then falling trajectory with time as was experienced in the industrialized countries, albeit with much lower peak values because of technical progress since the earlier era when the industrialized countries experienced their maximum intensities. (See Fig. 1).

As indicated in Table 8, the "energy efficient" scenario differs from "business as usual" mainly in having a slowly declining rather than slowly growing energy use rate in the industrialized nations, and in having a considerably larger growth of energy use in developing countries. Nonetheless, the global totals are very similar in the two scenarios, reaching 24–26 terawatts in 2030; and the industrialized developing gap in *per capita* use of industrial energy forms, although narrowing more rapidly in the "energy efficient" case than under "business as usual", is still nearly a factor of three in 2030.

Table 8: *Conventional projections for use of industrial energy forms*

	Actual		Projection			
	1980	*1990*	*2000*	*2010*	*2020*	*2030*
Population (millions)						
industrialized	1075	1158	1215	1260	1295	1315
developing	3310	4085	5000	5900	6750	7575
Energy Use/Person (watts)						
"business as usual"						
industrialized	7170	7255	7360	7465	7570	7675
developing	615	770	965	1205	1500	1880
"energy efficient" (Anderson)						
industrialized	same		7435	7225	6325	6285
developing	as above		950	1340	1720	2300
Total Energy Use (terawatts)						
"business as usual"						
industrialized	7.7	8.4	8.9	9.4	9.8	10.1
developing	2.0	3.2	4.8	7.1	10.1	14.2
world total	9.7	11.5	13.8	16.5	19.9	24.3
"energy efficient" (Anderson)						
industrialized	same		9.0	9.1	8.2	8.3
developing	as above		4.8	7.9	11.6	17.2
World Total			13.8	17.0	19.8	25.7

Notes: "Business as usual" results obtained by extrapolating 1980–90 rates of increase in *per capita* use of industrial energy forms for industrialized and developing countries. One terawatt = 10^{12} W. Sources are World Bank (1990); Anderson (1991), and calculations by the authors.

Indicative of "business as usual" thinking about where the energy will come from 20 years hence is the "consensus view" described by Davis (1990), in which a total rate of use of industrial energy forms of 17.5 TW in 2010 is obtained in the form of 6.1 TW of oil, 5.3 TW of coal, 3.5 TW of natural gas, 1.0 TW of nuclear energy, and 1.6 TW of hydropower and other renewables. In this scenario, then, fossil fuels in 2010 are still supplying 85% of all use of industrial energy forms, and the total use rate of these fuels has increased by 46% over 1990 levels; the increase in emissions of carbon dioxide from fossil fuel combustion would be 48% over 1990 levels if the fossil fuels were burned, as they are today, with full release of the contained carbon.

Given the difficulties posed by world energy supply in 1990, these more or less conventional projections of energy patterns over the next 20 to 40 years ought to be considered alarming. The indicated increase from 11.7 terawatts of industrial energy forms in 1990 to 24-26 terawatts of the same in 2030 is more than enough to generate formidable environmental problems from expanded fossil fuel use – above all, potentially intolerable impacts on climate from the associated CO_2 emissions – and it is more than enough to provoke intensified disputes about siting new facilities and about the responsibility for trans-boundary environmental impacts, to increase dependence on oil from the Persian Gulf, and to exhaust the possibilities for financing new energy facilities in developing countries where most of the growth is expected to take place. But this increase in energy supply is not enough to provide a high material standard of living to the nine billion people likely to be alive in 2030, assuming that the energy intensity of economic activity decreases only at about the historical rate; and it is not enough to reduce, to tolerable dimensions, the gap in energy use per person between industrialized and developing countries without a far bigger redistribution than these scenarios project.

The financial aspect of this predicament is made more difficult by the likelihood that energy supplies will become costlier over time. This is so because, not surprisingly, industrializing and industrialized societies found and used the most convenient and least expensive energy resources

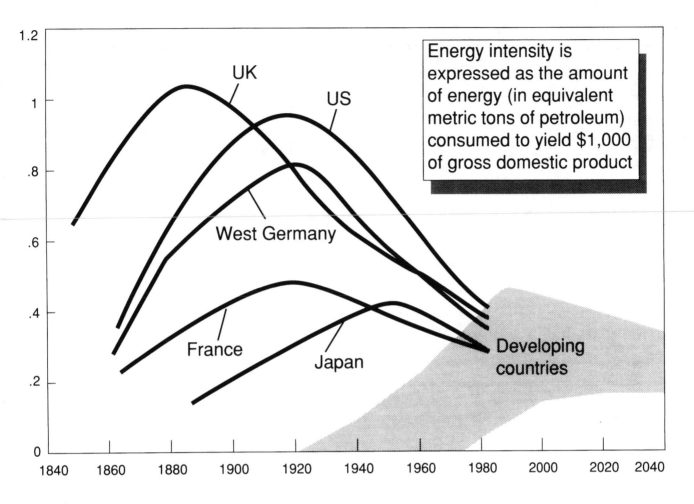

Figure 1: Energy intensity versus time in industrialized and developing countries. In industrialized countries the energy intensity (ratio of energy consumption to gross domestic product) rose, then fell. Because of improvements in materials science and energy efficiency, the maxima reached by countries during industrialization have progressively decreased over time. Developing nations can avoid repeating the history of the industrialized world by using energy efficiently. We would add, however, that it may be unrealistic to expect the developing countries to reach very quickly an energy-efficient development path, given the capital constraints and industrial weakness these countries confront (Pachauri, 1990). [From "Energy for the Developing World", by A.K.N. Reddy and J. Goldemberg. Copyright © 1990 by Scientific American, Inc. all rights reserved.]

first: the biggest, shallowest, closest deposits of oil and natural gas and the nearest and most cost-effective hydroelectricity sites. Cumulative depletion and rising demand now require resort, by most countries, to smaller, leaner, more distant, more difficult – and hence more expensive – resources of these kinds, or to more abundant resources, such as coal and uranium and solar energy, which are more capital-intensive to convert into the fluid fuels and electricity that industrialized countries now require. The problem is further compounded by the circumstance that controlling the environmental risks and impacts of some of the most abundant energy sources – above all coal and nuclear fission – tends to drive monetary costs still higher (Holdren, 1987).

4. INGREDIENTS OF AN ALTERNATIVE ENERGY STRATEGY

The data and argumentation in Sections 2 and 3 suggest

that an energy future based on "business as usual" – or the modest deviations from it represented by the "energy efficient" scenario shown in Table 8 – probably will not be manageable economically, environmentally, or politically. In order to overcome these difficulties, an alternative approach must be able to deliver: more affordable means of meeting energy needs; a more rapid narrowing of the rich-poor gap in energy use; a smaller increase (or a reduction) in fossil fuel use; and a reduction in environmental damage per unit of fossil and non-fossil energy supplied. Some ingredients of a strategy that could accomplish these things are described, along with some corresponding research priorities, in what follows.

4.1 Increasing the efficiency of energy use

The cornerstone of an alternative strategy is achievement of much more rapid increases in energy efficiency (i.e. reductions in energy intensity, or ratio of energy to economic

Table 9: *Efficiency-based scenario for narrowing the rich-poor gap*

	1990	2000	2010	2020	2030
Population (millions)					
industrialized	1158	1215	1260	1295	1315
developing	4085	5000	5900	6750	7575
Total energy per person (watts)					
industrialized	7465	6125	5025	4120	3380
developing	1085	1320	1610	1965	2395
GDP per person (1990 US $)					
industrialized	15 600	17 230	19 030	21 026	22 230
developing	750	1 220	1 990	3 240	5 280
Total energy (terawatts)					
industrialized	8.6	7.4	6.3	5.3	4.4
developing	4.4	6.6	9.5	13.3	18.1
Total world	*13.0*	*14.0*	*15.8*	*18.6*	*22.5*

Notes: Population projections are same as for Table 8 (World Bank, 1990). Energy figures include both industrial and traditional forms and come from authors' calculations based on assumptions stated in text.

product) than the 1% per year long-term historical average typically used in forecasting. Improvements in energy efficiency are the equivalent of new energy supply – the energy "saved" by increased efficiency in one application can be used in another – and expansion of the efficiency "supply" in the period after 1973 occurred in many countries at rates in the range of 2–3% per year (EIA, 1991; Kelly *et al.*, 1989). A large literature indicates that the potential for further efficiency improvements, at costs per gigajoule generally much lower than those of all other expandable energy sources, is vast in all sectors – residential and commercial buildings, industrial processes, agriculture, and transportation – and in industrialized and developing countries alike (see, e.g. Goldemberg *et al.*, 1987; Williams *et al.*, 1987; IEA, 1989; Kelly *et al.*, 1989; Carlsmith *et al.*, 1990; OTA, 1991; Lovins and Lovins, 1991). Exploiting this potential is essential for meeting increased requirements for energy services affordably and with tolerably low environmental impacts. The environmental impacts of measures for increasing efficiency, while sometimes non-negligible, are nonetheless generally much smaller – or can be made much smaller – than those of the energy sources displaced. See, e.g. Holdren (1987).

Rapid improvements in energy efficiency also hold the key to narrowing the rich-poor gap in energy use while avoiding unmanageably high aggregate energy demand. Suppose, for example, that the ratio of energy to real GDP could be reduced at an average rate of 3% per year in the

industrialized and developing nations alike for the next 40 years. (For comparison, the actual reduction in the United States between 1973 and 1986 was 2.4% per year (EIA, 1989). This would permit a real rate of *per capita* GDP growth of 1% per year in the industrialized countries while *per capita* energy use falls at 2% per year, and a real rate of *per capita* GDP growth of 5% per year in developing countries while *per capita* energy use grows at 2% per year (Holdren, 1991).

As indicated in Table 9, this scenario narrows the gaps in *per capita* energy use; and income much more rapidly than the more conventional scenarios presented in Table 8, and it does so within a substantially lower aggregate energy use. (The latter margin is even larger than it appears from comparing the two tables, since Table 9 is based on the total of industrial and traditional energy forms while Table 8 includes industrial forms only.) Slower rates of change in efficiency and *per capita* energy use after 2030 would still suffice to produce convergence in the "industrialized" and "developing" energy use per person at about 3 kW between 2040 and 2050; if the use rate could then be held at this level, with further increases in well-being coming from continuing improvements in efficiency, and if world population were stabilized at 10 billion, growth of aggregate energy use would stop at 30 terawatts (Holdren, 1991).

While many may question the feasibility of maintaining for the next 40 years a 3% per year rate of reduction in energy intensity, as assumed in this scenario, we think that the feasibility of scenarios that entail higher overall energy use and slower narrowing of the rich-poor gap is at least equally questionable. It is essential to proceed with the work of determining, by analysis and implementation, how much reduction in energy intensity can really be achieved.

This work should not ignore, as too many previous analysis have done, the energy-saving possibilities of changes in lifestyle away from that which now prevails in the most industrialized nations. Such changes, although they may be resisted by those who find the status quo agreeable, will not necessarily entail reductions in the energy services being provided to society. For example, it seems quite possible to use changes in commuting patterns and improvements in public transportation to reduce dependence on the private automobile while increasing the quality of transportation services available to the average citizen, reducing pollution, and saving energy and money.

4.1.1   Research priorities
Develop "supply curves for conserved energy" showing the most cost-effective approaches for decreasing energy intensity in all regions of the world.

Identify obstacles to the achievement of this potential and develop strategies for overcoming them.

4.2   Reducing the environmental impacts of current energy sources
Irrespective of progress in reducing energy intensity, civilization will remain substantially dependent for decades to come on the same fuels that supply most of the world's energy today. Thus, reducing the environmental impacts per unit of energy supply means, first of all, reducing these impacts from today's energy sources.

In the short term, the needed steps include: more widespread use of available control technologies to reduce emissions of sulfur and nitrogen oxides from fossil fuel combustion and emissions of hydrocarbons and particulate matter from fossil and biomass fuels alike; greater efforts to reduce the leakage associated with ocean drilling and transport of petroleum; increased resources and inspection powers for national and international nuclear-safety and anti-proliferation authorities; and, in less-developed countries especially, efforts to put fuelwood harvesting on a sustainable basis.

In the medium term, steps to be undertaken include: increased substitution of natural gas for coal and oil to gain further reductions in emissions of carbon dioxide and other pollutants; deployment of more efficient and cleaner-burning coal technologies, such as integrated-gasification combined-cycle power plants and pressurized fluidized-bed combustors (see, e.g. Johansson *et al.*, 1989); investigation of the potential of advanced nuclear reactor designs with more forgiving characteristics in accidents (Forsberg and

Weinberg, 1990); and replacing direct combustion of biomass fuels in developing countries with biogas and alcohol fuels that burn more cleanly and return some nutrients to the soil (Smith,1987).

It is likely that the reduction in greenhouse gas emissions resulting from the measures just described and from reductions in energy intensity will not be sufficient to ward off highly disruptive climatic changes. If that is so, pressure will grow to do more to reduce the carbon dioxide emissions from fossil fuel use. One option for accomplishing this would be to modify the largest CO_2-emitting facilities, such as coal-burning electric power plants, so that the carbon dioxide can be captured from the stack gases for sequestering in depleted natural gas wells or deep ocean waters (Okken *et al.*, 1989). This will be difficult and expensive because the volume and mass of the CO_2 involved are so large – nearly 3 tonnes of CO_2 for every tonne of coal burned, nearly 10 million tonnes of CO_2 from a 1-million kW coal-burning powerplant in a year.

4.2.1   Research priorities
Develop cost-effective and implementable means for reducing indoor concentrations of hydrocarbons and particulates from cooking, space heating, and water heating with traditional fuels.

Determine the most cost-effective means of controlling emissions of particulate matter and oxides of sulfur and nitrogen from existing uncontrolled coal-burning power plants in various parts of the world.

Investigate the feasibility and comparative costs of alternative means for capturing and sequestering carbon dioxide from large fossil fuel-burning facilities.

Investigate whether and how the difficulties of nuclear fission with respect to reactor safety, waste management, and weapons linkages can be overcome in a manner sufficiently convincing to enable this energy source to increase its current modest share of world energy supply.

4.3   Increasing the contributions from low-impact energy options
The principal energy-supply options that, in theory, could be used to replace or augment fossil fuels, nuclear fission, and today's high-impact/low-efficiency uses of biomass energy are readily listed: direct harnessing of sunlight; "indirect" harnessing of sunlight as hydropower, wind, ocean heat, and (with improved technology) biomass; geothermal energy; and nuclear fusion. The magnitudes of some of these resources are summarized in Table 10. All of these options have promise, but most of them have liabilities as well (see, e.g. National Research Council, 1980; Lashof and Tirpak, 1989; SERI, 1989; Brower, 1990; Flavin and Lennsen, 1990): high costs of solar collectors, competing uses for biomass and rivers,

Table 10 : *Resources for alternative energy options*

Stock Resources		Remaining Ultimately Recoverable Supplies (terawatt-years)
Geothermal, hot water and steam		4,000
Geothermal, hot dry rock (>45°C/km)		200,000
Lithium (for deuterium–tritium fusion)		140,000,000 (oceans)
Deuterium (for deuterium–deuterium fusion)		250,000,000,000 (oceans)
Flow Resources	**Total Flow (terawatts)**	**Plausibly Harnessable Flow**
Sunlight	88,000 at Earth's surface 26,000 on land	Converting insolation on 1% of land area at 20% efficiency yields 52 TW (electric)
Biomass	100 global net primary productivity, 65 on land	Biomass fuels from 10% of land area at 1% efficiency yields 26 TW (chemical)
Ocean heat	22,000 absorption of sunlight in oceans	Converting 1% of absorption at 2% efficiency yields 4 TW (electric)
Wind	1000–2000 driving winds worldwide	Using all cost-effective terrestrial sites may yield 1-2 TW (electric)

Notes: Estimates are the authors' based on a variety of sources; see, e.g. Hubbert (1969), Haefele (1981), World Energy Conference (1983); SERI (1989).

uncertain technology and economics for fusion and the most abundant forms of geothermal energy.

At the same time, there is much reason to believe that more systematic attention to carrying out economic comparisons on a consistent basis, and to correcting for subsidies now widely enjoyed by fossil fuels and nuclear fission, would increase the relative economic attractiveness of renewables. Beyond that, the necessity to diminish the overwhelming environmental burdens of today's energy mix makes it appropriate to begin now to promote the deployment of those renewable energy options – such as wind power, solar–thermal electricity generation, and biomass-derived alcohol fuels – that are close to economic competitiveness at current prices of conventional alternatives and that offer significantly smaller environmental or political risks than those alternatives.

Also appropriate is increased support for research and development on such longer-term non-fossil energy options as photovoltaics, solar-thermochemical hydrogen production, ocean-thermal energy conversion, hot-dry-rock geothermal energy, and fusion energy systems. These R

and D efforts should pursue not only the attainment of practical, economic ways to harness these possibilities, but also the prospects for minimizing their environmental costs. Finally, greatly increased research and development effort is warranted on the use of hydrogen (obtainable by using any primary energy source to split water) as a clean-burning, non-CO_2-emitting replacement for fossil fuels in dispersed applications; particular promise for high-efficiency dispersed uses of hydrogen (and, in the shorter term, methane) is offered by fuel cells (Ogden and Williams, 1989).

4.3.1   Research priorities
Development of more consistent economic comparisons between renewable energy options, on the one hand, and nuclear fission and fossil fuels on the other, taking into account current subsidies and the costs of modifying nuclear and fossil fuel technologies to meet appropriate environmental and safety standards.

Identification of other (non-monetary) barriers to the implementation of already practical renewable energy options, and development of means to overcome these barriers.

Expanded efforts to discover and develop additional affordable and low-environmental-impact approaches to the use of renewable, geothermal, and fusion energy resources.

Development of improved fuel cells and other elements of methane- and hydrogen-based energy systems to replace today's less efficient and more polluting dispersed uses of fossil fuels.

4.4 Increasing international co-operation and assistance

Greatly expanding programs of international co-operation and assistance in research, development, deployment, and financing of appropriate energy options will be necessary if the preceding elements are to be implemented not only in the richest industrial nations but also in the industrialized but cash-poor countries of eastern Europe and the former USSR, and in the less developed countries of the South. This will require not altruism from the richest countries but only responsibility and self-interest: responsibility, because the countries that are now richest became so by burning cheap fossil fuels and, in the process, consuming the absorptive capacity of the atmosphere for the resulting carbon dioxide (thus calling into question the feasibility of the rest of the world's following the same "inexpensive" path to prosperity); and self-interest, because the rich will not escape the consequences of the environmental and political disruptions that will follow if the peoples of the East and the South attempt to build their prosperity on the historical pattern of inefficient energy end use and conversion technologies, fueled by coal and oil.

Building East–West and North–South co-operation on energy could begin with increased co-operation on energy research and development. This would relieve some of the constraints imposed by inadequate energy R and D budgets world-wide by eliminating needless duplication, sharing diverse specialized strengths, and dividing the costs of the large projects that some avenues of energy R and D require. (Until now, nuclear fusion has been the only area of energy research that has enjoyed major international co-operation.) It is especially important that international co-operation in energy research be extended to include North–South collaborations on energy technologies designed for application in developing-country contexts. It is shameful that the industrialized countries – the only ones that can afford to do much energy research – have confined their efforts until now almost entirely to technologies tailored to industrial nation economic and cultural contexts. A step toward rectifying this deficiency could be the establishment of an international agency under UN auspices for research, development, and deployment of renewable energy options.

International co-operation on understanding and controlling the environmental impacts of energy supply is also extremely important, since many of the most threatening environmental problems are precisely those that respect no boundaries. Air and water pollution from Poland, Czechoslovakia, Hungary, Germany, and Russia reach across Europe and into the Arctic; and the environmental impacts of energy supply in China and India, locally debilitating at today's levels of energy use, could become globally devastating at tomorrow's. But pleas from the rich countries that global environmental problems require global energy restraint will fall on deaf ears in the less-developed and economically intermediate countries unless the rich find ways to help the rest achieve increased economic well-being and environmental protection at the same time.

In international and national planning for an energy transition, local factors will have to be kept in mind. In developing countries with large resources of coal, for example, the economic and social constraints inhibiting a rapid transition to environmentally more desirable fuels are formidable; overcoming those constraints will require large-scale education and training, as well as transfer of technology from North to South. Changes in methods of communication and management may also be required, such as creating and harnessing decentralized institutions for managing the energy sector and relying on a diversity of change agents appropriate to diverse circumstances. (Better management of household energy use, for example, will require emphasizing the role of women).

4.4.1 Research priorities

Co-operative and comparative assessment of energy/environmental problems and requirements.

Development of new approaches for transfer and financing of existing technologies for pollution control, cleaner energy supply, and increased energy efficiency, and for co-operative development of improved technologies in these categories.

Research on appropriate formulations for international agreements on standards and targets for energy efficiency and environmental performance.

4.5 STABILIZING WORLD POPULATION

All of the preceding measures together may not be enough to assure adequate energy supplies – at tolerable monetary, environmental, and political costs – to meet the aspirations of the population of 10 billion to which the planet is practically committed. (To halt population growth at that level would require fertility to fall to the replacement level of 2.1 children per woman – from today's world average of 3.5 – by the year 2025; fertility reduction at a faster pace seems unlikely, since, historically, declines in fertility usually have taken place only when a reasonable level of material well-being has been achieved, accompanied by widespread literacy and the availability of health care (Pachauri, 1991). Assuming that increases in energy efficiency make it possible to supply a high material standard of living to everyone based on average energy use per person around 3 kW (about 60% of the current figure for Japan and 25% of that for the United States), a population of 10 billion would require 30 terawatts. That is about 2.5 times the energy use rate that is so severely straining the planet's technological, managerial, and environmental resources today.

If replacement fertility is not achieved until 2060, world population will stabilize at 12.5 billion rather than at 10.0 billion; just the difference in energy use between these two population figures, at the hypothesized 3 kW per person, is equal to the world's 1970 use rate of all industrial energy forms. If a satisfactory standard of living in the long run turns out to require closer to half the 1990 US rate of energy use per person, say 5 kW, then a population of 12.5 billion would use energy at nearly five times the world's 1990 rate.

The higher the ultimate rate of energy use, the less flexibility society will have in its choices about how the energy is to be supplied and, in all likelihood, the greater will be the cost of energy supply – not only in the aggregate but also per gigajoule and hence *per capita*, and not only monetarily but also environmentally. If people are more numerous, therefore, they will probably be poorer. Bringing about the fertility reduction needed to stabilize world population at 10 billion will itself be an enormous challenge, requiring massive development assistance and other forms of international co-operation. But as difficult as that will be, it is likely to be easier, and cheaper, than providing extra billions of people with energy (and food and water and much else).

5. CONCLUSION

The current global energy-supply system is failing to meet the needs of the majority of the world's population and, at the same time, is generating massive environmental disruptions that imperil human well-being in rich countries and poor countries alike. Attempting to address the problems of the present and the needs of the future mainly through expansion of the current system – the "business as usual" scenario – will do too little for the poor and too much to the environment, and it will cost too much money.

In fact, the problem of financing is critical not only for bringing about a worldwide transition in the mix of energy and pattern of utilization, but also in promoting other aspects of the development needed to eliminate the poverty that afflicts over 2 billion people in today's world. To see how energy's capital requirements could squeeze capital availability for other purposes, it is only necessary to observe that:

(a) at least $14–22 \times 10^{12}$ will be required over the next 40 years to replace existing energy facilities at the ends of their useful lives and to expand energy outputs, and

(b) 70 to 80% of this total – hence $0.25–0.44 \times 10^{12}$ per year – needs to be spent in developing countries, where annual availability of investment capital for all purposes around 1990 was only about 0.8×10^{12} (World Bank, 1990).

Clearly, innovative means of financing will be necessary. These should rely to the extent possible on major institutional improvements in the management of the energy sector in developing countries, but in the final analysis significant transfers of resources from North to South will also be required. Such transfers could be funded by an appropriate level of tax imposed on carbon emissions from fossil fuel use above a certain *per capita* level.

What is required, more broadly, is an alternative approach that emphasizes cost-effective reductions in the energy intensity of economic activity, reductions in the environmental impacts of today's energy-supply technologies and the fastest practical transition toward new ones that are intrinsically less disruptive of the environment, and greatly increased international co-operation to achieve these transformations together with stabilized population and sustainable prosperity everywhere.

Alas, there is little evidence so far of the needed national and international commitments to move in these directions. The sharp reductions in energy intensities that characterized the period from 1973 to 1986 in many countries have slowed. Although numerous countries have announced their intention to stabilize or reduce greenhouse gas emissions over the next 10 to 20 years, all too few have devised or begun to implement plausible plans for achieving this. Investments in research and development toward improving existing energy sources and developing new ones have plummeted: government R and D on energy efficiency, fossil fuels, fission, fusion, renewables, and geothermal energy combined fell almost fourfold in the United States between 1980 and 1990 (corrected for inflation) and nearly twofold in the IEA (International Energy Agency) member countries as a group (Holdren, 1990; Anderson, 1991). Energy efficiency and renewables, both critical to the success of an alternative strategy, in 1990 were receiving only 25% of the public funds for energy-supply R and D in the United States and 11% in the IEA countries as a group. And net bilateral development assistance from OECD members to low-income economies (*per capita* GDP <$500) was only 0.09% of the donors' GDP in 1988, compared to 0.13% in 1970 and 0.20% in 1965 (World Bank, 1990).

This is not the path to the alternative energy future that society needs and, if it chooses, can attain. One must hope that the occasion of the 1992 UNCED will prove to be a turning point, at which the nations of the world agree collectively to reverse recent trends and begin to address, seriously and jointly, the energy/environment/development interactions that otherwise portend only trouble.

REFERENCES

Anderson, D. 1991. *Energy and the Environment.* Special Briefing Paper No. 1, Edinburgh, Scotland: The Wealth of Nations Foundation.

British Petroleum. 1989,1991. *BP Statistical Review of World Energy.* London.

Brower, M. 1990. Cool Energy: *The Renewable Solution to Global Warming.* Cambridge, MA: Union of Concerned Scientists.

Carlsmith, R.S., Chandler, W. U., McMahon, J. E. and Santini, D. J. 1990. *Energy Efficiency: How Far Can We Go?* Report ORNL/TM-11441. Oak Ridge, TN: Oak Ridge National Laboratory.

Davis, G. R. 1990. Energy for planet Earth. *Scientific American,* **263** (September), 55-62.

EIA (Energy Information Administration). 1989. *International Energy Annual 1988.* Washington, DC: Government Printing Office.

EIA. 1991. *Annual Energy Review 1990.* Washington, DC: Government Printing Office.

Flavin, C. and Lennsen, N. 1990. Beyond the Petroleum Age: Designing a Solar Economy Worldwatch Paper 100. Washington, DC: *Worldwatch.* December.

Forsberg, C. W. and Weinberg, A. M. 1990. Advanced reactors, passive safety, and acceptance of nuclear energy. *Annual Review of Energy,* Vol. 15, pp.133-152.

Gleick, P. H. 1989. The implications of global climatic changes for international security. *Climatic Change,* **15**, 309-325.

Gleick, P. H. 1990. Climate change and international politics. *Ambio,* **18**, 333-339.

Goldemberg, J. Johansson, T. B. Reddy, A. K. N. and Williams, R. H. 1987. *Energy for a Sustainable World.* Washington, DC. World Resources Institute.

Goldemberg, J. and Reddy, A. K. N. 1990. Energy for the developing world. *Scientific American,* September, 111-118.

Graedel, T. E. and Crutzen, P. J. 1989. The changing atmosphere. *Scientific American,* September, 58-68.

Haefele, W. 1981. *Energy in a Finite World: A Global Systems Analysis.* Cambridge, MA: Ballinger.

Hall, D. O. Barnard, G. W. and Moss, P. A. 1982. *Biomass for Energy in Developing Countries.* Oxford: Pergamon.

Harris, D. P. 1979. World uranium resources. *Annual Review of Energy,* **4**, 403-32.

Harte, J. 1988. Acid rain. In: *The Cassandra Conference: Resources and the Human Predicament.* P. Ehrlich and J. Holdren [eds.], pp. 125-146. College Station, TX: Texas A&M University Press.

Holdren, J. P. 1975. Uranium availability and the breeder decision. *Energy Systems and Policy,* **1**, 205-232.

Holdren, J. P. 1987. Global environmental issues related to energy supply. *Energy,* **12**, 975-992.

Holdren, J. P. 1989. Civilian nuclear technologies and nuclear weapons proliferation. In C. Schaerf, B. Holden-Reid, and D. Carlton [eds.]. *New Technologies and the Arms Race,* pp.161-198. London: MacMillan.

Holdren, J. P. 1990. Energy in transition. *Scientific American,* **263** (September), pp.156-163.

Holdren, J. P. 1991. Population and the energy problem. *Population and Environment,* **12**, 231-255.

Hubbert, M. K. 1969. Energy resources. In: *National Research Council, Resources and Man.* pp. 157-241. San Francisco: W. H. Freeman.

Hughart, D. 1979. *Prospects for Traditional and Non-conventional Energy Sources in Developing Countries.* Washington, DC. World Bank.

IEA (International Energy Agency). 1989. *Electricity Conservation.* Paris: Organization for Economic Co-operation and Development.

IPCC (Intergovernmental Panel on Climate Change). 1990. *Climate Change: the IPCC Scientific Assessment.* Cambridge University Press, Cambridge.

Johansson, T. B. Bodlund, B. and Williams, R. H. 1989. *Electricity: Efficient End Use and New Generation Technologies and Their Planning Implications.* Sweden: Lund University Press.

Kelly, H. C. Blair, P. D. and Gibbons, J. H. 1989. Energy use and productivity: current trends and policy implications. *Annual Review of Energy,* **14**, 321-352.

Lashof, D. A. and Tirpak, D. A. [eds.]. 1989. *Policy Options for Stabilizing Global Climate.* Washington, DC: Environmental Protection Agency.

Lovins, A.B. and Lovins, L.H. 1991. Least-cost climatic stabilization. *Annual Review of Energy,* **16**, pp. 433-531.

MacDonald, G. J. 1990. The future of methane as an energy resource. *Annual Review of Energy,* **15**, pp.53-84.

Masters, C. D. Root, D. H. and Attanasi, E. D. 1990. World oil and gas resources – future production realities. *Annual Review of Energy,* **15**, pp. 23-51.

Mintzer, I. M. 1990. Energy, greenhouse gases, and climate change. *Annual Review of Energy,* **15**, pp. 513-550.

NAS (National Academy of Sciences). 1979. *Risks Associated with Nuclear Power: A Critical Review of the Literature. Committee on Science and Public Policy.* Washington, DC.

National Research Council, Committee on Nuclear and Alternative Energy Systems. 1980. *Energy in transition 1985-2010.* San Francisco: W. H. Freeman.

Office of Management and Budget. Various years. *Budget of the US Government. Washington,* DC: Government Printing Office.

Ogden, J. M. and Williams, R. H. 1989. *Solar Hydrogen: Moving Beyond Fossil-Fuels.* Washington, DC: World Resources Institute. October.

Okken, P., Swart, R. and Zwerver, S. (eds.) 1989. *Climate and*

Energy: The Feasibility of Controlling CO₂ Emissions. Dodrecht, Holland: Kluwer Academic Publishers.

OTA. (Office of Technology Assessment, U.S. Congress) 1991. *Energy in Developing Countries.* Washington, DC: Government Printing Office.

Pachauri, R. K. 1985. *The Political Economy of Global Energy.* Baltimore: Johns Hopkins University Press.

Pachauri, R. K. 1990. Energy, environment, and sustainable development. In: R. K. Pachauri and L. Srivaslana, (eds.) *Energy, Environment, Development,* New Delhi: Vikas Publishers.

Pachauri, R. K. 1991. Energy efficiency in developing countries: policy options and the poverty dilemma. Natural Resources Forum.

Population Reference Bureau. 1990. *1990 World Population Data Sheet.* New York

Schneider, S. H. 1989. *Global Warming.* San Francisco: Sierra Club Books.

Smith, K. R. 1987. *Biofuels, Air pollution, and Health.* New York: Plenum.

Smith, K. R. 1988. Rural air pollution. *Environment,* **30**, No. 10 (December), 17-34.

SERI (Solar Energy Research Institute). 1989. The potential of renewable energy. [Prepared jointly with the Idaho National Engineering Laboratory, the Los Alamos National Laboratory, the Oak Ridge National Laboratory, and the Sandia National Laboratories]. Golden, CO.

Sweet, W. 1984. *The Nuclear Age: Power, Proliferation, and the Arms Race.* Washington, DC: Congressional Quarterly.

WEC (World Energy Conference). 1983. *Energy 2000-2020: World Prospects and Regional Stresses.* London: Graham and Trotman.

Williams, R. H., Larson, E. D., and Ross, M. 1987. Materials, affluence, and industrial energy use. *Annual Review of Energy,* **12**, 99-144.

World Bank. 1980. *Energy in the Developing Countries.* Washington, DC.

World Bank. 1983. The Energy Transition in Developing Countries. Washington, DC.

World Bank. 1990. *World Development Report 1990.* New York: Oxford University Press.

Chapter 5: Health

S.K.D. Bergstrom and V. Ramalingaswami

EXECUTIVE SUMMARY

Four fundamental concepts constitute the foundation of this chapter. They are:

(a) Human development must be sustainable and environmentally sound. Otherwise it is not human development;

(b) Environmental degradation results from *both* the *per capita* use of resources (consumption) and the number of people using resources (population);

(c) The state of the environment is a major determinant of health and good health is a fundamental human right;

(d) Humanity must strike a moral and material balance between the needs and wants of individuals and the common good. And the fulcrum of this balance is the consumption of resources by the rich and the fulfilment of the basic needs of the poor.

1. SETTING THE STAGE

The essence of development is the application of human capabilities towards achieving freedom, social justice, health, education and equitable living standards for *all people.*

1.1 Health, environment and development – general considerations

Good health is a major pathway to development and is one of its principal products. Environment is a crucial determinant of health. However, environment is linked together with other factors such as population density, education, economic status, shelter, sociocultural characteristics, lifestyles and the availability of health technologies (Fig. 1). These factors nearly always interact to produce a positive or negative impact in health. This interdependency underscores the need for multisectoral approaches to health and development. Today's environmental concerns stem from the growing realization that there are limits to the capacity of the natural systems to generate and yield resources and to absorb wastes arising out of human endeavors. The preservation and enhancement of the natural systems are the debt that the present generation owes to the future. While many of the effects of environmental degradation are felt locally and regionally, there are others that exert their effects through winds and waters over long distances, indeed the entire planet. Within countries and across countries, it is the more affluent groups with their high consumption lifestyles that generally succeed in creating better environmental conditions for themselves with the least polluted environments, but whose activities in the process lead to environmental degradation felt largely by the poor and defenceless. The latter through their own activities for sheer survival add further burdens on the environment. It is said in economic terms that the rich get the loans and the poor get the debts.

1.2 The concept of health – a holistic view

The purely medical connotation of health is yielding place to a holistic view of health being regarded as a state of complete physical, mental and social well-being and not merely as a state of absence of disease (WHO, 1988). The Alma Ata declaration regards health as a fundamental human right and has set the goal of Health for All through Primary Health Care. Since the declaration was adopted by the nations of the world in 1978, the Member States are attempting to give practical expression to its attainment through universal coverage of the population with health care services provided according to need. The care provided has to be comprehensive including preventive, promotive, curative and rehabilitation elements. It has to be culturally acceptable, economically affordable and scientifically sound. Countries should be involved in the process in a spirit of self-reliance and thorough multisectoral approaches. Through this declaration the nations of the world had attempted to bring health into the political and economic arena, accepting the principle of equity and access to services and bringing health to the center stage of the development process with deep ethical overtones.

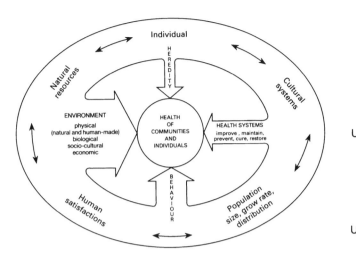

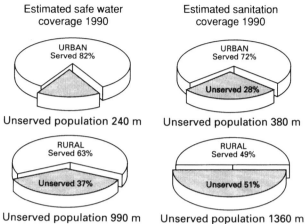

Figure 1: The width of the four . . . input-to-health arrows indicates . . . assumptions about the relative importance of the inputs to health. (Adapted from H.L. Blum, *Planning for Health*, Human Sciences Press, 1974, p3.)

Figure 2
[Source UNICEF estimates based on WHO data]

This view of health profoundly influences our approach to environment and enlarges the scope of environmental issues to be considered, not only to physical and biological but also social and psychological problems. Health for All implies the principle of improvement of health of all of the world's people, especially those underprivileged in developing and developed countries suffering from the worst effects of environmental degradation without further depletion of the world's environmental resources nor crossing the sink thresholds. This is the challenging task confronting science and technology well into the 21st Century. The challenge is to reduce inequalities and achieve equity in health and well-being and enhance access to services especially for the poor within the framework of sustainable development.

There is poverty and absolute poverty at the root of underdevelopment. Poverty expressed as income level cut-offs is inadequate. The criteria should be the extent of access to productive assets, to food and health promoting services and to education. A food and health security system is the need.

2. A HISTORICAL PERSPECTIVE AND THE UNFINISHED AGENDA

Historically speaking, the remarkable improvements in health that took place in the 19th and the first half of the 20th Centuries in the developed countries were due in large part to environmental improvements in the form of safe water, hygienic disposal of human wastes, prevention of food-borne infections and intoxications, personal hygiene, waste water and excess water management including clearing of marshes for vector control.

2.1 Population
No discussion of health and environment can be meaningful without a consideration of population growth rates and trends and their implications for sustainable environment and development.

Population growth is extensively treated in other contributions. Suffice it here to point out that we consider research and development on human reproduction and on better energy production systems as the two most important fields during the next half century. The prerequisite for a successful health development in a sustainable environment is decisive progress in bringing the world's population in equilibrium in respect to total numbers and *per capita* consumption as well as increased but safe energy supplies.

2.2 The epidemiological transition
It is the environment measures noted above applied at individual and whole population levels accompanied by education and improved nutrition that laid the foundation for the demographic, epidemiological and health transitions that took place in the developed countries. They were the products of rise in living standards which accompanied the Industrial Revolution. In the world of Lewis Thomas, these environmental measures are "complete technologies" which act through primary prevention. Each intervention, for example, clean water supplies or environmental sanitation, acts against multiple causal agents of disease. Unfortunately, developing countries have not been able to catch up with developed countries in the application of these classical environmental measures to their peoples even today, as Fig. 2 will show.

Environmental sanitation lags far behind safe water supplies, and rural areas are considerably behind urban areas. The international drinking water supply and sanitation decade of the 1980s has come and gone and with it has also gone its goal of 100% coverage of the entire population with safe water and sanitation. This is the unfinished agenda of safe water supply and environmental sanitation which needs to be addressed today with a sense of urgency. One is not waiting for a super-technology in this area. The decade of the 1980s has seen considerable progress; the technology is reasonably well developed: the India Mark II and the Tara hand pumps for example; the training is in place; community maintenance of installations is being achieved and yet a lot more needs to be done. It has been estimated that the implementation rates of the 1980s for urban water and sanitation and for rural water would have to be increased by two and a half times and that for rural sanitation about forty four times during the 1990s if 100% coverage is to be attained by the year 2000 (UN Children's Fund, 1989).

2.3 Vector-borne diseases and environmental control

Many vector-borne parasitic diseases are still widely prevalent especially in tropical areas despite progress in the discovery and use of insecticides and specific chemotherapies. The dream of the mid 1950s that tropical parasitic diseases would soon be brought under control is yet to materialize. Vector resistance to chemical insecticides and parasite resistance to chemotherapies had adversely affected the control programs coupled with management failures, cost escalations following oil price rises and failure to recognize the importance of community participation. Technology yields full benefits only when used within a framework of social development and community organization (WHO Press, 1991). The Tropical Diseases Research Programme of the World Health Organization is currently engaged in improving existing tools of diagnosis, treatment and prevention and discovering new ones and has already achieved substantial progress with great potential for the application of new tools in endemic countries. Through basic studies into the nature of immune responses to parasites, identification of new antigens and molecular genetics, new vaccine candidates against parasitic diseases are being developed. To overcome vector resistance to chemical insecticides, resources are being put into the development of biological control agents. The extensive use of *Bacillus thuringiensis* in the onchocerciasis (river blindness) control program has demonstrated that it is ecologically superior to conventional chemical insecticides (UNDP/WB/WHO, 1989). There is also a growing realization that the strategies of the pre-DDT era for environmental improvement aimed at control of vector proliferation at source sites need to be revived.

Environmental methods may prove to be effective in many parts of the developing world in combating malaria, filariasis, dengue and the general nuisance of mosquitoes and flies. Several demonstration projects recently carried out, for example, in India, using environmental methods with or without biological control agents give support to this approach which needs to be pursued with a sense of urgency and determination.

Meanwhile, the toll of death and disability from tropical diseases continues. Malaria is worsening or barely being held in check in many parts of the world according to the latest information released by the World Health Organization (WHO). Two billion people or two-fifths of the world's population are still exposed to malaria; 100 million clinical cases occur each year in Africa; there are at least one million deaths largely in children and mostly in Africa that are still taking place annually. The malaria situation is deteriorating in all frontier areas of economic development such as agriculture and mining in newly opened jungles and in areas of war, conflict, illegal trading and migration of refugees. In Brazilian Amazon, falciparum malaria is claiming the lives of some six to ten thousand people a year mainly young adults working in development projects. It is clear that malaria cannot be controlled by major campaign approaches but through an approach based on an understanding of local epidemiologic factors especially social, economic and ecological situations that encourage malaria and through designed specific strategies to address each ecological setting.

2.4 Health effects of developmental activities

Developing countries are attempting to overcome in a few decades their economic backwardness and social disabilities which took a longer time for developed countries to overcome. Well-intentioned development programs may not foresee the adverse health effects that might ensue in-short or long-term as a result leading to success in one domain coupled with setbacks in the other. For example, provision of water to rural areas for agriculture and human use is an essential part of development and must be pursued, but if the specific projects such as construction of dams, the digging of irrigation canals, the laying of railroads and highways are not accompanied by a water management system that takes potential breeding places of disease vectors into account, the benefits of material progress may be off-set by the adverse health consequences. Man-made malaria is all too real. In addition to problems of seepage, silting and salination that major dams and irrigation schemes may entail, there is the danger of encouraging the introduction, revival or spread of agricultural pests and diseases, of changing the insect pest complex especially when drainage and water management systems are not instituted simultaneously with the development projects. There are innumerable examples of the health consequences of labour migrating from non-endemic to endemic areas as the Brazilian Amazon malaria outbreak illustrates or vice versa from endemic to non-endemic areas introducing new infections into an area where ecological conditions for vector proliferation are conducive. Lymphatic filariasis is spreading in India on account of labour migration and unplanned urbanization. The periodic outbreaks of Japanese encephalitis that the region of South-East Asia is experiencing today is largely an expression of changes in ecology.

3.   SPECIFIC ISSUES

3.1   Health and the global climatic change

There is deep concern about the effects of climatic change, ozone shield depletion and ultraviolet radiation in the 300 nm range and global warming which will receive detailed consideration in other sections of the report. How these global events will affect health of the inhabitants of this planet is not all too clear although it is possible to visualise its impact on the under-nourished, impoverished and disabled populations of developing countries as a recent Study Group convened by the WHO indicated. The impact on these populations is expected to be a major one because they have no reserves and no effective adaptive responses. Cataracts which are already known to occur in younger age groups in developing country populations may increase as well as the incidence of many non-melanotic skin cancers. Heat illnesses may be accentuated and humidity increases may aggravate the spread of droplet infections. Dry, arid and cyclone prone areas may be worse off with adverse impact on food production systems, accentuating protein energy malnutrition in children. In addition, as a result of changes in the eco system, temperature and humidity and large water collections vector born diseases may increase. As already mentioned, two billion people are at risk of malaria, 900 million at risk of lymphatic filariasis and 600 million at risk of schistosomiasis.

3.2   Health and the work environment

Developing countries are naturally anxious to achieve rapid economic growth in a catch-up effort through industrialization. In the process, however, many of them are repeating the history of the earlier industrial revolution that occurred in developed countries in the form of health hazards in the internal work place environment and outside. On the basis of the earlier experiences of developed countries, a systematic corpus of knowledge exists in the sectors of industrial and occupational health which could guide industrial policies and practices of developing countries and avoid the mistakes that they had made. This is, however, not often the case and the scene of

syndromes, deaths and disabilities of industrialization of the past era are now being enacted on the stage of the developing world. It is often the case that regulations for the health protection of the work force have been enacted but nearly as often are not implemented fully and effectively. This is a vast area awaiting determined action.

The Bhopal disaster is a reminder of the need for policies that ensure location of hazardous industry away from congested human dwellings. Five years after a nuclear reactor in the small town of Chernobyl in the Ukrainian SSR exploded and burned, the long-term effects from this nuclear disaster still hover over the planet like a big question mark. The World Health Organization in collaboration with the Soviet Union is establishing an international program to study and deal with the long-term impact of the Chernobyl explosion.

3.3   The dilemmas of urban health

By the year 2000, a majority of the world's population will live in large cities or towns and the urban environment will play a major part in determining the health of most urban dwellers. 80% of urban health problems are connected with the environment, such as water supplies, safe disposal of human wastes, control of vectors and their breeding sites, control of rodents and atmospheric and noise pollution. The essential problem here is the extensive breakdown of infrastructure and services in slum areas creating conditions conducive to the resurgence of diseases that are believed to have been brought under control long since. An epidemic of cholera, one of the oldest diseases of mankind, is sweeping rapidly through parts of Latin America today causing in its wake death, suffering and economic dislocation, especially among the urban and rural poor. Until January of this year, this ancient scourge was virtually unknown in the Americas over the past 100 years. The current outbreak of cholera in Latin America is clearly the result of breakdown in health infrastructure due to lack of clean water and sanitation and safe food. And surely cholera is neither the last nor the least important of diseases to so behave. Crowding and microbial flourishing lie at the root of today's urban health crisis.

The growth of cities is no doubt a fundamental accompaniment of the process of economic growth. Cities have a dual structure, one which is modern and the other which is semi-rural or slum where most of the poor live. They have a dual pattern with the well-to-do showing disease and life expectancy pattern akin to that of the industrialized countries while the poor living in slums show higher morbidity and mortality rates due to past patterns of infectious disease, malnutrition and rapid population growth not yet conquered.

The problem of urban health is one of overwhelming urgency and perhaps there is not much time left to correct the prevailing deficiencies and hazards. Decentralization and strengthening of municipal structures is an obvious priority. The provision of safe drinking water and waste disposal, increasing the income of the poor, provision of sustainable health services with family planning as the center piece and large scale of mobilization of people themselves in the process of ascertaining their needs and structuring services are elements of a strategy that is urgently required. A balance between urban population and infrastructure must be ensured and urban and rural development policies must be so adjusted as to provide incentives for the public, industry, the private sector and government agencies to prevent excessive rural to urban migration. Leadership, organization and management are essential to bring about change. In the provision of services cost sharing may have to be resorted to, ensuring in the process that the fee for service pattern does not mitigate against the poor and the indigent. The key elements for a strategy for dealing with the urban health crisis and social breakdown are a stronger political commitment, better management, more resources mobilized from multiple sources and involvement of the urban communities. Building local capabilities to provide the services needed and developing criteria for urban health improvement and developing new and cost effective technologies are relevant to the current dilemmas of urban health. This topic is of such urgency that it formed the basis for technical discussions this year at the World Health Assembly which passed a resolution on the subject and some of the strands in the discussions are presented above.

4. CONCLUSION

4.1 The double burden of developing countries
In any discussion on health and development the situation of developing countries needs special attention. Developing countries are not a homogeneous entity; they are in different stages of development and are moving at varying rates. This holds true both between developing countries and within these countries. But they all carry both the burden of diseases arising out of malnutrition — infection complex aggravated by population pressure on the one hand and the evolving burden of diseases associated with increasing affluence. These diseases of affluence take up a disproportionate part of available resources.

At this stage in human history one must endeavour to find a balance between the biosphere of man's heritage and the technosphere of his creation (Ramalingaswami, 1981). Social and economic development are not two separate issues but are integral parts of the development process. Rise in GNP must be accompanied by a rise in those components that enhance the quality of life for all. What is envisaged is a trajectory of development which takes into account social and environmental issues as they affect all strata of the social systems.

4.2 Health for all and sustainable environment
The crucial question is whether the Health for All concept with its comprehensiveness, equity and access issues and development orientation is compatible with sustainable environment and development. We believe that this is possible and is indeed an imperative for the future. The phenomenon of significant overall economic growth on a national scale going hand in hand with accentuation of disparities between the affluent and the poor is no longer acceptable. Development must meet the basic needs of the poorest sections of society and it is futile to conceive of the health system functioning outside a developmental process.

Integrated programs of social and economic development make health services more effective even at low levels of economic growth. A health program and an economic program that are complementary will have a synergistic effect on human development. The mutual interaction between health and development forms the foundation of the Primary Health Care approach for the achievement of the goal of Health for All. Given favorable social and political factors, even comparatively limited economic growth can lead to perceptible improvements in the level of health. Improved standards of living as reflected in improved nutrition, education, healthy environment and fulfilment of basic needs are essential to a sustained improvement in a population's health and well-being. Much can be achieved by the incorporation within the health infrastructure of technologies for the control of specific diseases keeping the Primary Health Care approach as the basis. Tools of molecular epidemiology are being developed for early recognition of disease. They will also be immensely useful in surveillance of populations for health and disease.

4.3 The cost of new technologies
The cost factor of new health technologies effectively excludes their benefits from reaching the masses in developing countries.

In the remaining part of this century and well into the 21st Century, a problem of deep concern will be how to ensure that the fruits of science and technology are available to those in greatest need and especially to those that are unable to purchase those technologies. There is an important role, humanistic in nature, that the pharmaceutical industry, the governments of developed and developing countries, the international agencies including the UN system and the voluntary sector have to play together in resolving this issue.

4.4 The human factor in development
There is a growing consensus that people should be

positioned at the heart of development and that development plans must start with the human balance sheet rather than with macro-aggregates of GNP. Distribution objectives should be given as much emphasis as production objectives. Alongside Health for All is concern for people and equity. There is also developing a sense of unity in environment, economics and development thinking. There is growing support for preserving the ecological tapestry of our planet and an urge for environmentally sustainable economic systems. Science and technology have an important role to play in ensuring that the goal of Health for All is compatible with a sustainable environment and development.

REFERENCES

Ramalingaswami, V. 1981. Health aspects of developmental activities. *World Health Forum*, **2** (4), 563-566.

Ramalingaswami, V. 1990. Child health in the changing global economic and environmental situation — global responsibility. *International Child Health*, **1** (2), 69-74.

United Nations Children's Fund. 1989. Towards the 1990s: The water and sanitation sector workplan for 1990-1995. New York, UNICEF, WET/567/89.

United Nations Development Programme/World Bank/World Health Organization 1989. Tropical Diseases Progress in Research 1989-1990, Special Programme for Research and Training in Tropical Diseases. Tenth Program Report. pp. 135.

United Nations Development Programme, 1990. *Human Development Report*, Oxford University Press, Oxford. pp. 189

World Health Organization. 1988. From Alma-Ata to the year 2000, Reflections at the Midpoint. Geneva.

World Health Organization Press. 1991. World Health Assembly says make cities livable again. WHA/11.

World Health Organization Press. 1981. Malaria worsens in many areas. WHA/6.

II. Scientific Understanding of the Earth System

Introduction

J.W.M. la Rivière

For an analysis of the problems of environment and development (*Section I*) it is necessary to investigate the Earth System and how it operates so as to arrive at an assessment of its carrying capacity as well as its resistance to disturbances. This has become necessary because the development process is already being slowed down in many areas of the world by the constraints the global life-support system imposes on it. Similarly the best available knowledge about the Earth System is required as a basis for the formulation of policies and strategies for response (*Section III*) by prevention, mitigation and adaptation.

The physical and biotic components of the *Earth System* have evolved together in continuous interaction towards its present state of complexity. Over the past few decades scientific work has established that man's activities have caused *abrupt and unprecedented modifications* in the planetary life-support system. These signals are now leading to a transition from a blissful but unfounded state of certainty to a more realistic and sobering state of uncertainty about the future capacity of the planet to sustain human life of reasonable quality to all.

Thus a *new mandate for science* in these areas is being generated: what precisely constitutes this planetary life-support system? How does it work? To what extent are changes-reversible and irreversible-taking place? Which of these are dangerous? What could be their effects and on what time-scales? Can they be halted, reversed, ameliorated or adapted to? If so, how? This mandate presupposes that what man can damage, man can also repair, or in other words, that man, to some extent, will be capable of managing the Earth System to his own advantage.

Systematic investigation on a global scale has only recently become feasible. This section formulates the steps that need to be taken to develop gradually an understanding of the Earth System that can serve as basis for prediction, impact assessment and policy option design. This is being done, of course, on the basis of what is already known and also use is being made of the current stage of planning and priority setting by the relevant major international programs that are already underway.

Investigating the Earth System requires a *pragmatic separation in its component parts*: the atmosphere (Chapter 7), the marine (Chapter 8) and the terrestrial (Chapter 9) compartments. These are connected by fluxes of matter, i.e. the hydrological and the biogeochemical cycles (Chapter 6). The biotic part of the Earth System is dealt with in Chapter 11. Because of its special significance, a separate chapter has been devoted to freshwater resources (Chapter 10) which are part of the terrestrial compartment.

It should be stressed that the Earth System is in principle *one and indivisible* because all parts are interconnected by delicate control mechanisms operating on various space and time scales. Hence no part of the system can be understood in isolation. Therefore, while expediency dictates some degree of separation a *continuous focus on the interconnections* is essential if eventually under-standing of the functioning of the Earth System is to be achieved.

The framework of this Section obviously *does not permit comprehensive treatment* of the Earth System in all its aspects. Thus the human species was excluded from the treatment of the biota in Chapter 11 and dealt with in Chapter 1 in Section I. Other aspects could not be treated at all such as the relevant parts of the solar–terrestrial system, the deep ocean and the arctic and antarctic regions.

The six *chapters* that follow have been finalized by the authors on the basis of the discussions in the working groups which thus provided an intensive peer review of the original drafts for discussion sent to the working group participants before the meeting.

The *recommendations* developed in this Section are to be found in each of its six chapters; some of them also appear in the Conference Statement. Additional recommendations emanating from the working groups and the plenary sessions will be presented in this introduction. Attention should be drawn to the UNCED chapter for Agenda 21 on Science for Sustainable Development the draft of which is in general harmony with the ASCEND conclusions. This also holds for other, so called "sectoral" UNCED papers on more specific subjects like, biodiversity, marine systems, atmosphere, etc.

Conclusions that were *common* to almost all chapters were the following:

(a) proper attention to *time and space* scales; for practical purposes local and regional scales are particularly important while often most difficult to deal with scientifically.

(b) there is a strong need for *educating* scientists in the areas treated in this section, specialists as well as scientists capable of interdisciplinary research. The concern was expressed about a "brain drain" towards environmental studies at the expense of the pursuit of other scientific disciplines essential for the solution of environmental and developmental problems. The problems are most acute in developing countries, where major efforts for differentiated capacity building are called for as discussed in more detail in Chapter 14 of this volume.

(c) in many cases *co-operation between natural and social scientists* is required. It was stressed that this could best be done at the working level in concrete projects where the objectives clearly cannot be reached without such co-operation.

(d) intensified *co-operation with scientists in the Third World* was strongly advocated and its connection with the problems of capacity building recognized.

(e) in view of the *increased interests of governments* in Earth System Research the scientific community should develop effective mechanisms for clearly explaining their findings to policy-makers and for interacting with them.

(f) the scientific community should promote a better *understanding and appreciation* of research in general and Earth System research in particular among the *general public* at all levels.

In all chapters the view was expressed that for most purposes *adequate research programs are in place* and that strengthened support to existing programs rather than new initiatives or institutional arrangements were needed. An exception was made for the theme *Biodiversity*, for which National Biodiversity Institutes were recommended.

In most of the working group discussions the additions and amendments brought forward were incorporated in the revised texts of the papers. In the case of the chapter on *Freshwater Resources*, the working group realized that the subject is vast and complex, embracing scientific as well as

a considerable number of socio-economic and management aspects while at the same time it was of direct special significance to the development issues of the day. The working group was of the opinion that the constraints of space and the mandate for the Conference paper had made it difficult for the authors to do full justice to all of these aspects. In particular, the working group considered that in the absence of these constraints, more attention could have been given to such topics as: the importance of wetlands, reservoir management, water supply and waste water management in megacities, over- and under-exploitation as well as protection of ground water, potential impact of climate change on water availability on regional and local scales, improved water use efficiency. The authors as well as a number of working group participants will have had the opportunity to ensure that the contents of the paper as well as the above topics receive due attention in the International Conference on Water and the Environment (Dublin, January 1992), specifically devoted to the area of freshwater resources.

In the discussions of the chapter on Atmosphere and Climate it was emphasized that after doubling of the CO_2 concentration in the atmosphere, global warming would not stop if emissions continued. Furthermore, the point was made that more research is needed on impact assessments, techniques and methodologies, so as to clarify and quantify impacts.

Mention was made of the *rapid evolution* the *conduct of science* is now undergoing. Not only increased interaction with governments, industry and the public at large and a tendency towards reconvergence of the sciences (interdisciplinarity) are playing their role, it is also the rapid improvements in telecommunication, in data and information handling and in modelling that are changing the scientific endeavour more rapidly than ever before. Symptoms of this are the information explosion in the form of the growing number of publications and journals as well as the proliferation of scientific meetings large and small. It was felt that the scientific community should consider how to cope most effectively with these developments.

Although the discussions at the ASCEND Conference of the themes in this Section were devoted to the design of strategies for achieving an understanding of the Earth System, it can already be concluded that sustainable use of the planetary life-support system will require a rapid increase in the efficiency of the use of materials and energy sources.

Chapter 6: Global Cycles

J.W.B. Stewart, R.L. Victoria and G. Wolman

EXECUTIVE SUMMARY

Human influences are changing the amounts and manner in which elements cycle, are stored and interact. These changes in fluxes and sinks are observed at global, local and regional levels through increases in atmospheric CO_2, CH_4, Dimethyl sulphide (DMS), NO_x and CO, ozone depletion, acid precipitation, erosional processes, sediment transport, etc. Many of these changes threaten life on Earth in an unprecedented way. In this review of the hydrologic and biogeochemical global cycles, we outline the importance of terrestrial, atmospheric and oceanic processes and their interactions on the global cycles of carbon, nitrogen, sulfur, phosphorus and other elements. We stress that we have to understand these processes and interactions fully if we are to reverse, manage or mollify any of these effects as important components of global balances in maintaining atmospheric composition, sustaining food production or the quality of land, freshwater and sea resources:

(a) Important facts

(i) Trace gases in the atmosphere have increased steadily during the past 100 years and it is known that these gases in association with clouds regulate the radiation balance of the atmosphere. The important gases include CO_2, CH_4, DMS, NO_x and CO.

(ii) The fluxes and pools in the major cycles of elements are roughly but not precisely known. In the case of carbon, natural exchanges between air-sea and air–biosphere mask the relatively small anthropogenic inputs from fossil fuel and from land use changes. These anthropogenic additions show up in the atmosphere and in the oceans. However, it is not possible at present to fully document where all the carbon is stored – "missing carbon sinks", nor is it possible to accurately describe all the processes governing gaseous fluxes.

(iii) Water plays a central role in transportation of nutrients from land to sea and in the biogeochemical cycles. We stress that only a small fraction of the hydrologic cycle (less than 1%) contributes to the freshwater that is available for humans. The hydrologic cycle is of prime importance for this role alone, although its influence on transportation of sediments and nutrients must not be overlooked. Water also has a very important role in energy transport in oceans (cf. Chapter 7).

(iv) Interfaces between land, ocean and atmosphere are important zones in which to study cycles and feedback processes. Studies of the biogeochemical and hydrologic cycles in small catchments to large rivers provide a means of understanding spatial and temporal scales that will be required as we move to regional and global scales.

(b) Research needs

(i) Despite progress which has been made, specific gaps in knowledge remain which must be addressed. Thus the global sinks and fluxes of carbon, and other important biogeochemical elements (and their interactions) have to be assessed if we are going to use predictive modeling to assist decisions that impinge on energy use, land use, population pressure, acid rain, climate change and other environmental factors.

(ii) Both process studies and controls on processes will be required as will the collection of data at specific spots to verify model predictions as well as quantifying elemental sinks.

(iii) It will be necessary to maintain the continuity of comprehensive, coherent scientific studies and networks for observations of biogeochemical cycles and hydrologic cycles at global, regional and local scales.

(iv) Mechanisms must be developed for enhancing the cadre of talented environmental scientists, particularly in the developing countries, using communication networks of individuals, academic and other research centers in studies of major cycles related to coherent research efforts.

(v) Studies will have to be prioritized so as to target the most sensitive areas first. We anticipate that many will be concentrated on the interface between land and water, or oceans and atmosphere involving water, nutrient and other substance fluxes.

(c) Urgent actions

We recommend that:

(i) Integrated studies of the processes, controls and sinks associated with global elemental cycles along with simulation model predictions should be the tools that are used to complete our understanding of specific parts of the biogeochemical cycles. The most urgent problems should be targeted first such as the missing carbon sinks and disturbed fluxes of CO_2 and CH_4. This allows us to address problems associated with anthropogenic loading of the atmosphere through fossil fuels or through land use changes. Gaps in knowledge in the biogeochemical and hydrologic cycles and interactions must be addressed.

(ii) Integrative work is needed on key areas such as the interfaces between land, ocean, atmosphere and biosphere as well as the pertinent feedback processes. Studies of the biogeochemical and hydrologic cycles in small catchments as well as in large river basins provide the means of understanding spatial as well as temporal scales. These activities will allow us to address some of the spatial and temporal scales needed in moving to comprehension of global cycles.

(iii) Programs, which strengthen international co-operation, are the essential tools to achieve recommendations (i) and (ii). Such programs are among others:

- Research oriented programs (IGBP, including JGOFS, GCTE, and START, WCRP, HDGEC),
- Monitoring programs (GCOS, GEMS/GRID, WDC), and
- Assessment programs (SCOPE, SCOR, IPCC).

1. INTRODUCTION

The Earth System constitutes a web of interacting cycles involving atmosphere, biosphere, geosphere and hydrosphere. Together they form the Earth System that has evolved towards its present state of high complexity in continuous interaction with the biota that inhabit it. Each system includes transport and loci of storage. It is within these complex systems, land, oceans and atmosphere, that human activities manipulate and modify processes and constituents participating in a myriad of interacting systems. We know that the collective integrated mass of biota on Earth are the prime pumps in the major biogeochemical cycles which are a vital part of the planet's life-support system but we do not know how this role is affected by human activities, especially atmospheric C species changes or the rapid replacement of many species by the increasing human population. It is likely that the ultimate impact of human interventions in these cycles will not be adequately deduced until the global processes, as manifested in the steady flow of elements and water in the atmosphere, oceans and other aquatic systems, vegetation and soil, are understood. In this chapter we review progress in understanding some of the major elemental interactions starting with the hydrologic cycle as it is involved as a controller of biological processes as well as being a dominant factor in transport phenomena. We judge that the interactions between carbon, nitrogen, phosphorus and sulfur are important to an understanding of gaseous fluxes and therefore have concentrated mainly on them. It is realized that the processes and interrelationships among them are very important, and that information obtained at a small scale will be important as one attempts to synthesize information on regional and global scales.

One must at this stage distinguish between natural cycling in systems and anthropogenic effects, e.g. those induced by changing land use, fossil fuel use, urban pollution, etc. Cumulative data collected over time but especially since the start of the industrial revolution suggests that these effects are upsetting natural elemental balances and the cycling on nutrients and water in an irreversible manner, with potentially disastrous global effects (cf. Chapter 7, Atmosphere and Climate, this volume). We think that our ability to make any major progress towards reversing these effects lies in understanding the major cycles and will depend on a knowledge of elemental cycling processes in combination with a modeling approach (backed by adequate observational data collection) to verify hypotheses. Anthropogenic disturbance of global cycles such as increase of atmospheric CO_2 or CH_4, or acid precipitation will be observed at local and regional levels but understanding of global cycles will be fundamental to understanding causes and recommending remedial action.

2. THE HYDROLOGIC CYCLE

Water is essential to life. Water is the mediator of a variety of physical and chemical processes, the transformation of water from one state to another demands or releases large quantities of energy; the movement of water results thus in the transport of material as well as in the large-scale transport of energy (cf. Chapter 7, Atmosphere and Climate; Chapter 8, Marine and Coastal Systems, and Chapter 10, Freshwater Resources). Liquid water and water vapor are major determinants of climate. Moreover, biological processes in all organisms are controlled or mediated by water and water is a peculiar, nearly universal, solvent. A broad outline of the hydrologic cycle is as follows. Water moves from the oceans to the atmosphere to the land surface returning either to the ocean through runoff or to the atmosphere by evaporation or transpiration. The bulk of precipitation, following evaporation from the ocean, actually falls back upon the oceans.

Transport and disposition of water as liquid and vapor are determined by the radiation balance of the globe, driven by energy from the sun in patterns determined by differential heating at the Equator and the Poles and by the rotation of the Earth. Water is transported in the ocean and atmosphere. Climate is controlled by the dynamics of these global processes (cf. Chapter 7, Atmosphere and Climate). On a broad scale the dynamics of atmosphere can be modeled by the rapidly evolving field of global circulation models.

For the Earth as a whole, the volumes of water and the residence times within stored volumes are roughly but not precisely known. It is important to note that, because the oceans comprise 94% of water on Earth, the volume of freshwater is, in fact, exceedingly small. Moreover, the volume of freshwater customarily drawn upon for human activities, that is, the water in rivers and lakes constitutes less than a fraction of 1% of all available water. Knowledge of the temporal behavior of these freshwater stores is thus vitally important in human endeavors, both from the standpoint of direct use and from the standpoint of their influence on climate and the global water balance.

By far the largest fraction of water that falls on the continents in the form of precipitation is derived from oceanic sources. The amount of water recycled on the continents is, however, highly variable although in the interior of the North American continent roughly 80% to 90% of the precipitation derives directly from maritime air masses. On the other hand, some studies in the Amazon have indicated that 50% or more of the precipitation in the central part of the basin derives from evapotranspiration of the forest (Salati *et al.*, 1979).

2.1 Research needs and issues

While the broad outline of the hydrologic cycle in various parts of the globe is known in some detail, the ability to predict the behavior of the system under a variety of climatic and management assumptions remains uncertain. Given the rapidly increasing impact and interaction of

human activities with the hydrologic cycle, knowledge of processes within the cycle becomes increasingly important (NAS, 1992).

2.1.1 Land surface–atmosphere–ocean inter-actions

Vital interactions in the hydrologic cycle occurring at the interfaces between the land, ocean surfaces and the atmosphere exercise major climatic controls . Thus in modeling the potential climate impact of global warming induced by greenhouse gases, it is essential to know whether warming of the oceans will result in additional cloud cover which, in turn, will shield or reduce the radiative input to the lower atmosphere and Earth surface. Quantitative estimates of such consequences are fundamental to evaluating the impact of potential climate change. Similarly, because models of the general circulation currently have large spatial grids measuring hundreds of kilometers along degrees of latitude and longitude, characterizing the hydrology of soil moisture and vegetation to delineate exchange with the atmosphere in general circulation is critical to evaluating and modeling the impact of climate change. This characterization of both soil moisture and vegetation at different scales as noted above, is essential to better definition of the interactive processes of moisture exchange from the land, the biosphere and the atmosphere. It is essential that the dynamics of energy and moisture exchange processes be quantified in order to relate major features of the landscape and vegetation to general circulation models. Currently the Global Energy and Water Cycle Experiment (GEWEX) of the World Climate Research Programme (WCRP) is focussed on these problems.

While not an interfacial phenomenon, a major feature of the hydrologic cycle still inadequately quantified is the flux of energy and moisture between the oceans and the continents. Better information on the magnitude, timing and transformations involved in the flux of energy between oceans and continents is essential. These are being researched by components of the World Climate Research Programme; specifically the World Ocean Circulation Experiment (WOCE) and the Joint Global Ocean Flux Study (JGOFS) of the International Geosphere-Biosphere Programme have been designed to address these issues.

Much of the discussion about climatic change concerns the impact of human activities on vegetation as well as on atmospheric chemistry. Throughout history the major alteration of the terrestrial system by human beings has been through the initiation and expansion of farming and grazing. These have resulted in wholesale changes in land cover. Such changes in cover can have a major impact on the albedo of the land surface and in turn on the regional climate (cf. Chapter 9, Terrestrial Systems). Analyses of the margin of the southern Sahara in Africa indicate that changes in vegetation may induce feedback mechanisms which intensify aridity promoting further loss of vegetation and in turn more aridity.

At a continental scale, recent studies of the Amazon basin indicate the important relationship between forest cover and the dynamics of atmospheric processes and climate in the region. Some models suggest that, with complete substitution of grassland for forest, a conceivable result of progressive destruction of tropical forests, evapotranspiration and to a greater degree precipitation would be reduced resulting in reduced runoff.

2.1.2 Biogeochemistry in the hydrologic cycle

Water is the transporter of major chemical and biological materials throughout the hydrologic system. Though the medium of water biota are provided with nutrients, nutrients are removed from specific systems, and elements are made available to the biota and to the physical environment. The interaction and transformation of chemical and biological constituents in water constitute one of the most important features of the hydrologic system. Innumerable interactions with human activities alter the biogeochemical cycles at a local scale and at the global level (as discussed in more detail later).

The need for fundamental information about the chemistry and biology of water systems occurs at several different scales. The climate system itself is influenced by chemical constituents which interact with water in the atmosphere resulting in major transformation of the chemistry of the atmosphere. The greenhouse gas–climate change phenomena represent a single example of the interaction of chemistry, hydrology and climate, as does the transformation of sulfur and nitrous oxides in the atmosphere which result in acid deposition through both wet and dry precipitation.

Synthetic organic products are now ubiquitous on the globe. The chemical and biological reactions which may produce degradation products within the natural environment all take place in the presence of water. The ability to predict degradation in soil and water is needed to evaluate the capacity of the globe to absorb and transform residuals from human industrial and agricultural activities. The application of pesticides and herbicides in agriculture can result in long-term damage to soil and water unless techniques are developed to assure that such useful additives are biodegradable by design or through natural processes in the absence of design. The intimate relationship between the flow of water, its composition and chemical and biological transformations which take place within it must be better understood at all scales if we are to avoid contaminating the Earth with a myriad of materials of great longevity.

Because virtually all river systems carry both sediment and water, detailed studies of the type started on major world rivers (cf SCOPE 42) are needed of the source, transport, storage and fate of sediments and associated contaminants in a variety of biogeochemical environments. In contrast to the knowledge of the chemistry of the water column, the mechanics of chemical and biological degradation which alter the composition of constituents adsorbed on sediments are poorly understood. Evaluation of levels of quality deterioration in water bodies, as well as estimates of the response to water quality improvement programs are dependent on this understanding.

3. BIOGEOCHEMICAL CYCLES

3.1 Carbon

Studies on the amounts of trace gases in the atmosphere, notably carbon dioxide (CO_2), carbon monoxide (CO), methane (CH_4) and nitrous oxide (N_2O) are of comparatively recent origin (Scope 29). It is not surprising therefore that despite some excellent research, particularly in the last two decades, the carbon cycle remains poorly understood on a global scale. There is much integrated research into global oceanic carbon flux and some excellent studies are underway but these must be expanded and supported (JGOFS, SCOR, SCOPE, cf IGBP 1990), and much but unintegrated research into terrestrial fluxes (Mooney, 1991) but the global carbon cycle is still a mystery; even the relative importance of oceanic and terrestrial processes is still a topic of fierce controversy (Tans *et al.*, 1990).

It is essential that the carbon cycle be understood in fine detail. A high priority of future and on-going research should be to quantify fluxes, sources and sinks of CO_2, CH_4 and CO on a worldwide grid including both continents and oceans, and to quantify carbon transfers at various levels in the oceans and air. Fluxes of these gases are not increasing at similar rates; carbon dioxide has increased by 25% over approximately 150 years whereas methane concentrations have increased at twice this rate. Yet both these fluxes, although very important as forcing greenhouse gases for different reasons, are small in comparison to natural cycling. The exchanges of carbon between the ocean and atmosphere are many times larger than the anthropogenic emissions. Thus it becomes important to understand the fluxes of these gases to and from sinks to be able to predict future trends.

It has already been stated that the ocean plays an important role in the climate system through its storage and transport of heat. It also has an important role as a source and a sink of biogenic gases that have radiative properties in the atmosphere. The JGOFS core project to IGBP, started in 1989, has been designed to investigate the oceanic biogeochemical processes relating to the cycle of carbon in the ocean and to assess the capacity of the ocean and its circulation in the uptake of CO_2 produced from natural and anthropogenic processes such as the burning of fossil fuels. This uptake occurs by both physical and biological processes. Neither is well quantified at the global scale, and the regulation of the biological processes is poorly understood. In particular, the biological processes responsible for the long-term storage of a portion of the total primary production cannot at this time be resolved sufficiently in time and space to say how they might be affected by climatic change (IPCC,1990).

3.1.1 Research needs and issues
Specific problems and issues can be identified concerning each carbon gas.

Carbon dioxide sources and sinks are still not understood, and observational constraints are as yet inadequate to prove satisfactorily exactly how the carbon is transferred. The major fluxes in the system are from the forests, the grasslands, tundra, wetlands, and oceanic photic zones; the short-term stores in air, trees, soils, peats and shallow oceanic levels; the longer-term stores in deep water and sediments, and in fossil carbon and carbonate. The transfers between these systems are very poorly known. In particular, attempts to link short-term carbon transfers with general circulation models are only at a rudimentary stage. Programs, which are of relatively recent origin, have been designed to provide needed data and are described in the GCTE, JGOFS, and IGAC core programs of IGBP.

Methane sources and sinks remain very poorly characterized (Schimel and Andreae, 1989; Crutzen, 1991). The major sources are boreal wetland, rice fields, tropical wetland, ruminants, termites, fires, etc., but the individual fluxes are not accurately known, and there is much debate about the relative importance of tropical and northern wetland. The causes of methane seasonality are not understood, nor are the full reasons behind the relationship between latitude and mixing ratio. Sinks include OH, which is not yet properly monitored (Vaghjiani and Ravishankara, 1991), soils, and loss upward to the stratosphere: none is fully quantified. Finally, there are massive stores of methane in gas hydrates, in the Arctic and in marine sediment. There is a risk this will escape if tundra landscapes are to become warmer, but we do not know how great that risk is.

Carbon monoxide is little studied, in part because of the problems in keeping it in flask samples. Yet it is an important gas, both in its own right and in the methane cycle. Sources (e.g. fires, cars, industry) and sinks need to be quantified.

All three gases have been discussed in terms of sources and sinks. However, to understand their cycling properly, simple atmospheric monitoring is inadequate. What is needed is a full characterization of the biological processes involved in the fluxing — everything from the transfer by plankton to the release of methane from northern wetlands- and linkage of this information at appropriate spatial resolution, to a global climate model.

3.1.2 Needs for observation
There exists a rudimentary global atmospheric monitoring program for the study of CO_2, limited monitoring of CH_4 and some stations studying CO. These programs need substantial upgrading. At present, for example, the only intercalibrated CH_4 monitoring net is from the South Pole to Alaska, via Pacific islands; little data is available from the Atlantic and Indian oceans, and from the major landmasses. Furthermore, isotopic monitoring of carbon in carbon gases, to study seasonal variation and fossil input, is very limited despite the importance of this data in pinning down sources (e.g. in studying methane hydrate release with ^{14}C). A high priority is setting up a global intercalibrated net of atmospheric monitoring stations, measuring CO_2, CH_4, and CO mixing ratios and carbon, oxygen and hydrogen isotopic ratios on a continual basis. Methods of linking measurements from point sources to that observed from towers or from aircraft have a high priority but equal effort must be placed on measuring devices at enough resolution from satellite imagery. Oceanic fluxes also need further emphasis. Much of the detail of scientific approaches underway or planned are contained in components of the World Climate Research Programme, the International Geosphere-Biosphere Programme and related programs which have been designed to address these issue. The present programs are an excellent start, but will need much strong support in future.

Linked with the monitoring programs should be a series of experimental campaigns to determine biological controls on the carbon cycle. These should be designed not only to discover what pristine nature used to be like, but to investigate the real world, filled with humanity. In particular in terrestrial systems, more emphasis needs to be placed on processes in tropical landscapes, on soil carbon storage and turnover, on the boreal forest and wetlands of the northern latitudes, and on the relationship between present land management practice and the carbon cycle. Similarly in oceanic systems priority has to be given to both physical and biological processes and their controls. Other priority areas include a much better understanding of the effects of fossil fuel use both in terms of greenhouse forcing effects and on the balance of carbon between sinks and atmosphere. The aim of this work should be to build up a successful model, in fine detail (say 100 km x 100 km squares), of carbon transfers, and to integrate this model with global climatic models, both on a current basis and for use in prediction.

3.2 Nitrogen
A preliminary effort was made to develop a global budget for nitrogen and other elements in 1976 (SCOPE 7) and another UNEP/SCOPE project (1979–1981) dealt with

regional budgets of nitrogen cycling in tropical ecosystems. There also have been several early attempts to document the interaction of nitrogen with other biogeochemical cycles at the scale of the ecosystem (SCOPE 17, 21). A more recent study published in 1988 concentrated on detailed processes of nitrogen cycling and their controls in coastal oceans (SCOPE 33).

Our understanding of the nitrogen cycle processes, of sources and sinks of nitrogen, and of the effects of increased nitrogen cycling has since increased tremendously. However uncertainties exist and the current task is to re-examine the global budget of nitrogen and to analyze nitrogen budgets for key regions of the globe.

Ongoing work on Trace Gas Exchange (SCOPE/IGBP, IGAC) is concerned in part with the cycle of nitrogen oxides in the atmosphere, but the focus is with nitrogen oxides as greenhouse gases, and other nitrogen species are not considered. It is important to quantify and qualify the sources, transport through the atmosphere, and fate of nitrogen particles, ammonia gas, and nitrogen oxides (NO_x, NH_3, HCN, CH_3CN and other nitrites, N_2O, N_2) as well as hydrologically mediated fluxes of nitrogen. In particular, information is needed on the influence of burning of biomass due to human influences in the tropics on nitrogen compounds in the atmosphere (Kuhlbush *et al.*,1991).

3.2.1 Background information
The global cycle of nitrogen has been altered by human activity to a greater extent than that of any other element. The production of nitrogen fertilizer, cultivation of legumes and incidental nitrogen fixation in engines together transfer more nitrogen from the atmosphere into biologically available forms than is fixed by all natural processes combined. The past decade has witnessed a large increase in environmental problems attributable to increased nitrogen cycling through the atmosphere and waters of the world. Perhaps best documented among these is the eutrophication of coastal marine ecosystems, a problem largely caused by increased nitrogen inputs (at least in most of the temperate zones). As a result of this eutrophication, estuarine and coastal marine resources are being lost world-wide at an alarming rate as the volume of anoxic water grows and nuisance algal blooms become more prevalent (SCOPE 42).

Nitrogen inputs from the atmosphere to the open ocean oligotrophic gyres of the world have also increased due to human influence; the potential consequences on the ocean as a global carbon sink have received some, but insufficient, attention. The oceans are the largest known sink of carbon dioxide, and by increasing oceanic production increased nitrogen inputs may increase the magnitude of this carbon sink. The potential impact of increased nitrogen deposition on the ocean as a source of dimethyl sulphide (DMS) to the atmosphere has received

even less attention despite the fact that dimethyl sulphide is a major controller of cloud formation and therefore climate over the world oceans.

Terrestrial systems have been receiving increased input of nitrogen as air-borne ammonium and nitrate. This has sometimes been invoked as a contributing factor to forest die-back in parts of northern temperate zones Whether this is true or not, it seems certain that forests in many parts of the world are inadvertently being fertilized with nitrogen, with unknown consequences in terms of carbon storage within the forests.

3.2.2 Research needs and issues
These ecosystem-level changes could be, and often are, treated as separate problems. However, more rapid progress in understanding and solving these problems could be achieved by recognizing their commonality. They are all caused by an increased rate of nitrogen cycling at the scale of regions and by an increasingly important control of the nitrogen cycle by humans. Their understanding can only result from continued interdisciplinary efforts.

Consider the case of coastal marine eutrophication as an example of a required synergistic study. Although increased nitrogen inputs are now known to be the major cause of eutrophication in most temperate-zone estuaries, the factors responsible for these increased nitrogen inputs are not as well understood. What is the relative contribution of atmospheric deposition? How much of the nitrogen deposited on terrestrial ecosystems in precipitation and dry deposition is retained in the ecosystem, and how much is exported down stream to estuaries? How do changes in land use patterns affect the down stream export of nitrogen? How important are losses of fertilizer nitrogen from agricultural lands in coastal eutrophication? These questions, critical to the management of the world's estuaries and coastal seas, have not yet been answered. Also, although nitrogen inputs usually regulate coastal eutrophication in temperate zone marine ecosystems, phosphorus is more commonly hypothesized to control marine eutrophication in tropical regions.

It is important to develop a global budget and refine regional budgets for nitrogen for selected key regions of the world; to fully understand the problems stemming from accelerated nitrogen cycling and formulate operational recommendations for solving them. Specific questions include:- How much of the current rate of terrestrial, bacterial nitrogen fixation (110-170Tg yr^{-1}) is retained and how is it distributed or lost? What is the extent to which various marine systems really are nitrogen limited? We need to know since society is dumping large amounts of N into offshore waters in acid nitrate deposition (Howarth, 1988). Another question of whether forest growth and production (and thereby C storage) is really N limited or not (Vitousek and Howarth, 1991). The question of N limitation in forests seems important in determining how forests will change as C sinks as N deposition changes.

3.3 Phosphorus

Phosphorus occupies a key place among the major nutrients because of its relative scarcity among the light elements and its essential role in energy transformations in all life forms. Man's use of phosphate reserves has produced both desirable and undesirable effects on the environment. Widespread fertilizer phosphate applications have greatly increased food and fibre supply for an expanding world population. On the other hand, phosphorus associated with eroded sediments from agricultural lands, as well as phosphorus discharges from urban and industrial areas in sewage effluents and other wastes, are major causes of eutrophication of water bodies.

Scientific information is increasingly needed to guide the use of phosphorus to obtain maximum benefits without producing undesirable impacts on the environment. To this end, information on the phosphorus cycle needs to be summarized and then integrated with knowledge of other nutrient elements and their interactions. Phosphorus differs from carbon, nitrogen and sulfur, since it does not have a significant gaseous atmospheric transfer. However, phosphorus has significant indirect global effects on the environment through its effects on carbon, nitrogen and sulfur transfers. Extensive data are available in a number of regions of the world on:

(a) mineral phosphate deposits;
(b) fertilizer production and usage;
(c) detergent, pesticides and other industrial production and usage;
(d) soils and plant composition in various ecosystems;
(e) river transport, and
(f) lake sediments.

There have been few attempts to integrate this information.

Recent studies on phosphorus transformations focus on microbial activity and the importance of both inorganic and organic forms, as organic phosphorus forms are both a significant source and sink for biologically active phosphorus in ecosystems (Tate, 1984; Stewart and Tiessen, 1987). New methods have helped quantify levels of biologically active phosphorus and relatively inert physically and chemically occluded forms in ecosystems. Key processes of phosphorus interactions with carbon, nitrogen and sulfur have been identified and incorporated into computer models to guide interpretation of phosphorus data (Parton *et al.*, 1988). These models need to be extended to a wider range of ecosystems so that, in addition to the data base, they will provide a mechanism

for the evaluation of the short- and long-term impact of man's manipulation of phosphorus in the biosphere. A realistic understanding of elemental cycles is not possible in isolation from other nutrients and a holistic approach to global biogeochemistry is needed.

3.3.1 Research needs and issues

Currently major studies are underway (examples are SCOPE 42, IRRI 1990) which are aimed at better understanding the nature, sources and fluxes of phosphorus in the terrestrial and aquatic ecosystems and its global environmental effects through interactions with cycles of other elements. The primary focus of these projects are to integrate and synthesize information on phosphorus in diverse environments with emphasis on the flows of phosphorus between terrestrial systems, ground waters, rivers, lakes, estuaries, and oceans. Both natural and anthropogenic fluxes in the phosphorus cycle are being assessed in a study of the biogeochemical processes. Particular attention is being paid to phosphorus interactions with other elements (carbon, nitrogen, sulfur and metals).

Typical questions that need resolution include: What are the effects on the natural processes of phosphorus translocation and movement? What type of phosphorus transformation and translocation occur in landscapes where erosional and soil development cause differences in the accumulation of phosphorus forms? What are the sources of phosphate and how long will they last — very little is known of P budgets in weathered tropical soils or in Amazon rain forests? What is the relative importance of natural and anthropogenic phosphorus flows? Can one quantify the natural processes and the changes brought upon these processes by different management systems? What is the acceptable loading capacity of phosphorus in rivers (SCOPE 42 data on world rivers documents anthropogenic induced differences in N and P concentrations in rivers of two to three orders of magnitude) and lakes? What are the processes involved in lake eutrophication and to what extent can control of phosphorus or other inputs affect this? In estuaries and river deltas, what do we know about phosphorus movement and transformation in these areas? And finally, in an enclosed sea area such as as the Baltic, what are the current loading rates of phosphorus and what is the net result of these rates on the marine environment?

3.4 Sulfur

Current evidence indicates that biogenic sources of sulfur to the atmosphere, with the exception of marine sources of dimethyl sulphide (Andreae, 1986) are much smaller than once thought (50–120Tg S per year), and so anthropogenic sources of atmospheric sulfur are of the same or slighter larger magnitude (60–110 Tg S per year). These anthropogenic sulfur emissions have started to decrease in

North America and Europe and are predicted to continue to decrease (SCOPE 48), but large increases in Asia, Africa, and South America seem likely (SCOPE 36). A knowledge of sulfur cycling at the scale of the ecosystem is required to evaluate the recovery of areas already polluted by acid deposition (SCOPE 19) and the potential impact of further sulfur deposition and acidification in tropical countries (SCOPE 36).

3.4.1 Research needs and issues

Detailed and relatively accurate information on sulfur processes is generally now available for a variety of specific sites in a variety of types of ecosystems. However, measurements are usually made at a fine areal scale (a few square meters), and a challenge exists in developing the ability to integrate these to regions, to continents, and to global scale (SCOPE 35). One approach may be to synthesize process information into credible mathematical simulation models that accurately mimic the behavior of sulfur in natural ecosystems.

Sulfur and sulfur compounds are not easy to determine analytically. Despite much progress in analytical methodology, further method development is needed. The analytical challenge of sulfur has been further complicated by the number of additional sulphides that have been discovered in the atmosphere during the last 15 years, such as dimethyl sulphide, carbonyl sulphide and carbon disulphide.

Substantial progress has been made during the past two decades, especially through intensive studies to understand the effects of anthropogenic deposition on lakes, wetland ecosystems, and upland forests (SCOPE 48). In marine wetlands, the tidal fluxes of sulfur are very large in comparison to the rates of sulfur processing within the sediment, and rapid cycling occurs primarily through inorganic reduced forms of sulfur. Sulphate reduction is the dominant form of decomposition in these systems. Gaseous fluxes are high in comparison to any other ecosystems, but represent only a tiny fraction of a cycle within the wetlands. Also, because of the relatively limited area of these marine wetlands, their influence on the global atmospheric sulfur cycle is limited. The freshwater wetlands have a much lower sulfur input and the cycling of organic sulfur is much more important. One recent study suggested that S deposition may increase rates of anoxic decomposition and thereby decrease storage of organic carbon in the peat. Increased inputs of sulfur to a wetland also clearly reduce the fluxes of methane from the wetland to the atmosphere.

Research of the past two decades on sulfur cycling in lakes and other inland water bodies has revealed that freshwater lakes are among the ecosystems most obviously damaged by acid deposition. The attendant changes are in acid-base chemistry of lakes, the alteration of interactions with carbon, nitrogen, phosphorus, and other major elements, and the ability of the lakes to partially counteract acidification. Key features in freshwater lakes are assimilatory reduction and uptake of sulphate in the water column by algae and bacteria, dissimilatory sulphate reduction by bacteria in anoxic zones, reaction of reduced S species with organic matter and reduced metals, and the production of sulfur gases likely to accompany decomposition of sulfur-bearing organic matter.

In terrestrial ecosystems much of the information has been gained from studying the impacts of acid-sulphate deposition on forest vegetation and soils (SCOPE 48). The role that sulfur retention or release plays in affecting the flux of acidity to downstream ecosystems is now clear, as is the interaction of sulfur with other elemental cycles, especially nitrogen and basic cations. There are still inconsistencies in the measurement of dry and cloud-water deposition, and there are still major unanswered questions with regard to sulfur retention and absorption in soil.

Significant progress also has been made in understanding sulfur cycling in upland agriculture systems. Managed lowland systems, specifically wetlands used in rice culture, offer many opportunities for expansion of knowledge. Also left relatively understudied are the potential toxic effects of sulfur on rice and the effect of sulfur cycling on methane fluxes from rice paddies.

Finally, society has greatly altered sulfur cycles in a variety of ways, but the full magnitude of these is still unknown. Acid sulfur decomposition has damaged forest and lakes, and the challenge has been accepted in industrialized temperate regions to begin to ameliorate the problem and restore damaged areas. Tropical areas are also known to be vulnerable to a greater or lesser extent, but questions arise as to our ability to control the magnitude of anthropogenic sulfur inputs to these systems and to limit their deleterious effects. Eutrophication of continental and coastal seas has increased the abundance of sulphides in the water column, to the detriment of life in these waters bodies. And by developing high yield agriculture, we have stressed natural systems of sulfur supply and developed a need to add fertilizer sulfur in a manner that supplies plant needs without altering important soil quality characteristics and properties. Atmospheric sulfur comes principally from anthropogenic sources, and biogenic fluxes of sulfur from terrestrial ecosystems are thought to be relatively small. However, we do not have the capability to predict how they will change in response to added inputs of sulfur or other human influences. The major biogenic source of atmospheric sulfur appears to be oceanic dimethyl sulphide, but the exact magnitude of this flux and its controls remain poorly known. We are unable at present to predict how this flux may change in response to climate change or changing nutrient inputs to the oceans.

3.5 Other important elemental interactions

Other elements have significant interactions with the biogeochemical elements already discussed. Calcium and magnesium are important in the precipitation of carbonates in oceans and on land in precipitates of carbonate and in the rate of carbonate dissolution. (cf review by Degens *et al.*, 1984). The weathering of silicates is also an effective sink for CO_2. Iron and molybdenum are also key elements in C, N and S interactions (Howarth and Stewart, 1992) and iron deficiency has been shown to limit phytoplankton growth in oceans (Martin and Fitzwater, 1986). Cadmium has also been shown to behave like a nutrient in oceanic waters (Bruland *et al.*, 1978). A more comprehensive discussion of marine geochemistry and interactions is discussed in Chester (1990).

4. SPATIAL AND TEMPORAL SCALES

4.1 Hydrologic and Biogeochemical Phenomena

Hydrologic and biogeochemical phenomena at both the global scale and over the full range of drainage basin scales are relevant to development. Because of its unique continental and global importance the hydrologic behavior of the Amazon basin is highlighted here. Hydrologic and biogeochemical processes are equally important at a variety of different scales. Drainage basins of successive size nest within one another and are linked together by channels comprising the drainage net. From the smallest to the largest, all are subject to major change by human activities. These influences range from accelerated erosion induced by grazing or cropping practices to runoff of polluted water from urban land surfaces.

The relevant drainage region may dominate the landscape of entire nations or constitute critical water resources, for example, adjacent to urban pollution centers. Observations over a period of time, of stream flow and constituents including sediment and other potential pollutants such as nutrients, metals, or synthetic organics, are essential to achieve two purposes. First, because of the linkage of the hydrologic network, flows and constituents derived from small areas constitute contributions to a larger system. Each part of the system may serve as a source, loci of intermediate storage, or sink. Thus the knowledge of phenomena at the range of scales contributes to an understanding of the larger whole. Second, such information on water flow and its composition is an essential tool in managing land use and water quality. A comprehensive understanding of natural and manmade influences on both streamflow and constituents is fundamental to effective management. For example, urbanization increases both peak discharges from frequent intense rainfalls and urban surface runoff contributes significant quantities of pollutants, such as metals and hydrocarbons as well as nutrients and oxygen demanding

wastes to receiving waters. Monitoring of ambient water quantity and quality are important in evaluating potential drinking water sources, fisheries and recreational waters in the urban scene.

Hydrologic processes occur at all spatial and temporal scales. At one extreme, the infiltration of soil moisture following rainfall operates not only at the surface of the particles but within capillary and smaller openings. The forces involved, including surface tension, are affected by the chemical and biological constituents in soil and water and, in turn, these constituents are influenced by the dynamics of the movement of water. Time scales in ground-water flow may involve thousands of years while those in cloud physics span microseconds. In-between, the movement of water through drainage basins ranges over time scales of minutes, days and months. In predicting the behavior of hydrologic systems it is essential to be able to transpose spatial and temporal scales. For example, the permeability of soil and the subsurface is not only highly variable, but varies with scale. While progress is currently being made in characterizing different spatial scales by means such as fractal dimensions, much remains unknown.

Simple long recognized spatial and temporal phenomena remain to be explained. Rain is measured at a point. The integration of point rainfalls in space is essential in predicting runoff and floods. Similarly, the temporal features of nearby measuring points cannot be simply integrated in order to specify the likely temporal frequency of events such as rainfall or runoff in a particular area.

The distribution of vegetation is climatically controlled within the geologic framework of a given region. Much, however, remains to be learned about those temporal features of the climate, such as the distribution in time of temperature and precipitation (and their sequencing) that may control the adaptation of biota in a given region to a given site. Moreover, the relative significance of extremes in relation to smaller events, whether of precipitation or temperature, in controlling landforms and vegetation, continues unresolved. The processes are likely to be much influenced by human activities and in turn, to significantly influence such activities. Thus, alternating dessication and filling of river valleys in semi-arid regions, an important global phenomenon directly influencing human habitation, remains unexplained.

Because global circulation models employ large grid scales, problems of aggregating measurements of hydrologic variations at ground level are of particular importance. Techniques for doing so are essential in relating climate models to land use and other changes made by man. Monitoring programs currently being set up use a combination of satellite imagery with surface-based measurements are an essential means of achieving this end (GCOS, GEMS/GRID and WDC).

4.2 Integration of processes across regions to global scales

In the last two or three decades scientists have begun to look at ecosystems with a more holistic approach, trying to understand the complex interrelations between their geological, biological and chemical properties, in order to understand the processes through which energy and materials are cycled at the Earth's surface. This is a complex task due to the enormous range of time and space scales involved. We will illustrate this with an example of an ongoing attempt to understand the biogeochemistry of the Amazon Basin.

The Amazon basin comprises about 30% of what is left of tropical forests on Earth. Because of its importance as a tropical forest ecosystem, and because within its boundaries undisturbed regions coexist with others of changing land use, it may serve as a template for understanding other great tropical river basins. Major changes in land use, especially deforestation have already occurred in developed countries. Therefore, the control of processes affecting the biogeochemical cycles are much more linked to anthropogenic activity. Environmentalist pressure in developed countries is driven toward a minimum restoration of their lost natural ecosystems, and management of whatever is left of the already modified environment. In developing countries, on the contrary, the struggle is to stop or at least delay the currently strong changes in land use. Despite the importance of tropical forests for Global Climate Change issues, major deficiencies still exist in the knowledge of their water and biogeochemical cycles. There is no way to preserve tropical forests and their biodiversity using ecosystem sustainability concepts without a sound knowledge of its basic functioning mechanisms.

A key challenge for the science of global change is to understand how hydrologic and biogeochemical cycles function and interrelate on regional to continental scales. The properties of an ecosystem that best correlate with materials being transported in the river are precipitation and runoff. At the scales above, rivers act as integrators of processes at their basins, and are the main pathway for the transfer of organic matter from the land to the oceans. Carbon and nutrient fluxes in large rivers are reflections of their watersheds and floodplain, and some extension of the properties of the smaller rivers that form them. Organic and inorganic materials measured in the main channel are a mixture of materials originating from upland regions, as well as material introduced continuously from the adjacent floodplain (SCOPE 42).

An impressive amount of information has accumulated over the last decade on the biogeochemistry and hydrology of the Amazon. Information has been collected on sources and routing, chemical distributions and fluxes, analyses of particulate and dissolved organic biomarkers including $\partial^{13}C$, $\partial^{18}O$ and $\partial^{14}C$ analyses and dissolved gas regimes of the Amazon floodwave (Martinelli *et al.*, 1991; Richey and Victoria, 1992). Trophic food chains and mercury contamination have also been addressed as have the possible water recycling mechanisms in the region and their importance. There is now an urgency in identifying the research necessary to integrate the information of the kind above in order to develop models to assist in the self-sustainable development of the Amazon basin. Within these guidelines, a major challenge is to evaluate the potential impact of development alternatives on the water and chemical cycles of the basin. These cycles are of fundamental importance not only to the maintenance of the natural forest but also to any human occupation. The critical question is, "To what extent can the land use of a particular region be altered without impacting the overall regime of rainfall and runoff or of organic matter production and decomposition?"

The sheer physical size and logistics of problems posed by the Amazon challenge the effort to answer this question. Models must deal with the issue of "scale"; the ability to transfer understanding from very small scales (where it is possible to do field research), to regional scales (a river basin), and ultimately to the Amazon as a whole and its interactions with the global atmosphere. Combinations of field and remotely sensed data for analysis of water and chemical cycles across different scales are essential. Such an approach should allow definition of the critical scales of water, organic matter, and nutrient cycles relative to the needs of self-sustainable development of the Amazon.

4.3 Data and Observations on an International Scale

While models are essential to an understanding of the global cycles, the need for global observations and for detailed measurements of specific hydrologic and biogeochemical processes remains acute. Few models exist that are not in need of verification, or at the least, calibration of a variety of parameters. Large uncertainties remain. It is essential that we develop the means to observe and document the fundamental aspects of the climate and Earth biospheric system and changes occurring within them. Existing observational capabilities must be sustained and expanded to accomplish this goal.

It is essential that the present activities at the international level involving the development and enhancement of global observations of the hydrologic and biogeosphere cycles be maintained and expanded. Measurements are required throughout the globe of the flux of water and energy and of the reservoirs of water described above. These measurements can only be properly made through continuing international co-operation represented by activities such as the World Climate Data and Research Programmes and the Global

Energy and Water Cycle Experiment. Moreover, it is essential that long-term observations be made of the components of the hydrologic and biogeochemical cycles at specific places with carefully controlled instrumentation. While century or longer precipitation records exist at some locations, long-term records of precipitation in much of the world are scarce. Carefully designed networks are essential in assuring the most cost effective distribution of both long-term and model observations.

Remote sensing and global information systems provide unique opportunities for global observations of hydrologic and biogeochemical phenomena. At the same time, much work must be done to calibrate and interpret images obtained by remote sensing. For such important areas as soil moisture or soil carbon, measurable with great difficulty even at discrete points, improvement of the ability to characterize soil moisture states over large areas will be a major contribution to understanding and modeling the cycles.

Constant and rapid improvement in technology produces a burgeoning increase in the amount of data collected to describe facets of the hydrologic cycle. These masses of data require improved data management if mere data is to be converted into useful information. This need for data management is not unique to the field of hydrology or biogeochemical cycles. Rather, it is the combination of observations of the Earth's many interacting systems, often in real time scales, which demands systems of management that can help elucidate complex environmental processes. Because the information demanded is global, techniques to assure that the observations made are effectively used is a matter of major international importance.

5. INTERNATIONAL CO-OPERATION AND INSTITUTIONAL ARRANGEMENTS

The program suggested above demands major international co-operation on a heretofore unprecedented scale. Substantial planning and implementation of linked co-operative programs has already taken place and has given rise to the establishment of an Intergovernmental Panel on Climate Change (IPCC) which has the responsibility of assessing the state of scientific knowledge of climate changes due to human influences. The World Climate Research Programme (WCRP) and the International Geosphere–Biosphere Programme (IGBP) together constitute the international framework of the quest for scientific understanding of climate and global change. The proposed scientific strategy to achieve prediction of the climate system must be based on a combination of process studies, observation and modeling. IPCC considers the following areas as the most critical:

(a) control of greenhouse gases by the Earth System;

(b) control of radiation by clouds;
(c) precipitation and evaporation;
(d) ocean transport and storage of heat; and
(e) ecosystem processes.

Within the WCRP, the Global Energy and Water Cycle Experiment (GEWEX) is addressing (b) and (c) while the World Ocean Circulation Experiment (WOCE) is addressing (d) and part of (a). Two core activities of IGBP, the Joint Global Flux Study (JGOFS) and the International Global Atmospheric Chemistry Programme (IGAC) are designed to investigate the control of greenhouse gases by the oceanic and terrestrial processes while the Biospheric Aspects of the Hydrological Cycle (BAHC) is a complement to GEWEX and also addresses (c). An additional core project of the IGBP focusses on Global Change and Terrestrial Ecosystems (GCTE). Both WCRP and IGBP have other essential core activities that contribute to these efforts to reduce uncertainties in climate predictions (IPCC, 1990; IGBP, 1990). These internationally coordinated programs depend on strong national participation and support. World-wide networks of monitoring stations will need to be greatly expanded, especially in the eastern hemisphere. Experimental campaigns, particularly those into the biology and interrelationships of the biogeochemical cycles, will need active collaboration between scientists from tropical and non-tropical nations, and between researchers of all nations.

5.1 Communication of research results and public reaction

There is a great need for education and dissemination of information at all levels in society: an informed public will, perhaps, find ways of reducing the rate of deforestation, fossil fuel use and enact laws governing pollution problems. An excellent way of improving communication is through the schools, but it is essential that we have means of communication to scientists in all nations especially in developing countries, and to the public at large perhaps via books, magazines and TV programs. Similarly we hope that this will result in a greater will to address legal aspects of global change and that environmental ethics will strengthen public perception.

5.2 Means to attract and train scientists

The above discussion demonstrates the need for scientists in many disciplines on a world-wide basis. In many nations, scientific resources are thin, especially in appropriate scientific disciplines such as soil science and wetland biology, or atmospheric chemistry. Emphasis should be placed on encouraging training of scientists; in these areas, especially if major experimental campaigns take place on their territory. More generally, it is important to develop a career structure for scientists working on biogeochemical cycle problems as long-term research programs should also be started in relevant tropical sub-tropical areas. Education of scientists is a key aspect of IGBP and WCRP programs and should be supported.

ACKNOWLEDGMENTS

We should like to acknowledge assistance obtained from Euan G. Nisbet and Robert W. Howarth in the preparation of parts of this article as well as the numerous contributors to SCOPE biogeochemical workshops.

REFERENCES

Andreae, M. O. 1986. The ocean as a source of atmospheric sulfur compounds. In: (ed.) P. Buat-Menard. *The Role of Air–Sea Exchange in Geochemical Cycling.* pp. 331-362. D. Reidel, Boston.

Bruland, K.W. et al., 1978. Cadmium in Northeast Pacific waters. *Limnology and Oceanography,* **23**, 618-625.

Chester, R. 1990. *Marine Geochemistry.* Unwin Hyman, London.

Crutzen, P.J. 1991. Methane's sinks and sources. *Nature,* **350**, 380-381.

Degens, E.T., Kempe, S. and Spitzy, A. Carbon dioxide: a biogeochemical portrait. n: O. Hutzinger (ed.) *The Handbook of Environmental Chemistry.* part c p 127-215

Howarth, R.W. et al. 1991. Inputs of sediment and carbon to an estuarine ecosystem: the influence of land use. *Ecological Applications,* **1**, 27-39.

Howarth, R.W. 1988. Nutrient limitation in net primary production in marine ecosystems. *Ann. Rev. Ecol.,* **19**, 89-110.

Howarth, R.W. and Stewart, J.W.B. 1992. The interaction of sulfur with other elements in ecosystems. Chapter 4 in *SCOPE 48.*

IGBP, 1990. *The International Geosphere–Biosphere Programme: A Study of Global Change.* The Initial Core Projects. Report No. 12 .

IPCC, 1990. Scientific Assessment of Climate Change. Report prepared by the Intergovernmental Panel on Climate Change.

IRRI, 1990. Phosphorus Requirements for Sustainable Agriculture in Asia and Oceania. Proc. Symp International Rice Research Institute.

Kuhlbusch, T. A., Lobert, J. M., Crutzen, P. J., and Warneck, P. 1991. Molecular nitrogen emissions from denitrification during biomass burning. *Nature,* **351**, 135-137.

Likens, G.E. 1985. *An Ecosystem Approach to Aquatic Ecology.* Springer-Verlag.

Martin, J.H. and Fitzwater, S.E. 1986. Iron deficiency limits phytophankton growth in the North-East Pacific sub artic. *Nature,* **331**, 344.

Martinelli, L.A., Devol, A.H., Victoria, R.L. and Richey, J.E. 1991. Stable carbon isotope variation in C3 and C4 plants along the Amazon River. *Nature,* **353**, 57-59.

Mooney, H.A. 1991. Emergence of the study of global ecology:

Is terrestrial ecology an impediment to progress? *Ecol. Appl.* **1**, 2-5.

NAS, 1992. *Opportunities in the Hydrologic Sciences.* National Academy of Science, Washington, DC.

Parton, W.J., Stewart, J.W.B. and Cole, C.V.1988. Dynamics of carbon, nitrogen, phosphorus and sulfur in cultivated soils: a model. *Biogeochemistry* **5**, 109-131.

Salati, E.; Dall Ollio, A.; Gat, J. and Matsui, E.1979. Recycling of water in the Amazon basin: an isotope study. *Wat. Resour. Res.* **15**, 1250-1258 .

Richey, J.E. and Victoria, R.L. 1992. C, N and P export dynamics in the Amazon river. In R. Wollast (ed.), *Interactions of C, N, P and S Biogeochemical Cycles.* Springer-Verlag, Berlin. (in press).

SCOPE 7, 1976. *Nitrogen, Phosphorus and Sulfur – Global Cycles.* B.H. Svensson and R.Soderlund. (eds.) Ecol.Bull. (Stockholm) No 22.

SCOPE 17, 1981. *Some Perspectives of the Major Biogeochemical Cycles.* G.E. Liken. (ed.) John Wiley and Sons Ltd., Chichester.

SCOPE 19, 1993. *The Global Biogeochemical Sulfur Cycle.* M.V.Ivanov and J.R. Freney. (eds.) John Wiley and Sons Ltd., Chichester.

SCOPE 21, 1983. *The Major Biogeochemical Cycles and their Interactions.* B. Bolin and R.B.Cook. (eds.) John Wiley and Sons Ltd., Chichester.

SCOPE 29, 1986. *The Greenhouse Effect, Climatic Change and Ecosystems.* B. Bolin, B.R. Döös, J. Jaeger and R. A. Warwick. (eds.) John Wiley and Sons Ltd., Chichester.

SCOPE 33, 1988. *Nitrogen Cycling in Coastal Marine Environments.* T.H. Blackburn and J. Sorensen. (eds.) John Wiley and Sons Ltd., Chichester.

SCOPE 35, 1988. *Scales and Global Change: Spatial and Temporal Variability in Biospheric and Geospheric Processes.* John Wiley and Sons Ltd., Chichester.

SCOPE 36, 1988. *Acidification in Tropical Countries.* H.Rhode et al. (eds.) John Wiley and Sons Ltd., Chichester.

SCOPE 42, 1991. *Biogeochemistry of Major World Rivers.* E.T. Degens, S. Kempe and J. Richey. (eds.) John Wiley and Sons Ltd., Chichester.

SCOPE 48, 1992. *Sulfur Cycling on the Continents: Wetlands, Terrestrial Ecosystems and Associated Water Bodies.* R.W. Howarth, J.W.B.Stewart and M.V. Ivanov. (eds.) John Wiley and Sons Ltd., Chichester.

Schimel, D.S. and Andreae, M.O. (eds.) 1989. *Trace Gas Exchange between Terrestrial Ecosystems and the Atmosphere.* Pub John Wiley and Sons Ltd., Chichester.

Stewart, J.W.B. and Tiessen, H. 1987. Dynamics of Soil Organic Phosphorus. *Biogeochemistry,* **4**, 41-60.

Tans, P.P., Fung, I.Y. and Takhashi, T. 1990. Observational constraints on the global atmospheric CO_2 budget. *Science,* **247**, 1431-8.

Tate, K.R. 1984. The biological transformation of phosphorus in soil. *Plant and Soil* **76**, 245-256.

Vaghjiani, G.L. and Ravishankara, A.R. 1991. *Nature,* **350**, 406-408.

Vitousek, P. M. and Howarth, R. W. 1991. Nitrogen limitation on land in the sea: How can it occur? *Biogeochemistry,* **13**, 87-115.

Chapter 7: Atmosphere and Climate

G.A. McBean, G.S. Golitsyn and E. Sanhueza

EXECUTIVE SUMMARY

The atmosphere is a thin blanket of air surrounding the Earth with special characteristics that make life on Earth possible. Biological processes have provided us with an atmosphere of 21% oxygen; the ozone layer shields us from harmful ultraviolet radiation; water cycles through the atmosphere, nourishes us and the biosphere; and the greenhouse gases warm us up. The atmosphere is a dynamic system which links the global biogeochemical cycles. The atmosphere and the climate system are in states of continuous evolution on a wide range of time and space scales. Most of these variations are natural but human activities are now having demonstrable impact on the atmosphere. Natural disasters such as cyclones, tornadoes and floods, among other weather phenomena, cause widespread destruction and loss of life every year. Improved weather forecasts and warning services are helping to reduce the losses but more needs to be done. Longer-term forecasts of El Niño and droughts, for example, appear possible in the future with improved scientific understanding and better observations and prediction models.

(a) **Important Facts:**

(i) The atmosphere and climate systems are complex, interacting and dynamic, involving physical, chemical and biological processes. We need to understand this complex natural system as a basis for predicting its evolution and the role of humans in altering it.

(ii) Human activities are causing depletion of the ozone layer and increasing the pollution of urban areas, the acidity of precipitation and the atmospheric concentration of long-lived greenhouse gases.

(iii) The stresses on the atmosphere and climate are increasing and the rates of change are unprecedented.

(b) **Research Needs:**

(i) Nations are encouraged to contribute to the World Climate Research Programme and the International Geosphere-Biosphere Programme to provide the understanding, predictions and information base essential to determine climatic change impacts and to develop climate response strategies.

(ii) Support is needed for research studies in atmospheric chemistry (and related oceanic chemistry), including the sources and sinks of atmospheric pollutants, recognizing that problems of urban air pollution, acidic deposition, ozone depletion, climate change and others are strongly linked to atmospheric chemistry and that it is an area that is considered most likely to generate surprises.

(c) **Urgent Actions:**

(i) Nations are urged to contribute to the Global Climate Observing System (GCOS).

(ii) Improved models need to be developed to provide the basis for prediction of future evolution of the atmosphere and climate system.

(iii) All nations should review options for and take steps to reduce emissions of gases and particulates to the atmosphere. It seems appropriate that developed countries should lead in these reductions.

(iv) The scientific and technical capacity of nations needs to be improved, particularly in developing countries to address the regional aspects of these issues and to contribute to global programs.

1. INTRODUCTION

The atmosphere is a thin blanket of air that makes life on Earth possible. Although it extends far out into space, about half of the atmosphere's mass is within 5.5 km of the

Earth's surface and about 99% is within 30 km. The composition of the atmosphere is unique in our solar system. Whereas the outer planets have atmospheres primarily of hydrogen and helium and those of Mars and Venus are primarily carbon dioxide, the Earth's atmosphere is dominantly nitrogen (78% N_2 by volume), oxygen (21% O_2) and argon (1%), with many trace species of highly variable concentration (Budyko *et al.*, 1987). Oxygen is about 10^{13} times more abundant than normal chemical equilibrium theory would expect (Wayne, 1985) which is a consequence of, and at the same time, a prerequisite for, life. Our atmosphere's oxygen must be maintained in steady-state disequilibrium, being continuously replenished through biological processes with a residence time of about 10^4 y. Solar radiation provides the energy that allows the biosphere to provide not only the fuel for life (hydrogen, carbon and nitrogen) but the oxygen as well, through the process of photosynthesis that consumes CO_2 and releases O_2. Intricate connections between life and atmospheric composition provide a major critical example of the complex cyclical interactions that characterize our planet and make it habitable.

Among other components of our atmosphere are highly variable concentrations of water vapour (H_2O, up to a few per cent), carbon dioxide (CO_2, now 0.035% or 350 ppmv and increasing) and other trace species with much smaller abundances which play a major role in atmospheric dynamics, chemistry and air pollution. The hydroxyl radical is present in minuscule concentrations but acts as a powerful oxidant in cleansing the atmosphere of many pollutants. Atmospheric concentrations of many of these gases are increasing, but some, such as stratospheric ozone (O_3), are decreasing.

Solar radiation is the source of energy, not only for the biosphere and photochemical reactions, but also for all motions within the atmosphere and oceans which arise on the spherical planet due to inhomogeneous heating of its surface. The general circulation of the atmosphere and ocean is influenced by planetary rotation and by topography and experiences instabilities which manifest themselves as vortices on various scales, such as discontinuities and fronts, and other atmospheric phenomena which we recognize as weather.

The lowest layer of the atmosphere, which is well mixed and contains about 80% of the total mass, is called the troposphere. The troposphere is dominated by weather and the water cycle. Water is a critical part of natural weather and climate variations and it cycles through the troposphere in about 10 days, between evaporation, transportation and transformation, and then precipitation, usually some distance from source. In addition to nourishing natural and managed ecosystems, providing animals and humans with essential water, and refilling our lakes, rivers and aquifers, water, as vapour, serves as the main greenhouse gas, and, as clouds, provides the venetian blinds that control climate and it cleanses the atmosphere by precipitating out some pollutants.

Enveloping the troposphere is the stably stratified stratosphere with strong horizontal motions but little vertical motion so that pollutants, once reaching the stratosphere, remain for long periods and spread around the globe. A series of layers of increasing ionic complexity extend outward from the stratosphere. These layers, with their particular gaseous and electromagnetic structure shield us from short ultraviolet radiation and from other unfriendly invaders (such as solar energetic particles called solar wind) from outer space. The ozone layer lies mainly within the stratosphere, extending from 15–45 km from the surface (although there are important ozone concentrations, about 10% of the total, in the lower troposphere, as well). Although the total amount of ozone is very small, occupying a column only 3 mm high if collected and brought to the surface, it plays a major role in the atmosphere and for life on Earth.

Weather is usually thought of in terms of temperatures, precipitation and wind, with their extreme events such as frosts, floods and hurricanes. In middle and high latitudes, weather systems pass through every few days, while in the tropical regions diurnal cycles of precipitation and sunshine, mixed with monsoons and longer period oscillations, dominate. Climate is the statistics of weather, or an averaged weather for a particular region and time period. It has been traditional to consider the climate in terms of 30-y norms (e.g. the 1950–80 period) and this approach is still much in use. However, it is important to recognize that climate is not static but dynamic, varying on a multitude of time scales. The climate also, more than weather, involves the ocean, ice and land surfaces, including vegetation; they function as an integral whole.

The mean temperature of the Earth is determined by the intensity of the solar radiation reaching our planet, by the Earth's albedo or fraction reflected back to space and by the thermal radiation of the atmosphere that comes back to the surface. Albedo depends on cloud amount, aerosols and other composition of the atmosphere and characteristics of the surface (such as vegetation, snow/ice, etc). Thermal radiation back to the surface is determined by the atmospheric concentrations of greenhouse gases, primarily water vapour, carbon dioxide, methane, nitrous oxide, ozone and now chlorofluorocarbons. Without them, and for the same Earth's albedo, the Earth's temperature would be some 30°C colder than its present value of 15°C (Houghton *et al.*, 1990; the report of the Intergovernmental Panel on Climate Change (IPCC) Working Group I, Scientific Assessment). Note that the natural greenhouse effect raises the average temperature from well below freezing to well above. The distribution of temperatures

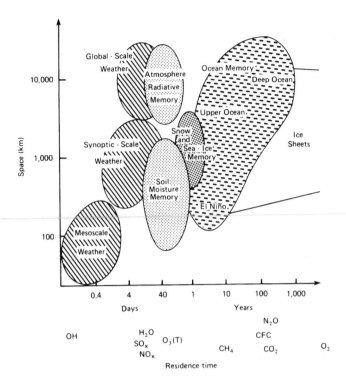

Figure 1: Schematic representation of the time- and space-scales of the atmosphere and climate (modified from Dickinson, 1986). Approximate atmospheric residence times of selected chemicals are indicated along the lower part.

over the planet, as well as the precipitation, is determined by atmospheric circulation, Earth's topography and albedo and interactions of the atmosphere with the oceans, ice and land surface. The oceans and atmosphere transport heat from the tropical regions to the polar regions, reducing global temperature gradients. The oceans have a tremendous heat capacity and long adjustment time and play a major role in climate.

Important concepts for the atmosphere (and the oceans that interact with the atmosphere to determine climate) are representative time and space scales. There is a full range of natural variability of the atmosphere, from the milli second and millimetre scales of turbulence through the global and decadal scales of climate variations (Fig. 1). For natural variability there is generally a correspondence between time and space scales: e.g. small spatial scale events have short time scales; large spatial scales have long time-scales. When we deal with the varying constituents of the atmosphere, this correspondence is less consistent and it is appropriate to speak of residence time: i.e. what is the typical time that a pollutant or some other constituent remains in the atmosphere. A similar concept is that of adjustment time: i.e. how rapidly does the atmosphere adjust to a significant change.

Dynamically, the atmosphere adjusts in a few days; it will re-establish its radiative equilibrium in about a month. Water cycles through the atmosphere in about 10 days. The upper layers of the ocean adjust in months through to the

seasonal cycle while the deep ocean may take centuries to respond to a change in external forcing. Connected to this concept of time and space scales is the concept of predictability. It is not possible to predict the evolution of a weather phenomenon beyond a few of its time scales. Statistical forecasts may be made for those elements of weather that are dependent on longer time-scale components of the coupled atmosphere–ocean system. Similarly, we can conduct sensitivity studies of slowly varying components of the climate system to changes in external forcing, such as the increasing atmospheric concentration of greenhouse gases. Within such modeling studies, the statistical character of events can be studied, but the emphasis is usually on the larger time and space scale features.

When we deal with pollutants added to the atmosphere, we need to consider the interactions of the atmosphere's natural time and space scales and those of the pollutant (including its time-scales for chemical transformation). A pollutant injected into the troposphere, which is generally well mixed, is spread horizontally by winds and mixed upward, perhaps to the tropopause (the level separating the troposphere and the stratosphere). Due to both wet and dry deposition, water-soluble or particulate pollutants usually do not stay in the troposphere for extended periods. Atmospheric concentrations are then a balance between emissions and removal processes. If a pollutant is injected into, or reaches, the stratosphere, where there is no rain and little vertical mixing, its residence time will be greatly lengthened due to the inefficiency of the removal processes. Chemical transformations may result in products that are more or less hazardous and more or less efficiently removed.

For many chemical species, transformation rates are the determining factor. Carbon monoxide, for example, has a photochemical lifetime of about 1 month in the tropics but it becomes about 6 months in winter high latitudes, where transport becomes the determining factor. The global average is 2–3 months (Houghton *et al.*, 1990). Carbon monoxide is mainly lost through reactions with hydroxyl radicals (which have a very short lifetime, about 1 second). Carbon dioxide, on the other hand, does not transform in most of the atmosphere (although it dissociates in the thermosphere) and its residence time is determined by complex interactions with the biosphere and the oceans. Houghton *et al.*, (1990) considered a range (50–200 years) as appropriate. For most chlorofluorocarbons, values around a century are appropriate. The lifetime for methane is primarily determined by reaction with the hydroxyl radical (Vaghjiani and Ravishankara, 1991) and is estimated to be about 10 ± 2 years. From a policy point of view, it is important to recognize that, while remedial actions for short-lived constituents will result in quick responses, the consequences of actions for long-lived species will not be seen for many years.

The main concern today, not only of scientists but also of society as a whole, is the rate of global change of our environment, related to human activities. Atmospheric composition is changing, global warming threatens, ozone concentrations decrease in the stratosphere and increase in the Northern Hemisphere troposphere, all accompanied by other changes such as deforestation and degradation of soils, waters and air. The Earth's environment is an integral whole and the topics dealt with in this chapter must be seen in the context of, and interacting with, the topics of other chapters.

2.   WEATHER PREDICTION AND WARNINGS

Almost all activities of humans are influenced by weather and climate, ranging from the inconvenience of an afternoon rain shower to the devastation of typhoons, storm surges and floods. Weather services around the world are co-operating to improve global prediction of weather elements and to provide, in particular, warnings of extreme events. Each year thousands of people perish in hurricanes, floods, tornadoes, storm surges and other atmospheric and atmospheric-driven phenomena. More than 2.8 million people have lost their lives through natural disasters over the past 20 years. The total cost to societies is immense. Tropical cyclones alone are estimated to have cost $6-7 billion annually (National Academy of Sciences, 1987). For some countries, these natural disasters destroy the work of many years to develop their socio-economic infrastructure. Although the energy amounts in these events are enormous and unlikely to be influenced by humans, better predictions of their occurrence would lead to greatly reduced loss of life and some reduction in material damage.

In a sense, the atmosphere has a short memory and its prediction from an initial state is limited according to the spatial scale under consideration. The predictability for thunderstorms and other dangerous mesoscale events is only a few hours. For major weather systems of mid-latitude weather, the predictability limit is thought to be about 2 weeks (Lorenz, 1969). Through improved prediction models, better observations and faster computers, there has been a steady increase in the skill of weather forecasts. This success can, in part, be attributed to the success of the ICSU/WMO Global Atmospheric Research Programme (GARP, 1967-80). Beyond this limit of predictability (which pertains to specific time and place forecasts), there is possibility of general predictions of longer time-scale features. For example, if a particular type of weather is related to upper ocean heat content, then statistical forecasts may be made on the basis of ocean parameters.   Skill in predicting the occurrence of rainfall anomalies, in a statistical sense, based on oceanic sea surface temperature anomalies has

been demonstrated for periods of several months (Folland *et al.*, 1990).

3.   HUMAN INFLUENCES ON THE ATMOSPHERE

Human influences on the atmosphere have a common origin: expanding human population and their increasing need for energy (usually based on consumption of fossil fuels), materials and food. Many other human activities, related to expanding population, lead to creation of vast amounts of waste, some of which is emitted directly to the atmosphere and some indirectly. Industrial production to meet the desires of the expanding population is another major source of pollutants to the atmosphere. Each of these activities contribute to atmospheric pollution. Whereas sources of these pollutants are concentrated in the developed countries, they are becoming much more widely spread so that few parts of the land surface are devoid of pollutant sources.

For many years, it was assumed that the atmosphere had a large assimilative capacity that could be used as a local resource. This was the "solution to pollution is dilution" philosophy. It is now not clear whether the concept of assimilative capacity has much meaning.

Three major issues where human intervention is causing or may cause significant atmospheric modification: urban air pollution; acidic deposition; and depletion of the ozone layer will be discussed first. Climate variability and change, another major concern where human intervention is important, will then be discussed. These interrelated issues were chosen for their importance and are representative, but not all inclusive, of atmospheric problems.

4.   URBAN AIR POLLUTION

Urban environments were the first to suffer from air pollution due to the congestion of people, industry, transportation systems and other activities in a small geographical area. Meteorological conditions play an important role; for example, very stable conditions in the Los Angeles basin make this urban area one of the most polluted in the United States. The main pollutants of concern in urban environments are sulfur dioxide, nitrogen oxides, carbon monoxide, ozone, volatile organic compounds, heavy metals and suspended particulate matter. The main basis for concern has been the adverse effects of pollution on human health. Sulfur dioxide levels exceed World Health Organization annual standards in many cities (Whelpdale, 1991). Better understanding of the sources of urban air pollution has led to the implementation of a variety of control strategies in most developed countries. Ambient urban sulfur dioxide and suspended particulate matter levels have declined substantially in several developed countries but in most developing countries decreases are not evident. In many large cities of the developing world, such as Calcutta, Mexico City and Sao Paulo, exposure to air pollutants is increasing. Growing urban populations and economic difficulties are making it very difficult to address air pollution (and other) problems in urban areas of developing countries.

Sulfur dioxide and particulate emissions are being reduced in many developed countries (which have the largest emissions *per capita*), through the imposition of progressively stricter emission standards, use of cleaner fuels and control technologies and less dependence on heavy industry (Whelpdale, 1991). Improvements are not as clear for nitrogen oxides because emissions are more difficult to control, in part because their sources are much more numerous and smaller (motorized vehicles). Photochemical smog is formed from chemical reactions among nitrogen oxides and reactive hydrocarbons in the presence of sunlight. The main photochemical products are ozone, hydroxyl radical, hydrogen peroxide and peroxyacetyl nitrate. Some of these species are very reactive chemically and cause damage to vegetation, materials and human health. Oxidant levels have increased over the past two decades as the result of increased urban nitrogen oxide emissions and fairly constant hydrocarbon emissions. Urban ozone concentrations regularly exceed established standards in countries which have routine monitoring. One of the main dilemmas in reducing urban photochemical pollution is the relative degree of control which must be placed on nitrogen oxide and hydrocarbon emission sources.

Unlike other atmospheric issues being discussed in this chapter, urban air pollution problems can be addressed at local or at least national level, although there are examples of trans-boundary atmospheric pollution transport. The level of scientific understanding has increased substantially over the past few decades and impacts on human health and ecosystems are becoming better documented. In many cases, urban air pollution is apparent to the eyes and noses of the population. The main difficulties are social issues and economic costs. However, it should be stressed that addressing the sources of urban air pollution problems will generally aid in addressing the other air pollution issues discussed below.

5. ATMOSPHERIC TRANSPORT AND ACIDIC DEPOSITION

Concern about damage to ecosystems due to the acidification of precipitation was first brought forth at the 1972 Stockholm Conference. Although precipitation can be naturally acidic, human activities were causing an increase in the acidity. Through the efforts of several national and international programs in North America and Europe, irrefutable evidence of pollutant transport over long distances in the atmosphere and across national boundaries was available and the resulting acidification of precipitation and damage to aquatic and terrestrial ecosystems and to materials has now been documented (Swedish Ministry of Agriculture, 1982). By the late 1980s international agreements on emission controls for both sulfur and nitrogen oxides were in place for Europe and North America. The global budget of sulfur emissions is an example where anthropogenic emissions (60–110 Tg sulfur per year) are about equal to or larger than natural emissions (50–120 Tg sulfur per year) (Whelpdale, 1991). Sources and processes of acidic deposition and long-range transport are reasonably well known. In both North America and Europe, total sulfur emissions peaked in the 1970s and are now decreasing; there is now evidence that the downturn in emissions over the past 10 years appears in the acid deposition record. However, severe acidification damage to soils, forests and lakes still exists in many countries and it will be some time before the overall impact of current reductions becomes known. Critical load and critical level values of acidic depositions (Brydges and Neary, 1984; Nilsson and Grennfelt, 1988; Chadwick and Kuylenstierna, 1991; Zhao and Seip, 1991), below which it is thought damage will not occur, are now being established for many ecosystems.

Although there is strong evidence of transport of acidifying pollutants on regional scales, it is not likely that emissions from northern mid-latitudes reach the tropics in significant amounts. Residence times for oxides of sulfur and nitrogen, and for their secondary acids (sulfuric and nitric) in the lower troposphere are less than a week whereas the travel times are at least several weeks. Available data indicate that there are significant differences between the Northern and Southern Hemispheres; much lower concentrations of sulphate and nitrate are observed in the Southern Hemisphere rains (for example, in southern Chile and New Zealand). However, values are likely to be high in certain areas of India and China. During this century a significant increase of sulphate and nitrate was found in Greenland ice cores (Neftel *et al.*, 1985; Mayewsky *et al.*, 1986) whereas in Antarctic ice cores no trends have been detected (Legrand and Delmas, 1986, 1987). These results indicate that, in contrast with the Northern Hemisphere, the Southern Hemisphere is being little affected by large-scale regional air pollution at present. On the other hand, since there are more efficient vertical mixing processes in the tropics and chemical lifetimes are longer in the upper troposphere, it is possible that significant quantities of pollutants emitted or formed in the tropics may be transported to higher latitudes.

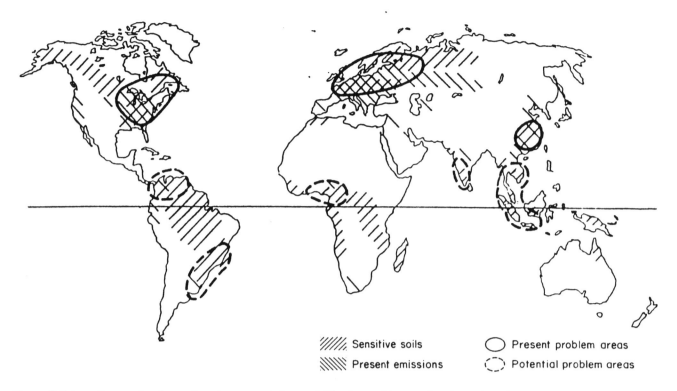

Figure 2: Schematic map showing regions that currently have acidification problems and regions where, based on soil sensitivity, expected future emissions and population density, acidification might become severe in the future (Rodhe *et al.*, 1988).

It seems that the acidification phenomenon could spread to the developing world. The combination of sensitive soils and increasing industrialization, with corresponding increases in emissions, has resulted in a study by SCOPE (Rodhe and Herrera, 1988) identifying southern China and other areas of southeast Asia, southwestern India, equatorial Africa including Nigeria, southeastern Brazil and northern Venezuela as being areas of probable sensitivity to soil and surface water acidification (Fig. 2) and a recent initiative has resulted in a study on Acid Rain in Asia (Bhatti *et al.*, 1992). In tropical regions, soil emissions and biomass burning are major contributors to the total emissions of both sulfur and nitrogen oxides and also for vapour phase and particulate organic compounds. Emissions due to human activities (industrial processes, fossil fuel consumption, biomass burning and volatilization of livestock and human wastes) already match or even possibly outweigh, emissions from natural sources (Rodhe *et al.*, 1988). Limited data on the chemical composition of rainfall have made it difficult to document the impact of these increased emissions. The pH of rainwater in tropical regions does seem to be consistently quite low (pH = 4.5-5.0) but this may be due to naturally occurring organic acids (Keene *et al.*, 1983; Sanhueza *et al.*, 1991). The importance of this area is reflected in several major activities in the International Global Atmospheric Chemistry Project (IGAC) of the IGBP.

The atmosphere also transports a wide variety of pollutants, including trace metals and other toxic contaminants. The atmosphere is also an important pathway for certain pollutants into the ocean (see Duce in Mantoura *et al.*, 1991).

6.   DEPLETION OF THE OZONE LAYER

The fragility of the ozone layer was first debated in the early 1970s, based on concern that supersonic aircraft, flying well into the stratosphere, would add pollutants to that layer and destroy the ozone layer (Crutzen, 1971; Johnston, 1971). Although the number of such aircraft did not become large enough to impact the ozone layer, scientific interest and increased observations of the stratosphere led to the realization that other pollutants, such as chlorofluorocarbons (CFC), may affect the ozone layer (Molina and Rowland, 1974). Chlorofluorocarbons had been deliberately designed to be non-toxic, non-flammable and inert to chemical change in use. This inert property, however, allowed them to remain in the atmosphere long enough to be transported into the stratosphere. Chlorofluorocarbons have lifetimes of 60 to 130 years and are potent greenhouse gases (about 15,000 times more efficient, on a per molecule basis than carbon dioxide). Their action in reducing ozone, also a greenhouse gas, acts in the opposite sense.

One of the most dramatic discoveries of our time was the identification of the hole in the Antarctic ozone layer in the second half of the 1980s. In 1985, UK scientists (Farman *et al.*, 1985) reported that during Antarctic spring time (September–October), large reductions of total

Table 1 *Total Ozone Trends (% per decade with 95% confidence limits; UNEP-WMO, 1991).*

Season	TOMS: 1979-91			Ground-based: 26-64°N	
	45°S	Equator	45°N	1979-91	1970-91
Dec-Mar	-5.2±1.5	+0.3±4.5	-5.6±3.5	-4.7±0.1	-2.7±0.7
May-Aug	-6.2±3.0	+0.1±5.2	-2.9±2.1	-3.3±1.2	-1.3±0.4
Sep-Nov	-4.4±3.2	+0.3±5.0	-1.7±1.6	-1.2±1.6	-1.2±0.6

ozone content were observed over the continent of Antarctica. The maximum decrease was in October, and by November the ozone concentrations were re-established. In the mid-1980s, there was a quasi-biennial cycle in the ozone hole intensity; however, for 4 of the past 5 years, the Antarctic ozone hole has been deep and extensive in area (UNEP-WMO, 1991). The first satellite measurements over the globe started in the mid-1960s, first from Soviet and then from American satellites. Measurements over Antarctica are on the edge of the view of satellites and though they were always revealing some peculiarities the results were rejected because data problems made them suspect. Retrospective analysis of past satellite and other data performed after 1985 shows that statistically significant spring changes in stratospheric ozone started around 1980. Special high-altitude aircraft observations in 1987 revealed very large amounts of CFC photodissociation products in the Antarctic stratosphere and the presence of stratospheric clouds which are formed only at very low temperatures. It is believed the ozone destruction is instigated by reactions which occur on the surface of these cloud particles releasing destructive catalysts. The strong polar vortex inhibits the exchange of heat between the Antarctic and warmer lower latitudes. With spring, sunlight appears and the released destructive catalysts destroy ozone. Scientific evidence now indicates that halogenated compounds are responsible for the Antarctic ozone hole.

Recent observations in the Arctic reveal depressions in the total ozone content in the Arctic as well, but these depressions are shallower, smaller in extent and duration and are not bonded to topographic features as their Antarctic counterpart. It is likely that future ozone depletion will depend on the atmospheric levels of halogenated compounds and on the particular meteorology of each Arctic winter.

Observations of ozone amounts from both ground based instruments and satellite (TOMS) now show significant decreases in total-column ozone in both the northern and southern hemispheres, but not in the tropics (UNEP-WMO, 1991; see Table 1). The heaviest losses are in the winter in the southern hemisphere. These decreases cannot be attributed to known natural processes.

Destruction of stratospheric ozone will lead to enhancement of solar ultraviolet radiation at the Earth's surface which suppresses plants and immune systems and causes skin cancer, eye damage and other health effects. If the process of destruction continues it will pose severe problems for the whole biosphere including people. Damage to US agriculture is estimated to be 1 to 5 billion US dollars (MacKenzie and El-Ashri, 1989; Heaton et al., 1991).

The physics and chemistry of the stratosphere have become reasonably well understood and the sources of the pollutant, CFCs, are quite limited in number and geographical coverage. The concept of an Ozone Depletion Potential was developed through the WMO/UNEP Ozone Assessments and allows for semi-quantitative assessment of the impact of various chemicals. These facts allowed for the negotiation of the Vienna Convention of 1985, which declared a need for protection of the ozone layer. The first practical step was the Montreal Protocol of 1987 requiring phasing out of certain CFCs. Now there is an understanding of a need to accelerate this phasing out process and to add new substances to the list. The new substances replacing old CFCs are about ten times less stable with lifetimes of the order of 1-10 years instead of 100 years. In particular, because they are destroyed largely in the troposphere, they have less effect on the ozone layer. Included in the agreements on the ozone layer are technology transfers to help less-developed countries take actions so as to not increase their emissions. Scientific studies and review, also included in the Protocol, allowed for modification of the Protocol based on new scientific information.

It should be noted that concentrations of tropospheric ozone in the western Europe have been increasing from values in the range 5-15 ppbv around the turn of the century to over 40 ppbv in the 1980s (Volz and Kley, 1988; Crutzen and Zimmermann, 1991), although the distribution is not homogeneous.

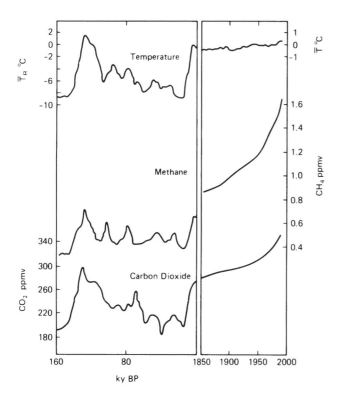

Figure 3: Time histories of temperature, carbon dioxide and methane for the past 160,000 years from the Vostok ice core (left) and for the past 140 years from ice cores and direct measurements (right). It is important to note that the time scales are quite different and the rates of change over the past century have been very large. The temperature scale for the Vostok core (left), representing a regional temperature, has been matched to that on the right assuming the global mean temperature change during an ice age was about $-5°C$.

7. NATURAL CLIMATE VARIABILITY AND ANTHROPOGENIC CLIMATE CHANGE

The Earth's climate has varied between glacial and interglacial states over the past million years, due to only small imbalances in the planetary radiation budget that have probably been amplified by changes in atmospheric concentrations of greenhouse gases. The history of atmospheric composition over the past 160,000 years has been established through analysis of air bubbles in ice cores drilled at the Soviet Antarctic station Vostok (Barnola et al., 1987; Chappellaz et al., 1988) and for rapid changes during glacial and deglaciation from Greenland ice cores (Staffelbach et al., 1988). Ice core data over the last 200 years (Siegenthaler and Oeschger, 1987) coupled with the precise direct atmospheric measurements of the past three decades make it possible to map the time variations in atmospheric composition (carbon dioxide and methane, for example) and temperature (Fig. 3). During warm periods, carbon dioxide and methane concentrations were generally 1.5 or 2 times larger than during cold periods, although cause and effect cannot be determined and for some periods the changes are negatively correlated.

The same ice cores also show that during the last two or three centuries, the concentration of these two gases started to increase, at first slowly, and then during the last century, much more rapidly, much faster than at any other time during the last 160,000 years. During the Eemian period (120–130 ky before present), the last interglacial, the global mean temperature was about 2°C warmer than present. At the time of maximum glaciation (about 18 ky before present), the global mean temperature was several degrees colder, while during the early Holocene (about 6–9 ky before present), temperatures were about 1° C warmer than present, the so-called "climatic optimum". Evidence is that these changes were global in extent. For several other warm or cold periods, such as the Medieval warm period or the Little Ice Age, it is not clear whether these phenomena were only regional or global. The ice cores do not show any marked changes in the carbon dioxide or methane concentrations for these periods (Khalil and Rasmussen, 1987). Volcanoes and other natural phenomena also play roles in determining climate. Further study of these paleoclimatic data is organized through the Past Global Changes Project (PAGES) of the IGBP.

The Earth's climate has warmed 0.3 to 0.6 °C during the last century (Fig. 3) and the 1980s has been the warmest decade on record. However, this warming could be entirely due to natural variability and cannot be attributed to the impact of increasing greenhouse gas concentrations (Houghton et al., 1990). The warming over the century has been mainly due to rapid increases from 1910–1944 and after the mid-1970s (both about 0.1°C/decade or 1°C/century). Warming rates coming out of the last glaciation were 0.2–0.3°C/century. Stratospheric temperatures have decreased in the past decade or two (UNEP-WMO, 1991), which may be due to ozone depletion and an increase of carbon dioxide. There is a possibility that this decrease may cause climatic change through dynamical linkages. The sea-level has risen 10–20 cm over the same period, mainly through thermal expansion of the ocean waters due to warming and melting of small land glaciers. The climate system may be thought of as being controlled by a "fast system" and a "slow system". The fast system is mainly determined by the atmospheric heat engine which drives the whole Earth environment and determines the ultimate amplitude and geographical patterns of climate change. Adjustments in the fast climate system are very complicated and involve feedback processes which amplify or reduce the primary greenhouse effect (Houghton et al., 1990). Water, in the form of vapour and clouds, plays the dominant role in the greenhouse effect. However, the water content of the atmosphere adjusts very rapidly and its atmospheric concentration is not being significantly modified, on a global basis, by human activities. The long-lived greenhouse gases whose concentrations are being changed,

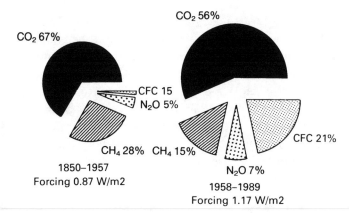

Figure 4: Accumulated climate forcings due to the long-lived greenhouse gases (except ozone) for the periods 1850-1957 and 1958-1989 (modified from Hansen and Lacis, 1990; reproduced from IGBP Rpt # 13).

are carbon dioxide, methane, nitrous dioxide, ozone and chlorofluorocarbons. Although carbon dioxide is the major contributor to the radiative forcing, the other greenhouse gases are also important (Fig. 4). The contribution of ozone has not been included in this Figure. The slow climate system is controlled by the global ocean which sets the pace for climatic change and may introduce a delay of decades in the transient response of the Earth's climate to greenhouse forcing. The temperature change of the past few decades is less than could be inferred directly from the imbalance of the radiation budget due to the observed increase in greenhouse gases.

This may be due to our underestimating the lag of the oceans, overestimating positive feedbacks of the fast climate system or other influences that could be counterbalancing increasing greenhouse gases. For example, recent studies have suggested that increasing northern hemispheric concentrations of anthropogenic aerosols may be increasing the Earth's albedo and hence acting to cool the planet (Charlson *et al.*, 1991). There is also the natural process whereby dimethyl sulphide released from the ocean may cause changes in cloud albedo and influence the climate (Charlson *et al.*, 1987). It is difficult to detect greenhouse gas-induced global warming because of natural variability and other factors. It is important that the origin and nature of natural climate variations be understood.

The problem of predicting future climate change, induced by human's activities, cannot be separated from that of understanding and predicting natural variations. Typically every 4–5 years the eastern tropical Pacific Ocean warms leading to disastrous consequences for the local fishery. We now know that the El Niño is part of a much larger natural variability of the coupled ocean–atmosphere–land climate system. In 1982-83, the strongest El Niño of the century occurred with extreme droughts in many regions and floods in others. We can

expect that this type of event will occur again and the scientific results of the Tropical Ocean–Global Atmosphere Programme (TOGA) of the World Climate Research Programme will aid in the prediction. For much of the past 20 years, the Sahel region of Africa has been impacted by drought, causing great suffering to this region. Although human activities are compounding these problems, it is clear that natural variability must be understood and better predicted to enable societies to better respond to its impacts.

The concentration of carbon dioxide, which accounts for over half the enhanced greenhouse effect due to long-lived gases in the past, has increased by 25% since the beginning of industrialization. The dominant importance of carbon dioxide is expected to continue in the future. Methane concentrations have been increasing at a rate more than twice as fast as carbon dioxide. The atmospheric concentration of nitrous oxide has increased by almost 10% in the past century. Changing ozone concentrations are also important for the radiative forcing of climate (Lacis *et al.*, 1990). In the Northern Hemisphere, tropospheric ozone has increased by a factor of two or three; in contrast, no increase or even a decrease (at Samoa and South Pole stations) is indicated by scanty data series in the Southern Hemisphere (i.e. Janach, 1989). Chlorofluorocarbons, virtually absent only 40 years ago, are the fastest growing greenhouse gas (by 5–10% per year) and already contribute about 20% of the greenhouse effect. It should be noted that the atmospheric concentrations of the longer-lived greenhouse gases (carbon dioxide, CFCs and nitrous oxide) adjust only slowly to changes in emissions. Because of uncertainties in future emissions, projections on climate warming rates will also be uncertain. It should also be emphasized that the anthropogenic emissions are but small increments on the large exchanges between the atmosphere and the oceans or land. Relatively minor adjustments in the natural system could significantly affect future atmospheric greenhouse gas concentrations.

The oxidizing capacity of the atmosphere plays a key role in the biogeochemical cycles of many compounds. Reduced species emitted from natural (i.e. methane) or anthropogenic activities (i.e. hydrocarbons) are oxidized in the atmosphere by the hydroxyl radical and subsequently removed through wet and dry deposition. Due to increasing anthropogenic emissions (i.e. carbon monoxide, nitrogen oxides) the oxidizing capacity of the atmosphere should be changing; some authors (e.g. Rotmans and Swart, 1990; Crutzen and Zimmermann, 1991) indicate that it is likely that an overall decrease in the reactivity of the atmosphere is occurring, leading to a build up of trace gases that would have been removed by hydroxyl reactions. On the other hand, recent analysis of methyl chloroform (CH_3CCl_3) data from the Atmospheric Lifetime Experiment–Global

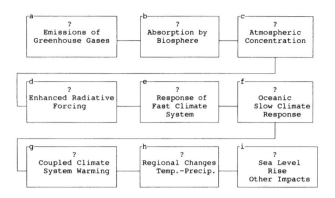

Figure 5: Schematic diagram of the factors determining the change in climate due to increasing greenhouse gas concentrations. Each has major uncertainties. Anthropogenic greenhouse gas emissions (a), some of which are absorbed by the biosphere (b) lead to atmospheric concentrations (c) which produce enhanced radiative forcing (d). The climate system responds on both fast (e) and slow (f) time scales, resulting in coupled climate system warming (g). The climate warming will have regional variations in both temperature and precipitation (h) and result in sea-level rise and other impacts (i).

Atmospheric Gases Experiment (ALE/GAGE) implies an increase of the OH concentration in tropical latitudes (Prinn *et al.*, 1991). The spatial inhomogeneity of tropospheric chemistry makes generalizations difficult. Since this feedback mechanism could significantly affect future concentrations of greenhouse gases, this subject (oxidizing capacity of the atmosphere) must be a priority area of future research.

To understand the uncertainties in climate simulations for enhanced atmospheric concentrations of greenhouse gases, it is useful to consider the following segments of the problem (Fig. 5). Over the next decades to century, anthropogenic emissions of greenhouse gases to the atmosphere are likely to increase, but their actual variations are difficult to predict. Presently about 50% of the anthropogenic emissions of CO_2 remain in the atmosphere and the remainder are absorbed in the oceans or on land; the qualification of this ratio is a major uncertainty. The Joint Global Ocean Fluxes Study (JGOFS), the International Global Atmospheric Chemistry (IGAC) Project, and the Global Change and Terrestrial Ecosystems (GCTE) Project, all of the IGBP, and other projects will make contributions towards understanding these biogeochemical cycles. The total response of the climate system to these additional greenhouse gases will depend on the climate system feedbacks or amplifiers, which are quite uncertain (Houghton *et al.*, 1990). Present climate models of the fast system (the atmosphere coupled to a shallow ocean) indicate that the global mean warming for doubled concentration of carbon dioxide will be 1.5 to 4.5°C (Houghton *et al.*, 1990). A major unresolved problem is that of predicting accurately the role of clouds

and generally water in its various phases (McBean, 1991a). Also, the models need to be able to predict the regional changes in the global water cycle, including precipitation. The Global Energy and Water Cycle Experiment (GEWEX) of the World Climate Research Programme is aimed at these problems. An additional complication that is now becoming more apparent is the role of aerosols in climate (Charlson *et al.*, 1987, 1991). The time-dependent response of the climate system to this additional radiative forcing will depend on the response of the slowest component, i.e. the oceans. This will result in a lag and/or reduction in climate change (McBean, 1991b). The formation of deep water, the thermohaline circulation that links the global oceans and the oceanic eddies that transport heat, salt and other properties are examples of key ingredients, as yet poorly understood, of the ocean's role in climate. The World Ocean Circulation Experiment (WOCE), also a component of the WCRP, is addressing these issues. The ocean also plays an important role in the cycling of greenhouse gases, such as carbon dioxide. The exchanges of carbon between the ocean and the atmosphere are many times larger than the anthropogenic emissions. How these exchanges are controlled and how they will alter in the future (as they probably altered in the past) is not well understood.

In the report of the IPCC, a scientific strategy, based on a combination of process studies, observation and modeling, to achieve effective prediction of the climate system was outlined. There is still significant scientific uncertainty and to narrow these uncertainties, substantial scientific activities need to be undertaken.

The following areas were considered most critical:
(a) control of the greenhouse gases by the Earth System;
(b) control of radiation by clouds (and aerosols);
(c) precipitation and evaporation;
(d) ocean transport and storage of heat;
(e) mass balance of polar ice sheets (for sea-level rise), and
(f) ecosystems processes.

Components of the World Climate Research Programme, the International Geosphere–Biosphere Programme and related programs have been designed to address these issues. To observe and document the fundamental aspects of the climate system and changes occurring within it, improvements are needed in the global atmosphere, ocean, ice and land surface observing systems. The Second World Climate Conference recommended that a comprehensive system for climate monitoring (the Global Climate Observing System) be implemented.

Within the United Nations International Negotiating Committee discussions have already started for a Framework Convention on Climate Change. Although scientific evidence for possible global warming and its potential impacts justify such international discussions, it is important to maintain strong scientific programs to reduce our uncertainties about climate variability and change and to provide a firmer basis for policy development. The Convention should build in scientific review and re-assessment so that targets and strategies can be modified as appropriate, as was done for the Montreal Protocol on Depletion of the Ozone Layer. Further, since no plausible emission reduction strategy will entirely prevent climate change, it is essential that predictions of climate variability and change be improved as a basis for adaptation, as much as for response.

8.   SUMMARY OF ISSUES

It is important to recognize that although these issues cover a range of spatial scales from local/regional to global, they are clearly interrelated. Urban air pollution is in many places an immediate issue with documented impacts on human health and on ecosystems. The solution to the problem is usually evident and the response time, in terms of reducing the atmospheric loading, is quite short. In Europe and North America there have been political decisions made to address the problem of acidic precipitation, and evidence for improvement is being seen in some chemical monitoring data. There is the potential for problems to develop in other parts of the globe. For the ozone layer, depletion is being observed and countries have responded through international agreements. Unfortunately, the response time of adjustments to reductions in CFC emissions is decades or more so the impact of positive actions will not be observed for some time. Climate change due to increasing atmospheric concentrations of greenhouse gases is a different kind of issue. As yet, we cannot detect, in a strict statistical sense, the imprint of human intervention on the physical climate of the globe, despite the clear evidence of human intervention in the chemical composition of the globe. Further, the response time of the climate system, as determined primarily by the response of the oceans, is long (at least decades) so that changes in what humans do will take time to show themselves.

For our improved understanding of the Earth's atmosphere and climate, it is clear that we must treat it in a more integrated way than has been our practice in the past. In all cases, it is the increased loading of the atmosphere by emissions due to human activities that are the concern. In some cases, such as the carbon cycle, human intervention has been a small part of very large natural cycles; but usually these natural cycles are in equilibrium, at least on time-scales of decades to centuries, so that the small anthropogenic addition has a major impact in creating an imbalance. In other cases, such as the sulfur cycle, anthropogenic sources are about the same as natural sources; while for CFCs, human activities are the only source of the chemical.

It is important to recognize the strong interconnections between these atmospheric environmental issues. For urban air quality, acid rain and increasing greenhouse gas concentrations, there are common sources of pollution: generally, the burning of fossil fuels, the waste emissions of industrial processes and agricultural practices. For example, fossil fuel combustion results in emissions of carbon and nitrogen oxides which result in photochemical smog in cities, acidify the rain and increase the greenhouse effect. They also have an impact on ozone, both tropospheric and stratospheric. The benefits of emission strategies for reducing urban air pollution will aid in responding to the increasing greenhouse effect and other environmental problems. This synergism adds weight to the argument that it is appropriate to take action on fossil fuel emissions, for example, because the benefits will accrue in several places.

9.   A LOOK AHEAD

As we look ahead 30 years, we see an increasing interplay of the atmospheric environment, climate and development. Severe weather events will continue to happen and it is unlikely that their general properties will change noticeably. We will still have devastating hurricanes, tornadoes and storm-driven coastal ocean surges. We can expect improvements in our ability to observe and predict such weather events over most of the globe. Improvements in our observations will mainly come from better understanding and observations from satellites and automated systems on land and in the oceans. Unfortunately, surface-based meteorological observations have deteriorated over the past decade; major commitments from all countries are needed to reverse this trend. Data on sea-level and river flow are needed for storm and flood warnings. As computer capacity increases, numerical weather forecast models will be run on finer and finer scales, allowing simulation of smaller scale phenomena and improving the overall forecast skill. Improved observations will, however, be needed to justify these models. Conversion of improved numerical forecasts for 10 days into weather predictions and warnings will be of great economic benefit in those parts of the world where the economic infrastructure is such that they can take advantage of them.

Dynamic extended range forecasts now show some skill for time-average forecasts well beyond 10 days. As our understanding of these predictions improves, better monthly to seasonal forecasts will be possible. On seasonal to interannual time scales, the oceans become critical. The Tropical Ocean–Global Atmosphere Programme of the WCRP (1985–95) is leading to increased confidence that the slow oscillations of the coupled El Niño-Southern Oscillation (ENSO) can be modelled and eventually predicted. Experimental predictions are now being done and it is expected that prediction of major interannual changes associated with ENSO will be possible by the end of the century.

Since the basis for atmospheric forecasting beyond a month or so lies in the longer term memory of the land-surface (perhaps for season or so) and of the oceans, it is imperative that we observe, understand and model those components of the coupled climate system, as well as the atmosphere. The World Ocean Circulation Experiment will finish its observing phase about 1997 and fully understanding the data will take up to another decade. New satellite systems will provide improved global coverage of ocean surface characteristics (wind, temperature and topography), but we will still need measurements within the oceans, as well as on their surfaces to provide ground truth for satellite calibration purposes.

To provide the observational basis for prediction of natural climate variability, it is essential that the Global Climate Observing System (GCOS, as proposed by WMO, IOC, ICSU and others), including operational programs such as the World Weather Watch (WWW), Global Ocean Observing System and the Global Atmosphere Watch (GAW) and research programs, be fully implemented. Support for the Global Energy and Water Cycles Experiment (GEWEX) which will continue until about 2010 and for the follow-on scientific programs for TOGA and WOCE on coupled climate system variability is also required. Assuming the nations of the world take up this challenge, one can be optimistic that seasonal through interannual climate predictions will be in place in the outlook period.

It is always necessary to mention the possibility of surprises. Depletion of the Antarctic ozone layer was unpredicted. The level of the Caspian Sea, which had been falling for many years, started to rise, quite unexpectedly (Golitsyn and Panin, 1989; Golitsyn *et al.*, 1990) and has risen by almost 2 metres in the last 15 years. The atmosphere and the climate system are very non-linear systems and perturbations can lead to quite unexpected results. Further, there are other natural phenomena such as volcanoes, that can significantly perturb the atmosphere. From the history of the climate record, we know that major changes have happened in the past and hence are likely to happen again in the future. This only stresses the necessity for global monitoring and further studies of the climate system.

Changes in urban air quality and acidic precipitation will be very dependent on local and regional emission control strategies. In much of the developed world, emissions from stationary sources are being reduced; some progress is also being made in the transportation sector. Sulfur emissions and observed atmospheric concentrations of sulfur dioxide and sulphate aerosols

actually decreased in western Europe between 1979 and 1986 by 15–19% (Mylona, 1989). Increases in sulfur and nitrogen emissions in North America and western Europe are likely to be small over the next 30 years. For Asia, Africa and Latin America, emissions *per capita* are expected to increase (to levels closer to those in the developed world), which will lead to increased total emissions. With major increases in emissions, the already serious urban air pollution problems in several large cities in developing nations will become more severe and many more cities will join the critical list. As noted earlier, there is also a potential problem for acidification of lakes, streams and soils in parts of Asia, Africa and South America. These increases in emissions of nitrogen and sulfur compounds to the atmosphere could change potential into actual.

The evolution of the Earth's climate over the next 30 years will result from ongoing natural variations combined with a warming trend forced by anthropogenic modifications of the atmospheric concentrations of greenhouse gases and aerosols. Based on current model results, the IPCC predicted that the rate of increase of global mean temperature during the next century will be about 0.3 °C/decade or 3°C/century (with an uncertainty range of 0.2 to 0.5°C/decade), assuming emissions continuing to increase as they have over the recent past. This temperature change will be greater and faster than that seen over the past 10,000 years. Hence, the global average temperature would be about 1oC warmer by 2020 and about 3oC warmer by 2100. High northern latitudes would warm by several times more during the winter and generally land areas will warm more than the oceans. The sea-level would rise another 10 cm. The temperature and sea-level rises would not be steady because other, natural factors will modulate the change.

This magnitude of climate change and sea-level rise would seriously threaten low-lying islands and coastal zones (Second World Climate Conference, 1990; Jaeger and Ferguson, 1991). There would be impacts on water resources, agriculture (especially in arid or semi-arid areas), forests and fisheries. Climate change impacts will range from the inundation of low-lying areas to reducing the snow-covered areas in Austrian Alps. Climate change, at the rate and magnitude currently estimated over the next 40 years, may exceed the "critical loads" for certain ecosystems and economic sectors and may further the gap between developing and developed countries, because the impacts in many cases will be felt most severely in regions already under stress. An important and developing area of research is on climatic impact analyses and improved methods for assessing the potential effect of changes on the biosphere and an aspects of the world economy are needed. Since the impacts will be felt primarily on the regional or smaller scale, climate modellers and impact assessment scientists must work together to interpret climate change scenarios on these scales. The impacts of climate change are discussed further in other chapters (e.g. 2 and 9) of this volume and also in the report of IPCC Working Groups II and III (1991).

If international and national actions are taken to reduce the rate of increase of emissions, or actually reduce the emissions, then the temperature increase would be correspondingly less. We again must stress the possibility of surprises. The role of the biosphere has been neglected in these calculations and could amplify or possibly reduce the change. The direct effect of aerosols could be to reduce the change, while the indirect impacts through clouds is less clear. The IPCC also presented results for other emission scenarios. It is important to note that, even if, the carbon dioxide emissions are reduced to 50% of 1985 levels by 2050, the atmospheric concentration of carbon dioxide will still increase through the next century. Actual greenhouse gas-induced warming will be delayed by thermal inertia of the oceans, but it will also continue long after the composition of the atmosphere is stabilized.

10. SCIENTISTS AND SCIENTIFIC CO-OPERATION

When dealing with global environmental issues such as ozone depletion and climate change, it is essential that there be strong international co-operation. Through international independent organizations such as the International Council of Scientific Unions and through intergovernmental agencies like the World Meteorological Organization, scientific meetings and internationally co-ordinated research programs have proven an effective way of involving large numbers of scientists from a wide range of countries. Unfortunately, the co-operation is not always optimum. In many countries economic conditions have led to the underfunding of science so that some scientific programs have participation only from a small number of countries. In particular there is need for more participation by scientists in developing countries, who need support from developed countries in order to carry out first-rate research. In this respect, an important initiative is the IGBP project called Global Change – System for Analysis, Research and Training (GC-START) (IGBP Report #15) which will lead to the creation of a world-wide system of research networks to implement the study of global change on a regional level. Equatorial South America, northern Africa and the tropical Asian monsoon region have been identified as three areas of highest priority for network development.

11. RECOMMENDATIONS FOR ACTION

In consideration of the widespread effects and very long recovery times, the following have been identified as priorities.

(a) Research Needs:
  (i) Nations are encouraged to contribute to the World Climate Research Programme and the International Geosphere-Biosphere Programme to provide the understanding, predictions and information base essential to determine climatic change impacts and to develop climate response strategies.
  (ii) Support is needed for research studies in atmospheric chemistry (and related oceanic chemistry), including the sources and sinks of atmospheric pollutants, recognizing that problems of urban air pollution, acidic deposition, ozone depletion, climate change and others are strongly linked to atmospheric chemistry and that it is an area that is considered most likely to generate surprises.

(b) Urgent Actions:
  (i) Nations are urged to contribute to the Global Climate Observing System (GCOS):
    to provide the data needed as a basis for understanding and prediction. The GCOS, sponsored by ICSU, the World Meteorological Organization and the Intergovernmental Oceanographic Commission of Unesco must be a globally integrated observational network with appropriate means for establishing data standards and quality control mechanisms and with free and scheduled international data exchange.
  (ii) Improved models need to be developed to provide the basis for prediction of future evolution of the atmosphere and climate system.
    Model resolution and process representation must be improved to address regional variations in the atmosphere and climate.
  (iii) All nations should review options for, and take steps to reduce, emissions of gases and particulates to the atmosphere. It seems appropriate that developed countries should lead in these reductions.
    Significant reduction of fossil fuel consumption will have positive effects on reducing urban air pollution, acidic deposition and transport of toxic metals, and in slowing the accumulation of greenhouse gases and reducing the rate of climate change.
  (iv) The scientific and technical capacity of nations needs to be improved, particularly in developing countries to address the regional aspects of these issues and to contribute to global programs.

For example, assistance could be provided to developing countries through the World Climate Programme to enable them to apply available climatic data to improve the efficiency of a wide range of economic and social activities and to assess actual and potential impacts of climatic change and variability.

REFERENCES

Barnola, J.M., Raynaud, D., Korotkevich, Y.S. and Lorius, C. 1987. Vostok ice core: a 160,000 year record of atmospheric CO2. *Nature*, **329**, 408-414.

Bhatti, N., Streets, D.G. and Foell, W.K. 1992. Acid rain in Asia. *Environmental Management* (in press).

Brydges, T.G. and Neary, B.P. 1984. Target loadings to protect surface waters. Ministry of Environment, Ontario, Canada.

Budyko, M.I., Ronov, A.B. and Yanshin, A.L. 1987. *The History of the Earth's Atmosphere*. Springer-Verlag, Berlin, pp.139.

Chadwick, M.J., and Kuylenstierna, J.C.I. 1991. Critical loads and critical levels for the effects of sulfur and nitrogen compounds. In: *Acid Deposition*, J.W.S. Longhurst (ed.), Springer-Verlag, Berlin.

Chappellaz, J., Barnola, J.M., Raynaud, D., Korotkevich, Y.S. and Lorius, C. 1988. Ice core record of atmospheric methane over the past 160,000 years. *Nature*, **345**, 127-131.

Charlson, R.J., Lovelock, J.E., Andreae, M.O. and Warren, S.G. 1987. Oceanic phytoplankton, atmospheric sulfur, cloud albedo and climate. *Nature*, **326**, 655-661.

Charlson, R.J., Langner, J., Rodhe, H., Leovy, C.B. and Warren, S.G. 1991. Perturbation of the Northern Hemisphere radiative balance by backscattering from anthropogenic aerosols. *Tellus*, **43AB**, 152-163.

Crutzen, P.J. 1971. Ozone production rates in an oxygen-hydrogen-nitrogen atmosphere. *J. Geophys. Res.*, **76**, 7311-7327.

Crutzen, P.J., and Zimmermann, P.H. 1991. The changing photochemistry of the troposphere. *Tellus*, **43AB**, 136-151.

Dickinson, R.E. 1986. Impact of human activities on climate – a framework. In: *Sustainable Development of the Biosphere*, W. Clark and R. Munn (eds.), 252-289.

Farman, J.C., Gardner, B.G., and Shanklin, J. 1985. Large losses of total ozone in Antarctica reveal seasonal ClO$_x$/NO$_x$ interaction. *Nature*, **315**, 207-210.

Folland, C.K., Owen, J., Ward, M.N. and Colman, A. 1990. Prediction of seasonal rainfall in the Sahel region using empirical and dynamical methods. *J. Forecasting*.

Golitsyn, G.S. and Panin, G.N. 1989. Contemporary changes of the Caspian sea-level. *Soviet Meteorologya and Hydrologya*, No. 1, 57-64.

Golitsyn, G.S., Dzuba, A.V., Osipov, A.G. and Panin, G.N. 1990. Regional climate changes and their impact on the Caspian Sea-level rise. Doklady, *USSR Ac. Sci.*, **313**(5), 1224-1227.

Hansen, J.E., and Lacis, A.A. 1990. Sun and dust versus greenhouse gases: an assessment of their relative roles in global climate change. *Nature*, **346**, 713-719.

Heaton, G., Repetto, R. and Sobin, R. 1991. *Transforming*

Technology: An Agenda for Environmentally Sustainable Growth in the 21st Century. WRI Report, April 1991, pp. 36.

Houghton, J.T., Jenkins, G.J., and Ephraums, J.J. (eds.) 1990. *Climate Change: The Scientific Assessment*. The IPCC Report. Cambridge University Press, Cambridge, pp. 364.

IGBP, 1990. *Terrestrial Biosphere Exchange with Global Atmospheric Chemistry. International Geosphere–Biosphere Programme*, Stockholm, Rpt 13, pp. 103.

IPCC Working Group II, Impacts 1991. Climate Change. The IPCC Impacts Assessment.

IPCC Working Group III, Response Strategies 1991. Climate Change. *The IPCC Response Strategies*. Island Press, Washington, pp. 272.

Jaeger, J., and Ferguson, H.L. (ed.) 1991. *Climate Change: Science, Impacts and Policy*. Cambridge University Press, Cambridge, pp. 591.

Janach, W.E. 1989. Surface ozone: trend details, seasonal variations and interpretation. *J. Geophys. Res.*, **94**(18)289-295.

Johnston, H.S. 1971. Reduction of stratospheric ozone by nitrogen oxide catalysts from supersonic transport. *Science*, **173**, 517-522.

Keene, W.C., Galloway, J.N. and Holden, J.D. Jr. 1983. Measurements of weak organic acidity in precipitation from remote areas of the world. *J. Geophys. Res.*, **88**, 5122-5130.

Khalil, M.A.K. and Rasmussen, R.A. 1987. Atmospheric methane: trends over the last 10,000 years. *Atmos. Env.*, **21**, 2445-2452.

Lacis, A.A., Wuebbles, D.J. and Logan, J.A. 1990. Radiative forcing of global climate by vertical distribution of change of atmospheric ozone. *J. Geophys. Res.*, **95**, 9971-9981.

Legrand, M.R. and Delmas, R.J. 1986. Relative contributions of tropospheric and stratospheric source to nitrate in Antarctic snow. *Tellus*, **38B**, 236-249.

Legrand, M. and Delmas, R.J. 1987. A 220-yr continuous record of volcanic H$_2$SO$_4$ in the Antarctic ice sheet. *Nature*, **327**, 671.

Lorenz, E.N. 1969. The predictability of a flow which possesses many scales of motion. *Tellus*, **21**, 289-307.

MacKenzie, J.J. and El-Ashri, M.T. 1989. Tree and crop injury: a summary of the evidence. In: *Air Pollution's Toll on Forests and Crops*. MacKenzie, J.J. and El-Ashri, M.T. (eds.), Yale University Press, New Haven, Connecticut.

Mantoura, R.F.C., Martin, J.-M. and Wollast, R. (eds.) 1991. *Ocean Margin Processes in Global Change*. J. Wiley and Sons, Chichester, 469 pp.

Mayewsky, P.A., Lyons, W.B., Spencer, M.J., Twickler, M., Dansgaard, W., Koci, B., Davidson, C.I. and Honrath, R.E. 1986. Sulfate and nitrate concentrations from a South Greenland ice core. *Science*, **232**, 975-977.

McBean, G.A. 1991a. Global energy and water cycles. In: *Climate Change: Science, Impacts and Policy*. J. Jaeger and H.L. Ferguson (eds.), Cambridge University Press, Cambridge, pp. 591.

McBean, G.A. 1991b. Possible impacts of climate change on marine resources. *Trans. Roy. Soc. Canada*. (in press).

Molina, M.J. and Rowland, F.S. 1974. Stratospheric sink for chlorofluoro-methanes: Chlorine-atom catalyzed destruction of ozone. *Nature*, **249**, 810-812.

Mylona, S.N. 1989. Detection of sulfur emission reductions in

Europe during the period 1979-1986. EMEP MSC-W Report 1/89, Norwegian Meteor. Inst., Oslo, pp.149.

National Academy of Sciences, 1987. *Confronting Natural Disasters. An International Decade for Natural Hazard Reduction.* US National Academy of Sciences, Washington.

Neftel, A., Beer, J., Oeschger, H., Zurcher, F., and Finkel, R.C. 1985. Sulfate and nitrate concentrations in snow from South Greenland 1985-1978. *Nature*, **314**, 611-613.

Nilsson, J. and Grennfelt, P. (eds.) 1988. Critical loads for sulfur and nitrogen. Miljorapport, 15, Nordic Council of Ministers, Copenhagen.

Prinn, R., Cunnold, D., Simmonds, P., Alyea, F., Boldi, R., Crawford, A., Fraser, P., Gutzler, D., Hartley, D., Rosen, R., and Rasmussen, R. 1991 Global average concentration and trend for hydroxyl radicals deduced from ALE/GAGE trichloroethane (methyl chloroform) data for 1978-1990. *J. Geophys. Res.*, (in press).

Rodhe, H. and Herrera, R. (eds.) 1988. *Acidification in Tropical Countries.* John Wiley and Sons, Chichester, pp.405.

Rodhe, H., Cowling, E., Galbally, I.E. Galloway, J.N. and Herrera, R. 1988. Acidification and regional air pollution in the tropics. In: *Acidification in Tropical Countries.* H. Rodhe and R. Herrera (eds.), John Wiley and Sons, Chichester, pp.3-39.

Rotmans, J. and Swart, R.J. 1990. The role of the CH_4-CO-OH cycle in the greenhouse problem. *Sci. Total Environ.*, **94**, 233-252.

Rowland, F.S. 1990. Stratospheric ozone depletion by chlorofluorocarbons. *Ambio*, **19**, 281-292.

Sanhueza, E., Ferrer, Z., Romero, J. and Santana, M. 1991. HCHO and HCOOH in tropical rains. *Ambio*, **20**, 115-118.

Second World Climate Conference Statement 1990. Available from World Meteorological Organization, Geneva, pp.10.

Siegenthaler, H. and Oeschger, H. 1987. Biospheric CO_2 emissions during the past 200 years reconstructed by deconvolution of ice core data. *Tellus*, **39B**, 140-154.

Staffelbach, T., Stauffer, B., and Oeschger, H. 1988. A detailed analysis of the rapid changes in ice-core parameters during the last ice age. *Annals Glaciology*, **10**, 167-170.

Swedish Ministry of Agriculture 1982. *Acidification Today and Tomorrow.* Swedish Ministry of Agriculture, Stockholm, pp.231.

UNEP-WMO. 1991. Executive Summary: Scientific Assessment of Stratospheric Ozone. *United Nations Environment Programme — World Meteorological Organization*, Geneva, pp.7.

Vaghjiani, G. and Ravishankara, A. 1991. New measurements of the rate coefficient for the reaction of OH with methane. *Nature*, **350**, 406-409.

Volz, A. and Kley, D. 1988. Evaluation of the Montsouris series of ozone measurements made in the 19th Century. *Nature*, **332**, 240-242.

Wayne, R.P. 1985. *Chemistry of Atmospheres.* Clarendon Press, Oxford, pp.361.

Whelpdale, D.M. 1991. Atmospheric Pollution. *State of the Environment Report*, United Nations Environment Programme, (to be published).

Zhao, D. and Seip, H.M. 1991. Assessing the effect of acid deposition on soil in southwestern China. *Water, Air and Soil Poll.*, (in press).

Chapter 8: Marine and Coastal Systems

P. Bernal and P.M. Holligan

EXECUTIVE SUMMARY

Marine and coastal systems provide the world's population with important food resources and with facilities for recreation, transportation and waste disposal. The proportion of people living close to the coast continues to increase. The different uses of the coastal zone give rise to conflicts over resource exploitation, conservation and pollution issues and, with generally weak regulation policies, to continuing severe degradation of coastal ecosystems.

Three key issues underlying this situation are discussed in this chapter: the exploitation and sustainability of living resources, the degradation of marine and coastal environments, and the dynamic properties of the land–ocean interface. The latter is of particular importance since marine environments are significantly affected by climate change and associated variations in sea-level, and models to predict future states of marine and coastal systems must take into account the interactions of both climatic and human effects.

In considering what is required to promote sustainable utilization of marine resources, three actions are recommended: to initiate a global system for observing the coastal and open oceans, to construct and test simulation models of the responses of marine and coastal systems to environmental change, and to develop further the scientific basis for the management of living marine resources and for the conservation of coastal ecosystems. These actions will require maintaining and strengthening support for the ongoing and planned large scale research programs on coastal and marine environments sponsored by UN agencies and ICSU, in particular under the WRCP and IGBP.

Three particular needs are recognized: efficient technologies for the long-term observation of marine and coastal systems must be made operational as soon as possible, objective methods for assessing the socio-economic values of the coastal zone need to be established as a basis for determining the resources and priorities for

new research, and improved facilities are required especially in less developed countries for education and training in marine sciences. Progress towards sustainable utilization of marine resources will depend on establishing strong international partnerships in research and management between developed and developing countries, and on a close linkage between scientific, environmental and socio-economic policies for the future use of the coastal zone.

1. INTRODUCTION

The oceans play a major role in determining the climate of the Earth through effects on global albedo, on the storage and poleward transport of heat, and on the composition of the atmosphere. Major international research initiatives such as the Tropical Ocean Global Atmosphere (TOGA) experiment, the World Ocean Circulation Experiment (WOCE) and the Joint Global Ocean Flux Study (JGOFS) under the auspices of the WCRP and the IGBP are addressing the nature and causes of interdecadal and seasonal variability in the global climate system and ocean circulation, and the ways the ocean-atmosphere system responds to changes in the hydrologic cycle and in ocean biology through effects on the water vapour and trace gas content of the atmosphere. In general the major interactions between ocean processes and climate occur over timescales $>10^2$ years, although ice core records demonstrate decadal variability during glacial–interglacial transitions, and significant short-term global fluctuations are recognized from recent studies of El Niño-Southern Oscillation (ENSO) events which have strong impacts on the climate and fisheries of coastal regions (Johnson and O'Brien, 1990). These issues are considered in more detail under Chapters 6 and 7. Here we wish just to reinforce the concepts that the oceans are an essential component of the Earth's support system for life as we know it, that the composition of the atmosphere, and therefore the climate, would be very different in the absence of the biogeochemical activity of the marine biota, and that the main uncertainties in predicting climate concern interactions with the oceans (heat transport, cloud cover).

In this chapter we focus on issues of immediate concern that stem from the indirect (climate, sea-level changes — see Tegart *et al.*, 1990; Jaeger and Ferguson, 1991) and direct impacts of humans on the ocean margins (Mantoura *et al.*, 1991). We consider in the context of UNCED the steps needed to enable "prudent management of the marine environment for the survival of humanity", and the scientific basis for developing policies for the sustainable utilization of marine resources. At present too little attention is being given to studying the coastal oceans which provide the major proportion of living marine resources and, unlike the open oceans, are already strongly affected by human activities, and to assessing and predicting large scale ecological changes in the oceans which have important implications for the exploitation of marine resources but may not require a detailed understanding of the associated complex chemical and biological processes.

Estuaries, shelf seas and the continental slopes are extensively used by people in a wide variety of ways (recreation, energy production, transportation, etc.), provide more than 95% of the total marine fish catch, and receive much of the waste materials that result from urban and industrial development. Recent demographic forecasts indicate that pressures on coastal oceans will worsen as the world population grows and a greater proportion wishes to live near the sea it is estimated that by the year 2000 more than 75% of people will be living within 60km of the coast. The combined effects of humans and of variations and trends in climate and sea-level on coastal ecosystems are potentially severe (Bardach, 1989; UNESCO, 1990; Warrick and Farmer, 1990; Holligan and Reiners, 1991), and changes are already occurring that have profound and long-term implications for coastal management policies and threaten sustainable utilization of marine living resources at a global scale (Glasby, 1988; Ray, 1989). Furthermore, marine ecosystems under stress from humans are likely to be more sensitive to variations in climate and sea-level.

Any attempt to study and predict change in coastal oceans requires a global synthesis of the complex interactive and feedback processes occurring at the boundary between land and ocean as a basis for designing robust and statistically sound sampling strategies to test hypotheses about the nature and consequences of environmental change (see discussion on pp. 349-363 in Mantoura *et al.*, 1991). Such an approach is needed to understand how fluxes of energy (tides, waves, buoyancy), nutrients and suspended matter from continents and oceans affect biological processes on a global scale. The coastal plains and coastal oceans occupy only 3% and 5% respectively of the Earth's surface (Ray, 1989), but

account for perhaps 25% of global biological production and include the most productive ecosystems on Earth. Changes in coastal zones not only affect habitats and finite resources of great social importance, but also the way in which biogeochemical and biogeomorphological feedback processes determine the nature of the land–ocean boundary and, to some degree, the composition of the atmosphere. For example, large quantities of organic matter are being transferred from land to sea as a result of agricultural development and deforestation, but relatively little is known at a global scale about the fate of this material and the impacts it has on marine productivity, on the properties of coastal sediments, and on the CO_2 budget of the oceans.

Studies of global change at the ocean margins must depend on the application of appropriate observation and modeling methods, and of systems for classifying coastal oceans in terms of relevant physical, chemical and biological parameters. At present, not only are the resources to carry out these tasks inadequate but also the conceptual basis and special techniques needed for studying the dynamic behavior of the land–ocean boundary still need to be developed, taking account of the critical temporal and spatial scales of key processes and episodic events that characterise many coastal systems. Any global program on marine environmental change must clearly prioritize scientific objectives, and be based on observing methods that have the necessary resolution and coverage to provide a proper global database. Observation programs designed primarily for the open oceans such as the Global Ocean Observing System (GOOS) need to take account of the different scales of ocean and atmospheric forcing for shallow and deep waters, and include specific coastal ocean components that address land–sea interactions and the influence of land drainage basin dynamics on the ocean margins.

It is reasonable to assume that changes in coastal oceans are predictable to some degree in relation both to assessing the consequences of certain types of environmental forcing and to designing management policies needed to achieve certain environmental or ecological aims. Such optimism is based on new advances in understanding the structure and functioning of marine ecosystems and in modeling techniques. However, sustainable harvesting and the conservation of living resources in the context of multiple and conflicting uses by humans of coastal habitats remain an enormous challenge, and one that can only be met by acquiring a much better knowledge of how the coastal oceans work, how populations of marine animals are determined, and how coastal ecosystems respond to the anthropogenic impacts.

There are two major barriers to developing sensible policies for the use of the coastal oceans. The first concerns the lack of an objective basis for assessing the social and economic values of coastal environments and resources. Information for the fisheries is relatively good despite the lack of understanding of marine ecosystem dynamics market revenues are known, quotas for catches within Exclusive Economic Zones (EEZs) are being established, the ecological importance of traditional fishing methods especially in tropical regions is recognised, and attempts to reach global agreements on conserving stocks of certain species (in particular, marine mammals) are being made. By contrast, it is still very hard to place a true value on the use of coastal regions for tourism, waste disposal and transportation that takes into consideration the long-term economic and social costs of degrading marine environments as well as the immediate financial returns. The costs of developing and using alternative methods to get rid of untreated sewage and toxic materials that are presently being released into the sea would probably exceed by many times the total value of marine fisheries. Reefs, mangrove, salt marsh and other coastal habitats act as a natural defence against sea-level rise, and the economic implications of the losses of such habitats to urban and agricultural development in terms of constructing replacement walls and dykes to prevent coastal erosion are very difficult to assess (Broadus, 1989). The development of coastal regions along sensible lines must take account of the economic implications of over-exploitation of resources and of environmental damage (Pearce and Maler, 1991), giving particular attention to the relative time-scales of the benefits of resource use, the impacts of environmental change, and the capacity for environmental recovery.

The second barrier is a lack of reliable, quantitative data on the types, rates and causes of change in marine environments. There is much information, largely anecdotal, on the states of fish stocks and of coastal habitats before extensive interference by humans, but the ways in which marine food chains and environmental conditions have been altered even over the last few decades is poorly understood. There are many reasons for this situation — a lack of suitable techniques for observing change, shortage of financial support for monitoring programs, uncertainties about what ecological parameters to measure, and difficulties in distinguishing trend from natural variability — and it is a sad indictment of our attitude to marine environmental problems that ignorance is so often used as an excuse to take no action. One particular concern for the future is that coastal ecosystems will show rapid, threshold responses to the combined effects of human disturbance, of variations in climate and of a rise in sea-level, leading to irreversible declines in living resources and in ecosystem health and biodiversity. Developing the capability of predicting the behavior of coastal ecosystems subjected simultaneously to several types of environmental change is a major scientific challenge, but we must not make the mistake of thinking that each type of change can be considered in isolation; non-linear, interactive responses are to be expected.

In this chapter we review briefly three scientific issues that are of central importance in attempts to use marine and coastal environments effectively. The first covers the scientific principles underlying the control and exploitation of populations of commercially valuable marine organisms, including those amenable to mariculture, which are a major source of protein for the peoples of many less developed countries (Bardach, 1989). The second issue focuses on the wide range of environmental problems that stem from human disturbance and pollution of nearshore marine ecosystems (GESAMP, 1990); understanding both the causes and consequences of ecological change in marine systems is a responsibility we must accept immediately for the sake of the interests of future generations. And the third concerns the ocean margins which are sensitive to climate and sea-level variations and which are already significantly altered by human activities; a better understanding of the complex processes occurring in the boundary region between land and ocean is fundamental to the development of long-term plans for the sustainable use of coastal oceans (Ray, 1989).

2. REVIEW OF MAJOR SCIENTIFIC AND ENVIRONMENTAL ISSUES

2.1 Exploitation of living marine resources

The world annual fish landings are now about 100 million tonnes of which about 70% are consumed directly by humans (FAO, 1989). Fish protein represents about 16% of the animal protein intake of the world population (Bardach, 1989), and is of greatest dietary importance to the people of developing countries for whom the total daily protein intake is, on average, 60% of that for people in developed countries (Table 1). Other major living resources include shellfish, crustaceans, molluscs, seaweeds and, largely in the past, whales.

Exploitation by man has by far the greatest impact on stocks of living marine resources, with catches often exceeding the annual production and leading to collapse of populations. Man acts as a generalist top predator that develops new strategies of resource exploitation according to demand and technological abilities. Fishing methods are now so efficient that large-scale changes are taking place not only in the stocks of most marketable species of fish and shellfish but also in the structure and dynamics of the supporting food webs (e.g. Oliva and Castilla, 1986; Duran and Castilla, 1989).

Over 90% of the world fish catch is reported from coastal oceans where biological productivity in the surface illuminated layers is maintained by the efficient replenishment of inorganic nutrients. Important mechanisms for nutrient inputs include upwelling of ocean

Table 1: *Daily per capita protein consumption in grams (1979–1981)* (adapted from Bardach, 1989)

	Total protein	Animal protein	Fish protein
		(A)	(%A)
Developed countries	98.9	56.2	7.5 (13.3)
Developing countries	57.5	12.3	2.6 (21.1)

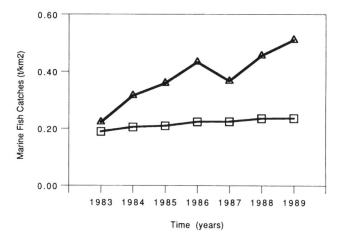

Figure 1: Changes in marine fish catches (tonnes km^{-2}) from world (▢) and from southeast Pacific (▲) fishing areas between 1983 and 1989. From FAO (1989), by permission.

water associated in particular with eastern boundary currents, and land sources via riverine and atmospheric transport. There are also significant, at least in economic terms, ocean pelagic fisheries for species such as tuna and squid.

The high level of exploitation of fish resources around the world leaves little opportunity for further growth especially with the recent expansion of relatively under-exploited fisheries, such as those in SE Asia (Fig. 1). The dramatic rise in the export value of the marine fish catch from about US$170 to 320 million between 1985 and 1988 (FAO, 1989) is a measure of the high demand for fish products and the high prices that certain countries are prepared to pay. New harvesting technologies are being introduced, often with uncertain environmental consequences especially for bottom spawning and feeding species.

The challenge for the next few decades is to develop sustainable strategies for managing the world fish resources without hindering the economic benefits of the industry. Improved management will depend on a better understanding of the natural system that supports the resource and of the economic and social restraints that control the market, as well as on improving the scientific basis for establishing catch quotas. It will be necessary to review the methods currently used to predict natural and man-induced changes in the abundance and availability of fish stocks. Fish populations show inherent levels of variability associated with ecological and environmental factors (e.g. Southward *et al.*, 1988; Corten, 1990). Prediction of stocks under conditions of strong exploitation by humans will only be possible if the mechanisms by which climate affects fisheries are better understood.

Recruitment efficiency, or the incorporation of new offspring to the adult population, has long been identified as a key process determining stock size (Shepherd and Cushing, 1990) and is generally regarded as independent of stock size above some low threshold value. The survival rates of the eggs, larvae and young of marine fish, as for most other marine animals, are determined by the various environmental conditions that affect predation and food

supply. Since most commercially valuable fish and shellfish inhabit coastal waters, at least at some time during their life history, much effort has recently been placed on understanding how physical conditions, including the effects of climate, in coastal oceans affect the dispersal of the planktonic juvenile stages with respect to predators and food (see Sharp, 1988). However, the nature of the regulation mechanisms that determine long-term changes in fish populations remains very uncertain (Shepherd and Cushing, 1990), so that quantitive observations on how fish populations respond to over-exploitation (or climate change) will remain an important component of fisheries science.

Some progress has been made recently in elucidating the links between climate variations, recruitment success, and fluctuations in the size of fish stocks. However, in order to detect the effects of climate change on fisheries and validate the results of predictive models (Sharp, 1987), a new mode of monitoring incorporating in situ measurements and remotely sensed data will be required which takes account both natural variability of the ocean system and the large scales of interaction between the physical environment and the biota (see McGowan, 1990). The results of fisheries models must continue to be treated with great caution until such processes are better understood.

The effects of climate change upon the oceans can be ascertained from General Circulation Models (GCMs) that take account of the dynamics of the ocean interior (Bernal, 1991). Model results indicate that the deep ocean (>1000 m) will take much longer (up to 3500 y) to reach equilibrium conditions than the shallow ocean in response to climate forcing due to a doubling of atmospheric CO_2. Also amplification of the climatic effects at the poles will lead to larger changes at high latitudes but with significant asymmetry between the northern and southern hemispheres due to the distributions of land masses and to the Drake

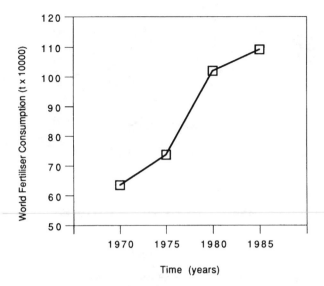

Figure 2: Changes in world fertilizer consumption (10^4 tonnes y^{-1}) between 1970 and 1985. From IOC Meeting of Experts on Land Sources of Marine Pollution, Canada, March 1991, by permission.

Passage which permits the existence of the circumpolar current around Antarctica.

After a few decades of increased CO_2, a general warming of the upper ocean (<500 m) is predicted, with the maximum anomaly around 60°N and the minimum close to the Antarctic continent. The warming is likely to be accompanied by increased rainfall and decreased surface salinity at mid-latitudes (30–50°), with possible enhancement of coastal inputs of nutrients due to land runoff (Bernal, 1991) and local/regional intensification of coastal upwelling (Bakun, 1990). The implications for marine resources are uncertain, but changes in the biogeography of both planktonic and benthic organisms and in the extent of specific pelagic and coastal habitat types are expected. Such changes will affect the survival of juvenile stages and the distributions of many marine organisms. Although the overall potential productivity of marine fisheries may not be altered, the distributions of fish stocks are likely to change and stronger stratification of the surface waters may favor the growth of pelagic species compared to demersal ones (e.g. Frank *et al.*, 1990).

Thus, a general picture is emerging that climate change is likely to have significant effects on marine fisheries (Tegart *et al.*, 1990; Jaeger and Ferguson, 1991), some good and some bad in relation to human needs. Indeed, as shown by El Niño events, adverse environmental and resource impacts in one part of the world may be accompanied by beneficial ones in another. The main implications for humans, therefore, are social ones (Francis, 1990; Glantz, 1990) concerning the adjustment of exploitation and marketing procedures to a new situation. International agreements on monitoring the state of fish

populations and agreeing to catch quotas will continue to be an integral part of sustainable utilization.

Mariculture has grown in importance rapidly over the past two decades, and now supplies more than 10% of living marine resources (Goldberg, 1990). Although there are restraints to the industry concerned with new diseases in farmed populations of marine organisms and with environmental impacts associated with feeding and disease control (e.g. Ackefors and Enell, 1990), expansion is likely to continue as new technological advances are made. However, much more careful evaluation of large-scale operations, such as the removal of mangroves to create ponds for shrimp, must be undertaken if long-term harm to coastal systems is to be avoided.

2.2 Degradation of marine environments

A wide range of human activities on land and at sea influence marine ecosystems (Goldberg, 1990). On a global scale the most important are (a) land use practices which affect the riverine and aeolian transport to the sea of particulate matter, organic matter, inorganic nutrients (Fig. 2), and pollutants; (b) freshwater use and storage on land which, in general, reduces the total and episodic (spring melt, floods) discharges of water and suspended and dissolved constituents from land to sea; (c) coastal development, engineering (building of harbours and sea defences, dredging, etc.), tourism and mariculture which are all associated with the modification and loss of shoreline habitats and the direct discharge and dumping of urban and industrial wastes into the sea; and (d) the exploitation of living and non-living marine resources leading to disturbance of both pelagic and benthic food chains.

Changes in patterns of biological productivity resulting from eutrophication (Fig. 2) and associated environmental effects (Turner and Rabalais, 1991) and giving rise to increases in rates of production or to modifications in food web structure, are now widely observed in inland and semi-enclosed sea areas such as the Caspian, Baltic and Adriatic Seas, and along densely populated coasts (e.g. Rosenberg *et al.*, 1990; Radach *et al.*, 1990; Fisher and Oppenheimer, 1991).

Blooms of nuisance algae (Fig. 3) appear to have become more frequent and widespread with a variety of adverse effects on animal communities and human practices, but they are not always a direct consequence of excess nutrient loading (Richardson, 1989). The occurrence of phytoplankton blooms is to some degree predictable on the basis of hydrographic conditions, but the precise factors that favour the growth of one species as opposed to another and that enable the rapid accumulation of plant biomass are often unknown. Certain algal species, in particular dinoflagellates, produce substances toxic to marine animals and to man which tend to be accumulated

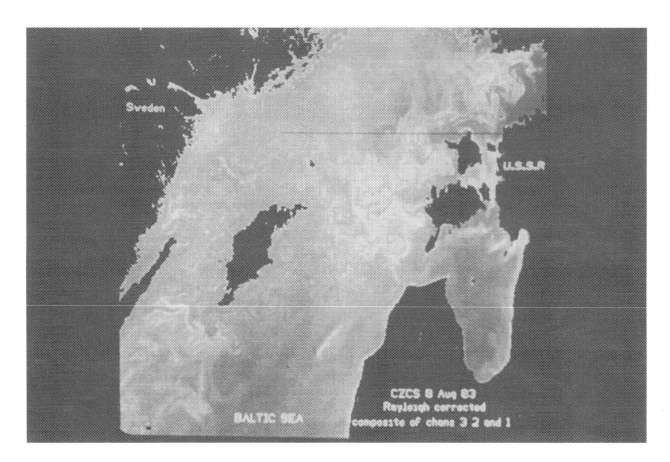

Figure 3: Image from the Coastal Zone Color Scanner on the Nimbus-7 satellite of a phytoplankton bloom in the Baltic Sea, 8 August, 1983. The bright, wavy features are probably caused by accumulations of surface foam.

by filter feeding organisms such as shellfish. Others release organic matter leading to the formation of surface foams (Fig. 3) which accumulate on beaches and to the emission of volatile compounds such as dimethyl sulphide. Models to predict the occurrence and species composition of marine phytoplankton blooms would have wide applications.

Under certain conditions large quantities of organic matter may accumulate below the pycnocline and lead to extensive de-oxygenation with severe impacts on benthic animals and demersal fish. Long-term observations (Justic et al., 1987) show that bottom anoxic conditions are increasing in frequency and extent in both shallow and deep water environments, apparently due to a combination of enhanced primary production in surface waters and fluxes of organic matter from land. The most severe case is the Black Sea where oxygen depleted water is continuing to reach closer to the surface and threatens the survival of important fisheries (see Halim in Mantoura et al., 1991).

A recent review of the health of the world's oceans (GESAMP, 1990) has summarized the present state of knowledge about the distributions and impacts of toxic pollutants in the sea. Although the public remains very conscious of oil pollution and the release of radioactive substances, the very wide range of synthetic organic compounds such as chlorinated hydrocarbons

accumulating in the sea are a greater threat to the marine biota. Such substances are often present at very low concentrations, and long-term exposure to sublethal levels is thought likely to cause significant damage to marine ecosystems at both the community and species levels. Concern is also expressed about the great increase in plastic litter in the marine environment and the accumulation of tar along shorelines and beaches.

New techniques based on a combination of ecological and physiological approaches (see Bayne et al., 1988) are now being developed for the early detection of the effects of toxic substances in the sea, and these are likely to be of considerable practical value for environmental monitoring (Gray et al., 1990) especially if linked to advanced chemical methods for assessing the impacts of hazardous chemicals (Landner, 1988). However, our understanding of the precise effects of toxic substances on marine life is very imperfect, particularly in relation to reproduction and recruitment processes and to the ways in which secondary concentration and dispersal mechanisms affect the time and space scales of exposure to toxicants. Recent pathological studies on the young stages of demersal fish in the North Sea (Dethlefsen, 1988) indicate relatively severe local impacts that have been previously overlooked, but the interpretation of the observational data remains controversial, emphasizing the very considerable difficulties of relating cause and effect in studies of marine ecosystems.

There is a rapidly increasing awareness of the significance of microbiological contamination of marine environments. Human pathogenic bacteria can survive for long periods in sea water (Grimes *et al.*, 1986). The GESAMP (1990) report states that "Early views, that there is no demonstrable link between human disease and bathing in contaminated sea water, can no longer be supported". A related problem is that consumption of seafood from contaminated waters can cause infectious diseases such as hepatitis and cholera (Goldberg, 1990). Apart from direct human interest in such problems, especially in the context of tourism, they also highlight an area of general ignorance about the microbiology of marine ecosystems; for example, the abundance of marine viruses has only recently been demonstrated (Bergh *et al.*, 1989).

Shoreline development which has led to the extensive modification and destruction of mangrove, coral reef, lagoon, seagrass and salt marsh habitats (see Chapman, 1977) has perhaps been even more damaging than eutrophication and pollution. Such damage is serious in itself but has broader implications for marine ecosystems due to the dependence of populations and communities of organisms on the integral functioning and health of the various ecosystems that constitute the land–ocean boundary. For example, mangroves are sensitive to salinity variations related to freshwater inputs, stabilize coastal sediments, are a major contributor to coastal biological productivity, and provide important spawning and nursery areas for various animals (Fortes, 1988). Their loss due to cutting, drainage, or changes in salinity distributions has far-reaching ecological impacts which reduce the overall long-term value of coastal regions to man. Against a background of urgent, short-term resource needs, however, management schemes to conserve mangrove forests and other types of coastal habitat (Batisse, 1990) will be difficult to implement effectively.

Another significant ecological disturbance is the introduction of new species. Many examples from marine environments are now known, including seaweeds and phytoplankton as well as various invertebrates (Rueness, 1989; Carlton *et al.*, 1990), and rapid colonization can lead to marked changes to the structure, diversity, functioning and commercial value of native communities. Although much is now known about how such introductions occur, it remains difficult to apply preventative control measures. There are also constant dangers for the mariculture industry from new diseases of cultured organisms.

There are many well-documented examples of biological change in marine systems, some extending back through historical time (Southward *et al.*, 1988), but most including those caused by human activities which go undetected. Not only is there far too little baseline information about the "natural" state of coastal ecosystems, but long-term observation projects are given low priority within marine research programs and are usually the first to be stopped at times of financial cutbacks. Scientific attitudes must be changed both towards the value of long-term data on the marine environment and about the way such information is used to justify and plan experimental work on the causes, prediction and economic implications of ecological change.

Not all environmental changes are adverse in terms of the health, diversity and human use of coastal ecosystems and, for this reason, management policies need to distinguish between ecological impacts that are acceptable or non-acceptable. The concept of environmental capacity is used as a basis for determining the degree of environmental disturbance, for example, related to pollutant discharges, that can be tolerated in relation to certain objectives. However, as discussed by Krom and Cohen (in Mantoura *et al.*, 1991), it is likely to be too complex and expensive to apply for legislative purposes and will always be open to different interpretations. The precautionary principle (Gray, 1990), which advocates the need for information about environmental impacts before potentially harmful activities are allowed, appears to be a more practical framework for considering pollution control measures at least as an initial step.

Ultimately objective decisions will have to be made about limits to environmental degradation that can be tolerated without undue damage to marine resources based on a global assessment of societal needs, on an understanding of how marine ecosystems respond to different types of disturbance, and on realistic predictions of the short and long-term impacts. The base of scientific knowledge is presently inadequate to be confident in most predictions about how marine ecosystems will be further altered by continuing human activities and respond to attempts of ecological restoration, especially in the context of natural variability related to the dynamic properties of biological systems and their response to fluctuations in climate (e.g. Southward *et al.*, 1988; Aebischer *et al.*, 1990). However, immediate action is needed to prevent further irreversible losses of coastal resources in heavily populated and developed regions, especially around deltas and estuaries.

2.3 Understanding the dynamics of the land–ocean interface

The boundary between the land and the oceans is characterized by variable geomorphological features – emergent and trailing coastlines associated respectively with narrow and wide continental shelves, deltas and lagoons, islands and reefs, smooth and canyoned

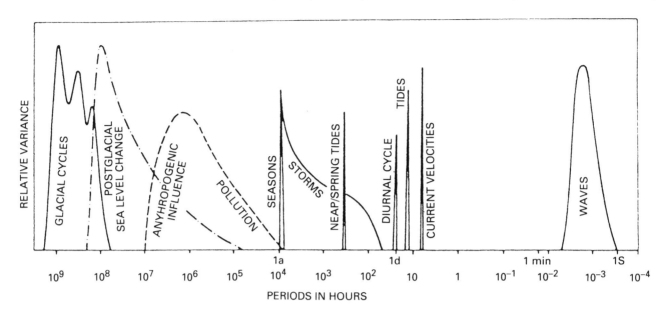

Figure 4: Timescales of variability in estuaries due to environmental factors. From Kempe (1988), by permission.

continental slopes and so on – and variable energy states determined by tidal and wind mixing, upwelling, the influence of ocean eddies and currents, and the effects of freshwater inputs on coastal circulation (see Postma and Zijlstra, 1988; Mantoura *et al.*, 1991). The very wide range of coastal habitats, reflecting different combinations of substrate type, water depth and motion, light and nutrient availability, and temperature, is associated with a high degree of biological diversity (Grassle *et al.*, 1991; Ray, 1991). Relatively little effort has been given to developing methods for classifying coastal ecosystems at regional or global scales (Hayden *et al.*, 1984). This situation represents a significant obstacle to attempts to compare the structural and functional properties of marine ecosystems, and to predict the ecological impacts of man or of changes in climate, sea-level at the ocean margins.

All marine and coastal habitats show strong periodic and episodic variations in physical forcing (Fig. 4) with close coupling between land catchment basin, coastal ocean and open ocean processes. The ecological implications depend both on the type and frequency of such forcing; for example, changes in patterns of river flow due to management of freshwater, or of storminess due to climatic variability have quite different impacts. Good progress has been made over the last two decades in improving our understanding of how physical conditions in the water column affect sedimentary, chemical and biological processes, but the ability to predict with confidence how large scale factors such as ENSO events (Johnson and O'Brien, 1990), global warming (Tegart *et al.*, 1990; Jaeger and Ferguson, 1991), and changes in windiness and the flushing characteristics of shelf seas are likely to affect coastal ecosystems regionally and locally remains crude.

Both the land-sea and shelf edge boundaries are active sites for the exchange and transformation of particulate and dissolved materials (Eisma, 1988, Kempe, 1988; Degens *et al.*, 1990; Milliman, 1992). Inventories for such exchanges are very uncertain, although riverine, ground-water and atmospheric inputs of particles, organic matter, nutrient and trace elements do modify strongly the chemical environment of coastal waters as shown by the strong gradients in properties between estuarine, shelf and open ocean waters. Recent studies have shown that atmospheric transport pathways for certain materials are considerably more important than previously thought (Dorten *et al.*, 1991; Duce in Mantoura *et al.*, 1991). The ocean is the major source of inorganic nutrients for most shelf regions (Walsh, 1991), and local variations in upwelling and shelf edge exchange are linked closely with rates of biological productivity and fishing activity. Sites of sediment accumulation, in particular within estuaries and on the upper continental slope, are important regions for biogeochemical transformations due to various processes involving particle–water reactions, benthic and microbial activity, and sediment diagenesis. Quantitative measurements of such processes are still sparse so that global estimates for coastal sources and sinks of compounds of climatic and ecological importance (e.g. CO_2, nitrous oxide, methane, organic carbon, nutrients) remain uncertain.

The varied physical and chemical nature of coastal habitats is the basis of combined high biological diversity and high biological productivity, a distinctive feature of coastal marine ecosystems such as tropical reef and temperate nearshore benthic communities. Understanding the close coupling between physics and biology (e.g. Nihoul, 1986; Nixon, 1988) is fundamental to any attempt to predict how environmental change will affect coastal ecosystems, and must take account of physical forcing due both to the climate system and to the human activities (Fig.

4). The significance of the latter is well recognized in terms of the impacts of such practices as coastal engineering and river management schemes, but the more general implications for coastal habitats of physical perturbation related to energy wastage by man (Glasby, 1988) have not been studied in detail. The effects of physical and dependent chemical (e.g. nutrient inputs) processes on biological productivity and diversity are likely to be different for pelagic, benthic and coastal habitats (Ray, 1991), and there is a need for comparative field and modeling studies to elucidate how, and over what critical time and space scales, different types of marine ecosystems respond to changes in environmental conditions. However, the limitations of present ecological models need to be clearly recognised, especially with respect to their applicability to different climatic conditions.

Important functional attributes of marine ecosystems include biogeochemical and biogeomorphological processes that, in essence, represent mechanisms of biological feedback on the marine and global environment. Estuarine, deltaic and slope sediments are major sinks for both marine- and terrestrially derived organic matter as well as marine calcium carbonate (Kempe, 1988). The efficiency of burial of organic carbon directly affects the concentration of CO_2 in the overlying water and, therefore, the air–sea exchange of CO_2. Coastal ecosystems also contribute signicantly to fluxes of other trace gases such as N_2O and dimethyl sulphide.

The marine biota affect coastal geomorphological processes in a wide variety of ways (Spencer, 1987) – building of reefs, accumulation of carbonate sediments, erosion by boring organisms, shoreline stabilization by mangrove and salt marsh communities. One aspect of coastal biogeomorphology which continues to be overlooked concerns the stability of soft sediments; benthic phytoplankton, microbes and shellfish serve to filter, package and bind fine sediment particles, reducing the turbidity of overlying water and allowing colonization by other benthic organisms. Disturbance of soft sediment habitats through shellfish harvesting, dredging, introduction of toxic pollutants, and even eutrophication of the overlying water tends to cause large scale changes in the functioning of estuaries (Goldberg, 1990) which are likely to promote sediment erosion in regions exposed to wave action and tidal resuspension and to be difficult to reverse.

Various mechanisms by which external changes in sea-level and climate (Warrick and Farmer, 1990; UNESCO, 1990; Tegart *et al.*, 1990; Jaeger and Ferguson, 1991) and in land usage (Kempe, 1988; Milliman *et al.*, 1989; Bardach, 1989; Degens *et al.*, 1990) affect coastal and marine systems are recognized. Those concerning the impacts of climate change on fisheries and of toxic pollutants on coastal ecosystems have been dealt with in the preceding two sections. Variations in relative sea-level, due to a combination of eustatic and isostatic effects and coastal subsidence, directly threaten towns and cities and will be felt most severely along low-lying coasts and on islands (Titus, 1990; Paw and Thia-Eng, 1991) especially where natural ecosystems such as salt marsh and mangroves have been removed by man and no longer provide a natural buffer against any mean rise. The causes are only partly understood although improvements in the ability to distinguish variations from trends and in extending interannual predictions to interdecadal ones are being made. It is important to recognize that the effects of episodic climatic events such as ENSO on sea-level and the amplification of seasonal sea-level differences are likely to have greater impacts on coastal regions than a gradual rise in mean sea-level.

Arctic coastlines which are expected to experience the greatest warming trends are likely to be the most affected especially if there is any significant subsidence associated with permafrost melting and the release of methane clathrates. By contrast at mid- and low latitudes the worst problems are associated with the effects of agricultural development, deforestation, wetland reclamation (Ruddle, 1987), and river management on the fluxes of water, suspended matter and dissolved matter to the coastal zone (e.g. Turner and Rabalais, 1991). Their severity will continue to vary regionally depending on the combination of environmental changes occurring and on the nature of the coastline.

The marine habitats most threatened today by global change include deltas (Milliman *et al.*, 1989), estuaries (Kempe, 1988), coral reefs (Warwick *et al.*, 1990), lagoons, low-lying islands, inland seas, and coastal waters close to densely populated regions. Such regions must remain the focus of long-term environmental protection as human demands for resources will continue to increase. Environmental management strategies must focus on steps to ameliorate the most severe adverse effects, and develop procedures for habitat restoration (Jansson and Jansson, 1988) based on objective scientific criteria. Ecosystem and landscape modeling procedures (Lindeboom *et al.*, 1989; Costanza *et al.*, 1990) are likely to be important tools for evaluating the effectiveness of particular management procedures, but must themselves be based on a sound understanding of how marine ecosystems work and be supplemented by new long-term observational programs as the best means of both defining change and detecting responses to new environmental conditions.

At each stage, our ability to make sound environmental decisions will be limited by knowledge of the complex physical, chemical and biological interactions that characterize coastal ecosystems. Although good progress has been made over the last decade in initiating new multidisciplinary studies, their scientific objectives are often not fully met due to limitations of resources and of trained personnel in key scientific disciplines. This restraint is particularly severe for studies of transitional environments such as the land–sea or shelf break boundaries as it requires investigations of two or more physically and geomorphologically distinct systems. The coastline is the meeting place of land, sea, freshwater and atmosphere, but most coastal research is funded by agencies or departments that have responsibility only for marine issues. The removal of such demarcations in responsibility and in resource allocation will open the way to more effective progress.

3. FUTURE ACTIONS AND NEEDS

The case for new actions to conserve the resources of marine and coastal ecosystems and to prevent further environmental degradation must be based on objective evaluations of such ecosystems. The costs of the subsidence of deltas, of declines in water quality near all major coastal cities, of various health hazards at seaside resorts, of over-fishing in less developed countries dependent on the sea as a major source of food protein, and of enhanced erosion of the land margin due to the destruction of coastal habitats and sea-level rise must be properly counted, and set against the investment required for proactive, anticipatory management policies to ameliorate such situations. We have the technical capacity and knowledge to avoid these problems provided an holistic, integrated approach to coastal management is adopted. All long-term economic and social planning for coastal regions must take account of ongoing and likely future environmental changes related both to human activities and to expected trends in sea-level and climatic conditions. Early and continuing investment to resolve a range of global marine environmental problems is not a choice but a necessity if future generations are to benefit from living near the coast.

Knowledge and education are vital if good environmental decisions are to be made. In this sense marine scientists have an important responsibility both to work on problems of immediate relevance to the needs of society and to provide accurate information to the public about the nature and implications of environmental change.

3.1 In relation to the major issues discussed in this chapter there are three priorities for future action

3.1.1 **Action 1**. *To initiate a global observing program dedicated to studying dynamics of the ocean margins and land–sea interface*
This will require satellite, aircraft and in situ remote sensing techniques, coupled with ship observations. Apart

from long-term measurements of basic physical and chemical parameters, including sea-level, water temperature, salinity, levels of nutrients and certain pollutants, attention must be given to changes in the abundance and distributions of key species of planktonic and benthic organisms which reflect broad-scale variations in environmental parameters and are likely to be accompanied by fluctuations in fish stocks. The state of estuaries and of ecosystems along the land margin must also be carefully monitored with respect to human activities and to sea-level rise. Baseline studies of the present distributions of threatened coastal habitats (mangroves, reefs, seagrasses, etc.) will be needed in order to document any further destruction of these communities. An initiative to acquire compatible information on relevant socio-economic variables is needed.

The immediate priorities must be to implement ongoing and planned programs for global observations, in particular the Global Ocean Observing System (GOOS) being developed by IOC/WMO/UNEP/IUCN which includes a coastal ocean module, and to define minimum targets for such programs. Appropriate low-cost and robust technologies are now available even for chemical and biological parameters (e.g. the Continuous Plankton Recorder — see Aebischer *et al.*, 1990, McGowan, 1990), and important contributions can be made by all coastal nations maintaining time series measurements of basic properties in nearshore waters.

Care must be taken with the design of the coastal module of GOOS to ensure that the time and space scales of sampling are appropriate for detecting the significant variability, trends and episodic events in coastal environments which affect human activities and policies for sustainable utilization of marine resources. A well balanced program is vital as coastal data interpretation will always be dependent on having reliable information on ocean dynamics, but ocean measurements alone will have a very limited value for coastal environmental problems.

3.1.2 **Action 2**. *To construct and test simulation models of the responses of marine and coastal ecosystems to environmental change*
The ultimate aim will be to use predictive models for fishery and coastal management (see Fig. 5), but new techniques to reduce the uncertainties related to the non-linear behavior of biological systems and the complex physical, chemical and biological interactions at the boundary between land and ocean must first be developed. Such simulation models should take account of biological feedback effects on fundamental properties such as the cohesiveness of sediments which affects coastal geomorphology, and the fate and transformation of contaminants. Confidence in the models will depend on experimental measurements of rates of key processes under

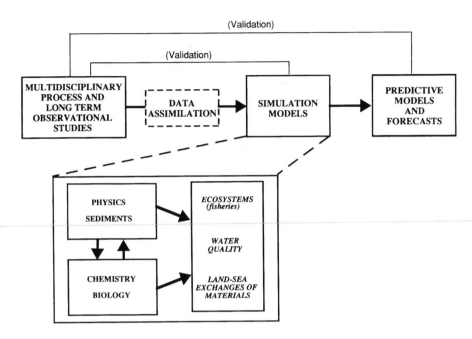

Figure 5: Development of models for coastal zone management.

controlled field or laboratory conditions, and on validation by careful time-series observations of the impacts of environmental change on natural ecosystems (Fig. 5) which take account of the broad nature of interactions between land and ocean.

Implementation of this action should also initially be through existing programs such as the IOC/FAO Ocean Science in Relation to Living Resources (OSLR) and the proposed IGBP (1990) Land–Ocean Interactions in the Coastal Zone (LOICZ), as well as various national efforts. Great care must be taken to validate the results of models of ecosystem structure and function, as demonstrated in the work on understanding environmental changes in the Baltic Sea (Jansson and Jansson, 1988; Elmgren, 1989; Rosenberg *et al.*, 1990). Although such modeling work must have a strongly multidisciplinary approach, the quality and validity of the models is likely to be strongly dependent on maintaining a structure based on the best available disciplinary skills.

3.1.3 **Action 3**. *To develop further the scientific basis for the management of marine living resources and for the conservation of coastal ecosystems; within a changing environment*

Inadequate planning for the sustainable utilization of both living and non-living marine resources represents a serious threat to the coastal marine environment in terms of physical degradation and the loss of ecosystem integrity. Within the framework of the 1982 UN Law of the Sea Convention, in particular parts IX, XII and XIII dealing with the protection and preservation of the marine environment, enclosed and semi-enclosed seas, and marine scientific research, there is now an urgent global need to reduce pollution (see GESAMP, 1990), to protect

biodiversity, and to control coastal development in a way that allows long-term optimal use of marine and coastal resources. Important work has been started under the Global Investigation of Pollution in the Marine Environment (GIPME) program sponsored by IOC and UNEP, but more attention must be given as soon as possible to understanding and predicting the impacts of global change on marine ecosystems, to the prevention of resource exploitation practices (including certain traditional ones) which damage the environment, to the restoration of seriously damaged coastal environments, and to methods of matching projected uses of the coastal zone over the next few decades to the distributions of particular resources.

The main tool for coastal management will be integrated models based on a scientific understanding of the environment and on social needs. These are presently not limited in the developed countries by the availability of scientific information or technology, but by a lack of commitment by scientists and funding agencies to invest time and money in such work. The case to invest in model development must be based on a more realistic long-term assessment of the value of coastal and marine systems to society. Models of population dynamics of marine organisms are needed which take account of new discoveries about the recruitment control and about the combined effects of man and climate on stock size and on food web structure. Mariculture will continue to increase in importance (see Goldberg, 1990), and require further research on the physiology and genetics of cultured species especially in relation to yield efficiencies and to the control of associated adverse effects on water quality. Healthy coastal ecosystems are not only of inherent value to man, but also play a vital role in maintaining the integrity of the coastline against erosion and in providing nursery grounds for the young of many commercially valuable species of fish and shellfish.

One aspect of coastal management that should not be overlooked is preparation for exploiting new opportunities in the marine environment that will emerge from technological developments related to ocean energy sources, coastal engineering, marine waste disposal, harvesting of living resources, etc. Constructive thinking about new uses of marine and coastal systems will be important in order to meet the needs of society, especially if further environmental degradation is to be avoided.

3.2 In order to carry out these actions at a global scale there are three major needs related to technological, financial and human resources:

3.2.1 Need 1. *Deployment of efficient technologies for the long-term observation of marine and coastal systems*

The development of platforms, instruments, and data analysis methods for long-term studies of the marine environment has reached the stage where global programs such as GOOS, and relevant national components (some of which may already be operational), should be initiated as soon as possible. There is an immediate need for baseline data in relation to a wide range of development and management issues. Efficient co-ordination of satellite missions needs to be maintained, and for measurements that require expensive equipment, such as gravity studies in relation to sea-level changes, the possibilities of moving such equipment from one site to another should be considered. Continuing attention must be given to the problems of data quality control and of the compatibility and exchange of data sets for different regions/programs. In certain areas, such as ecosystem dynamics and modeling, there is a need for further investment into the development of new techniques for improving understanding both of underlying key processes and of interactions between organisms at the species and community levels.

3.2.2 Need 2. *Objective methods of assessing the socio-economic values of marine and coastal ecosystems as a basis for determining the resources and priorities for future research*

A dialogue between environmental scientists, social scientists and economists concerned with marine and coastal issues must be fostered as a basis for determining optimal policies for the future use of resources and, therefore, what the levels and priorities should be for environmental research and management. Attributing objective, long-term values to marine resources is in itself a research problem requiring new information about the environment and social needs in both developing and developed countries.

3.2.3 Need 3. *Improved facilities, especially in less developed countries, for education and training in marine science and technology*

Sustainable utilization of coastal and marine resources is a global issue which can only be tackled through the involvement of all coastal states. National differences in policy need to be resolved since most environmental and resource use problems extend across national and regional boundaries. In countries where the capabilities and skills to deal with urgent environmental problems are inadequate, education and research capabilities in marine sciences must be improved with the help of richer nations. Important activities include the development of partnerships through joint research and educational programs(taking place when possible within the less developed countries), fellowship training schemes, regional co-operation in environmental management, and the establishment of research centers in parts of the world where marine environmental problems represent a danger to life and security. Measures to prevent the loss of trained, skilled scientists to other types of employment may also be required.

4. LINKAGES TO OTHER CONFERENCE THEMES

Being at the junction between land and ocean where rivers discharge agricultural, urban and industrial wastes, the coastal zone bears the brunt of the combined effects of environmental pollution and of climate and sea-level change (Oppenheimer, 1989). As the needs of human society will continue to have to be met, the management of marine and coastal systems must focus on amelioration rather than prevention policies (Glasby, 1988). Strategies for the sustainable utilization of resources and for resolving the environmental problems will depend on treating the relevant scientific, social and economic issues together, and on recognising that each type of coastline is amenable only to certain types of use if unacceptable and irreversible environmental damage is to be avoided. The global nature of the present impacts of human activities on coastal ecosystems means that a global policy is urgently required for resource exploitation and conservation which, in turn, must be linked directly to global and regional forecasts of the ways in which the coastal zone is likely to be developed over the next few decades.

Agreements on how to improve the use and management of the coastal zone will have to take account of the global significance of ecological and biogeochemical processes at the land–ocean interface and of the full range of societal needs now and in the future. Achieving the optimal strategies will depend on improved public awareness of the need for appropriate action, on improved scientific knowledge and technical capacity to undertake such action, and on strengthened national and international institutional arrangements to prioritise and execute what needs to be done. The most urgent environmental issues at the global scale are the effects of changes in land use, the lack of appropriate technology to reduce industrial and urban discharges of toxic wastes, and the pressures of population increase on coastal ecosystems in less developed countries. However, just as important is the present unwillingness of developed nations to invest beyond immediate national needs in long-term global policies for the sustainable utilization of coastal resources.

5. CONCLUSIONS

Marine and coastal systems are an invaluable, but poorly managed and over-exploited, resource to the human population which is further threatened by global changes in climate, sea-level and environmental pollution. The main scientific priorities in relation to establishing policies of sustainable utilization are to provide more accurate information about the changes now taking place at the land–ocean interface, to develop reliable methods of ecological prediction for the coastal zone given a range of scenarios for future human needs, climate and sea-level, and to develop new methods for the long-term management of coastal resources by applying the expertise of the scientific community to situations where it is really needed. The technological and programmatic means are now largely in place through the initiatives of UN agencies and ICSU so that, with the proper level of investment reflecting the true value of these resources to society, these tasks defining the scientific agenda for the next few decades are achievable.

There is an urgent need to continue developing both regional and global perspectives of change in the coastal oceans which take account of the large spatial and temporal scales of the natural and anthropogenic episodic events (see Fig. 4) that affect coastal and marine resources. At present, views and decisions are generally based on parochial national needs, whereas most of the adverse changes now occurring to the marine environment and its resources transcend national boundaries. Several barriers need to be broken down: inappropriate economic policies, often based on the demands of developed countries, have given rise to an attitude of "take what you can while you can" without accounting for environmental damage; regulatory and administrative systems for dealing with marine environmental issues are usually weak; and many of our poor attitudes and ways of using marine resources have strong societal roots and need to be modified.

A key issue will be the promotion of education and research in marine sciences in less developed countries

where most of the resource and environmental problems lie. The national differences in technological capabilities and in attitudes to environmental issues which must be overcome will require intergovernmental co-operation and financial support by the richer nations. Establishing scientifically sound practices for the sustainable utilization of marine resources, and for preventing irreversible damage to marine habitats and the diversity and health of marine ecosystems are global needs of the highest priority. In many regions a great deal can be achieved immediately even with rather modest investment since, as pointed out by Sharp (1988), our greatest ignorance is often in processes that occur near to the shore. Also it is the ecosystems at the boundary between land and sea that are of crucial importance in maintaining the stability of the coastal zone under conditions of change in the hydrologic cycle, in sea-level, and in storm frequency and intensity.

REFERENCES

Ackefors, H. and Enell, M. 1990. Discharge of nutrients from Swedish fish farming to adjacent sea areas. *Ambio*, **19**, 28-35.

Aebischer, N.J., Coulson, J.C. and Colebrook, J.M. 1990. Parallel long-term trends across four marine trophic levels and weather. *Nature*, **347**, 753-755.

Bakun, A. 1990. Global climate change and intensification of coastal upwelling. *Science*, **247**, 198-201.

Bardach, J.E. 1989. Global warming and the coastal zone. *Clim. Change*, **15**, 117-150.

Batisse, M. 1990. Development and implementation of the biosphere reserve concept and its applicability to coastal regions. *Env. Conserv.*, **17**, 111-116.

Bayne, B.L., Clarke, K.R. and Gray, J.S. (eds). 1988. Biological effects of pollutants. Results of a workshop. *Mar. Ecol. Prog. Ser.*, **46**, 1-278.

Bergh, O., Borsheim, K.Y., Bratbak, G. and Heldal, M. 1989. High abundance of viruses found in aquatic environments. *Nature*, **340**, 467-468.

Bernal, P.A. 1991. Consequences of global change for oceans. A review. *Clim. Change*, **18**, 339-359.

Broadus, J.M. 1989. Impacts of future sea-level rise. In: *Global Change and Our Common Future*. Papers from a Forum. R. S. DeFries and T. F. Malone. (eds.) pp. 125-138, National Academy Press, Washington DC.

Carlton J.T., Thompson, J.K., Schemel, L.E. and Nichols, F.H. 1990. Remarkable invasion of San Francisco Bay (California, USA) by the Asian clam *Potamocorbula amurensis*. *I. Introduction and dispersal. Mar. Ecol. Prog. Ser.*, **66**, 81-94.

Chapman, V.J. (ed.) 1977. *Ecosystems of the World. 1. Wet Coastal Ecosystems*. Elsevier, Amsterdam. pp. 428.

Corten, A. 1990. Long-term trends in pelagic fish stocks of the North Sea and adjacent waters and their possible connection to hydrographic changes. Neth. *J. Sea Res.*, **25**, 227-235.

Costanza, R., Sklar, F.H. and White, M.L. 1990. Modeling coastal landscape dynamics. *BioScience*, **40**, 91-107.

Degens, E.T., Kempe, S. and Richey, J.E. 1990. Summary:

Biogeochemistry of major world rivers. In: *Biogeochemistry of Major World Rivers*. SCOPE 42. (eds.), E.T. Degens, S. Kempe and J.E. Richey. (eds.), John Wiley and Sons, Chichester.

Dethlefsen, V. 1988. German Bight. In: *Pollution of the North Sea. An Assessment*. W. Salomons, B.L. Bayne, E.K. Duursma and U. Forstner. (eds.) pp. 425-440, Springer-Verlag, Berlin.

Dorten, W.S., Elbaz-Poulichet, F., Mart, L.R. and Martin, J.-M. 1991. Reassessment of the river input of trace metals into the Mediterranean Sea. *Ambio*, **20**, 2-6.

Duran L.R. and Castilla, J.C. 1989. Variation and persistence of the middle rocky intertidal community of central Chile, with and without human harvesting. *Mar. Biol.*, **103**, 555-562.

Eisma, D. 1988. The terrestrial influence on tropical coastal seas. *Mitt. Geol.-Palaont. Inst. Hamburg*, **66**, 289-317.

Elmgren, R. 1989. Man's impact on the ecosystem of the Baltic Sea: Energy flows today and at the turn of the century. *Ambio*, **18**, 326-332.

FAO Yearbook. 1989. *Fishery Statistics, Catches and Landings*. Vol. 68. pp.516.

Fisher, D.C. and Oppenheimer, M. 1991. Atmospheric nitrogen deposition and the Chesapeake Bay Estuary. *Ambio*, **20**, 102-108.

Fortes, M.D. 1988. Mangrove and seagrass beds of East Asia: habitats under stress. *Ambio*, **17**, 207-213.

Francis, R.C. 1990. Climate change and marine fisheries. *Fisheries*, **15**(6), 7-9.

Frank, K.T., Perry, R.I. and Drinkwater, K.F. 1990. The predicted response of northwest Atlantic invertebrate and fish stocks to CO_2-induced climate change. *Trans. Am. Fisheries Soc.*, **119**, 353-365.

GESAMP. 1990. *The State of the Marine Environment*. Rep. Stud. GESAMP No. 39, UNEP. pp.111.

Glantz, M.H. 1990. Does history have a future? Forecasting climate change effects on fisheries by analogy. *Fisheries*, **15** (6), 39-44.

Glasby, G.P. 1988. Entropy, pollution and environmental degradation. *Ambio*, **17**, 330-335.

Goldberg, E.D. 1990. Protecting the wet commons. *Environ. Sci. Tech.*, **24**, 450-454.

Grassle, J.F., Lasserre, P., McIntyre, A.D. and Ray, G.C. 1991. *Marine Biodiversity and Ecosystem Function. A Proposal for an International Research Programme*. Biology International Spec. Issue No. 23. IUBS. pp.19

Gray, J.S. 1990. Statistics and the precautionary principle. *Mar. Poll. Bull.*, **21**, 174-176.

Gray, J.S., Clarke, K.R., Warwick, R.M. and Hobbs, G. 1990. Detection of initial effects of pollution on marine benthos: an example from the Ekofisk and Eldfisk oilfields, North Sea. *Mar. Ecol. Prog. Ser.*, **66**, 285-299.

Grimes, D.J., Atwell, R.W., Brayton, P.R., Palmer, L.M., Rollins, D.M., Roszak, D.B., Singleton, F.L., Tamplin, M.L. and Colwell, R.R. 1986. The fate of enteric pathogenic bacteria in estuarine and marine environments. *Microbiol. Sci.*, **3**, 324-329.

Hayden, B.P., Ray, G.C. and Dolan, R. 1984. Classification of coastal and marine environments. *Environ. Conserv.*, **11**, 199-207.

Holligan, P.M. and Reiners, W.A. 1991. Predicting the responses of the coastal zone to global change. *Adv. Ecol. Res.* **22**, 211-255.

IGBP. 1990. *The International Geosphere-Biosphere Programme: A Study of Global Change. The Initial Core Projects*. Ch. 4. How changes in land use affect the resources of the coastal zone, and how changes in sea-level and climate alter coastal ecosystems. Report No. 12. Stockholm.

Jaeger, J. and Ferguson, H.L. (eds.) 1991. Climatic Change: Science, Impacts and Policy. Task Group 3, pp 443-446. *Proc. 2nd World Climate Conference*. Cambridge University Press, Cambridge.

Jansson, A.M. and Jansson, B.-O. 1988. Energy analysis approach to ecosystem redevelopment in the Baltic Sea and Great Lakes. *Ambio*, **17**, 131-136.

Johnson, M.A. and O'Brien, J.J. 1990. The Northeast Pacific Ocean response to the 1982-1983 El Niño. *J. Geophys. Res.*, **95**, 7155-7166.

Justic, D., Legovic, T. and Rottini-Sandrini, L. 1987. Trends in oxygen content 1911-1984 and occurrence of benthic mortality in the northern Adriatic Sea. *Estuar. Cstl. Shelf Sci.*, **25**, 435-445.

Kempe, S. 1988. Estuaries - Their natural and anthropogenic changes. In *Scales and Global Change*. T. Rosswall, R.G. Woodmansee and P.G. Risser. (eds.) pp 251-285. John Wiley and Sons Ltd., Chichester.

Landner, L. 1988. Hazardous chemicals in the environment - Some new approaches to advanced assessment. *Ambio*, **17**, 360-366.

Lindeboom, H.J., van Raaphorst, W., Ridderinkhof, H. and van der Veer, H.W. 1989. Ecosystem model of the western Wadden Sea: a bridge between science and management. *Helgol. Meeresunt.*, **43**, 549-564.

Mantoura, R.F.C., Martin, J.-M. and Wollast, R. (eds.) 1991. *Ocean Margin Processes in Global Change*. John Wiley and Sons, Chichester. pp.469.

McGowan, J. 1990. Climate and change in oceanic ecosystems: the value of time-series data. *Trends Ecol. Evol.*, **5**, 293-299.

Meybeck, M. and Helmer, R. 1989. The quality of rivers: From pristine stage to global pollution. *Palaeogeogr., Palaeoclim., Palaeoecol*, **75**, 283-309.

Milliman, J.D. 1992. Fluvial sediment discharge to the oceans. Re-evaluating the importance of small mountainous rivers. (in press).

Milliman, J.D., Broadus, J.M. and Gable, F. 1989. Environmental and economic implications of rising sea-level and subsiding deltas: the Nile and Bengal examples. *Ambio*, **18**, 340-345.

Nihoul, J.C.J. (ed.) 1986. *Marine Interfaces Ecohydrodynamics*. Elsevier, Amsterdam. pp.670.

Nixon, S.W. 1988. Physical energy inputs and the comparative ecology of lake and marine ecosystems. *Limnol. Oceanogr.* **33**, 1005-1025.

Oliva D. and Castilla, J.C. 1986. The effect of human exclusion on the population structure of key-hole limpets *Fissurella crassa and F. limbata* on the coast of central Chile. P.S.Z.N. I: *Mar. Ecol.*, **7**, 201-217.

Oppenheimer, M. 1989. Climate change and environmental

pollution: Physical and biological interactions. *Clim. Change*, **15**, 255-270.

Paw, J.N. and Thia-Eng, C. 1991. Climate changes and sea-level rise: Implications on coastal area utilization and management in south-east Asia. *Ocean and Shoreline Management*, **15**, 205-232.

Pearce, D. and Maler, K.-G. 1991. Environmental economics and the developing world. *Ambio*, **20**, 52-54.

Postma, H and Zijlstra, J.J. (eds). 1988. *Ecosystems of the World 27. Continental Shelves*. Elsevier, Amsterdam, pp.421.

Radach, G., Berg, J. and Hagmeier, E. 1990. Long-term changes of the annual cycles of meteorological, hydrographic, nutrient and phytoplankton time series at Helgoland and at LV ELBE 1 in the German Bight. *Contin. Shelf Res.*, **10**, 305-328.

Ray, G.C. 1989. Sustainable use of the coastal ocean. In: *Changing the Global Environment. Perspectives on Human Involvement*. D.B. Botkin, M.F. Caswell, J.E. Estes and A.A. Orio. (eds.) pp. 71-87. Academic Press, Boston.

Ray, G.C. 1991. Coastal-zone biodiversity patterns. *BioScience* **41**, 490-498.

Richardson, K. 1989. Algal blooms in the North Sea: the good, the bad and the ugly. *Dana*, **8**, 83-93.

Rosenberg, R., Elmgren, R., Fleischer, S., Jonsson, P., Persson, G. and Dahlin, H. 1990. Marine eutrophication case studies in Sweden. *Ambio*, **19**, 102-108.

Ruddle, K. 1987. The impact of wetland reclamation. In: *Land Transformation in Agriculture. SCOPE 32*. M.G. Wolman and F.G.A. Fournier. (eds.) pp.171-201. John Wiley and Sons Ltd., Chichester.

Rueness, J. 1989. *Sargassum muticum* and other introduced Japanese macroalgae: Biological pollution of European coasts. *Mar. Poll. Bull.*, **20**, 173-176.

Salomons, W., Bayne, B.L., Duursma, E.K. and Forstner, U. (eds). 1988. *Pollution of the North Sea. An Assessment*. Springer-Verlag, Berlin. pp.687.

Sharp, G.D. 1987. Climate and fisheries: cause and effect or managing the long and short of it all. *S. Afr. J. Mar. Sci.*, **5**, 811-838.

Sharp, G.D. 1988. Fish populations and fisheries: Their perturbations, natural and man-induced. In: *Ecosystems of the World 27. Continental Shelves*. H. Postma and J.J. Zijstra. (eds.) Elsevier, Amsterdam. pp.155-200.

Shepherd, J.G. and Cushing, D.H. 1990. Regulation in fish populations: myth or mirage? *Phil. Trans. R. Soc. Lond. B*, **330**, 151-164.

Southward, A.J., Boalch, G.T. and Maddock, L. 1988. Fluctuations in the herring and pilchard fisheries of Devon and Cornwall linked to change in climate since the 16th Century. *J. Mar. Biol. Ass. UK.* **68**, 423-445.

Spencer, T. 1987. Coastal biogeomorphology. In: *Biogeomorphology*. H.A.Viles, (ed.) pp.255-318.

Tegart, W.J.McG., Sheldon, G.W. ands Griffiths, D.C. 1990. *Climate Change. The IPCC Impacts Assessment*. Chap. 6. World oceans and coastal zones. Austral. Govt. Publ. Service, Canberra. pp.28.

Titus, J.G. 1990. Greenhouse effect and sea-level rise. In: *Handbook of Coastal and Ocean Engineering*. Vol. 1. Wave phenomena and coastal structures. J.B. Herbich. (ed), pp. 673-702. Gulf Publ. Co., Houston.

Turner, R.E. and Rabalais, N.N. 1991. Changes in Mississippi River water quality this century. Implications for coastal food webs. *BioScience*, **41**, 140-147.

UNESCO. 1990. Relative sea-level change: a critical evaluation. *Unesco Rep. Mar. Sci.* No. 54, pp.22.

Walsh, J.J. 1991. Importance of the continental margins in the marine biogeochemical cycling of carbon and nitrogen. *Nature*, **350**, 53-55.

Warrick, R. and Farmer, G. 1990. The greenhouse effect, climatic change and rising sea-level: Implications for development. *Trans. Inst. Br. Geogr. N.S.*, **15**, 5-20.

Warwick R.M., Clarke, K.R. and Suharsono. 1990. A statistical analysis of coral community responses to the 1982-83 El Niño in the Thousand Islands, Indonesia. *Coral Reefs*, **8**, 171-179.

Chapter 9: Terrestrial Systems

H.A. Mooney and W.G. Sombroek

EXECUTIVE SUMMARY

An assessment of what elements constitute terrestrial systems, our current state of knowledge of their spatial distribution and interactions, and how humans are impacting these systems lead us to the following statements:

(a) Important Facts
  (i) Terrestrial systems are being highly modified by additions and subtractions of system elements.
  (ii) Environmental stresses on terrestrial communities are increasing.
  (iii) The physical environment is changing faster than ever in historical times and to states that are unique in evolutionary history.

(b) Research Needs
  (i) Develop new methods to study whole ecosystem response to changing conditions at differing time and space scales.
  (ii) Develop techniques for restoring sustainable systems (restoration ecology).
  (iii) Develop the capacity (concepts, models, basic data, etc.) for predicting responses of terrestrial systems (species distribution, soil processes, etc.) to short-and long-term perturbations.

(c) Urgent Actions
  (i) Support major scientific initiatives directed toward understanding terrestrial system dynamics and predicting their response to change (e.g. International Geosphere–Biosphere Programme), and their sustainable use under increased population pressures (Sustainable Biosphere Initiative). Particularly important is the quantification of terrestrial carbon and nitrogen pools and fluxes and how modification of ecosystem structure alters these fluxes, and stemming the loss of the Earth's primary productive capacity.

  (ii) Identify and analyze current agro-ecosystems as to their inherent sustainability and their adaptability to climatic change.
  (iii) Identify and analyze environmentally critical areas (high expected impact of climate change, unusual diversity, inherent fragility, high population impact).
  (iv) Develop a global digital data base of terrestrial system components (landforms, soils, bioclimates, biomes, land use, etc.) in a compatible and readily accessible form and a program for monitoring their changing states as part of a global observing system.

1. INTRODUCTION

We as humans depend on the atmosphere, the oceans, and the land to maintain the environment in which we live. It is the land, however, that provides us with most of the food, fuel, fibre, and shelter and with 80% of the global biodiversity. Here we examine the biotic and certain physical resources that the land provides and how these resources are structured and how they are changing. We examine how well we know the components that make up our natural realm since fundamental to any understanding of the functioning or management of natural systems is knowledge of the nature of the component pieces. We show that in fact we have rather poor information on the basic building blocks of the terrestrial system. We do this in order to assess what the gaps are in our understanding of terrestrial systems, to develop a plan for eliminating these deficiencies (research needs), and to indicate priority actions.

Soil, water, and sunlight are the fundamental resources driving the productivity of the land surface. Soils affect human beings by providing the medium by which essentially all plants obtain their nutrients and water. Soils, interacting with plant cover, serve as the interface for minerals, water, and gases between the land, the atmosphere, and the oceans. They affect humans indirectly by emitting and absorbing greenhouse gases, by serving as a major reservoir and controller of the rates of movement of elements, and as a major reservoir and pump for water. Thus for managing the biosphere we need to know the character and distribution of soils and the modifications that they are undergoing due to global change, including the direct effects of human activity.

Human beings have had a major impact on the productive potential of the soil resource. The pathway from a natural ecosystem to, where possible, a robust and sustainable agro-ecosystem is difficult. Humankind has often not found this pathway, with serious degradation of the soil and land resources as a consequence. This is exemplified in the recent United Nations Environment Programme/International Soil Reference and Information Center (UNEP/ISRIC) assessment of global soil degradation (Oldeman *et al.*, 1990), as discussed in detail in Chapter 2. That study refers to the various forms of degradation that have occurred over the past 50 years or so. One should add the amounts of originally human-induced but now arrested soil degradation that occurred during the periods of colonial expansion and the agricultural occupation of North America and Australia (100–200 years ago), as well as those lands that were degraded during the times of early Mediterranean and Asian civilizations (2000–3000 years ago).

It may be reiterated that there is a clear and urgent need to further quantify and to monitor the extent, degree and the causes of the various types of human-induced soil and land degradation that are occurring globally, per region and per country; to translate this information into socio-economic and ecologic terms, and to develop practical programs to control and reverse degradation trends at local levels. The International Scheme for the Conservation and Rehabilitation of African Lands (ISCRAL) as proposed by the Food and Agricultural Organization (FAO, 1990b) may constitute a useful mechanism for such programs.

2.   THE DYNAMICS OF SOILS AND LANDFORMS

The attributes of the various soils in the world can be grouped by "response times" or speed of adjustment to changing environmental conditions (Arnold, Szabolcs, and Targulian, 1990; Scharpenseel, Schomaker, and Ayoub, 1990). Response times range from rapid monthly changes to changes that are noticeable only over millennia. Many of the soil attributes that affect plant growth, the hydrological cycle and surface energy balance (transfer of sensible heat, albedo) can change within a few years. The respective response times of landforms, versus soils per se, to environmental change need to be quantified.

The moisture, temperature, nutrient/salts and organic matter dynamics of soils are easily changed. In the time perspective of global change, say 50–100 years, these

changes in these parameters would not normally affect the major pedological types. There are, however, some fragile or "threshold" soils where a slight change in one of the soil forming factors – viz., temperature, rainfall, surface hydrological conditions, and direct human influence — would induce a major change in the pathway of soil formation, such as ferralitization vs. podsolization, illuviation vs. homogenization, salinization vs. leaching. Sombroek (1990a) gives some examples for the tropical and subtropical regions, where a transition to a different genetic soil group may take place within a time span of only tens of years, upon only slight changes of prevailing climatic conditions or somewhat higher frequency of extreme weather conditions. For the boreal and the subpolar regions, where global change impacts may be large, Goryachkin and Targulian (1990) predict substantial changes in soil-forming processes and shifts in zonal soil geographic belts.

Slight changes in climatic conditions may cause a substantial change in the rate of emission or sequestering of greenhouse gases; such as methane and nitrous oxides, which are predominantly of terrestrial biotic origin (see also the discussion on global biogeochemical cycles in Chapter 6). At present there is insufficient information to ascertain the factors controlling flux changes. Many point measurements – chosen in relation to their representativeness for major land units – are necessary to understand the dynamics of these fluxes, and indeed several research programs have started to do so (an example is the Terrestrial Initiative in Global Environmental Research (TIGER) Programme of the Natural Environmental Research Council (NERC), UK. The relatively few actual measurements of the moisture and temperature conditions in soil profiles, and their variation over the seasons and years, need to be extended and systematically compared and related to fluxes of soil-related methane, nitrous oxide and sulfur.

Special attention needs to be given to the quantity, quality and dynamics of the organic matter of the world's soils. This not only because organic matter has a large bearing on the conditions of the soils in relation to plant growth, (micro)faunal activity, and human life-support systems, but also because it represents a major carbon storage pool. Soil respiration represents about 25% of the total annual release of CO_2 to the atmosphere. The amount of carbon stored in the world soils, as fresh organic matter, stable humus or charcoal is supposedly two to three times higher than the carbon stored in the natural vegetation and in standing crops (Houghton, Jenkins, and Ephraums, 1990). Where the plant biomass is very luxuriant such as in tropical rain forests there is still as much carbon in the soil as above ground; the soils of grasslands and farmland store up to ten times as much carbon as the plants growing on them (Goudriaan, 1990). Most of this soil carbon is not fixed, and thereby an important link in the global carbon cycle; it may enter directly into the atmosphere by combustion or microbial transformations, be transported overland by wind and water erosion and then be stored in sediments of floodplains and sea or lake bottoms; it may also move down into the soil profile and be transformed into near permanent occlusions with iron and aluminium compounds, or be instrumental in the dissolution of rock carbonate and the re-allocation of the products in the soil profile or downslope.

Soils of present or former aridic regions contain carbon locked up in the form of pedogenic carbonates ("calcrete", "petrocalcic horizons"). This inorganic soil carbon pool is estimated to be as large as the stock of organic carbon in the world's soils, but has a much longer residence time.

Further quantification of the global stock of carbon in soils (topsoils and subsoils) is needed, including potential sources as well as sinks of soil carbon. We need to know how present levels of surface organic matter, and of soil micro-biomass in the various biomes and agro-ecosystems, influence the annual input of organic matter in the soil; what is the residence time of organic matter of the various soils of the world; which compounds are the most stable and at which depth; which are the microbial processes that break down soil organic matter, and how they influence the release or sequestering of greenhouse gases; and how can the dynamics of soil organic matter be modeled?

Close co-operation is needed on methodologies, site selection, and the extrapolation of studies in order to derive an improved geo-referenced data base of the world's soil organic matter status. There are national initiatives either in-country or in a network of research sites elsewhere, such as TIGER and the Stimulation Programme for Research in Tropical Forest Areas (TROPENBOS, Holland); regional programs in developing countries on soil organic matter and soil fertility such as FAO's Integrated Plant Nutrition Systems (IPNS), and those of the International Board for Soil Research and Management (IBSRAM), the International Council for Research in Agroforestry (ICRAF), the International Fertilizer Development Center (FDC) and the Tropical Soil Biology and Fertility (TSBF); soil geography-related programs such as those of FAO's World Soil Resources Office, ISRIC and the Soil Conservation Service (US–SCS). Mechanisms need to be developed to integrate these important individual efforts.

About 70% of the atmospheric methane (CH_4) comes from terrestrial sources such as wetland rice fields, marshes and swamps, ruminants (cattle) and termites. Neither the processes of microbial methanogenesis in the various soils of the world and their cover, nor those of methane oxidation in soils are as yet well understood. Indications are that the drainage of wetlands and increased nitrogen inputs are decreasing the capacity of soils to consume methane (Melillo *et al.*, 1989). Early estimates of

methane production by ruminants need to be quantified over the various rangelands and pasture lands. Termites and other soil fauna as producers of methane are also poorly quantified over the different biomes.

The budget for the nitrous oxide (N_2O) exchange between ecosystems and the atmosphere has many uncertainties. At least 90% of the emissions are provisionally estimated to be of terrestrial biotic origin — either from tropical rain forests, from heavily fertilized arable lands and grasslands, or from subtropical and topical savanna soils. Soil microbiological activity forms the basis of nitrification, denitrification and diazotrophic processes that result in the production of nitrous oxides, but actual measurements on N_2O emissions on representative sites are very few.

The surface attributes of soils – color, crusting vs. porosity, degree of plant cover – influence the transfer of sensible heat and the reflection of sunlight (albedo). Soils are, moreover, a major determinant of the fluxes of water vapor, a major greenhouse gas, and a number of soil attributes influence the runoff, storage and transmittance of water, and thereby the world hydrological cycle.

In order to better quantify and understand the processes of terrestrial sources and sinks of greenhouse gases, there is a need for more measurements, at representative sites, of soil moisture and soil temperature conditions and dynamics; of the soil microbiological composition and activity, and of the quantity and dynamics of soil organic matter. Such measurements should be performed as a part of integrated or interdisciplinary research on the hydrological and soil chemical and physical effects of conversion of natural vegetation into human use (tropical forest clearing for shifting cultivation, or permanent agriculture such as perennials, grasslands or exotic trees; shrub savanna conversion into cropland by high-input soil management such as in the cerrado area of Brazil; tundra and taiga vegetation conversion into large-scale mechanized annual cropping, etc.). This research should be done with a rigorous experimental designs that are processes-oriented, on transects, paired sites and experimental catchments as is being proposed by the International Geosphere-Biosphere Programme - Global Change and Terrestrial Ecosystems (IGBP GCTE) Programme.

3. THE GLOBAL SOIL DATABASE – WHAT DO WE HAVE?

There is a confusingly large variation in soil classification systems that are utilized at national levels. There is, however, an overview inventory of the world's soil resources at a scale of 1 : 5 million (FAO, 1971–1978) prepared under the auspices of FAO and UNESCO. The map is mainly a compilation of a multitude of national soil maps available at the time, of differing detail and accuracy. The map utilizes 106 units based on "diagnostic (soil) horizons" and non-horizon related "diagnostic properties". The topsoil textural class of the predominant soil unit is indicated on the map, as is the average slope of the mapping unit as a whole. Areas of "non-soil" are indicated as miscellaneous land units: dunes or shifting sands, glaciers and snow caps, salt flats, rock debris, and desert detritus. Boundaries of permafrost or intermittent permafrost are indicated separately, but no other soil moisture and soil temperature characteristics are used, except implicitly in hydromorphic and desert soils. To estimate moisture and temperature characteristics, FAO devised special maps on "agro-ecologic zoning" (FAO, 1978–1981) which can be used as overlays. These zones are in fact agro-climatic regions, based on prevailing climatic conditions, and expressed in annual "length-of-growing periods". The above soil and agro-climatic data have been combined in an assessment of the population carrying capacity per country, at three different levels of land management and external inputs (FAO/UNFPA/IIASA, 1982).

There is a need to update the global soils map at the same scale of 1:5 million and to work toward a more detailed presentation that would be in digital form and compatible with biome mapping as discussed below. Recently a systematic revision of the world-level FAO/UNESCO map legend was completed (FAO, 1988), for use at the original 1:5 million scale. Digitizing of all 1:5 million map sheets, either by polygon vectoring or by gridding, has been undertaken by several organizations including UNEP-GRID in Geneva, but quantitative analysis of the polygon's composition will yield much more information.

Most existing soil classification systems are pedogenetically oriented, with varying degree of quantification of the characteristics of the soil ("morphometrics"). They take the subsurface and subsoil features as starting point, because of their supposedly more stable character than the surface and topsoil features. The topsoil features, and the associated land cover characteristics, are however important in the study of natural ecosystems and the atmospheric greenhouse functioning, and thus must be accommodated in soil mapping efforts.

There is, with the advance of modeling techniques, a growing need for both local and global geo-referenced information on individual attributes of soils and terrain units for use in the assessment of agricultural production potential, for identifying alternative sustainable uses of the land including forest use and silviculture, for identifying soil pollutant buffering capacity, and for global change studies. The kinds of information needed are: surface roughness, surface sealing/crusting, surface color in relation to reflectance; amount, type, stability and vertical distribution of organic matter; soil biologic activity; surface runoff, soil moisture storage and transmittal properties; soil temperature régimes, drainage and flooding conditions; soil nutrient status and storage capacity; but also relatively simple attributes such as bulk density per layer horizon. The latter are also needed to express soil characteristics and properties, as commonly given on a weight basis, in units-of-volume as used by modelers of climate and plant growth.

Soils do, however, not stand alone on the surface of the Earth; they are closely related to landforms: mountains, hills, plateaus, terraces, undulating uplands, plains, bottomlands, etc. Within the broad context of geological materials, climatic conditions and geomorphologic history, each facet of the present-day landforms has its own soil and associated hydrological régime. Soil formation has also a significant landscape-lateral element (Ruellan, 1985). Experienced soil geographers are aware of these landform–soil relationships and use it to delineate the spatial patterns, both in detail and scaling to large areas – often with remote sensing imagery as auxiliary tools.

No program has, however, been developed at the continental or global level to combine the nature and distribution of landforms (and their surficial geologic materials and hydrological régimes) with soils in a single integrated quantitative database. The only effort, in a generalized and therefore rather qualitative way, is a recent 1:15 million scale wall chart on the "Present Status of Landscapes of the World" prepared by the Institute of Geography of Moscow State University. The lack of a systematic landform–soils inventory at global level is partly due to the absence of an accepted multi-categorical hierarchical system of describing and distinguishing landforms. Nevertheless, a degree of consensus exists on "orders of land relief": primary continental subdivisions; physiographic provinces; relief sections; repetitive landscape patterns; relief facets; relief sites. Bridges (1991) in a recent book on the main landforms per continent, applies such orders of relief.

The technique of Digital Terrain elevation Modeling (DTM), translating topographic map features and remotely sensed altitudinal information into a digital form suitable for computer analysis, is now in an advanced stage of development and can be very helpful in generating global landform information.

Hydrological features of the landscape also need to be considered in an inventory and evaluation of soils and landforms. Hydrological attributes of the land are: the degree of surface runoff in relation to slope, land cover and surface crusting or sealing; infiltration, storage, percolation of water, and horizontal hydraulic conductivity; phreatic levels; spring line and seepage situations; areas with saline/sodic ground-water subject to rise; and the

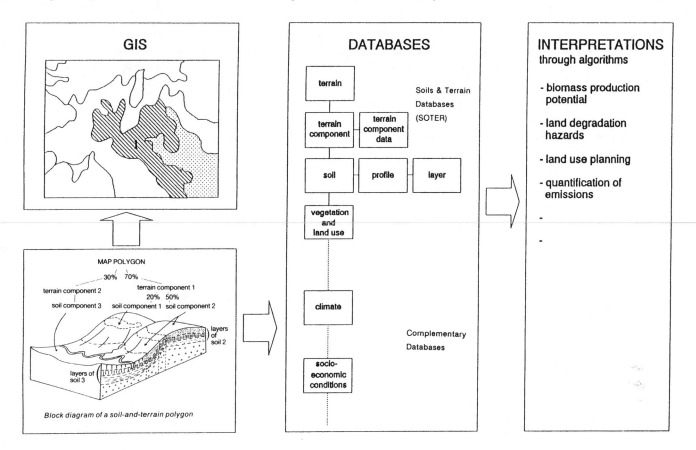

Figure 1: Structure of a computerized land resource information system.

characteristics of flooding and rainwater submergence régimes of low terraces, bottomlands, floodplains and deltas. For the latter, a classification scheme is needed, encompassing length, depth, velocity and regularity of flooding or overland flow within and outside the temperature determined annual plant growing period; as well as the sediment load and the chemical characteristics of the waters, upon which a systematic global inventory can be undertaken (see Aselmann and Crutzen, 1989 for a first effort in relation to methane emissions). International action plans on better water use for sustainable agricultural development, such as the International Action Programme on Water and Sustainable Agricultural Development (WASAD) of FAO (1990c) need to utilize both global and local geo-referenced data bases on hydrological conditions, as described above, linked to associated natural resources and to diffuse or point sources of pollution (see Chapter 10 for detailed discussion of the world's freshwater resource).

A Working Group of the International Society of Soil Science (ISSS) has developed a proposal for a world soils and terrain digital database at an average scale of 1:1 million, acronymed SOTER (World Soils and Terrain Digital Database), to be completed in a 10 to 20 year period. A number of pilot and priority areas were identified, and UNEP provided funding to the International Soil Reference and Information Center in Wageningen (ISRIC) for a first pilot area in South America, viz.

adjoining portions of Argentina, Brazil and Uruguay. Since then a procedures manual for map and database compilation has been prepared (van Engelen and Pulles, 1991) and training at the national level has started. A database structure using a relational data base management system was developed, and an international panel assessed various Geographic Information Systems (GIS) suitable for SOTER.

The essence of the approach is to screen all existing data in an area, whether or not registered on official soil and landform maps, and where necessary to complement these with remote sensing information. The data are then rearranged for the database, going from landform and terrain component features to soil layer attributes (Fig. 1). For all soil attributes, quantitative data or class limits are used. Those attributes that cannot yet reliably be quantified from available data, or from limited additional field and laboratory measurements, are flagged as such, for amendment at a later stage. Subfiles give the details of very representative individual soil profiles (International Soil Information System, ISIS) and will complete the information. Algorithms and computer programs are being developed for the use of such a comprehensive database for a number of purposes, for instance the risk assessment of several forms of soil degradation.

The SOTER initiative has already obtained the conceptual support of UN organizations such as FAO and

UNEP; the Consultative Group of International Agricultural Research Centers such as the International Service for National Agricultural Research (ISNAR); the IGBP Data and Information Systems unit (DIS); and many national institutions on land resources assessment. Priority areas to be entered in the data base in advanced state of planning are: Central America, parts of West Africa and Eastern Africa, the Danube catchment area, central USSR, and South-eastern China. The sequence of other areas can be related to the needs of IGBP core programs, in particular with respect to the envisaged 100 land system research sites of the Global Change and Terrestrial Ecosystems (GCTE) and the planned Regional Research Centers of the Global Change System for Analysis, Research and Training (START).

In summary, in spite of the vital link between soils and human welfare we do not have an adequate information base from which to evaluate the soil resources available and how they are changing. Our global soil database, as for most natural resources, is uneven and poor in quality. We call here for an integrated effort to remedy this through a concerted program. This program would provide a uniform, readily accessible database of soils and terrains that would be compatible with comparable databases for climate, hydrology, vegetation, and current agricultural land use. Such a database would be a vital tool for defining the interaction among these realms in developing global Earth System models as well as providing information for policies and strategies relating to the management of regional and global soil resources. This database of soils would involve improving the global soils resources inventory utilizing modern techniques. It would further involve studies of soil genesis as well as a global assessment of soil productivity and degradation. Periodic updating of the system will allow monitoring of the changes that will occur as a consequence of expected climatic change and population increase.

4. BIOCLIMATIC RESOURCES

Data bases on landforms and soils information are useful in assessing the biotic functions of the land only in combination with information on local bioclimatic conditions; amounts and temporal distribution of precipitation, temperature (means and extremes), solar inputs, evapotranspiration, wind speeds, and other biologically sensitive climatic parameters.

Although we now have techniques for modeling the world's climates in response to changing forcing functions this information is not yet sufficiently regionally specific to give us spatially detailed bioclimatic information (Dickinson *et al.*, 1989).

Because of its inability to move in short time-scales, the natural vegetation integrates the overall effects of individual climatic parameters acting on plant growth. It is therefore no surprise that most of the traditional climatic classifications at a global level have used the occurrence of natural vegetation types to identify class limits for the various climatic zones. Since, in addition to light and nutrients, plant growth requires moisture and warmth in varying amounts, such limits have been related mainly to hydric and thermic régimes.

There are a number of classifications of the world's bioclimates available including Köppen, (1931) that is based on thermal and hydric criteria and Thornthwaite, (1931) which utilizes precipitation and thermal efficiency indices. The Holdridge (1959) system, which has been generalized to a global level (Leemans, 1989), takes into account biotemperature (the sum of monthly temperatures in excess of freezing and averaged over the years) and the ratio of annual potential evapotranspiration over annual precipitation. Also available are continental maps on calculated soil moisture and temperature régimes for crop growth suitability of the US Soil Taxonomy system (Van Wambeke, 1981, 1982, 1985).

Even using the same climatic station databases, such as those applying the World Meteorological Organization's (WMO) Climate Data Management System (CLICOM), the results of these classification approaches are different because of the different combinations of the criteria utilized. Their simplicity is attractive for modeling purposes, but not necessarily adequate. At regional and local levels more climatic detail needs to be taken into account. The bioclimatic scheme developed by Papadakis (1961) can be considered to be more complete for application at a regional level, because it includes various essential features such as winter severity, summer heat, duration of the frost-free season, and duration, intensity and annual position of the humid and dry seasons. The concepts of Papadakis were taken up and elaborated by FAO for its agro-ecological zones project (AEZ) for developing countries, for application both at continental and at country level, and with food crop suitability assessment as its main aim. The major agro-climates in the AEZ approach are delineated on the basis of the mean daily or monthly temperatures during the growing period. Moisture régimes are quantified through the concept of reference length-of-growing period defined as the duration, in consecutive days, when moisture supply from precipitation and soil moisture storage permits crop growth. In the regional climatic inventories 20 reference length-of-growing period zones are mapped with 30-day intervals, for the whole of Africa, Asia and Latin America, and interpreted for all important food crops. Country-level detailed zoning has been carried out for Kenya, Bangladesh and China, allowing the definition of more specific land use options. The AEZ approach of FAO has not yet been systematically extended to temperate and cool regions, although examples are available for Canada, northern China and western Australia.

It may be evident that a more complete and more uniform effort for bioclimatic zoning at both global and regional or country levels is called for. The availability of world climatic data in digital form now makes it possible to tailor studies on climate–soils–vegetation/land use interactions to a much more specific level. Detailing of the (agro)-ecological zoning approach of FAO, and its extension to temperate and cold regions, may be the quickest way to accomplish this. The results will also offer reference material for the spatial and temporal dynamics of isolines of global biomass production potential as inferred from repetitive remote sensing imagery provided by NOAA-AVHRR satellite (the National Oceanic and Atmospheric Administration Agency – Advanced Very High Resolution Radiometer), as discussed below.

5. BIOTA

5.1 Biotic diversity – numbers and kinds

Surprisingly, in spite of systematics being one of the oldest natural science disciplines, we have an inadequate assessment of the biotic richness of the Earth. What we do know is uneven across groups. It is estimated that we have described less than 15% of the world's estimated over 1,000,000 microorganisms (di Castri and Younès, 1990) whereas we probably know over 99% of the world's bird species, which number somewhere about 9000 species (Clements, 1978). Estimates, which are only that, of the number of species on Earth are in the realm of ten million (May, 1988) (see Chapter 11). The rapid destruction of tropical forests (as high as 200,000 km^2 y^{-1}), where biotic richness is centered, is threatening the principal storehouse of the world's biodiversity. Elsewhere in this volume these losses, and the social-economic drivers of this devastation are treated in detail. Here we focus not on diversity as an issue per se, or as a human resource, but rather on the consequences of species losses to the functioning of ecosystems. Biotic diversity encompasses not only the numbers of species but also numbers within populations. Needless to say the loss of one half of the elephants of Africa in a single decade (Matthiessen, 1991) or of over 80% of the forests of the Ivory Coast in the last 35 years (Myers, 1990) is an enormous biotic impoverishment of the biosphere. Also being reduced or modified, as discussed below, are other components of biodiversity: races within species, varieties of habitats as well as landscapes.

The world's species richness can be collapsed considerably when considered functionally. At the most general level, for example, all organisms can be categorized as either producers, consumers, or decomposers. Finer level categorizations can be utilized for examining ecosystem function such as nitrogen fixers, photosynthetic pathway, particular energy source utilization, pollinator, etc. Similarly, organisms can be grouped based on structural traits related to size and shape, such as trees, shrubs, and herbs for plants, or at finer structural levels, such as canopy or understory, evergreen or deciduous, etc. As one moves from the grossest to the finest categories the numbers of species involved in each decline. This means that at the higher levels of categorization there is a great amount of apparent replication of function. We do not know whether this replication of function represents redundancy or whether critical functional roles are assumed by the various components of a functional group under normal environmental fluctuations on time scales of decades to centuries.

It is important to resolve these issues. For example, in order to model the Earth's global metabolism, or its altered biotic structure with changing climate, we must categorize the Earth's biota into meaningful functional groups. Equally urgent is assessing the impact of losses of biodiversity, in all its dimensions on Earth System functioning. The International Geosphere-Biosphere Programme (IGBP) and the Scientific Committee on Problems of the Environment (SCOPE) in conjunction with the International Union of Biological Sciences (IUBS) are presently addressing these concerns.

Organisms interact not only with their physical environment but also with their biotic milieu. This is an important consideration in viewing the impact of species additions or deletions to ecosystems as is noted below.

6. BIOTIC INTERACTIONS

The biota of any region display varying degrees of interdependencies, the most basic being energy related as noted above. Food webs describe the energy dependency in an ecosystem. Any system, although it may contain thousands of species, will be tied together by these basic relationships. At a finer level though complex linkages may be seen. Specific energy-source interactions between species are called links. The average number of linkages in a given energy chain is about four with the maximum being about ten (Cohen, 1990). Although many species in the foodweb will be generalists, there are specialists that serve to multiply the numbers of chains in any given food web.

In addition to energy transfer interactions, there are interactions that are involved in the transfer of genetic information. These mostly relate to pollination systems – the transfer of pollen among flowering plants through an animal vector. Of the approximately 250,000 species of vascular plants a large percentage are pollinated by animal vectors, preponderantly insects, many of which are generalists and pollinate many different kinds of plants.

In any given community there will also be a wide range of biotic interactions that may be classified as mutualistic – something is given for something gained. Most plants (75–80%) (Hawksworth, 1991), for example, harbor fungi associated with their roots (mycorrhizal fungi) that derive energy from their host while giving in return nutrients. Again, while many of these fungal species are generalists, indiscriminately affecting plants irrespective of their taxonomic status, there are some that are species specific.

Another vital mutualistic relation is between nitrogen-fixing bacteria and their hosts, generally leguminous plants. These interactions are crucial to ecosystem functioning since they provide the preponderant pathway for atmospheric nitrogen to be transferred to the soil. Strains of the nitrogen-fixing bacterium *Rhizobium*, tend to have somewhat narrow host specificities, in some cases extremely so. Isolates of *Rhizobium leguminosarum*, for example, may infect some pea (*Pisum sativum*) cultivars, but not others (Young and Johnston, 1989)

Thus in the categorizations of biotic capital of our world we can view species as independent entities, and additionally as components of a web of interactions. These webs have both temporal and spatial dimensions. Migrating birds may have vast geographic ranges and be involved in numerous species, and even different ecosystem interactions, whereas some mutualistic relations between a plant root and a fungus can be species specific and at a cellular level.

The terrestrial system can thus be viewed as a tapestry of many species, woven into complex patterns representing biotic interactions found in the living landscape. These patterns fall into assemblages of organisms and their interactions that are called ecosystems, or in a geographic context, biomes. We now view what we know of these biomes, or landscape units, and subsequently how humans are affecting them by modifying the biotic interactions as well as even the species present.

7.   BIOMES – THE GREAT MULTIPLIERS

In the simplest form the biotic communities of the world have been described by the dominance of the principal plant growth form, e.g. grassland, shrubland, or forest, either evergreen or deciduous. A next level of categorization divides these basic forms among various generalized climatic types, such as temperate or tropical grasslands, Mediterranean or desert scrublands, and so forth. These units have been further subdivided by the occurrence in a particular biogeographic province, which begins to capture the species peculiarities of the biotic community since these are generally few common species among biogeographic provinces. Categorizations below this level become increasingly localized and species

dependent. Knowledge of the distribution and nature of the large-scale units is important for understanding global metabolism, for example, and the finer categories for localized land management.

We have a fairly precise knowledge of the broad limits of the higher units of biome categorization on Earth and have had for a long time. This information is now available in digital form for 1 degree latitude grid sizes (Matthews, 1983). Also available, in digital form, for 1/2 degree grid units is the theoretical distribution of biomes based on bioclimatic constraints of vegetation types (Emanuel, Shugart, and Stevenson, 1985) as well as world ecosystem complexes (Olson, Watts, and Allison, 1983). There are currently efforts to bring these and climate and soils databases together into functional models for understanding vegetation dynamics of the globe (Prentice *et al.*, 1989). It is essential to do this in order to achieve the capability for predicting how biomes will respond to climate change as well as how climate change will affect feedbacks of terrestrial systems to the atmosphere.

The above information on geographic extent of particular biome types is used as multipliers for point measurements of system function, such as primary productivity, in order to assess, for example, global fluxes of carbon. Measurements of primary production are generally only made on tens of square meters of land surface and then multiplied by 10^6 km^2 to give global fluxes for various biome types. Obviously small errors in areal process measurements can have large effects on global estimates of fluxes. Estimates of the global living biomass have gone from over 1000 billion tonnes to about half this amount over several decades due primarily to the use of average rather than extreme values of point measurements of standing biomass (Olson *et al.*, 1983) but also due to real loss of living carbon stores. Thus there is a need to provide better estimates on both sides of this equation – the process and the biome extent.

The data sets referred to above have been assembled from many different sources and are not of high resolution. Resources, both financial and human, are not adequate to evaluate rapidly the extent as well as the nature of the Earth's biome types. We do, however, now have the capability of utilizing satellite imagery for obtaining both functional and structural information on vegetation. This information is global in extent and is gathered at repeated time intervals giving an assessment of the changing properties of the global terrestrial system. The AVHRR on the NOAA satellite collects information on the spectral properties of the Earth's surface. This information is being utilized to obtain measures of the absorption of photosynthetically active radiation, and hence photosynthesis (Sellers, 1985). Temporal measures give seasonally changing productivity of the Earth's vegetation types and bioclimate as discussed earlier. Information can also be derived on the exchange of water vapor of the Earth's various vegetation types.

One km^2 resolution AVHRR images compiled at monthly intervals are now available for a number of places on the globe. An effort to implement a global mapping project should be made at a centralized facility, as is being proposed by the IGBP. This information, which would be accessible in digital form, should be made available to users via mass storage media. To complement these measurements data should be assembled at the same resolution for soils, land cover and land use change, and topography. The 1 km^2 resolution represents probably the finest detail that is feasible at this time for global studies. This resolution involves 1.49×10^8 grid cells and a formidable challenge for the remote sensing community; however, this number is only a fraction of the number of nucleotides being sequenced in the human genome project.

Of course, neither the relatively fine global resolution of the 1 km^2 satellite measurements nor the large-scale ground based measurements of landscape types and changes will tell us all we need to know about the functioning and alterations of terrestrial systems. With the resources and time that are available there needs to be intensive attention given to in-depth studies, and protection schemes evolved, for particularly sensitive and critical ecosystem types. For example, (Myers, 1990) has identified 18 global hot-spots in tropical and Mediterranean climatic areas that encompass only 0.5% of the Earth's land surface but yet which contain 20% of the Earth's plant species. Habitats in all of these areas are threatened by destruction. Efforts need to be made to identify and analyze regions of high fragility and that are particularly sensitive to climate change and to a growing human population. The International Geographic Union (IGU) has launched a program to identify and assess such critical environmental zones. Looking at soils, landforms, hydrological and bioclimatic conditions in conjunction, on the basis of quantified geographic databases of sufficient detail (resolutions of 1–10 km), one can identify, delineate and analyze environmentally critical areas. Some already obvious examples of areas that are particularly sensitive to human-induced changes in land cover, and/or to climatic change are:

(a) mountain slopes in tropical areas, especially those covered with soft volcanic materials (land- and mudslides);

(b) floodplains and low terraces of river-catchments that are subject to deforestation in their upper reaches (flashflooding; salinization under aridic conditions; white sand podzolization or laterite/plinthite formation and hardening, in humid tropical regions);

(c) wetlands, delta's and coastal lowlands with high population pressure (salinization, acid sulphate soil formation, transformation of peat and mangrove

areas);

(d) upland areas of hitherto stabilized dunes and cover sands in desert fringe areas (wind and water erosion);

(e) hummocky terrains in permafrost fringe areas (solifluction processes; loss of soil organic matter);

(f) areas with sandy soils in regions with high degree of pollution by chemical compounds and heavy metals, especially when acidification processes are active (chemical-time-bomb hazards).

8. SYSTEM PROCESSES

Knowledge of the global extent of the Earth's terrestrial biomes provides, as stated, the multipliers for describing the storage and fluxes of energy, minerals, and water at regional and global dimensions. The terrestrial biota serve as regulators and pumps of these stores and fluxes. Although the distribution and structure of the Earth's biota is controlled by the environment, at the same time the biota has a major influence on environmental, features through ecosystem feedbacks. For example, vegetation structure and water use influences the transfer of energy and mass, directly influencing the physical climate system. Deforested areas will have a different influence on local climate than forested zones. In addition, the biota influences atmospheric properties through the direct emissions of gases that in turn alter atmospheric radiative properties and hence influence climate (Mooney *et al.*, 1987).

In addition to the coupling between the land biota and the atmosphere there are many coupled processes between land and the water. The high productivity of coastal marsh systems is fed by the nutrients moving in from upland systems. The coupled land and water systems have been one of the most impacted on the Earth by human activities, primarily through water diversions or land fill. In addition to land–water and land–air interactions, there are important linkages among various landscape units. An extreme case may be the fertilization of Amazonian forests by minerals transported by air from Africa.

The past few years has seen the active development of a new science, Earth system science, as exemplified through the International Geosphere–Biosphere Programme, that is attempting to quantify these linkages between the functioning of the terrestrial system with that of the atmosphere and ocean, through biogeochemistry and climate systems. This effort has shown, for example, the close ties between climate and plant water balance as noted above. It has further shown how altering the vegetative cover of the Earth has also changed the CO_2 content of the atmosphere.

Thus far we have viewed briefly the Earth's biotic building blocks and how they interact and function together as large Earth-system units. We now turn to how

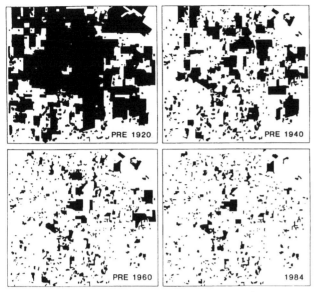

KELLERBERRIN STUDY AREA 1600km^2
UNCLEARED LAND

Figure 2: Fragmentation of a western Australia landscape. From (Saunders, Hobbs, and Arnold, in press). Such fragmentation alters the radiative, water and nutrient fluxes across the landscapes as well as the characteristics of the residual populations in the residual natural "islands" (Saunders, Hobbs, and Margules, 1991).

these units are being stressed and restructured by human activities.

8.1 Modification of Biotic Complexity of Natural Systems

8.1.1 Breaking the links – Humans as editors

Superimposed upon the natural geographic patterns of biodiversity is localized biotic variability due to disturbance and microsite characteristics. Another driver of pattern is human action, either directly, or indirectly. In most parts of the world we are seeing losses of species, losses of genetic variability within a species, dramatic reductions in the population sizes of certain organisms, and a reconfiguration of landscapes (Figs. 2 and 3). In short we are seeing a simplification of the Earth's biotic richness by most measures. At the same time there are certain activities, driven by humans, that are increasing diversity of certain components of the Earth. The diversity of natural primeval landscapes was driven by local variability in substrates (e.g. riparian corridors) and natural disturbance régimes. Human activity, at low levels, increased the landscape variability by increasing local disturbances and creating biotic-rich edges. For example, the human-influenced Mediterranean landscapes have gone from being rather simple in diversity, to ones which were quite rich under early human development, to ones that are becoming simplified again, but now due to land

degradation related to high population usage. In short, humans are acting as editors to terrestrial systems – they have added and deleted elements. We are modifying the overall structure of nature without, however, a comprehensive understanding of the consequences of these modifications.

Accompanying the increase in human activity was first the evolution of new races of organisms that could co-occur with humans (various weedy species) and as well as the domestication of organisms. As land use intensified and human commerce became intercontinental species of organism from one biogeographic region were moved to another, both purposefully and accidently. The disturbed landscapes of the world, which are ever increasing in extent, have become the habitat for a rather small number of highly successful invaders. Thus, although, a local biota may be enriched by the addition of these newcomers, the landscape has become depauperized since the population sizes of the endemics have often become greatly reduced. The intensification of agriculture has also resulted in the biotic impoverishment of regions since most often local crops are rather uniform genetically.

8.1.2 The Consequences of biotic alteration

We have a multifaceted problem. First, we have inadequate information of our biotic capital, as noted earlier, at virtually any level, and particularly for tropical regions. Secondly, we do not really have any good quantitative measures of how the Earth's biotic assemblages are being altered, either by species additions or subtractions. Worse yet, we have not developed adequate information to assess the consequences of these changes in terms of ecosystem functioning. There are indications that the removal of top carnivores from many of the Earth's ecosystems has profoundly altered their entire structure by changing species interactions at all levels. Many systems are still changing to a new equilibrium state as a result of these species removals (Terborgh, 1988).

What can we do to better our knowledge? We can mount an intensive effort to describe the biotic richness of the globe, in relation to geographic position, at least at a certain level and for certain groups, for example birds, butterflies, mammals, and higher plants. This could be done for 1 degree grid sizes equivalent to that now being utilized for global vegetation assessment. In time more difficult groups can be added to the global assessment. We can better estimate species and system losses by a comprehensive satellite monitoring program of land cover change as is being proposed by the IGBP. We can evaluate species additions by co-ordinating listings of regional invasive species into a global data base, as IUCN is beginning to do. Finally, we can support the program of SCOPE/IUBS/MAB to evaluate the ecosystem function of

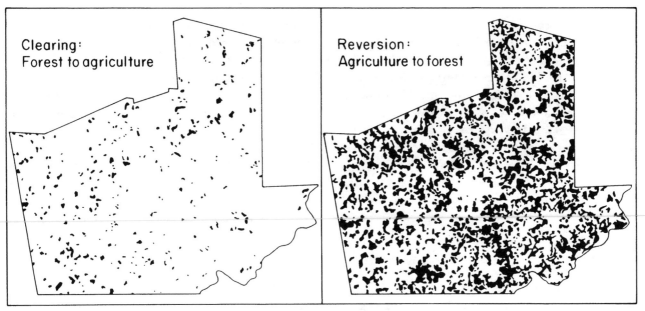

Carroll County, Georgia, 1937-74

Figure 3: Reversion of agricultural land to forests in Georgia, USA. (Williams, 1989) Changing economic and social factors can result in dramatic changes in the state of terrestrial systems.

biodiversity. This program would investigate the impact of species additions and subtractions over temporal and spatial scales on ecosystem functional properties utilizing explicit controlled manipulations as well as "natural" and "inadvertent" experiments.

8.1.3 Putting the pieces back together

It is now axiomatic that no place on Earth is untouched by the influence of humans. There is no such thing as a pristine environment. As human intervention increases in pulling the fabric of terrestrial systems apart so increasing amounts of human management will be needed for either mitigating against undesirable environmental effects locally, or even globally through geo-engineering. In addition to mitigation there will be increasing needs to rebuild or regenerate natural systems that have been destroyed, including streams, tropical forests, coastal marshes and so forth. The new field of restoration ecology has been developed to meet this need. Success in this endeavor has been called the true test of how well we understand natural systems. Do we have the knowledge, not only to reconstruct ecosystems, but also to make them self-sustaining? If we do not then we must put more effort into this important endeavor.

8.1.4 Modification of biotic complexity: Agroecosystems

At least half of the global land surface is no longer covered by natural ecosystems, but has been modified by humankind for agricultural purposes: cropland, animal husbandry and plantations (see Chapter 2). Many of these managed landscapes are stable under present-day climatic and cultural conditions, although others are not. In many

areas land management has resulted in an enlargement of the intrinsic biomass production potential and in some situations an increased biodiversity. Humans have made local landscapes more suitable for productive purposes by regulating runoff, terracing, clearing of stones, irrigating or draining, and by supplementing soils with lime, nutrients, and organic material. At the same time, agriculturally important plant and animal species have been improved by mass selection to become suited to local conditions resulting in a wealth of "land races". Traditional smallholder mixed farming in particular has produced diverse landscapes where wild species, crop cultivars, and domesticated animals co-occur. The sustainability of these agro-ecosystems is being threatened at present because of strong population growth and changing cultural practices that lead to the excessive use of chemical amendments and overmechanization.

There is a need to analyze the various present-day land use systems in terms of their degree of modification from pre-existing natural systems in order to determine how features of soil, microclimate, hydrology, and biodiversity and bioproductivity have been altered, and how these changes affect stability under contemporary socialeconomic conditions. Altered systems that are inherently unstable need to be identified along with what land management factor or climatic change element could lead to their sudden collapse. The spatial dimensions of these "collapse hazards" would require the matching of a geo-referenced data base on land use systems, as advocated in Chapter 2, with similar data bases on abiotic resources as discussed earlier in the present chapter, along with modeling of regional or local effects of climatic change.

In summary, multi- and interdisciplinary research is

needed on agro-ecosystem functioning at different intensities of human impact and climatological stress as advocated by UNESCO's Man and the Biosphere Programme (MAB), and which could be incorporated into the farming systems research of the international agricultural research institutes of the CGIAR system. The results of such research would be sustainable land use options that could support modifications to varying degrees. Hopefully, such an effort would permit us to avoid the environmental mistakes that we have made in the past that have resulted in intense land degradation and accompanying human misery.

9.   TERRESTRIAL SYSTEMS IN A NEW WORLD

9.1 Terrestrial systems under stress

Virtually all dimensions of our globe are changing at unprecedented rates due to the activities of a rapidly growing human population. Land cover, water and atmospheric chemistry, and species numbers and distributions are being greatly modified on a global scale. Not only is there rapid change but some of the features we see now have never existed in past history. We are tending toward a new world. This means that to a certain extent our knowledge of how the Earth operates now and has operated in the past will not be a totally reliable guide for understanding the future. There is fair agreement that our climate will be changing at a rapid rate to a warmer world. The atmosphere is becoming enriched in CO_2 going from 280 ppm in the pre-industrial atmosphere to over 350 ppm today (Houghton *et al.*, 1990). Tropospheric ozone have doubled in rural European sites this past century (Ashmore and Bell, 1991) and increases are even seen in non-industrialized tropical regions due to accelerated biomass burning (Fishman *et al.*, 1991). UV-B levels in spring in Antarctica are now far in excess than has ever been previously measured whereas at the same time there have been reductions in ground level-irradiances in the northern hemisphere due to increases in atmospheric pollutants (Frederick *et al.*, 1989). Increased atmospheric deposition of sulfur, nitrate and ammonium is significantly modifying soil chemistry and plant nutrition of European forests (Schulze, 1989). About one-third of all irrigated arable land is now salt affected (Reeve and Fireman, 1967). We do not know which systems are the most sensitive to these changing stress factors. Obtaining this knowledge is important as well as learning how to mitigate against or prepare for these changes.

Not only is the physical environment of the Earth changing but so is the biotic configuration of the continents as noted earlier. The breakdown of biogeographic barriers, both inadvertently as well as intentionally, through world commerce, is resulting in a homogenization of the world's biota. In New Zealand well over 50% of the flora is non-

native and in Hawaii 40% of the plants and 95% of the mammals are non-indigenous (Mooney and Drake, 1990).

Thus in looking to the future, we cannot entirely utilize the information from the past – the world is changing very rapidly, and to states that have not previously existed. It is predicted that future climate change may be greater than that which accompanied deglacial warming (Huntley, 1991).

We do not know fully what the consequences of these changes will be on terrestrial systems. We do know that increases in tropospheric ozone has reduced agricultural productivity in the United States (Heck *et al.*, 1988), that enhanced atmospheric deposition of nutrients has reduced forest growth in Germany (Schulze, 1989) and that increased CO_2 will probably promote greater plant growth in non-nutrient and water-limited conditions (crops and marshes) but not in nutrient or temperature-limited environments (Mooney *et al.*, 1991). Biotic diversity is predicted to decrease under present climate change scenarios (Woodward and Rochefort, 1991); however, the effects of land use change will probably have a far greater detrimental effect on biodiversity in the near term.

9.2 New knowledge for the new world

How then can we plan for the future on the basis of our present knowledge? We can refine our understanding of how natural systems operate and link this knowledge together into a new class of Earth System models as being developed by the IGBP. Thus we should be able to predict to a better degree how a given change in any part of the natural world will affect other parts. We must be able to consider landscapes as operational units – with important links between the units. The knowledge that we will need comes in large part from understanding how the larger units of the landscape operate. In order to gain this knowledge we will have to magnify meaningfully information derived from small scales to larger units. We must also develop a new experimental approach built on assessment of the response of whole ecosystems to change, for example, CO_2 enrichment. It is only through such experiments that we will be able to examine the feedbacks that will occur between plants, herbivores, and decomposers in a carbon-rich environment, for example. We will further have to show how organisms respond to multiple stresses building on our experimental base of single stress responses. We will also have to learn how certain organisms will respond to unprecedentedly rapid changes. This is a very large research agenda, but one which is needed to prepare for the uncertain future.

9.3 Training the new citizens and scientists

If you plan for one year, plant rice
If you plan for ten, plant trees

If you plan for one hundred years, educate mankind.

Kuan-Tzu (as cited in Holzner *et al.*, 1983)

As terrestrial systems become increasingly impacted by the influences of a growing human population, and in the process lose the building blocks and the connections, resilience and productive capacities will decrease. The activities of individuals and their effects on biotic systems will have to be evaluated even to a greater degree. Educational efforts on interactions between humans and the environment will need to be expanded to an ever greater extent and at all age levels. Additionally we will need more scientists, of more kinds, to study, understand, and manage terrestrial systems. This societal need, which surely will increase, comes at a time when training for resource scientists (foresters, taxonomists, range managers, etc.) is actually declining. We must reverse this trend and at the same time we need to develop new kinds of scientists; scientists, for example, who understand how both natural and human social systems operate and interact, scientists that understand how terrestrial systems interact with the atmosphere and hydrosphere, scientists who can think globally. This need will have to be fulfilled by restructuring some of our traditional educational institutions as well as for the establishment of new international research and training centers as proposed by IGBP and the WCRP. We must produce the stewards and visionaries that will guide us, and the Earth Systems on the planet that support our existence, into a future filled with the uncertainties that global change will bring.

ACKNOWLEDGMENTS

We thank the many people who have provided input into this brief overview either directly, or indirectly through participation in the formulation of a number of the emerging international programs referred to here.

REFERENCES

Arnold, R.W., Szabolcs, I., and Targulian, V.O. 1990. Global Soil Change. Report of a IIASA-ISSS-UNEP task force on the role of soils in global change. IIASA, Laxenburg, Austria.

Aselmann, L., and Crutzen, P.J. 1989. *Freshwater Wetlands: Global Distribution of Natural Wetlands and Rice Paddies, Their Net Primary Production, Seasonality and Possible Methane Emissions.* Mainz, Germany: Max Planck Inst. Chemistry.

Ashmore, M.R., and Bell, J.N.B. 1991. The role of ozone in global change. *Ann. Bot.*, **67**, 39-48.

Bridges, E.M. 1991. *World Geomorphology*. Cambridge: Cambridge University Press.

Clements, J.F. 1978. *Birds of the World: A Check List*. New York: The Two Continents Publ. Group.

Cohen, J.E. 1990. *Community Food Webs: Data and Theory.*

Berlin: Springer Verlag.

di Castri, F., and Younès, T. 1990. Ecosystem function of biodiversty. *Biology International*, Special Issue 22. International Union of Biological Sciences, Paris.

Dickinson, R.E., Errico, R.M., Giorgi, F., and Bates, G.T. 1989. A regional climate model for the western United States. *Climate Change*, **15**, 383-422.

Engelen, Van. V.W.P., and Pulles, J.H.M. 1991. The SOTER manual; procedures for small scale map and database compilation of soil and terrain conditions. Working paper and preprint 91/3, ISRIC, Wageningen, pp.92.

Emanuel, W.R., Shugart, H.H., and Stevenson, M.P. 1985. Climatic change and the broad-scale distribution of terrestrial ecosystem complexes. *Climate Change* **7**, 29-43.

FAO 1971-1978. *FAO/UNESCO Soil Map of the World 1:5.000.000.* Volumes II-X. Maps per (sub)continent and explanatory texts. UNESCO, Paris.

FAO 1978-1981. Report on the Agro-ecological Zones Project. Volume I-IV. *World Soil Resources Report* 48/1-4. FAO, Rome.

FAO 1988. Soil Map of the World, Revised Legend. *World Soil Resources Report* 60. FAO, Rome.

FAO 1990a. World Soil Resources map at 1:25.000.000 and explanatory text. *Soils Bulletin*. FAO, Rome.

FAO 1990b. The conservation and rehabilitation of African lands, an international scheme. ARC/90/4. FAO, Rome

FAO 1990c. An international action programme on water and sustainable agricultural development: a strategy for the implementation of the Mar del Plata action programme for the 1990's. FAO, Rome.

FAO/UNFPA/IIASA 1982. Potential population supporting capacities of lands in the developing world. Technical Report. FPA/INT/513, with maps at scale 1:10000 000 - FAO, Rome

Fishman, J. Fakhruzzaman, K., Cros, B. and Nganga, D. 1991. Identification of widespread pollution in the Southern Hemisphere deduced from satellite analyses. *Science*, **252**, 1693-1686.

Frederick, J.E., Snell, H.E., and Haywood, E.K. 1989. Solar ultraviolet radiation at the Earth's surface. *Photochemistry and Photobiology*, **50**, 443-450.

Goryachkin, S.V., and Targulian, V.O. 1990. Climate-induced changes of the boreal and subpolar soils. In H.W. Scharpenseel, M. Schomaker, and A. Ayoub (eds.), *Soils on a Warmer Earth*. Elsevier, Amsterdam. (pp. 191-210).

Goudriaan, J. 1990. Atmospheric CO_2, global carbon fluxes and the biosphere. In: R.E.A. Rabbinge (ed.), *Theoretical Production Ecology: Reflections and Prospects*. pp.17-40. PUDOC, Wageningen.

Hawksworth, D.L. 1991. The fungal dimension of biodiversity: magnitude, significance, and conservation. *Mycol. Res.* **95**, 641-655.

Heck, W.W., Taylor, O.C., and Tingey, D.T. (ed.). 1988. *Assessment of Crop Loss from Air Pollutants*. Elsevier Applied Science, London.

Holdridge, L.R. 1959. A simple method for determining potential evapotranspiration from temperature data. *Science*, **130**, 572.

Holzner, W., Werger, M.J.A., and Ikusima, I. 1983. *Man's Impact on Vegetation*. Dr. W. Junk, The Hague.

Houghton, J.T., Jenkins, G.J., and Ephraums, J.J. (eds.). 1990. *Climate Change: The IPCC Scientific Assessment*. Cambridge University Press, Cambridge.

Huntley, B. 1991. How plants respond to climate change: migration rates, individualism and the consequences for plant communities. *Ann. Bot.* **67**, 15-22.

Köppen, W. 1931. *Grundriss der Klimakunde*. Walter de Gruter and Co, Berlin.

Leemans, L. 1989. *World Map of Holdridge Life Zones*. IIASA. Laxenburg, Austria.

Matthews, E. 1983. Global vegetation and land use: new high resolution data bases for climate studies. *Journal of Climate and Applied Meteorology*, **22**, 474-487.

Matthiessen, P. (1991). *African Silences*. Random House, New York.

May, R.M. 1988. How many species are there on Earth? *Science*, **241**, 1441-1449.

Melillo, J.M., Steudler, P.A., Aber, J.D., and Bowden, R.D. 1989. Atmospheric deposition and nutrient cycling. In: M.O. Andreae and D.S. Schimel (eds.), *Exchange of Trace Gases between Terrestrial Ecosystems and the Atmosphere*. pp. 263-280. John Wiley, Chichester.

Mooney, H.A.,Drake, B.G.,Luxmoore, R.J.,Oechel, W.C., and Pitelka, L.F. 1991. Predicting ecosystem responses to elevated CO_2 concentrations. *BioScience*, **41**, 96-104.

Mooney, H.A., and Drake, J.A. 1990. The release of genetically designed organisms in the environment: lessons from the study of the ecology of biological invasions. In H.A. Mooney and G. Bernardi (eds.), *Introduction of Genetically Modified Organisms into the Environment*. John Wiley and Sons, Chichester. pp. 201.

Mooney, H.A.,Vitousek, P.M., and Matson, P.A. 1987. Exchange of materials between terrestrial ecosystems and the atmosphere. *Science*, **238**, 926-932.

Myers, N. 1990. The biodiversity challenge: expanded hot-spots analysis. *The Environmentalist*, **10**, 243-256.

Oldeman, L.R.,Hakkeling, R.T.A., and Sombroek, W.G. 1990. World map of the status of human-induced soil degradation: maps and explanatory note. UNEP, Nairobi, and ISRIC, Wageningen.

Olson, J.S.,Watts, J.A., and Allison, L J. 1983. Carbon in live vegetation of major world ecosystems. Environmental Sciences Division, Oak Ridge National Laboratory. Oak Ridge, Tennessee.

Papadakis, J. 1961. Climatic tables of the world. Private publication. Buenos Aires.

Prentice, I.C., *et al.* 1989. Developing a global vegetation dynamics model: results of an IIASA summer workshop. IIASA, Laxenburg, Austria.

Reeve, R.C., and Fireman, M. 1967. Salt problems in relation to irrigation. *Agronomy*,**11,** 988-1008.

Ruellan, A. 1985. Les apports de la connaissance des sols intertropicaux au développement de la pédologie: la contribution des pédologues français. *CATENA* **12**(1): 87-98.

Saunders, D.A.,Hobbs, R.J., and Arnold, G.W. (in press). The Kellerberrin project on fragmented landscapes: a review of current information. Biological Conservation.

Saunders, D.S., Hobbs, R.J., and Margules, C.R. 1991. Biological consequences of ecosystem fragmentation: a review. *Conservation Biology*, **5**, 18-32.

Scharpenseel, H.W.,Schomaker, M., and Ayoub, A. 1990. *Soils on a Warmer Earth*. Elsevier, Amsterdam.

Schulze, E.-D. 1989. Air pollution and forest decline in a spruce (Picea abies) forest. *Science*, **244,** 776-783.

Sellers, P.J. 1985. Canopy reflectance, photosynthesis and transpiration. *Int. J. Remote Sensing*, **8**, 1335-1372.

Sombroek, W.G. 1990a. Soils on a warmer Earth: tropical and subtropical regions. In: H.W. Scharpenseel, M. Schomaker, and A. Ayoub (eds.), *Soils on a Warmer Earth*. Elsevier, Amsterdam.

Sombroek, W.G. 1990b. Global Change: Do Soils Matter? ISSS, Wageningen.

Terborgh, J. (1988). The big things that run the world – a sequel to E.O. Wilson. *Conservation Biology* **2**, 402-403.

Thornthwaite, C.W. 1931. The climates of North America according to a new classification. *Geogr. Rev.* **21**(4).

Van Wambeke, A. 1981. Soil moisture and soil temperature régimes, South America. *SMSS Technical Monograph*, 2. Cornell University. Ithaca, New York.

Van Wambeke, A. 1982. Soil moisture and soil temperature régimes, Africa. *SMSS Technical Monograph*, 3. Cornell University. Ithaca, New York.

Van Wambeke, A. 1985. Calculated Soil moisture and soil temperature régimes of Asia. *SMSS Technical Monograph*, 4. Cornell University. Ithaca, New York.

Williams, M. 1989. *Americans and Their Forests*. Cambridge University Press, New York

Woodward, F.W., and Rochefort, L. 1991. Sensitivity analysis of vegetation diversity to environmental change. *Global Ecology and Biogeograpy Letters*, **1**, 7-23.

Young, J.P. W., and Johnston, A.W.B. 1989. The evolution of specificity in the legume-rhizobium symbioses. *Trends in Ecology and Evolution* **4**, 341-345.

Chapter 10: Freshwater Resources

N.B. Ayibotele and M. Falkenmark

EXECUTIVE SUMMARY

The chapter analyzes the scientific understanding of the freshwater cycling on the continents and water's involvement in environmental degradation, and identifies the research needed to address crucial issues related to environment and development.

The *three most important facts* stressed in the chapter are:
(a) the important environmental linkages between land use and water;
(b) a growing imbalance between water availability and growing water demands, driven by rapid population growth and increasing *per capita* demands;
(c) the growing problem of water pollution, reducing the usability of water.

Close hydroclimatically based *links between water, poverty, and environmental degradation* are highlighted. Both dry and humid tropics suffer from a particular environmental vulnerability related to the climate, which complicates the fact that development depends on a multitude of interventions with the natural rural landscape in order to make accessible and harvest its water, energy and biomass resources. Development strategies needed to cope with environmentally related poverty will have to include protection of environmental productivity as an essential ingredient. This involves an integrated management, where water management is integrated with land use management, since land fertility degradation is often water-related, and land use is generally at the same time water-dependent and water-impacting.

Development in poverty-stricken dry climate countries will depend on the possibility to secure easy access to safe water, and the drought-proofing of rainfed agriculture for livelihood security. In the water-deficient areas there is already a water crisis which will exacerbate as increasing population will increase the pressure on the finite water availability. A more and more firm scientific basis will be needed for development planning under aggravating water scarcity. Water quality degradation adds to the development problems, because the degradation reduces the usability of natural water. Water pollution reaches humans along various pathways and produces biological disturbances and biodiversity reduction.

The present level of hydrologic understanding emerges largely from *temperate zone*, research, characterized by a strong fluid mechanistic reductionism. Problems in the humid tropics have to be analyzed in a broad atmosphere–ocean–land context. Influence of thermodynamics and biochemical processes have a dominant control over the water fluxes. The dynamic features represent a wide spectrum of scales in terms of time as well as space: from diurnal convection up to the annual El Niño-Southern Oscillation (ENSO)-phenomenon. Much less interest has yet gone into the *dry tropics and subtropics*, the region now hosting the majority of the rural poor. Development in this zone is a question of skilful balancing between water consumed in rainfed biomass production, and the limited surplus of rainfall left to recharge aquifers and rivers and support societal water needs. Thus, also in this zone will vertical and horizontal water flows have to be studied together.

A fundamental fact in current research is that progress *in many areas of hydrologic science is limited by lack of data*. Fundamental gaps in knowledge relating to the use of hydrologic modeling include:

(a) inadequate representation in the present physically based models of the relative influence of precipitation and land surface parameters;
(b) problems of coupling outputs from General Circulation Models (GCMs) as inputs into land-based hydrologic models;
(c) difficulty in accounting for influence of the different climates on atmosphere and land-based model parameters;
(d) difficulty in deriving areal values from point measurements of hydrologic parameters.

The *most important research needs* in the field of

freshwater as related to environment and development include:

(a) development of General Hydrologic Models in support of analysis of climate change impact and of macroscale water resources assessment;

(b) closing the gap between terrestrial hydrology and ecology including the critical water-related processes behind land degradation and its restoration;

(c) study of the key processes in water quality genesis, closing the gap between hydrologic flows and biogeochemical processes;

(d) development of planning and management methodologies to cope with greater uncertainty.

Finally, the *most urgent actions* are the following:

(a) establish data bases to support both research and water resources assessment;

(b) education and training with a content responding to current problems and approaches;

(c) developing/strengthening of national and regional institutes for research on water resources in order to make possible a science-based planning and management of available freshwater resources;

(d) review and improve the legal and institutional mechanisms for planning and managing water resources.

1. INTRODUCTION

This chapter will focus on the scientific understanding of the freshwater cycle and water's multiple roles and functions as related to the issue of environment and development. It will highlight the most important facts, the most important research needs, and the most urgent actions. It will, however, basically leave the way in which water is managed and used in society outside the discussion. As global scale issues have already been addressed in Chapter 6, this chapter will *concentrate on regional, subregional, national and landscape scales*, i.e. the insights and understanding necessary to provide the scientific underpinning for development and for protection of environmental productivity as related to water.

Water is – besides carbon – the most essential molecule on Planet Earth (Hartman *et al.*, 1985) – the only planet where water exists in its liquid form. Due to its versatile characteristics and great chemical reactivity, water has a multitude of parallel functions both in the natural landscape and in society:

(a) water is the unifying agent of natural ecosystems with functions similar to the blood and the lymph of the human body;

(b) water is consumed in biomass production which is therefore limited by local water availability;

(c) water circulation is an important part of the global cycles (cf Chapter 6), and is in this sense intimately related to the climate;

(d) water is a fundamental resource on which depend the life-support systems, and which has to be equitably shared between all those living in a particular river basin;

(e) water is a crucial substance in the causality chains producing biodiversity disturbances.

Life on planet Earth takes place at the mercy of the water cycle. Water moves through the Earth System in an endless cycle that forms the framework of hydrologic science. Basically, land gets its life through the wettening from the atmosphere. Water operates plant production, thereby being returned to the atmosphere: plant growth is possible only as long as a column of water can pass through the plants from the roots up to the leaf surfaces. The non-evaporated part of the precipitation goes to recharge aquifers and rivers. The river water is a mixture of water fractions arriving along different pathways with different chemical history. Water bodies and wetlands are a noticeable part of the natural environment and habitats for biota.

Water, in other words, plays a pivotal role in many physical, chemical and biological processes regulating the Earth System, where man's activity is now inseparable in its effects from natural events. In the land phase of the cycle, water through fluvial erosion and sedimentation contributes in shaping the land surface. It is, moreover, the universal solvent and the medium in which most changes of matter take place. The fact that water is the basis for water supply of human life and activities makes many forms of land use *water-dependent*. At the same time due to their interaction with water partitioning and water quality genesis, land use is often also *water-impacting*.

With all those different functions and roles, water is closely linked to almost all other sectors related to the issues of environment and development addressed in this publication:

(a) *health:* persistent supply of safe water is a livelihood necessity. Water is needed both for drinking purposes, and for securing personal and household hygiene. Polluted water is also a central carrier of disease vectors, and in this function responsible for a dominating part of all hospital patients in Third World countries;

(b) *agriculture:* maximized crop yields depend on adequate access to soil moisture to compensate the water losses when the plants take in carbon dioxide for photosynthesis. At the same time the flux of surplus water through the soil leaches soluble agricultural chemicals, carrying them to ground-water aquifers and drainage system;

(c) *energy:* the flow of water down the rivers is important as an energy resource, highly influential in supporting economic development in some countries well endowed with hydropower like Norway and Sweden. In a world where fossil fuels will have to be outphased by renewable energy resources, hydropower – small scale as well as large scale – may eventually regain interest;

(d) *industry:* industrial production is dependent on water for a number of different production functions. Water is at the same time used as a conveyor of liquid waste, causing high pollution levels wherever the dilution flow is small;

(e) *transportation:* water has always played a fundamental role for transportation and the rivers and coasts have traditionally been highly influential in attracting the location of cities;

(f) *recreation:* both physical, biological and psychological factors contribute to the importance of wetlands and water bodies for human recreation.

2. ENVIRONMENT AND DEVELOPMENT AS SEEN FROM A WATER PERSPECTIVE

The *most important facts* as regards the linkages between water, environment and development are the following:

(a) a growing imbalance between water availability, later to be influenced by climate change, and growing water demands, driven by population growth and increasing *per capita* demands;

(b) the growing problem of water pollution;

(c) the important environmental linkages between land use and water: land use impacts on water, water impacts on land (e.g. floods, erosion, sedimentation).

2.1 Environment and development meet in the landscape

A recent analysis of the key environmental issues, as reported to the World Bank in a large-scale enquiry from more than 70 borrowing countries, showed that they mainly originated from human activities in the landscape (Falkenmark, 1991a). These activities were principally driven by social needs and waste production, and exacerbated by both perverse economic incentives and the particular environmental vulnerability, that is typical in tropical and subtropical climates.

Thus, environment and development can be said to meet in the landscape, where socio-economic development is reflected in more or less intensive human production activities. Since man depends on the water, the energy, the minerals and the biomass produced by natural processes in

the landscape, development generally calls for an array of more or less unavoidable landscape manipulations of the soil, the vegetation, and the water pathways (Falkenmark and Suprapto, 1992). Due to physical, chemical and biological processes going on in the landscape, man thereby produces – besides the intended benefits – also secundary impacts. Some of these impacts are avoidable, others unavoidable. Water-related changes are propagated in the water cycle, finally reaching biota due to their water-based life processes.

Basically, two parallel types of water-cycle related phenomena are involved in the *genesis of environmental problems* (Falkenmark, 1991b):

(a) *water-partitioning* is easily altered through interventions with soil and vegetation. Altered land permeability has consequential effects both on the rainfall infiltration, the fertility of the soil and on the leaching of excess nutrients and chemical components of the underground matrix; on the percolation to ground-water and its composition; and on the quality of water bodies into which the ground-water seeps. It also produces changes in surface runoff with consequential changes both in land erosion and silt transport and in flood régimes and risk for downstream inundations;

(b) water is a unique *solvent, chemically active and on continuous move* above and below the ground. It easily picks up soluble substances, carrying them to ground-water aquifers, to discharge (seepage) areas and wetlands, to the water bodies, and onwards to the coastal waters.

Due to the fact that life of every plant, every animal, every human being is based on micro-scale water flows, changes in these fundamental water processes will generate higher-order biological effects, in the long run influencing biological diversity.

2.2 Poverty, water and environmental degradation

In the past, environmental changes have been rapidly expanding in scale and number. The lack of interdisciplinary approaches has retarded the necessary scientific understanding. Problems have been met mainly on a cause-by-cause basis and by an *impact-thinking approach*. Environmental research has to a large degree been problem-driven and effect-oriented. The scientists have been pushed by reality rather than pulled by their own desire to understand the overall Earth System.

Increasing interest is now going into the link between poverty and environmental degradation. The dry tropical climate as such may in fact have contributed to the persistent development problems in that zone (Falkenmark and Suprapto, 1992). It has been pointed out that most (*not* all) of the poverty-stricken countries with low human capability index share the challenge of seasonal and/or recurrent droughts on the one hand, and low rain efficiency due to high evaporative demand of the atmosphere on the other. However, also the humid tropics exhibit tremendous challenges for human populations. Leonard *et al.*, (1990) have shown that the great majority of the poorest segments of society are dwelling in drylands and on hillsides. Also the tropical forests and the urban periphery are housing hundreds of millions of poor.

2.3 Development-generated perturbations and environmental sustainability

The accumulated results of past manipulations are now threatening the sustainability of life. At present a whole range of perturbations are operating in parallel, all affecting the same processes in the Earth System: climate change due to altered energy balance of the atmosphere; rapid population growth forcing intensification of man's manipulations of the land and water systems; delayed responses to pollutants already discharged into soil and water bodies (socalled chemical time bombs); ongoing output of exhausts to the atmosphere; ongoing output of waste water to the water bodies; etc . The consequences of all these perturbations are being superimposed on each other. The system may be full of synergisms and other interactions, all characterizing today´s *multicause syndrom of environmental change* (Clark, 1990).

The whole Earth System has an integrity which involves the seeking of equilibrium states after perturbations. New equilibrium states may be forced which may be adverse to the survival and productivity of the biological species. The low latitude countries will have an even harder time to meet their development needs and maintain the integrity of the environment. *There is a need for understanding of how disturbances are propagated forwards and backwards in the system, and the way they affect processes along the pathways of water through the hydrologic cycle, to seek new equilibrium states. The time and space scales of the propagations must also be elucidated.*

Thus, if development is seen as equivalent to perturbations of the Earth System, then *sustainable development is an issue of balancing the needed perturbations against unavoidable negative side effects so that certain basic sustainability criteria are satisfied* (Falkenmark and Subrapto, 1992). These criteria might include such conditions as the protection of land productivity, of ground-water potability, of fish edibility, and of biological diversity. *It follows that environmental protection is a necessary condition for development in poverty-stricken countries where land degradation and water pollution are part of the problems generating or conserving the present poverty.*

2.4 Integrated land–water management

The Preparatory Committee for the UN Conference on Environment and Development (1990) stresses that an essential strategy in addressing the present water-related problems is an integrated water resources management. It is worth stressing, however, that in its conventional meaning, integrated water management involves an integration across all water uses and all water-dependent sectors in society.

As land use – besides being water-depending – is also water-impacting, *water management will, however, have to be integrated also with land use.* Falkenmark and Lundqvist (1992) have for instance suggested that essential strategy components under dry-climate conditions would have to include a minimizing of the non-productive evaporation losses from a certain landscape, and a maximizing of the amount of water-related production – biomass production as well as other societal production – from each liter of water on the move through that landscape.

The land/water integration should include environmental protection of uplands which – as pointed out by Leonard *et al.* (1990) – has a dual scope: on the one hand to secure an acceptable livelihood for the local upland inhabitants, on the other to protect the fertile lowlands from degradation under silt flows and floods emerging from the water-driven degradation of the slopes in the upland basin. The fact that – as already stressed – most water in the rivers has earlier passed land, is another argument for integrating both land use and water management. The natural unit for such integration is the watershed.

Summarizing, the development strategies needed to cope with the hydroclimatic challenges and environmentally related vulnerability and poverty in the tropics and subtropics should in other words be based on an integrated land-water conservation and management approach, implemented in the context of the catchment or river basin.

3. WATER AVAILABILITY AND USE

3.1 Freshwater resources

Freshwater resources is a concept used for the freshwater potentially available for societal use in aquifers, rivers, lakes, wetlands and glaciers. Their characteristics are affected both by geophysical processes and human activities. They are renewed annually through precipitation processes typical of the various climatic regions. This renewal sustains human development and biogeochemical processes in the terrestrial and aquatic ecosystems.

Atmospheric, land and ocean processes operating at different time scales interact to endow various geographical areas – tropical (low-latitude), temperate (mid-latitude) and cold (high-latitude) – with water in an

Table 1: *Annual surface runoff and specific discharge over continents*

Territory:	Annual river runoff		Portion in total runoff in %	Area thous. km^2	Specific discharge l/s/km^2
	mm	km^2			
Europe	306	3210	7	10500	9.7
Asia *	332	14410	31	43475	10.5
Africa**	151	4570	10	30120	4.8
N and C America	339	8200	17	24200	10.7
S America	661	11760	25	17800	21
Australia and Tasmania	45.3	348	1	7683	1.44
Oceania	1610	2040	4	1267	51.1
Antarctica	160	2230	5	13977	5.1
Total land area	314	46770	100	149000	10.0

* with Japan, Philippines and Indonesia   ** and Madagascar

uneven manner, generally described as humid, semi-humid, semi-arid and arid. The arid (water deficit) areas cover about 30% of South America, 33% of Europe, 60% of Asia, 85% of Africa and most of Australia and Western areas of North America (UNESCO, 1978). The present knowledge about global amounts of freshwater resources in rivers, and the distributions over the continents are presented in Table 1. It can be seen that for surface water runoff which is the most accessible part of freshwater resources used by man, 56% of the annual global total is contributed by Asia and the South American continent, 10% by Africa and the remaining 34% by Europe, North America, Australia, Oceania and Antarctica .

In the case of freshwater reserves in large lakes, Africa and Asia account for 67% of the total world reserves while North America, Europe, South America and Australia and Oceania account for the remaining 33%. For man-made reservoirs as at 1972, Africa and Asia had 60% of the reserve with the remaining 40% shared by the rest of the continents.

Ground-water runoff which is annually renewed had South America and Asia contributing 31% and 28% of the global total respectively. The remaining 41% is accounted for by Europe, Africa, North America, Australia and Oceania.

Wetlands form another important type of freshwater resources. South America has the highest percentage of the total global wetland surface area with 46%. This is followed by Europe and Asia with 34%. The remaining 20% is accounted for by the rest of the continents.

89.7% of the freshwater reserves held in surface ice/glaciers are found in Antarctica. Europe accounts for 10% and the remaining 0.3% by Asia, North America and Africa. The amount of Africa is insignificant.

3.2 Variability of water resources

In addition to the uneven distribution of freshwater on Earth, the precipitation from which it derives is characterized by variability. This results in corresponding variability of surface runoff both in the long-term annual and seasonal means. These variations also differ among continents, within regions, sub-regions and countries.

In the water surplus areas (*humid*) of both the temperate and tropical climates, variabilities are less than in the water deficient (*arid and semi arid*) areas with similar temperature. However in the semi arid and arid regions of the tropical climates, the temporal and spatial variabilities of precipitation and runoff are considerable to the extent that the dependability of runoff is often seriously undermined. In the temperate climates, even though variability of precipitation may be appreciable, the variability in runoff is dampened by snow melt in the warmer months. Hence runoff is more dependable than in the tropics.

3.2.1 Extremes – Floods and Droughts

The precipitation that annually renews water resources may occur several percentage points above or below the long-term annual or seasonal mean. Depending on the direction and extent of the departure from the mean, the intensity and duration, floods and droughts of different magnitudes are caused. All climatic regions are prone to floods and droughts, but there are some which are more prone to one type of extreme than others. For instance the tropical water deficient African Sahel, western part of North America and most parts of Australia are more drought prone than other areas. Similarly the tropical water surplus South East Asian region (the Indonesian archipelago, India, Bangladesh) are more prone to floods.

Table 2: Dynamics of water consumption in the world according to various kinds of human activity (km³/s)

Years: Water uses:	1900	1940	1950	1960	1970	1975	1980 km³/year	%	1990	2000 km³/year	%
Irrigated lands mln ha	47.3	75.8	101	142	173	192	217		272	347	
Agriculture	525 409	893 679	1130 859	1550 1180	1850 1400	2050 1570	2290 1730	68.9 88.7	2680 2050	3250 2500	62.6 86.2
Industry	37.2 3.5	124 9.7	178 14.5	330 24.9	540 38.0	612 47.2	710 61.9	21.4 3.1	973 88.5	1280 117	4.7 4.0
Municipal supply	16.1 4.0	36.3 9.0	52.0 14	82.0 20.3	130 29.2	161 4.2	200 41.1	6.1 2.1	300 52.4	441 64.5	8.5 2.2
Reservoirs	0.3 0.3	3.7 3.7	6.5 6.5	23.0 23.0	66.0 66.0	103 103	120 120	3.6 6.1	170 170	220 220	4.2 7.55
Total (rounded off)	579 417	1060 701	1360 894	1990 1250	2590 1540	2930 1760	3320 1950	100 100	4130 2360	5190 2900	100 100

Note: Total water consumption is shown in the numerator; irretrievable water losses are given in the denominator. Source: Shiklomanov (1990).

In case of drought, this may involve the late arrival of rain, or insufficient rainfall during the crop-growing season resulting in insufficient soil moisture. The deficiency could also last some months resulting in low flows, lowering of reservoirs and ground-water levels. It could also persist for a few years like the Sahel drought (1968–1973) which later spread to the whole of Africa (1981–1986).

3.2.2 Causes of variability

The causes of variability of the annual precipitation that renews the freshwater resources come first from natural long-term geophysical processes and secondly from man-induced climate change. Evidence gathered from paleo data shows that the Earth's climate is dynamic at the geophysical scale, because various regions have been found to have experienced other climates different from those of today (Oeschger, 1991). As such the water endowments of the various climatic regions and ecosystems cannot be static (stationary) as assumed in the past in estimating the long-term means of hydrologic parameters (Abu Zeid and Biswas, 1991).

3.3 Water quality

Freshwater resources cannot be fully described without knowledge about the quality characteristics. The state of quality of freshwater resources is assessed from their physical, chemical and biological constituents. The latter may originate naturally from the environment (e.g. soils and geological formation), or from wastes discharged as a result of agricultural, human settlements and industrial activities. They are introduced either from point sources (mostly industrial) which are manageable, or from non-point sources (mainly agricultural) in which cases management is more difficult.

The concentrations of the constituents simply express the state in which the water is in physical, chemical and biological terms. Quality can only be discussed meaningfully when it is related to a use. In such cases guidelines on the concentrations of various constituents that should not be exceeded so as not to impair the water for particular use must be known (WHO/UNEP, 1989). The human activities presently impacting on water quality are discussed in subsequent paragraphs.

3.4 Estimates of freshwater resources

It must be pointed out that the estimates of freshwater resources distribution at global and continental levels as presented in Table 1, are based on present knowledge.

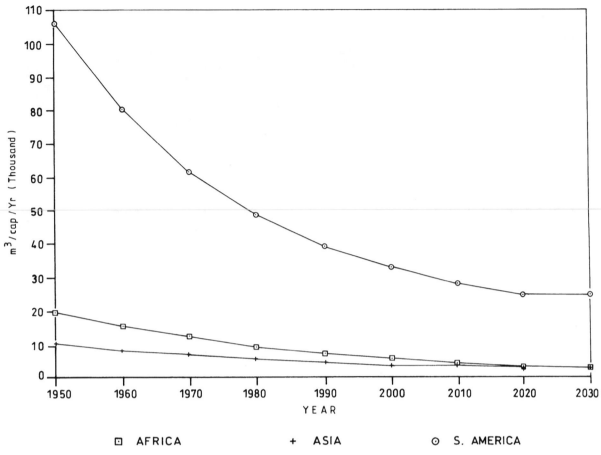

Figure 1: Decreasing trends in annual *per capita* freshwater availability in developing regions

Note: Population taken from UN World Population Prospects

There are significant differences in estimates obtained from different sources. For instance, for the African region estimates of the surface water resources vary by a factor over 3 from 2280 to 7826 km³. These arise mainly out of the adequacy of the scientific bases used, the suitability of the methodologies and techniques, and the adequacy of data in terms of areal coverage, representativeness, continuity and quality (Ayibotele, 1991a). The problem equally affects sub-regional (international river, lake and ground-water basins shared by a number of countries) and national estimates of freshwater resources. *Hence considerably more work is required in the scientific, methodological and data fields to obtain more accurate and dependable estimates, of particular importance as a base for water resources planning in regions where water is scarce.*

3.5 Growing imbalance between demand and availability

Water demands are a function of the size of the population on the one hand and its lifestyle and activities on the other. It has to be realized that the world population will at least double in the next few decades. If population growth

cannot be brought to a stop by rapidly enough reaching two-children-per-woman, it may even triple before it finally stabilizes. Ever increasing water needs have to be met from available freshwater resources accessible in rivers, lakes/reservoirs, ground-water, wetlands. Over the years the amount of freshwater available *per capita* has been decreasing due to increasing population. The well-informed and intelligent approach needed to facilitate human development in a water-scarce region has to involve water security both for biomass production and for other societal water needs. *Securing easy access to safe water and the drought-proofing of agriculture are consequently two fundamental goals to be achieved with greatest possible urgency.*

The evolution of increasing global use of water to meet needs in agriculture, industry and human settlements over the years and projected to year 2000 is presented in Table 2 (Shiklomanov, 1990).The bulk of the population increase will take place in the Third World. What is envisaged is in other words rapidly increasing water needs in regions already now struggling with the problems of water scarcity (cf Fig. 1). Also increasing wastes will be discharged either on land, into water bodies, the atmosphere, and the oceans.

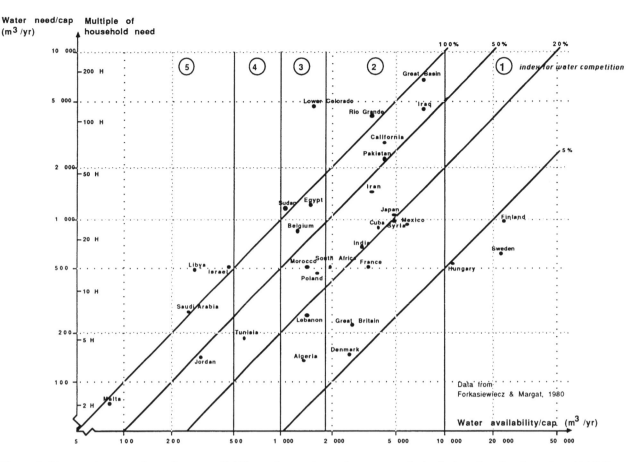

Figure 2: Macroscale comparison of water availability and water demand on a per capita basis, based on data from the mid-1970s. Demand also given as multiples of a household demand (H) assumed at 100l/p day. Crossing lines indicate degree to which potentially available water has been mobilized for use.

Figure 2 shows the variety of estimated situations in terms of *per capita* water availability on the one hand (*x*-axis) and manifested water demands on the other (*y*-axis). The sloping lines indicate the degree to which the potential water availability is already under use for withdrawal-based purposes. The ease in using high percentages of the potential water resource depends on its origin: whether it is *endogenous* water, recharged from national rainfall, or whether it is *exogenous* water, entering the country in international rivers or aquifers, recharged from rain falling over upstream countries.

In the water deficient areas there is already a water crisis (Falkenmark *et al.*, 1989; INWARDAM, 1990). This can be found in the countries of the Sahel of Africa, the Arab States of North Africa and the Middle East.

There is a shortage of freshwater that can be economically mobilized for use. Even within countries with adequate water resources there may be serious regional imbalances. Shortages may require expensive projects to be undertaken to transfer water from one basin to the other. Examples are surface water transfer projects in Algeria, or the ground-water transfer project Great Man Made River in Libya. In some cases the shortage of water has led to overexploitation of ground-water resources, particularly in coastal areas with the result that these areas

are threatened by saline water intrusion from the sea (Ayibotele, 1991b). These situations are worsened by prolonged droughts which hit the water scarce areas periodically as demonstrated in for example the falling levels of Lake Chad during the Sahel drought – 1968 to 1984.

3.6 Quality degradation of freshwater resources

In modern times, human activities have contributed increasing outputs of waste, unwillingly or deliberately transferred to the environment: as air exhausts to the atmosphere, as surplus agricultural chemicals to arable land, as dry waste deposits and landfills around cities, and as waste water to the rivers and lakes. Water-soluble pollutants along water pathways above and below the ground surface dissolve in the circulating water and are carried along with the water cycle.

As a result water quality is degrading all over the globe, in various ways disturbing the usability of water. Through co-ordinated observations within GEMS (GEMS MARC and WHO, 1991) a first overview has been achieved. In particular ten issues were highlighted:

(a) pathogen agents from faecal discharge, probable to become more severe as population increases more rapidly than waste water collection and effective

Table 3: *Limitations of water use owing to water quality degradation*

Use: / Pollutant type:	Drinking water	Aquatic wildlife & fisheries	Rec-reation	Agriculture (including irrigation)	Industrial uses	Power and cooling	Transport
Pathogens	++	0	++	+	++(1)	00	00
Suspended solids	+	++	++	+	+	+(2)	++(3)
Organic matter	+	+	++	0(9)	++(4)	+(5)	00
Algae	+(5)	+(6)	++	0(9)	++(4)	+(5)	+(7)
Nitrates	++	+	00	0(9)	++(1)	00	00
Salts	++	+	00	++	++(8)	00	00
Metals	++	++	+	+	+	00	00
Industrial organics	++	++	+	+	?	00	00
Acidification	+	++	+	?	+	+	00

++ marked impairment requiring major treatment
  or precluding this use
+ minor impairment
0 no impairment
00 irrelevant

Notes: (1) Food industries
(2) Abrasion
(3) Sediment settling in channels
(4) Electronic industries
(5) Filter clogging
(6) In fish ponds higher algal blooms
    can be accepted
(7) Development of water hyacinth
    (Eichhornia crassipes)
(8) Fe, Mn in textile industries, etc.
(9) Water quality degraded in this way
    may be beneficial for this specific use

Source: GEMS MARC and WHO, 1991

treatment are implemented;

(b) organic pollution from the same sources, consuming the oxygen in receiving water bodies with consequential effects on biota;

(c) salinization, primarily by poorly managed irrigation systems in arid regions – one of the primary water quality problems in the next few decades – but also from natural evaporation in belts of acid rocks in overexploited coastal aquifers; and as a result of mining activities;

(d) nitrate pollution of aquifers from surplus fertilizers and from human and cattle waste in rural areas;

(e) eutrophication caused by increasing levels of nutrients, normally phosphate but sometimes nitrate;

(f) heavy metals – widespread in all industrialized regions and in mining areas, absorbed on particulate matter in soils, landfills, mine tailings – of which only a minor fraction has yet leached into water bodies but the fraction may be exacerbated by rain acidification;

(g) pesticides, of which there has been an exponential rise in use for the last three decades, but the contamination of which remains poorly documented due to analytical difficulties;

(h) industrial organic substances such as solvents, chlorinated hydrocarbons, and polycyclic aromatic hydrocarbons from air exhausts, dump site leakage, and waste waters;

(i) acidifying substances released in air exhausts as inorganic acids, and causing acidification of aquifers, rivers and lakes in poorly buffered regions like sandstone and crystalline rocks, and associated with increases in dissolved aluminium and some other metals;

(j) river sediment loads which have increased considerably through man-enhanced erosion, locally causing enormous turbidity problems, and in general disrupting natural biological processes in rivers and coastal areas.

Human activities also involve massive modifications of natural hydrologic régimes, with various direct effects on water bodies: salinization of rivers, lakes and soils, salt intrusion in coastal aquifers, and reduced dilution and self-purification capacity of rivers.

Water quality change alters the usability of water, reaches humans along various pathways (Table 3) and produces biological disturbances in the natural environment. Among these are fish kills, deforestation and general loss of biodiversity.

3.7 Water resources / management issues

The widespread problems of water shortages, inadequate management tools, pollution and land degradation call into question the appropriateness and adequacy of the planning and management tools being used in many countries. The situation in the countries of the African region are typical examples (Ayibotele, 1990, 1991b). The following points which cut across regional and national boundaries could be mentioned.

(a) In most countries the *economic and legal tools* for conserving the quantity and quality of water through the control of water abstraction/demand and the discharge of wastes into water bodies are either weak, non-existent, or if available, the capacity to implement them are not available. Economic incentives and disincentives are not properly applied to manage demand because it has not been fully accepted that water is an economic good, for which the resources consumed in making it accessible should be fully paid for by the beneficiaries. This leads to excessive demand for water and its wasteful use in meeting domestic, municipal, industrial and irrigation needs.

(b) The need has been recognized, particularly in those countries in North Africa and the Middle-East, where the water shortage situation is more serious, to mobilize additional water from *unconventional sources* like the drainage water, treated waste and saline water, and treated sea water. However, this is not being realized fast enough, owing to the economic, technical, social and cultural barriers that have to be overcome.

(c) Allocation of water under conditions of scarcity between upstream and downstream users within national boundaries often leads to conflict. However, when the water is *trans-boundary* the principles for allocating to meet current and rising need in the future have to be acceptable to all the riparian countries, otherwise conflicts might arise that could jeopardize

regional and even world peace. The situation with regard to the sharing of the Nile, Jordan, Tigris and Euphrates waters among the riparian countries in Eastern and Northern Africa and in the Middle East provide clear examples.

(d) The *institutional gap* set up for planning and managing water resources are many and varied. They also have different levels of adequacy and effectiveness. Common shortcomings include:

(i)   The absence in many countries of an overall authority to give central direction to the utilization and conservation of water resources.

(ii)   Duplication and overlapping of functions leading to inefficient use of scarce resources.

(iii) Ineffective linkages between water agencies and their sector ministries on the one hand, and on the other ineffective linkages betwen sector ministries and the agencies dealing with overall national development planning and finance.

(iv) Lack of mechanisms for integrating land and water resources planning and management in spite of the fact that it is recognized as desirable.

(v)   Absence of clearly designated and mandated agencies to deal with trans-boundary water issues in countries where river, lake or ground-water basins are shared with other countries.

(vi) The importance of water to the sustainability of the Earth System and to socioeconomic development is not reflected in the institutions. The decision-making processes at various levels of government have not been able to incorporate water as a key factor in decision-making.

3.8 Emerging development issues

3.8.1 Water supply

A number of issues depend on an adequate scientific understanding of underlying hydrologic phenomena:

Possibility to mobilize an increasing share of the potentially available water by water resources development projects:

(a) principal water resources development measures adapted to the hydroclimate;

(b) increased use of ground-water: what are the constraints posed by the recharge of individual aquifers?

(c) the dam issue: how important is the storage of water for activitites in a dry-climate country? What are the avoidable as opposed to the unavoidable side-effects? In view of the "conventional wisdom" among environmental activists that dams are "bad", it is essential to clarify the crucial need for water storages in dry climate regions in an interdisciplinary manner;

(d) how large a share of the potential water availability can ultimately be mobilized and made accessible for societal use – it is essential to reach a broad understanding among planners of the constraints posed by topograpy, climate and geology.

Possibilities to share exogenous water passing through several countries (international rivers):

(a) data exchange problem: what is referred to here is the fact that data are often considered secret in water-scarce regions, but also an unequal access to information caused by differences in access to national expertise;

(b) the concept of sharing endogenous as opposed to exogenous water: for example, how large a part of the endogenous rainfall is a country entitled to use for rainfed biomass production, in view of its obligations to leave a certain water surplus as runoff in the river for downstream countries ?

3.8.2 Biomass production
In a dry climate with unreliable rainfall it is essential to achieve water security for biomass production:

(a) root zone water security may be achieved by alternative means: irrigation by exogenous water or alternatively by soil and water conservation, benefiting from endogenous water;

(b) how high yields can be produced given the hydroclimatic constraints of Third World region: the necessity of extending agricultural research from the present monodisciplinary thinking to taking an ecohydrologic flow-based perspective, acknowledging the water partitioning changes produced;

(c) preparing the rest of the world for food export to expanding regions in the dry climate tropics and subtropics that will be lacking a long-term self-supporting potential.

3.8.3 Water resources assessment
Measures needed to secure access to water to satisfy societal needs depend on good data and information, and good understanding of a number of issues:

(a) endogenous water availability: locally formed ground-water and river runoff; yield, seasonality and interannual fluctuations;

(b) exogenous water availability: aquifer and river flow formed in distant upstream areas; yield, seasonality and interannual fluctuations.

A particular problem is *how to address water resources assessment while waiting for long enough time-series of adequate basic data*. The present situation is that – in spite of the recommendation in the Mar del Plata Action Plan on *increasing* the basic data networks to provide adequate

background data for water resources planning and management the development has in fact gone in the opposite direction by stations being closed down in response to economic difficulties. There is therefore a need to develop approximation methods based on various types of proxy data.

3.8.4 Preparedness for climate fluctuations and change
Through climatic change, decision-makers will be faced with increased uncertainty in evaluating future water availability and water demand. Methodology should be developed for generating internally coherent regional scenarios of water availability changes as a basis for water resources planning (Task Group 1 – Climate, Hydrology and Water Resources, 1990). Knowledge of basin sensitivities will be needed to identify and assess the regions most vulnerable to climate-induced stress.

4. PRESENT LEVEL OF UNDERSTANDING

4.1 General understanding
The integration of land/water management, called for in an earlier paragraph, critically depends on a good enough understanding of fundamental hydrologic processes and their interactions with terrestrial and aquatic ecosystems. The role of water in the various processes of the hydrologic cycle and its impact on the life-support systems has however not been well appreciated in the past. What is now emerging is a recognition of the fact that there are linkages in the processes that involve water on the one hand, and activities that affect the movement of water in the hydrologic cycle on the other. This then leads us to take an expanded view on development (human, settlements, agriculture and industry) and its interaction with water in the life-support systems of the Earth as found in the atmosphere, in the oceans and on land.

It has already been stressed that in the *land phase* the water is involved in numerous processes: morphological (land forms), transport (sediment), chemical (hydrochemical, geochemical, and biochemical), and in the health and productivity of plants and animal populations (on land, in rivers, lakes/reservoirs and wetlands). Furthermore, the land is involved in the processes by which precipitation is partitioned into soil moisture, evapotranspiration, ground-water recharge and river runoff. The *ocean phase* is also involved in biological, and chemical processes and exchange of water between ocean and atmosphere.

A general survey of the status of understanding was recently presented by the National Research Council in USA (NRC, 1991). The evolution of hydrologic science has been in the direction of ever-increasing scale, from small catchment to large river basin to the Earth System, and from storm event to seasonal cycle to climatic trend. Arising questions today include how to aggregate the dynamic behavior of hydrologic processes under great natural heterogeneity; how to trace water pathways and to understand the natural buffering of anthropogenic acids; and the equilibrium and stability of moisture states and vegetation patterns.The report finally stresses the fundamental fact that *progress in many areas of hydrologic science is currently limited by a lack of data.*

4.2 Present hydrologic methods

Basically, the present hydrologic methods are the results of 200 years of efforts to solve water management problems in Europe and North America (Klemes, 1990). Their development took place when Europe was already stripped of its forests and when intensive cultivation had already started. The methods were those needed in order to be able to utilize water resources for the immediate demands of the industrial revolution. Scientifically, the work was characterized by a strong fluid mechanistic reductionism, the main forces being gravity and friction. The resulting methods can be characterized as empirism, later formalized into a systems approach. The efforts to explain the processes have been weak – the work being merely an effort of description and curve-fitting. The *climatic bias* involved in the development of hydrologic methods is important to have in mind when addressing the research needs in the field of environment and development. Scientific understanding is a fundamental prerequisite to solve fundamental livelihood problems in the tropical and subtropical region, and address the probable consequences of a changing climate.

4.3 Regional similarities and differences

Even if the basic hydrologic processes and the water balance components are everywhere the same, their respective intensity and dominance differ between regions (Falkenmark and Chapman, 1989).

4.3.1 Temperate zone hydrology

The present hydrologic understanding is best developed for the temperate zone, both for slopelands and flatlands. The evaporative demand of the atmosphere is moderate so that the precipitation surplus is large enough to recharge aquifers and rivers, producing runoffs adequate for the support of water-dependent activities. Studies on ground-water recharge and the runoff production process indicate that the landscape is divided into recharge areas with infiltration surplus, and discharge areas where the precipitation surplus meets a rising ground-water. The ideas have been validated by tracer analyses in some Swedish rivers indicating that even during the snowmelt most of the flood water had earlier passed under the soil surface (Rodhe, 1984).

4.3.2 Humid tropics hydrology

The humid tropics is a region of environmental stress due to massive deforestation (Klemes, 1990). It is the region with maximum input of solar radiation, and the prime mover of atmosphere and ocean, in other words the location of maximum water and energy fluxes. The dynamic features originating in the tropics represent a wide spectrum of scales, in terms of time as well as space: from the diurnal convection all the way up to the annual ENSO-phenomenon. Thus, hydrologic conditions have to be analyzed in a broad atmosphere–ocean–land context. The key to understand hydrologic phenomena is often outside hydrology, introducing a strong interdisciplinary component into the research needed. The atmospheric processes are of major importance for the hydrologic processes in the tropics, in particular the precipitation characteristics.

Major methodological difficulties that will have to be overcome include the following three (Klemes, 1990):

(a) bridging the scale gaps: past ways of scaling up developed in the temperate zone are not readily transferable to the tropics;

(b) the dynamical context: the classical hydrologic models suffer – as already indicated – from a mechanistic bias which was indeed adequate in the temperate zone. In the tropics, however, also influence of thermodynamics and biochemical processes have a dominant control over the water fluxes;

(c) the lack of appropriate data bases.

4.3.3 Hydrology of dry regions

The dry regions of the tropics and subtropics and elsewhere arid, (semi-arid, subhumid) are another area of heavy environmental stress, mainly due to disturbances of the land surface with vaste implications for infiltration, and propagated onwards to other water cycle components. The stress goes back to a large human pressure in an area with high environmental vulnerability. Post-war hydrologic research has concentrated on rainfall–runoff relationships, following the tradition from the temperate zone. Recent scientific interest as reflected by scientific symposia has however moved into local scale phenomena related to land surface, land degradation and ground-water yield. To the atmospheric scientists this area has lower interest than the humid tropics due precisely to its dryness – the vertical exchange between atmosphere and land surface is low. As seen from a hydrologic perspective, however, *this is the region that hosts most of the poverty-stricken Third World countries and consequently a region of high scientific relevance in regard to environment and development.*

The high evaporative demand in this zone has the effect that the water vapour flux entering the continents with the winds evaporates within a limited distance so that deserts tend to develop in the downwind direction. The fact that most of the rain returns to the atmosphere makes the water partitioning crucial for the amount left to form runoff. Whereas it makes sense in the temperate zone to study the horizontal flows separately without large attention to the evaporation, the same is not true for this region. On the contrary, changes in biomass production intensity would be expected to alter the water partitioning and have impacts on the runoff which may be considerable on the relative scale. In this region vertical and horizontal flows have therefore to be studied in combination. Indeed, in the Murray-Darling basin in Australia plant production is used as a way of actively mitigating the waterlogging and salinity problems suffered (O'Laughlin, 1990).

As regards *hydrologic processes* the earlier mentioned US overview expressed a number of conclusions regarding what we need to know:

(a) *Reservoirs and fluxes of water:* how patterns of surface wetness, temperature, reflectivity, and vegetation influence the formation of clouds and precipitation on a wide range of space and time-scales; the rates and pathways of moisture through the soil in order to predict soil chemical reactions, solute responses, and water quality changes; the issue of the representation of spatially aggregated non-linear behavior in the presence of large spatial variability; new observational technique to study and monitor the rates of snow accumulation and snow and ice melt over remote areas;

(b) *Flux of sediments:* seek quantitative understanding of the dynamics of channel formation or of the causal relationship between the three-dimensional network structure and the precipitation driving the erosion;

(c) *Involvement of biota:* better understanding of the physical relationships among climate, soil and vegetation that determine the dominance and stability of specific vegetation types at particular geographic locations.

4.4 Fundamental gaps of knowledge

The fundamental gaps in knowledge related to *hydrologic modeling* on the other hand have been discussed recently by Schaake and others (Schaake, 1991; Rowntree, 1991; Mawdsley, 1991; Harding, 1991). They arise from:

(a) inadequate representation of the relative influences of precipitation and land surface parameters like vegetation, soil structure and composition, slope etc., and their spatial and temporal variations in present physically based models;

(b) the problems of coupling the outputs (particularly precipitation and energy) of atmospheric GCMs as inputs into the land based hydrologic models and how to take account of the spatial and temporal variations

in model parameters;

(c) the difficulties in accounting for the influence of the different climates on atmospheric and land-based model parameters;

(d) the difficulties in deriving areal values from point measurement of behavior parameters.

With regard to the particular case of gaps in *water quality knowedge*, the first Global Assessment Report (WHO/UNEP, 1989) suggested that:

A better understanding of the basic biogeochemical processes and interactions between particulate and dissolved forms of pollutants, biota and sediment is required to assess the fate and pathways of pollutants in the aquatic environment. Models need to be developed for this purpose and reliable data need to be generated to verify their validity and practical relevance. Various models of eutrophication are available and have proved to be successful in reducing the problem. Acidification modeling is also progressing rapidly, particularly through ecosystem manipulation.

Although adequate models exist for biodegradable organic matter and oxygen balance in rivers, there have been only a few models generated to describe the fate of metals and organic micropollutants. For ground-waters, models are still lacking which describe the transfer of miscible and non-miscible pollutants into the microporous matrix of aquifers, taking into account all the hydrodynamic components of the transport and the exchange reactions between the liquid, solid and gaseous phases.

In the field of planning and *management* there is the particular need to develop tools to take care of non-stationarity in water resources estimates, due to man-induced climate change. The effects are being superimposed on the natural long-term geophysical processes (Task Group 1 on Climate, Hydrology and Water Resources, 1991).

5. RESEARCH NEEDS

Based on the earlier discussion of present level of understanding on fundamental gaps of knowledge, a few areas have been selected for particular priority. Thus, the *most important research needs* in the field of freshwater as related to environment and development are the following four:

(a) development of General Hydrologic Models in support of analysis of climate change impact and of macroscale water resources assessment;

(b) closing the gap between hydrology and terrestrial ecology including the critical water-related processes behind land degradation and its restoration;

(c) water quality genesis, closing the gap between hydrologic flows and biogeochemical processes.

(d) development of planning and management methodologies to cope with greater uncertainty.

5.1 Development of general hydrologic models

Rationale: As stressed by the Second World Climate Conference climate change involves new demands on hydrologic science to improve techniques and models not only for evaluating the role of the water cycle but also to assess water availability, flood risk and water quality in a non-stationary world (Task Group 1 – on Climate, Hydrology and Water Resources, 1990). The most important impact of climate change will be its effects on the water cycle and water management systems. The scale of the problems involved makes it natural that international agencies continue to take the lead.

Research: Needs include upscaling-downscaling in hydrologic modeling, and parametrization between scales. A new generation of coupled hydrologic atmospheric models in support of GCMs has to be developed. Improvements in GCM parameterization through use of finer grids are of great importance for hydrologic investigations. Furthermore to get a better representation of hydrologic fluxes research on the role of the unsaturated zone should be pursued. Studies are needed to identify river basin responses and methodologies needed for building internally consistent non-stationary water availability and demand scenarios. It is highly desirable to advance beyond the coarse approximations of the water-related interactions atmosphere/soil/vegetation by replacing the socalled Budyko's bucket-model.

5.2 Bridging hydrology and terrestrial ecology

Rationale: A crucial issue in a broader world context is socio-economic development under water scarcity. It has therefore become essential to understand better the particular environmental vulnerability in regions close to the hydrologic margin. Moreover, intensifying interactions man–water–soil/vegetation–other species involve the need to develop a capability of planning for sustainability. This is of particular concern for vulnerable regions where population-driven manipulations of the land surface are called for in order to secure the biomass supply needed for self-reliance and are more hazardous than in less vulnerable regions.

Research: Needs include the interactions between land surface processes, water partitioning between vertical and horizontal flows, and atmospheric processes. Methodology has to be developed for ecohydrologic water flow analyses on different scales. Uphill/downhill linkages in the landscape should be studied in different landscape settings

as a basis for renovation of degraded lands and watershed management. Linkages should be developed between plant modeling and hydrologic modeling.

5.3 Water quality genesis

Rationale: The scale of water quality deterioration is alarming and calls for focus on the links to – beyond waste water habits – land use and atmospheric pollution. There is a need to understand and to manage water quality, to mitigate water quality deterioration, and to protect usability of water (incl. drinking purposes) and biodiversity for coming generations. It is particularly essential to reveal crucial causality chains and phenomena behind delayed responses in both ground-water and water bodies.

Research: Study interactions between physics of flow in geologic media, aquatic chemistry, and microbiology. Reveal essential solute transformations, biogeochemical functioning, and mechanisms for contamination/purification of soils and water. Address the issue of chemical time bombs. Clarify water exchange mechanisms and the way they contribute to response delays.

5.4 Development of improved planning and management tools

The research needed to take care of the above were proposed by the Task Group 1 on Climate, Hydrology and Water Resources (1991). The specific problems they are meant to cope with include water management under uncertainty, sensitivity and vulnerability of water resources to climate, and, adaptive and preventive measures to withstand changes.

The corresponding research needs are:

(a) Development of methodologies for generating internally coherent regional scenarios of future water supply and demand as a basis for water resources planning.

(b) Development of numerical indices of vulnerability and those that identify river basin responses to the combined effects of climate and social and economic changes.

(c) Studies on means for developing more resilient, robust and flexible water resource systems, able to be cheaply upgraded to cope with future demands.

5.5 International Research Programs

Water research is to a large degree international, especially climate-related issues and issues of a global or regional scale. The World Climate Programme (Water) within which the research is being handled is part of the World Climate Research Programme (WCRP) which is co-ordinated with the International Geosphere-Biosphere

Programme. An important component of the WCRP, which is going to make a significant contribution to the WCP (Water), is the Global Energy and Water Cycle Experiment (GEWEX). The Scientific Committee for GEWEX has proposed a GEWEX Continental International Project (GCIP) with the following objectives:

(a) to determine the time/space variability of the hydrologic and energy budgets over a continental scale;

(b) to develop and validate macro-scale hydrologic models, related high resolution atmospheric models, and coupled hydrologic atmospheric models;

(c) to develop and validate information retrieval schemes, incorporating existing and future satellite observations coupled with enhanced ground-based observations;

(d) to provide a capability to translate the effects of a future climate change into impacts on water resources and temperature on a regional base.

Other international projects that will contribute to the above research are UNESCO's International Hydrological Programme (IHP) Phase IV, with its subprogram H which addresses Hydrologic research in a changing environment; and WMO's Hydrology and Water Resources Programme (HWRP) with its relevant components. The International Association of Hydrological Sciences (IAHS) and other non-governmental organizations are involved with UNESCO and WMO in making contributions. Within the International Geosphere-Biosphere Programme there is a particular project addressing the Biospheric Aspects on the Hydrologic Cycle (BAHC).

6.   URGENT ACTION

The actions that are particularly urgent:

(a)   establish data bases for a water resources assessment;

(b)   education and training with a content responding to current problems and approaches;

(c)   developing/strengthening of national and regional institutes for research on water resources in order to make possible a science-based planning and management of available freshwater resources;

(d)   review and improve the legal and institutional mechanisms for planning and managing water resources.

6.1 Data bases

The data needs for executing the scientific agenda are enormous. They involve firstly data collection and monitoring of water as well as sediment and solute transport by benefitting from remote sensing and global scale observations from space. In addition to the space-gathered data, land surface data and monitoring of hydrologic parameters are required in every country. The analysis and interpretation of the data will be made at national, subregional, regional and global levels.

The specific primary data to be collected from land and space, and their secondary treatment to get derived data for gridded fields for the calibration and validation of both mesoscale Atmospheric General Circulations Models (AGCM) and macroscale hydrologic models were discussed in NATO (1991). The exchange of data from all countries for global and regional studies underscored the need of the Global Runoff Data Center and the Global Precipitation Climate Center in Koblenz and Offenbach (Germany) respectively. Other data centers on land surface characteristics, such as for soils and vegetation, need equal support.

The data to be collected from space will involve the use of high technology. WMO is expected to make arrangements for data collection to be made available to national and regional institutions. The data collection and monitoring from land are primarily the responsibility of each nation within its territorial boundaries. Unfortunately the hydrologic networks for data collection and monitoring are weak in the developing countries of Africa, Asia and Latin America (UNESCO/WMO,1990).

Assistance is therefore needed for the countries of these regions to strengthen and expand their networks and hydrologic services to be able to collect the requisite data and share with other regions. In the case of Africa the World Bank/UNDP, Sub-Saharan Africa Hydrologic Assessment Project offers the most appropriate basis to provide the assistance.

6.2 Education and Training

The education and training components of the various international programs in hydrology and water resources offer opportunities for training water engineers and scientists. The developing countries have been the greatest beneficiaries of these programs. However, to train high-quality research personnel will require that special attention be paid to offer more training to the PhD level within the framework of these programs. Further the education and training programs in hydrology and water resources will have to be expanded to increase knowledge of hydrologic processes in the atmospheric phase, and their interrelationship with processes on the land phase and the aquatic and terrestrial ecosystems (Nash *et al.*, 1990). Planning and management of water resources should also be included.

In order to retain the scientists to be trained from the developing countries to work in their own countries, and not be attracted to work outside for better conditions, incentives to support them in the form of better remuneration and research facilities will need to be provided.

6.3 Research capability

The national research capability needs to be built up and/or strengthened in the developing countries where the need is greatest. First this calls for awareness-making, so that policy-makers get aware of the need for research so that the required national resources can be allocated. Advancing knowledge is crucial to create the capability to solve the water crisis that is imminent. Secondly, it will involve training adequate numbers in the various disciplines of hydrology and water resources. This has been indicated in the paragraph under education and training. Thirdly, adequate funds should be provided to acquire facilities in terms of field, laboratory and office equipment needed for research. The funding resources of national governments should be supplemented with external resources (IHE, 1991). Fourthly, training should be given in the planning and management of national and subregional research projects and programs.

Current efforts to support scientists from developing countries to attend scientific meetings should be identified and widened to enable a much greater number to interact, share experiences and be involved in what is happening in their fields both regionally and globally.

The adoption and implementation by all countries of the *Delft Declaration* on water resources capacity building would assist in building national research capabilities in the developing countries (IHE, 1991).

6.4 Instruments for Planning and Management

(a) Review and update present policy instruments for managing demand like water tariffs, abstraction control, etc., for effective application.

(b) Review laws and regulations for the control of wastes discharged into water bodies, the land and the atmosphere. These should be updated in line with current socio-economic conditions and effectively implemented.

(c) For the countries that have to mobilize additional water resources from unconventional sources (e.g. treated waste water, drainage water, etc.), studies should be carried out to find out the socio-economic, technical and cultural barriers that must be overcome in order to facilitate their widespread adoption.

(d) Since irrigation is the biggest water user, strategies for saving water like use of more efficient irrigation techniques (sprinkler, drip), prevention of losses in irrigation systems, and improvement in the efficiency of water use at the farm level should be implemented.

(e) Review the institutional framework for planning and management and update or adapt them to suit current conditions, taking into account the important physical linkages between water uses such as surface water and ground-water, the economic linkages between different water uses such as irrigation and hydroelectric power, and the social linkages between water development and people who are benefitting or adversely affected (UNESCO, 1990).

7. CONCLUSION

This chapter has highlighted the importance of water both in the biogeochemical processes of the Earth System, and in the socio-economic development process. Water is in numerous ways involved in the efforts aimed at improving the standard of living and quality of life of the world's populations. Socio-economic development profoundly depends on the benefits of water resources development by which adequate amounts of water are made accessible when and where needed. Regional shortages of water, an expanding and already widespread pollution of water resources, and a multifaceted degradation of the environment, closely related to water phenomena, pose serious challenges for the future.

In some regions, water is increasingly turning into a life-and-death issue. The above problems have come about partly because the knowledge of freshwater resources in the hydrologic cycle in the Earth System is far from adequate. At the same time the planning and management of water to meet developmental needs is insufficient. Increased uncertainties brought about by man-induced climate change add to the problems referred to. It is essential that the gaps in knowledge needed to develop the different regions of the world in a sustainable manner are being bridged through research into climate, hydrology and water resources. The integrated planning and management of land and water resources has to be underpinned by an adequate scientific understanding of local and regional hydrology. Only so can policy- and decision-makers and the public be provided with a sound basis for action.

REFERENCES

Abu Zeid, M.A. and Biswas, A.K. 1991. Some major implications of climatic fluctuations on water management. *Water Resources Development*, pp. 74-81.

Ayibotele, N.B. 1990. Institutional and Legal Infrastructures for Planning and Management of Water Resources in African Region – Progress with Implementation of Mar del Plata Action Plan. UNDTCD and DIESA, New York.

Ayibotele, N.B. 1991a. The world's water: assessing the resource. *Proceedings of the International Conference on Water and the Environment*. Dublin, January 26-31 1992 (To be published by WMO).

Ayibotele, N.B. 1991b. Conservation and Rational Use of Water Resources in North African Countries. UN Economic Commission for Africa. Addis Ababa.

Clark, W.C. 1990. The Human Ecology of Global Change. Global Environmetal Policy Project. Discussion Paper G-90-01. John F. Kennedy School of Government, Harvard University.

Falkenmark, M. 1991a. Regional Environmental Managment: The Role of Man-Land-Water Interactions. World Bank, Environment Dept, Divisional Working Paper No. 1991-21. Washington DC.

Falkenmark, M. 1991b. Environment and Development – Urgent Need for a Water Perspective. Chow Memorial Lecture, IWRA Congress, May 1991. Water International, in press.

Falkenmark, M., and Chapman, T. 1989. *Comparative Hydrology. An Ecological Approach to Land and Water Resources*. UNESCO, Paris.

Falkenmark, M, and Lundqvist, J. 1992. Coping with Multilevel Environmental Challenges-Taking a Water Perspective. Keynote Paper, UN Conference on Water and Environment, Dublin January 1992 (In press).

Falkenmark, M ., Lundqvist, J., and Widstrand, C. 1989. Macro-Scale Water Scarcity Requires Micro-Scale Approaches. Aspects of Vulnerability in Semi-Arid Development. *Natural Resources Forum*, pp.258-267.

Falkenmark, M., and Suprapto, R. A. 1992. Population-Landscape Interactions in Development. A Water Perspective to Environmental Sustainability. *Ambio*, (In press).

GEMS MARC and WHO 1991. Water Quality. Progress in Implementing the Mar del Plata Action Plan and a Strategy for the 1990s.

Harding, R.J. 1991. The Estimation of Areal Evaporation. NATO Advanced Workshop in Opportunities for Hydrologic Data in support of Climate Change Studies. Lahnstein, 26-29 August 1991. In press.

Hartman, H., Lawless, J.G., and Morrison, P. 1985. Search for the Universal Ancestors, NASA SP-477.

IHE 1991. Delft Declaration. Statement from UNDP Symposium on A strategy for Water Resources Capacity Building. Delft.

INWARDAM 1990. Water Resources Assessment for Arab League Countries. Islamic Network for Resources Development and Management, Issue No. 10. Amman.

Klemes, V 1990. The Problems of the Humid Tropics - An Opportunity for Reassessment of Hydrologic Methodology. International Colloquium on the Development of Hydrologic and Water Management Strategies in the Humid Tropics. *International Hydrological Programme*. UNESCO, Paris.

Leonard, H.J., and contributors 1990. Environment and the Poor: Development Strategies for a Common Agenda. Transaction Books.

Mawdesley, J.A. 1991. Estimation of Areal Evapotranspiration. NATO Advanced Workshop on Opportunities for Hydrologic Data in Support of Climate Change Studies. Lahnstein, 1991 (In press).

Nash, J.E., *et al* 1990. The Education of Hydrologists. Report of an IAHS/UNESCO Panel on Hydrologic Education. IAHS, Delft.

National Research Council, 1991. Opportunities in the Hydrologic Sciences. Commission on Geosciences, Environment and Resources. National Academy Press, Washington DC.

Oeschger, H. 1991. Paleo Data, Paleoclimates and the Greenhouse Effect. In: J. Jaeger and H.L. Ferguson (eds.) *Climate Change: Science, Impacts and Policy. Proceedings of the Second World Climate Conference*. pp. 211-224. Cambridge University Press, Cambridge.

O'Laughlin, E.M. 1990. Modeling soil water status in complex terrain. *Agricultural and Forest Meteorology*, pp. 23-38.

Preparatory Committee for the UN Conference on Environment and Development, Third Session (August 1991). Protection of the Quality and Supply of Freshwater Resources: Application of Integrated Approaches to the Development, Management and Use of Water Resources. General Assembly, Document A/CONF.151/PC/42/Add 7. United Nations.

Rodhe, A. 1984. Ground-water Contribution to Streamflow in Swedish Forested Till Soils. Isotope Hydrology 1983, pp 55-66. IAEA, Vienna.

Rowntree, P.R. 1991. Hydrologic Aspects of Modeling Climate Change.

Schaake, J.C. 1991 A Strategy for Using Macro-Scale Hydrologic Models in the Development of Improved Global Climate Models (GCMs). NATO Advanced Workshop on Opportunities for Hydrologic Data in Support of Climate Change Studies. Lahnstein, 26-30 August 1991 (In press).

Shiklomanov, I.A. 1990. The World's Water Resources. International Symposium to commemorate the 25 years of IHD/IHP, Paris 15-17 March 1990. UNESCO.

Task Group 1 on Climate Hydrology and Water Resources 1991. In: Jaeger, J., and Ferguson, H.L. (eds.) *Climate Change: Science Impacts and Policy. Proceedings of the Second World Climate Conference*. Cambridge University Press, pp. 435-438.

UNESCO, 1978. World Water Balance and Water Resources of the Earth. UNESCO Series Studies and Reports, No. 25. Paris.

UNESCO, 1990. International Hydrological Programme (IHP) 4th Phase: Hydrology and Water Resources for Sustainable Development in a Changing Environment. Paris.

UNESCO/WMO 1990. Report of the 5th Planning Meeting on World Climate Programme - Water. Laxenburg.

WHO/UNEP 1989. *Global Freshwater Quality. A First Assessment*. Geneva, Nairobi.

Chapter 11: Biodiversity

M.T. Kalin Arroyo, P.H. Raven and J. Sarukhan

EXECUTIVE SUMMARY

The extinction of a major proportion of the species of plants, animals, fungi, and microorganisms, which could amount to 20% of the total or more within the next 30 years, constitutes a very serious problem for human beings. It is the most rapidly moving aspect of global change, and is completely irreversible. The species of a few groups of organisms, including plants, vertebrate animals, and butterflies, have been enumerated reasonably well, but names have been given to only about 1.4 million of the estimated minimum 10 million species of organisms. Even less is known about the functioning of these species in ecosystems, the principles that could be applied to their conservation, or the ways in which they could be used for human benefit in an era of biotechnology.

To remedy these problems, we recommend that the issue of the extinction of biodiversity be brought to the attention of world leaders and that the available manpower and infrastructure for dealing with biodiversity throughout the world be strengthened greatly. The formation of National Biological Resource Centers (National Biodiversity Institutes) in each country is advocated as a particularly effective way to develop effective management structures and plans for biodiversity; to be able to study it effectively; and to conserve it. Scientific and technical personnel need greatly to be increased in all developing countries, in part so that their own biodiversity can be used for their national development. The training of geneticists, taxonomists, and ecologists specialized in the study of biodiversity is a particularly critical need.

It is clear that biodiversity will be conserved effectively only in a stable world in which major ecological problems have been resolved. This will involve the attainment of a stable human population in all regions (a particularly important problem for industrialized countries, since the level of consumption *per capita* is so high) as well as the moderation of overconsumption by the inhabitants of the industrialized nations as well as by relatively wealthy citizens of developing nations.

Within the context of the development of a general theory of biodiversity, much more research is needed into its function, from genes to ecosystems, and into the role of species diversity in communities. The diversity of organisms needs to be explored in detail, and estimates of the abundance and patterns of distribution of relatively poorly known groups need to be developed. The economic value of organisms, related to improved economic models that take into account the exhaustion of natural resources, needs to be explored in detail. The knowledge held by indigenous and rural people throughout the world should be gathered, evaluated, and applied to the solution of current problems.

Networks of protected areas, selected because of their potential for preserving biodiversity, need to be established and funded adequately, as do networks of botanical gardens, stock culture centers, seed banks, and similar facilities for *ex situ* preservation.

1. INTRODUCTION

The local and global extinction of species and the consequent reduction of genetic diversity of plants, animals, fungi, and microorganisms that we are experiencing constitutes a major environmental problem that requires urgent attention. This problem has arisen as a result of pressures generated by the population growth and patterns of consumption of human beings; the consequent loss of species is irreversible. As a result of this loss, our ability to understand and control the global ecosystem and use many kinds of organisms sustainably for our benefit is being lost permanently (Ehrlich and Wilson, 1991).

Species vary in genetic uniqueness and play diverse functional roles of greater or lesser significance in ecosystems; consequently, the loss of some species will be more critical than others. Ecosystems regulate the flow of energy, ultimately derived from the sun, both on a local and on a global basis. The functioning of these ecosystems directly influences the concentrations of atmospheric gases, determines the nature of soils, and profoundly affects the condition of bodies of water. They also regulate the cycling of the nutrients on which life depends. Organisms are crucial for the functioning of the Earth System, and form the basis for the life-support system on which humanity depends. We depend on them completely as the source of our food and other materials critical to our lives.

As a result of human activities, the gradual increase in the total number of species that has occurred over the past 65 million years has been abruptly reversed. Major extinction events have been recorded at widely separated intervals during the course of the Earth's history, with the most recent having occurred at the close of the Cretaceous Period, 65 million years ago. As a result of these events, some evolutionary lineages were lost and replaced by others, and some expanded while others contracted. In the past, environmental conditions usually changed slowly, allowing a gradual re-establishment of the major biological communities. The current extinction event, estimated to be proceeding at a rate from 1000 to 10,000 times greater than the natural background rate (Wilson, 1988a), however, is associated with rapid climatic change and the wholesale local and often unpredictable destruction of particular communities. Numerous evolutionary lineages that provide the genetic material for future evolutionary diversification on Earth will be severely eroded. Although such extinction could be seen philosophically as comparable to the several similar events that are known to have occurred in the past, leading to another reorganization of the world's ecosystems, its outcome in the near future is likely to be one with which human beings, as a species, can hardly cope.

As we enter the current spiral of extinction and other aspects of global change, we can expect the substitution of species that are dominant now by others that are fast growing, have short life cycles, and are weedy (Arroyo et al., 1992). It is uncertain whether the communities that we are forming will be as productive as those on which we depend now for our survival (Vitousek et al., 1986; di Castri and Younès, 1990). In addition, both modern and traditional agricultural practices depend on the very climatic stability that we are destroying simultaneously, as well as on the predictability of cycles of pests and infections. The extinction of a major proportion of the species that currently exist, coupled with the partial substitution into ecosystems of species that are biologically radically different from those that they are replacing and assembled into ecosystems that are structurally simpler, threatens to disrupt the viability of the basic patterns by which we depend collectively on the functioning of the global ecosystem.

The knowledge currently available is insufficient to assess the ecosystem function of biodiversity in any reasonable depth (di Castri and Younès, 1990). This question, although of theoretical interest, is of more limited relevance practically, when it is considered that the most serious immediate effect of extinction in this respect will be the emergence of radically different and structurally simpler ecosystems. With the former kind of uncertainty, and the need to conserve functionally important species to maintain presently structurally diverse ecosystems, we cannot afford to let the "natural gene banks" on which the adjustment of any local ecosystem clearly depends, simply to slip through our hands into extinction. In general, genetically diverse populations are able to cope with environmental change much better than those that are genetically impoverished, making the problem all the more serious in view of the human-induced changes that are altering the world so rapidly.

The search for ways to conserve biodiversity becomes all the more urgent considering that the human population, even under the most optimistic scenarios, involving continued attention to family planning, will increase for at least another century. As our numbers grow, we shall come to rely even more heavily than we do now on our ability to manage biodiversity as the primary basis of our subsistence and survival. At the very least, the development of strategies and agendas to conserve biodiversity as we attain a more profound comprehension of its role in the global ecosystem affords a minimum guarantee for a viable common future. The study of biodiversity and its incorporation into the livelihood of local people thus must come to constitute a major area of research and concern for the biological and social sciences as we move into the 21st Century.

It is worth pointing out that changes in the world's atmosphere, leading among other effects to a reduction in stratospheric ozone, with serious biological effects, and to global warming, through an increase in "greenhouse gases", came to widespread attention during the 1980s and led to the initiation of concentrated international action to address these serious problems. In contrast, the loss of biodiversity, although it is proceeding more rapidly than the atmospheric changes, is irreversible, and has far more serious consequences for human beings in the future, and has still not received adequate attention.

This chapter constitutes an attempt to draw the attention of world leaders and that of the scientific community generally to the importance of understanding biodiversity and managing it properly. As many as possible of the plants, animals, fungi, and microorganisms that exist now should be conserved, as we seek to understand better the role of biodiversity in ecosystem functioning and the principles by which stable ecosystems can be restored.

2. WHAT IS BIODIVERSITY?

The US Congressional Office of Technology Assessment (OTA, 1987) defined biodiversity as follows:

Biological diversity refers to the variety and variability among living organisms and the ecological complexes in which they occur. Diversity can be defined as the number of different items and their relative frequency. For biological diversity, these items are organized at many levels, ranging from complete ecosystems to the chemical structures that are the molecular basis of heredity. Thus, the term encompasses different ecosystems, species, genes, and their relative abundance.

Life originated at least 3.5 billion years ago, within no more than a billion years of the origin of the Earth. Terrestrial communities have existed for more than 430 million years, with forests appearing about 300 million years ago. These communities have evolved and become more diverse over the course of time; thus species-rich ecosystems are particularly characteristic of tropical regions, species-poor ones of areas such as the Arctic, where conditions for growth are more limiting. The differences between ecosystems need to be measured in terms of the distinctiveness of the evolutionary lineages involved; thus, although perhaps only a tenth as many species exist in the sea as on land, the great majority of the major groups of organisms (phyla and divisions) occur only in the sea. Such relationships need to be taken into account when making choices for conservation at the local scale.

The simplification of ecosystems that is so evident throughout the world began with the development of agriculture by humans about 11,000 years ago. As a result of our activities, the destruction, reduction in area, and simplification of ecosystems is now proceeding rapidly everywhere. Tropical rain forests, coral reefs, mangroves, wetlands, and temperate forests and prairies have all been modified extensively as a result of human activities, and only fragments remain of these and other ecosystems in many regions.

In functional terms, the relationships between species in ecosystems are still poorly understood. As these properties are more clearly evaluated, and the functional redundancy hypothesis is tested (di Castri and Younès, 1990), the loss of a single species in a temperate ecosystem might well turn out to be more critical than the loss of one in a tropical ecosystem, where a given evolutionary lineage may be represented by numerous species. The evaluation of biodiversity is not merely a question of species number, but also must concern itself with the genetic and functional uniqueness of individual organisms. Since it will clearly be impossible to save all of the species that exist today, a major area of scientific inquiry should be the determination of the genetic and functional significance of species in different ecosystems, as proposed for possible implementation by joint IUBS/SCOPE/UNESCO action (Solbrig, 1991b). Few tools are available to evaluate these factors at present, and we have therefore presented the problem of the loss of biodiversity here primarily in terms of species loss and of the obvious reduction in the genetic diversity of individual species that is occurring throughout the world.

3. THE DIMENSIONS OF BIODIVERSITY

The total number of named species of organisms is approximately 1.4 million (Wilson, 1988b). Of these, about 250,000 are plants (defined as vascular plants and bryophytes only), 44,000 are vertebrate animals, and 750,000 are insects. Plants and vertebrates, as well as a few other groups such as butterflies, are relatively well known, with probably 90% of the species named. For fungi and microorganisms, which play vital roles in nutrient cycling, decomposition, soil functioning, detoxification, and a great variety of symbiotic relationships, our estimates of overall species numbers are woefully inadequate. For example, with 69,000 known species of fungi, it has recently been estimated that as many as 1.5 million species may actually exist (Hawksworth, 1991); the group is highly significant ecologically. Other groups, such as mites, nematodes, and bacteria are even more poorly known. It is very difficult to establish how many species of organisms there may actually be on Earth, but we now estimate conservatively that at least 10 million species may exist (May, 1988; Gaston, 1991; Raven, unpublished observations.)

For freshwater habitats, we can describe biodiversity accurately for perhaps only a fifth of the total area, with many groups of organisms extremely poorly known (Allen, Flecker and Smith, 1992; Firth and Werner, unpublished observations). Marine biodiversity is also poorly known, especially that of the deep sea (Ray, 1988). A particularly serious deficiency concerns our knowledge of marine microorganisms; we can culture and therefore classify probably less than 1% of marine species – a group of organisms that plays a series of key roles in controlling major biogeochemical cycles (R.R. Colwell pers. comm.). In ecological terms, the linkage between the biodiversity of marine ecosystems and their functioning is even less well understood than are the analogous relationships on land. Coastal zones, where a majority of human beings live, are especially critical. The biological exploration of such zones, important for their proper management, should, therefore, be a matter of high priority.

Summarizing these considerations, E.O. Wilson is clearly justified in emphasizing that we cannot confidently estimate the number of species of organisms on Earth even to an order of magnitude, an appalling situation in terms of knowledge and our ability to affect the human prospect positively. There are clearly few areas of science about which so little is known, and none of such direct relevance to human beings. One evident challenge for scientists, in the face of our ignorance of the basic facts, will be to define a theoretical context for biodiversity in order to address adequately emerging issues in this area and to develop realistic sampling procedures to collect the additional data that is so badly needed.

4. PROXIMATE CAUSES AND ESTIMATED RATES OF LOSS

The reasons that species are being lost so rapidly have been outlined in other themes discussed in this conference, and the overall patterns involved have been reviewed comprehensively by Wilson (1988b). The loss of habitats throughout the world, as a result of human activities; the overexploitation of particular species for their commercial value; pollution; and the introduction of species from one part of the world to another have all contributed substantially to the problem. Of particular importance is the reduction of the world's forests, which has proceeded at an ever-accelerating rate over the past 50 years. Tropical rain forests, for example, may have occupied about 12% of the Earth's surface at their maximum extent, and have now been reduced to approximately half of that area. These forests apparently are home to more than half of the world's species – among the most poorly known found anywhere. Of the estimated 7.7 million square kilometers of tropical moist forest remaining in the late 1980s (perhaps about half of the original amount; Myers, 1988), 160,000 to 200,000 square kilometers were estimated to have been cleared each year – a 50% increase in the rate of clearing over the past ten years (World Resources Institute, 1990). At such a rate, all tropical moist forests would be lost in less than 40 years. In fact, the prospects for most areas are even worse, as the number of people living in tropical countries increases rapidly toward a doubling over the next several decades.

It is important to develop more precise methods for evaluating forest loss than are generally available at present, because it is difficult to reverse or deal with trends that are poorly documented. As an example of the difficulty of evaluating data in this area, Myers (1989) estimated that 50,000 square kilometers of forest in the Brazilian Amazon were being lost each year, but Fearnside (1990) calculated that the actual figure was about 20,000 square kilometers. The wider use of satellite imagery to evaluate forest loss, coupled with observations on the ground, would assist greatly in our evaluation of this problem, and has long been advocated; but the nature and scope of human activities, clearly even more difficult to evaluate and to predict, will determine the fate of all

natural communities, including tropical forests, in the future.

At any rate, taking the overall patterns of forest loss into account and evaluating them conservatively, estimates of global rates of extinction can be made, based on our knowledge of relatively well-known groups of organisms and the development of the areas where they occur. Clearly, if the rates of forest destruction can be decelerated, for example by focussing many productive activities on lands that have already been deforested, these rates could be slowed down substantially. If current trends persist, however, the loss of species is certain to be drastic. Using vascular plants as an example, because they are so well known, it can be calculated that the forests of areas that contain about half of the world's total number of species could be reduced to a tenth or less of their original size over the next 25 or 30 years (Raven, 1987). Such destruction of habitats would, according to the principles of island biogeography (MacArthur and Wilson, 1967), which relate species number to area on an empirical basis, put half of those species at risk, and greatly reduce the genetic diversity of the survivors. Not only does fragmentation lead to a loss of species diversity by reducing the size and genetic diversity of the populations, it fundamentally changes the environmental conditions to which they are exposed. These relationships lead to estimates that 20 to 25% of the world's plant species, and more than a third of those in the tropics, could be at risk over the next three decades (Raven, 1987). Many of the species of these areas do not occur in the forests themselves, and their fate in the face of deforestation is therefore more difficult to project; the estimates could be improved greatly by a careful evaluation of the species of particular areas. Nonetheless, the estimate appears to provide a useful approximation of the degree of threat to the world's biodiversity during this period.

Other studies have come to conclusions that are roughly similar. For example, applying the principles of island biogeography and projections of forest loss, Simberloff (1986) has calculated that 15% of the plant species and 2% of the bird species in South America are likely to be headed for extinction by the year 2000, and that approximately two-thirds of the plant and bird species are likely to vanish during the course of the next century. Similar estimates have been developed in an excellent review by Reid (1992) for organisms generally.

A consideration of the situations in individual regions lends strength to these projections. Considering western Ecuador, for example, an estimated 75% of the area was still forested in 1945, as compared with a rapidly decreasing 4.4% now (Dodson and Gentry, 1991). About 6300 species of plants, some 1260 found nowhere else, occur in the region, with at least 12% likely to disappear during the 1990s alone. Unfortunately, western Ecuador is by no means unique. In Madagascar, the Atlantic forests of

Brazil, in the Philippines, and in many other regions, the trends are similar. By analyzing areas with rich concentrations of endemic plant species that are under threat, Myers (1990) has demonstrated that 20% of the world's plant species are confined to 0.5% of the world's surface – areas that he has termed "hot-spots." Not only are very large numbers of species endangered locally, but conservation efforts could be concentrated in such areas, which occur in both tropical and temperate (e.g. Mediterranean- climate) regions, to good advantage.

The role of disturbance and other factors in maintaining biodiversity in ecosystems should be more completely investigated, since these appear to be key factors that might offer numerous possibilities for the management of biodiversity in ecosystems and their restoration. Another important area obviously concerns the management of small populations and the management of small reserves, since many of those that do persist into the future are likely to be limited in size.

Freshwater habitats are under severe threat everywhere, but especially in the tropics (National Research Council, 1980; Woodwell, 1990; Allen, Flecker, and Smith, 1992; Firth and Werner, unpublished observations.). Overfishing and especially pollution, such as runoff from agriculture, and acidification, are becoming widespread; dams, channel dredging, sedimentation, overexploitation of water supplies – all are affecting the world's freshwater habitats, among the most productive sources of protein for human beings throughout the world. Inventories are rarely complete enough to estimate actual rates of extinction, but the problem of species loss is evidently severe virtually everywhere. The introduction of exotic fishes is the most serious cause of extinction of lake-dwelling fishes, and possibly of freshwater organisms generally, thoughout the world (Lowe-McConnell, 1987). The zebra mussel, now spreading aggressively in the interior waterways of the United States and Canada, provides a particularly striking example of an aggressive invader. Of exceptional importance are the faunas of ancient lakes, such as Lake Baikal in Siberia (where manipulation of the fish stocks and the construction of additional factories are being considered) and the major lakes of East Africa, which contain impressive series of endemic organisms – more than 1000 species of endemic cichlid fishes inhabit these lakes – that have evolved in isolation, and are now being threatened with extinction, especially by the activities of introduced species. As calculated for the United States recently by The Nature Conservancy, it is likely that freshwater organisms are becoming extinct even faster than terrestrial ones throughout the world.

For the sea, it is difficult to project rates of extinction. The open sea is still relatively clean, but coastal areas are affected by humans almost everywhere (GESAMP, 1990). The destruction of beaches, coral reefs and wetlands, including mangrove forests, and increasing erosion of the shore, are evident all over the world. Bleaching in corals has become evident over wide areas during the past several years, indicating the death of the coral animals and some of their associates. At the same time, enormous amounts of human-generated contaminants, including mostly untreated sewage, are being discharged onto the continental shelves, which are also being enriched by the discharge of nutrients. Although estimates of extinction have not yet been attempted for the sea, the heavily stressed and very productive ecosystems of the areas where the land and sea come together are clearly endangered on a global basis, and urgently need attention if they are to be stabilized and preserved. An effort to begin to frame the problem of understanding the process of extinction in marine habitats is now being made (e.g. papers in Woodwell, 1990; Thorne-Miller and Catena, 1991), and this area should be pursued actively so that it can be addressed more adequately.

As scientists, we should clearly strive to improve our estimates of current and probable future rates of extinction. By focussing our attention on groups of organisms that are well known, or employing accurate sampling methods with others, we can certainly do so. For example, a large number of genera of plants and vertebrates are well understood; information concerning them could be incorporated into comprehensive data bases. To the extent that patterns of deforestation are also well understood or can be predicted accurately, this information could yield increasingly precise information about current and projected rates of extinction. At the same time, it is obviously counterproductive to become overly immersed in differences of opinion about rates of extinction, when even the most conservative estimates are so alarming and when the most important issue is clearly that of saving biodiversity.

5. ULTIMATE CAUSES OF THE LOSS OF BIODIVERSITY

Unfortunately, the severe loss of habitats that is evident everywhere is likely to accelerate unless radical steps are taken to address a number of fundamental issues, reviewed in many chapters of this volume. These issues, overall, have to do with population growth and patterns of consumption. To achieve the global stability within which biodiversity could be preserved and human prosperity could be assured, it will be necessary to address both of these factors. A stable population is a necessary prerequisite for the preservation of biodiversity, and so is the adoption of moderate patterns of consumption throughout the world. Unfortunately, most human activities are carried out in the light of outmoded economic theories that simply assume that all resources are inexhaustible, and that if they do run out, others can be subsituted without loss to the functioning of the system (Solbrig, 1991b). New ways of thinking must be developed if we are to be able to attain global stability.

Taking population first, human numbers have more than doubled to their present level of 5.4 billion over the past 40 years. In doing so, we have reached a state at which we are consuming, diverting, or wasting an estimated 40% of total net terrestrial photosynthetic productivity (Vitousek *et al.*, 1986). Many signs of the deterioration of the global environment, coupled with substantial losses in productivity, are becoming evident in both agricultural and natural ecosystems (Brown *et al.*, 1990). In temperate regions throughout the world, biodiversity is being sharply reduced as a result of the intense use of agricultural and other lands and widespread pollution; thus, for example, the diversity of the macrofungi of Europe has declined markedly since World War II (Jaenike, 1991). Taken together, these relationships indicate severe challenges for the human future, with population stabilization under the most favorable assumptions (sustained, world-wide attention to family planning, for example) not likely to occur until the human population has reached two to three times its present size a century from now. In order to feed, clothe, and shelter 10–12 billion people adequately at any time, it will clearly be necessary to strengthen scientific, social, and economic research on the Earth's biodiversity. Our dependency on the world's organisms for our sustenance and for maintaining the stability of the places where people live will be far more evident then than is the case at present.

Resources-money, commodities of all kinds, and relevant infrastructure – are highly concentrated in industrialized countries, where a disproportionate amount of the world's productivity is being consumed by a rapidly shrinking minority of the global population. At present, about a billion people – a quarter of the total living in developing countries – exist in what the World Bank defines as absolute poverty. Half of these people, according to UN estimates, receive less than 80% of the UN-recommended minimum caloric intake. The 77% of the people in the world who inhabit developing countries now have an estimated 15% of the world's wealth (measured as GDP) and consume 10 to 15% of the world's industrial energy and other materials that contribute to their standard of living. An estimated 1.5 billion people, over a third of the total number who inhabit these countries, depend on firewood as their primary source of fuel, obtaining it almost entirely from natural plant communities. Because the 23% of the world's people who live in industrialized countries are consuming the world's resources at a rate individually 10 to 40 or more times that characteristic of developing countries, the attainment of stable populations industrialized countries should be, but is

not, currently, a matter of serious concern. At the same time, the wealthier people who live in developing countries must learn to limit their consumption in the same way that these levels must clearly be reduced in industrialized countries for the sake of global stability.

The economic relationships between nations have had serious negative impacts on global stability. Ever since what we now know as industrialized nations came into contact with developing countries, they have controlled their economies so that they would concentrate on primary productivity, selling their goods on the world market at low prices and highly dependent on the vagaries of the marketplace. Although there was an upsurge in development assistance in the years following World War II, during the late 1980s there was an annual positive cash flow, amounting to tens of billions of dollars per year, from poor countries to rich ones. The $1.3 trillion debt of developing countries to industrialized ones plays a major role in the draining of resources from the poor to the rich, a pattern that has existed in one form or another for hundreds of years.

In terms of their scientific and technological development, the poorer countries of the world are badly off. Developing countries have, according to Abdus Salam (1991), about 6% of the world's scientists and engineers, mostly concentrated in a few of the countries, such as China, India, and Brazil; the chances for most countries to improve the technological basis for sustainable productivity without outside assistance are therefore poor.

Against this background, a deceleration of deforestation rates will only be achieved with a major change in attitude, whereby the conservation of natural resources comes to be seen as an "insurance policy" for global stability and thus for our own survival. Presently, a significant proportion of deforestation stems from the universal desire to convert resources into money – the same force that is causing deforestation in the northwestern United States, in Siberia, and other regions of industrialized nations where ancient forests still persist. In this process, medium- and long-term values are characteristically neglected. Natural resources are characteristically undervalued, and we need to find new ways to calculate their true worth and build those values into our calculations of how to deal with them (McNeely, 1988; Barbier, Burgess, and Markandya, 1991). The maldistribution and misuse of farmlands, coupled with perverse land-use policies that offer incentives for the inefficient conversion of forests into agricultural systems, or often all too soon into wastelands, that contribute significantly to the loss of forests in the tropics and in some temperate areas as well are a further expression of undervaluing natural resources.

Another major factor in deforestation in developing countries is poverty – the legitimate desire to find food for oneself and one's family. Without attention to poverty throughout the world on a much more serious and sustained scale than anything yet experienced, the fate of not only tropical lowland forests but also most other natural communities in these regions seems clear.

6. SCIENTIFIC CONSEQUENCES OF THE LOSS

The simplest way to convey the scientific consequences of the world-wide loss of biodiversity that is occurring at present is to emphasize the opportunities for knowledge that we are losing. Certainly those opportunities are greater now than ever will be the case in the future, simply because more species exist now than are likely ever to be the case again. No sound approach can be developed to the restoration of biological communities, a topic that will clearly come to be viewed as increasingly important during the course of the next century, unless biodiversity, and its functioning in ecosystems, is better understood.

If we take into account the many hitherto unknown species that are being discovered even in well-known groups of organisms, the urgency of learning about them becomes obvious. A better knowledge of these groups is the most readily accessible key to understanding the abundance and patterns of diversity of life on Earth (National Research Council, 1980). The study of organisms that have obvious economic importance should also be emphasized. At the same time, precise sampling methods must be established to provide even the overall nature of most groups of organisms, extremely poorly known and also disappearing rapidly.

The knowledge possessed by indigenous peoples, or other rural peoples living in close contact with the relevant communities, must also be viewed as a precious and rapidly vanishing resource about which we must gain knowledge while there is still time. Emphasis should certainly be given to the conservation of useful species of organisms utilized by such people. Knowledge concerning the organisms and communities of interest to them should include greater emphasis than has been traditional on the ecological features and requirements of the species (e.g. population size, phenology, breeding system, time of seed germination, accompanying species, storage practices, and so forth). To make this knowledge further useful to plant breeders, and for application in other parts of the world, there is a need to develop and to adhere to standardized methods for characterizing habitats and other environmental features.

Of equal importance to learning about the kinds of organisms that exist is gaining an appreciation of the role of biodiversity at an ecosystem level (di Castri and Younès, 1990). This role needs to be evaluated by studying ecosystems from the species- poor to the species-rich regions of the world. We have given names to fewer than

3% of all tropical organisms, and unveiling their existence should be seen as the first step towards placing an organism into the arena of scientific knowledge. Once that has been accomplished, it will be evident that a great deal of additional information is highly desirable. Thus we need to explore biodiversity in greater depth in order to understand its relationship to ecosystem processes (National Research Council, 1980; di Castri and Younès, 1990). To assess the role of an organism at the ecosystem level and hence the significance of biodiversity in the global ecosystem, we need to understand the unique biological features of organisms, a task that goes well beyond a descriptive approach and requires a strong experimental base. Scientists have tended to build models of ecosystem functioning and have emphasized particular features of ecosystems based on the ecophysiology, reproductive biology, and biological rhythms of a minute proportion of the world's best known and most easily manipulated species, almost exclusively native to the relatively species-poor ecosystems of the temperate zone. Such organisms could not possibly be more distinct from the dominant species in the world's most threatened ecosystems, such as tropical forests and coral reefs.

At a practical level, this narrow approach has led to the employment of inadequate models for exploiting ecosystems from temperate regions into the tropics. The inability of scientists to provide sustainable solutions to the exploitation of different kinds of habitats in various regions of the Earth is a direct reflection of our insufficient biological knowledge of the key organisms in many ecosystems. Studies of the biological properties of individual organisms must be stimulated now, particularly in developing countries, which house so much of the world's biodiversity. Because of the rapid destruction of most of the relevant biological communities, steps to realize these opportunities must be undertaken urgently. The importance of gaining information about the functioning of organisms in ecosystems rapidly and efficiently cannot be overstated, for it is such information that will prove especially valuable for the conservation of biodiversity throughout the world.

Knowledge of the degree to which organisms may be substituted in any ecosystem by others is fundamental for developing the principles of restoration ecology, as the determination of their key biological properties is a prerequisite for using natural ecosystems sustainably. There will be little hope for sustainable development as a means for supporting the human population, and of conserving biodiversity, if we continue to rely primarily on information derived from a handful of the Earth's most conspicuous and easily manipulated organisms. Indeed, there are many organisms that grow in unusual, extreme, or marginal habitats, and whose properties, therefore, might be of extraordinary interest in relation to improving the human prospect. The achievement of these goals will depend in large measure on the availability of well-trained taxonomists and ecologists, especially in developing countries.

It is evident that studies of the functioning of biodiversity in ecosystems must be stimulated especially in developing countries. The knowledge gained should be integrated into a strong theoretical framework concerning biodiversity. The development of such a framework would constitute an important incentive for young scientists to carry out research in the area of biodiversity. An attempt to lay out the basic conditions for the development of the framework has recently been presented by Solbrig (1991b), and this approach should be pursued and expanded. The principles of the emerging discipline of conservation biology should be given special attention, including such matters as the design of natural reserves, the viability of populations with different genetic constitutions, the role of population size *per se*, and many related areas. Systematists, ecologists, and social scientists should be directly and routinely involved in the concept, design, and geographical placement of parks and other protected areas.

In addition to these considerations, we now have greatly improved techniques to deal with the relationships of organisms; both analytical methods (cladistics and vicariance biogeography) and macromolecular comparisons, involving proteins and nucleic acids, have become available during the past few decades. The rapid pace of extinction of both terrestrial organisms generally and some groups of marine organisms constitutes a severe potential handicap to achieving a more complete understanding of the relationships of organisms and the reconstruction of the history of life on Earth. The patterns that are being investigated are of great theoretical interest. For example, they have much to tell us about the constraints on adaptation to changing environmental conditions; thus they constitute a key to how the Earth's organisms, if they are conserved, and even though they may be severely depleted genetically, will be able to cope with global change.

7. ECONOMIC CONSEQUENCES OF SPECIES LOSS

Briefly stated, sustainable productivity is essentially biological productivity. We need to understand the ways in which individual organisms function in order to be able to understand the ways in which communities function. For example, the development of sustainable agriculture in many areas of the tropics, where the soils are often relatively infertile, will depend on the intelligent use of individual kinds of plants and animals individually or in combination. Collectively, organisms make up the ecosystems that build and protect soils, control hydrologic cycles, and regulate the characteristics of the atmosphere, thereby playing a major role in determining climate throughout the world. The effectiveness of our ability to utilize ecosystems on a sustainable basis will play the major role in determining our quality of life now and in the future.

Individually, we use organisms as sources of food, medicines, chemicals, fiber, clothing, structural materials, energy (biomass), and for many other purposes. We also need to understand the harmful properties of organisms, as weeds or invading animal pests; agents destructive to valuable kinds of plants and animals; or causative agents for diseases in humans, domestic animals, and crops. We should learn to understand them better, and to formulate more precise biocontrol mechanisms than are available now.

To illustrate this point further, consider that only about 100 kinds of plants provide the great majority of the world's food (R. and C. Prescott-Allen, 1990). These plants are precious individually, and their genetic diversity should be preserved and enhanced. In addition, however, there are tens of thousands of other plants, especially in the tropics, that have edible parts and might be used more extensively for food, and perhaps brought into cultivation, if we knew them better. Many of these are being used locally, but they have not yet achieved wide use because of cultural barriers, and because very little attention is being paid to their improvement in cultivation to make them more attractive for human consumption. Important tropical cultivated plants, frequently trees and shrubs, are not amenable to the standard methods for cultivating annual and perennial herbs that have been developed in temperate regions, and thus tend to be undervalued, and capital is generally not available for their development. At the same time, the loss of genetic diversity in crops and other plants, and the consequences of such loss to the possibilities for protecting the plants against disease, as well as the value of genetic diversity for developing valuable features in these crops, are well documented. The practical consequences of forest loss for human welfare in Southeast Asia are well discussed by Hurst (1990).

Beyond food, there are many other useful products that plants, which are natural biochemical factories, can provide (Myers, 1983; Oldfield, 1984). Plants are traditional sources of medicine, and still provide the major source of medicine for most people in the world. Most pharmaceuticals, in turn, have either been derived from plants, fungi, or bacteria, or are based on molecules first observed in these organisms. To give just a few examples, vinblastine and vincristine, derived from the rosy periwinkle (*Catharanthus roseus*), a native of Madagascar, are effective against certain forms of childhood leukaemia, raising the chances of survival past the age of 5 from about 10% to about 90%; they reached the market in 1971. The alkaloid artemisine, derived from an annual wormwood

(*Artemisia annua*), is now being investigated actively as an alternative to quinine, the usual prophylactic against the *Plasmodium* organisms (now often resistant to quinine) that cause malaria, a disease that afflicts about 250 million people at any one time. The two molecules are entirely different in structure, and neither the existence of artemisin nor its effectiveness against malaria could have been predicted from the existence of quinine, except for the fact that the Chinese people, who employ between 5100 and 5600 of the estimated 30,000 species of plants in their country for medicinal purposes, have been using extracts of the wormwood against malaria for more than 2000 years.

Given the possibility of genetic engineering, which provides the opportunity of transferring genes from one organism to another and thus enhancing the qualities of the recipient, the potential usefulness of individual organisms – even though they may have no economic value themselves – is obvious. Screening methods still need to be developed to determine the genetic basis of most features of interest, but there are no theoretical difficulties. In view of these relationships, we shall need to begin to think of individual kinds of organisms not only in terms of their direct potential usefulness, but in relation to their potential as sources of tens of thousands of genes that might be transferred to other organisms to enhance their valuable features. Unfortunately, trained personnel who can screen organisms for useful features easily, or develop organisms with new sets of characteristics by transferring genes, are very poorly represented in most developing countries. Although the majority of the world's species of organisms occur in these countries, their human inhabitants will be unable to exploit the economic value of these organisms unless steps are taken to strengthen the individual and institutional capacities of these nations. The application of biotechnology to the marine sciences generally, and marine productivity in particular, has been outlined very well by Colwell (1983).

8. A PLAN OF ACTION

How can we learn about the world's organisms, save as many as possible of them, and thereby improve our prospects for the future? In the most general terms, biodiversity will be preserved best in a stable world, characterized by sustainable productivity. Actions to conserve biodiversity, therefore, must be organized around the achievement of this goal, and will become more apparently complex and multidimensional as global change continues (Soulé, 1991). Our concern here is in part with that strategy, but especially with the scientific underpinning of the overall effort.

An excellent and comprehensive proposal for preserving global biodiversity, involving many of the features that have been stressed here, has been developed by the World Resources Institute (WRI), together with the International Union of the Conservation of Nature (IUCN) and United Nations Environment Programme (UNEP), (WRI, IUCN, and UNEP, 1992). This document comprises a set of guidelines for action to save, know, and sustainably and equitably use Earth's biodiversity. It is a blueprint for action, dependent on governments and many public and private agencies for success. The role of non-governmental agencies can be particularly important in affecting public opinion throughout the world. In the following remarks, we have emphasized some of the elements that we believe to be of the greatest importance in relation to the role of science in the preservation of biodiversity on a global scale.

At the onset, it must be stressed that since the conservation of biodiversity depends on the direct perception of the governments and people of individual countries of its value, strategies to save biodiversity globally must be based on understanding and commitment from the participating nations. This necessarily implies that scientific efforts of the same level of sophistication as in industrialized nations must be encouraged and allowed to prosper in the developing countries. Specific actions designed to conserve the world's remaining biodiversity must also be consistent with the present global dynamics and in particular with the developing world's strong economic dependence on natural resources, which is unlikely to change during the next century at least. In addition, the effective preservation of marine biodiversity (Thorne-Miller and Catena, 1991), and, to a large extent, that of all biodiversity will require much stronger international collaboration than has been undertaken yet.

Given this level of participation by developing countries in the process, the first prerequisite of a sustainable world is the attainment of a stable human population. Considering the unstable and destructive way in which the global ecosystem has been affected by a population of 5.4 billion people, it is a sobering thought that two to three times the present number of human beings may exist when population stability is finally achieved. Also of fundamental importance to the attainment of a global stability is the problem of poverty and social justice, which must be addressed effectively throughout the world. To the extent that the twin problems of population and poverty in the developing world and overconsumption in industrialized nations can be addressed adequately, opportunities will become available, at least locally, for managing biodiversity sustainably.

Questions of the ownership of biodiversity and the rewards to be gained from its commercialization must be considered carefully, in the context of the development of adequate institutional and human resources in tropical countries. The relevant strategies are the subject of a proposed Convention on the Conservation of Biological

Diversity, being developed under the auspices of UNEP; negotiations on that convention are underway at the present time.

First, the framework of national and international economic policies and legislation should be modified in such a way as to foster the sustainable use of biological resources and the conservation of biodiversity. The policy and legal frameworks established by governments create the incentives and obstacles that influence decisions about how to utilize and manage biological resources. Nationally, policies covering areas ranging from natural resource exploitation to incentives for technological innovation should be modified with this objective in mind. Internationally, the crushing debt burden on many developing countries absorbs a large portion of available public resources and makes producing cash commodities to generate foreign exchange irresistible. Development assistance transfers often contribute to projects that speed up the rate of biological extinction. Many transnational investment practices also drain resources from developing countries, and do nothing to help their host communities and countries develop their own technological, professional, and institutional capacities significantly.

Given that background, the conservation of biodiversity would be served best by the establishment and maintenance of a world-wide system of protected areas, perhaps based, at least in part, on the UNESCO Biosphere Reserve and Man and the Biosphere (MAB) Biosphere Reserve Programmes. The sites to be preserved should be selected systematically, in part by rapid surveys of endemism and species diversity across as many groups of organisms as feasible, so that they will include the greatest possible proportion of the existing global biodiversity. They must be managed in a regional context, taking into account modified and partly natural ecosystems and human interactions of all kinds, since it will clearly not be possible to protect all of the world's biodiversity by preserving samples of pristine ecosystems in their original condition permanently. Global change will also have an important effect on the sustainability of reserves in the future.

Concerning marine biodiversity, there is an urgent need to identify reference sites for biodiversity. Among others, coral reefs, mangroves, lagoons and estuaries, salt marshes, beds of algae and sea grasses, and rocky shores should be considered as priority areas in which to monitor marine biodiversity and to understand better how the pattern of environmental variation in the sea translated into functional priorities and the maintenance of various levels of biodiversity. An important way to help implement this priority is to utilize the existing network of marine laboratories world-wide (Grassle *et al.*, 1991).

The industrialized world, currently comprising of 23% of the global population but about 85% of the world's financial resources must provide the bulk of the resources for the maintenance of these networks of protected areas that are established. The Global Environment Facility, a $1.5 billion fund, was established by the World Bank, UNEP, and UNDP in late 1990 as a pilot project to address global environmental problems. The Facility makes direct grants to countries whose *per capita* annual income is less than $4000. It offers a model of how the funding for such programs might be organized, although the participation of developing countries in the decisions of the Facility should be encouraged strongly. The amount of money involved is still minute, however, in relation to the estimated $100 billion per year lost to developing-world farmers by export restrictions; but the effort is potentially worthwhile, and appears to point in the right direction.

Secondly, mechanisms must be established for the *ex situ* preservation of samples of selected organisms. Plants, for example, with 250,000 species, are a likely target for such efforts. The botanical gardens that are members of Botanic Gardens Conservation International should be encouraged to form an operational network to conserve plants throughout the world, and a world-wide network of seed banks should also be formed. For either of these networks to be effective, adequate funding must be secured – and it would need to come primarily from industrialized countries – to make possible the collection of genetically adequate samples of living plants and seeds and to ensure their perpetuation. Special attention should be paid to plants of actual or potential economic importance, and others should be selected on the basis of their scientific interest and unique characteristics. The Center for Plant Conservation, which operates as a network of botanical gardens in the United States for such purposes, maintaining a national collection of threatened and endangered plants, is the best-developed organization of this kind currently in operation. Such a model could, with adequate funding, be utilized on a world-wide basis. Clearly, the ultimate goal of preserving species is to use them in the restoration of ecosystems, an emerging subject that will grow in importance as the human population continues to devastate the global ecosystem but at the same time approaches stability.

Beyond plants, other groups of organisms, such as bacteria and fungi, lend themselves to preservation in culture centers, provided that funding is made available. The World Federation for Culture Collections has over the past 20 years established in the form of its World Data Center a computerized and constantly updated catalogue of the microbial strains held in over 400 culture collections in 65 countries. It also fosters quality control and the development of new preservation methods. One of its members, the American Type Culture Collection, provides a well-developed example of a repository for microorganisms, with more than 42,000 strains preserved in its collections. The preservation of animals, as in zoos and similar facilities, is expensive but will continue to receive emphasis according to the values that particular groups of people emphasize, and the availability of funds.

For the establishment and maintenance of the protected areas and for the implementation of effective *ex situ* strategies as well, the full participation of developing countries is necessary. Such countries contain at least 80% of global biodiversity. Their relevant infrastructure must, therefore, be strengthened as rapidly as possible. Such strengthening should include, among other factors, funds to acquire adequate library resources, the encouragement of direct collaboration between scientists in adjacent countries that share particular biomes, the provision of adequate computer facilities, and access to inexpensive and rapid communication.

The value of biological resources and the nature of the economic strategies that would need to be applied for their preservation have been well documented (for review see McNeely *et al.*, 1990). In contrast, little serious effort has been made to improve the prevalent conditions of limited scientific manpower, constricted institutional development, and inadequate access to biotechnology that exist in most developing countries; these factors have greatly limited their ability to use their own biodiversity. Abdus Salam (1991) and his colleagues at the Third World Academy of Sciences have properly stressed the necessity for radical and speedy improvement in these conditions of the large and rapidly growing majority of the world's people who live in developing countries are to enjoy the fruits of development and obtain access to morally just and fruitful lives. It is only through the development of strong scientific communities and well-equipped laboratories and other facilities in the developing countries themselves that the people who live in these countries will come to appreciate the importance of the proper management, sustainable use, and conservation of biodiversity, and to carry out the necessary steps to accomplish these objectives.

Developing countries must be given the chance to explore the multiple paths, consistent with their own social values, for making biodiversity an indispensable ingredient of socioeconomic, cultural, and scientific development (e.g. Colwell, 1986). As Francisco Sagasti (1990) has put it, "Our aim must be to enable individuals and social groups to choose what is best for them, and to assist them in acquiring the capacity to decide and bring about their own destinies with full knowledge of the possibilities, limitations and consequences of their actions for present and future generations. . . [The] generation, dissemination and use of knowledge will play an even more critical role in whatever we may call development, progress or empowerment in the years to come. . . The presence or absence of this capacity constitutes the crucial distinction between developed or developing nations; between those parts of the world in which individuals have the potential

to decide and act with autonomy, and those in which people are not yet empowered to realise their potential."

The development of national schemes for the management of biodiversity, including its preservation, appears to be one of the best strategies for the strengthening of the relevant capabilities in developing countries. A good example is provided by the Instituto Nacional de Biodiversidad (INBio) in Costa Rica. Bringing together conceptually the nation's scientists, biological collections, libraries, and other sources of information, INBio is exploring the biodiversity of Costa Rica comprehensively, and taking the first steps in putting this knowledge to work for the benefit of Costa Ricans. The information about biodiversity is organized in a major relational database, which can be used to identify gaps in the available information. Since biodiversity is, in effect, a nation's stock in trade – the ultimate basis of its productivity, it is clearly in the national interest that it be managed properly. National databases can be used to identify individual organisms for commercial use; groups of organisms for the construction of artificial communities to promote local ecological stability; as sources of information to assist in making decisions about appropriate, sustainable land use; for education about and appreciation of organisms; for scientific studies; and for conservation purposes, including both the management of reserves and their integration into their regions, and the ex situ preservation of individual organisms of interest, including the formation of botanical gardens and zoos, seed banks, and stock culture centers for microorganisms. Chemical prospecting can be carried out at such centers in such a way as to strengthen the country's abilities in this area, and to help it realize the economic potential of its own biodiversity (Eisner, 1990). Organizations like INBio, which might be called National Biological Resource Centers, would appropriately differ widely in administrative detail from country to country, but are of great potential significance in helping to solve the problems enumerated above.

Research on the biological properties of selected species in ecosystems under strong pressure as a result of economic development should be a major objective of any national biodiversity institute. The inevitable exploitation of native forests, of savanna ecosystems, and of other biomes and aquatic systems, must be made sustainable. This objective will be achieved only to the extent that we have access to information on the critical ecophysiological properties of individual species, their biological rhythms, the interdependence with other organisms, the conditions and requirements for their reproduction, migration and dispersal patterns, and other key factors. Knowing the number and kinds of organisms present, their names, distributions, habitats, and features, provides a necessary first step towards identifying the components and probable level of complexity of an endangered ecosystem and an indication of how it can be managed properly for sustainable development, but an understanding of the functioning of the ecosystems in which these species occur is ultimately necessary if they are to be managed well (Solbrig, 1991b).

If a government feels that it has an appropriate understanding of, and way of dealing with, its own plants, animals, fungi, and microorganisms, then that country will be able to use and conserve those organisms for its own benefit, and to participate willingly in international agreements concerning them. Without such a structure, many nations will continue to feel that their biodiversity is a matter of concern largely for wealthy foreign nations, and that they lack the means to use it properly. It is for these reasons that national biological surveys, as management strategies, have great importance. If nations are able to use their own biodiversity for their own purposes, they will not consider themselves victims of the industrialized world, but rather participants in a common effort to manage global resources efficiently and sustainably.

An enhanced appreciation of the properties of species has the potential of allowing hundreds of additional tropical species to be developed at a commercial level. Achieving this goal would ultimately involve the stimulation of relevant industries, as, for example, by the provision of tax incentives, to return to the use of new sources of natural products to a substantial degree. Such developments should take into account the benefits to the owners of the forests from which the products are being derived – a strategy that would lead directly to the preservation of those forests. The background against which such developments would have to take place, however, includes an annual outlay of $200 billion by the developed countries of North America and Western Europe to protect domestic agriculture against foreign products, rather than to encourage their production.

For improving the properties of species that are important to people now, and to take advantage of additional species that are not utilized at present, all nations should have access to the relevant biotechnology. The intelligent use of biotechnology, which helps to make possible the incorporation of biodiversity into everyday living and thus the stimulation of economic growth, leads directly to reduced pressures on natural ecosystems. For these reasons, young scientists in the developing world should be encouraged to master the principles of biotechnology and to apply them to indigenous organisms.

At the same time, traditional knowledge concerning resource utilization by native and other rural people in developing countries provides many important indications about the ways in which biodiversity can be understood, managed, and preserved. For example, an understanding of the kinds of high-diversity "backyard orchards" that are well developed in many parts of the tropics, as in southern Mexico and southeast Asia, could be very important foundations for building sustainable agroforestry systems in their regions and elsewhere. They are attractive means of preserving and improving many economically important species. It is certainly highly desirable and extremely urgent to preserve and extend the ethnobiological knowledge of tropical areas very broadly, while such knowledge, which is often highly endangered, still exists.

The implementation of educational programs to stress the intrinsic value of biodiversity to society should be a major objective in all countries, and especially in those where biodiversity is under the greatest threat. Innovative techniques should be developed, with a strong emphasis on hands-on efforts. Educational and public facilities in developing countries suffer greatly from a general lack of resources, expressed in such terms as inadequate infrastructure, poor maintenance of facilities, and a lack of access to appropriate technology. The theme of the importance of biodiversity can be addressed very effectively through the development of local botanical and zoological collections in national educational institutions, the development of educational materials based on locally familiar organisms, and similar strategies.

The actions needed will ultimately be taken by individuals, villages, and nations, yet these actions can be catalyzed and supported by the international community. The Global Biodiversity Strategy will recommend the establishment of an international panel of scientists, community leaders, and representatives of governments and non-governmental organizations to help mobilize and orient the international response. This panel would document the magnitude of the problem, the impacts on society, and provide guidance in setting priorities at the international level. Many of its functions would ultimately be assumed by the Biodiversity Convention when it is ratified, an essential step for progress in the maintance of biodiversity. Such a panel could play a key role in stimulating the kind of response needed to deal with a threat of this magnitude.

In conclusion, we stress the importance of strengthening the scientific infrastructure in developing countries; so that we can collectively sample the biodiversity that exists now and work out the best possible strategies for managing it properly. The next few decades, with their projected doubling of a human population that already consumes, diverts, or wastes an estimated 40% of total net photosynthetic productivity on land, can only be a time of catastrophic extinction. In many ways, we have an opportunity now comparable to that of living in the final decades of the Cretaceous period, 65 million years ago; the opportunities that exist to sample the full range of biodiversity with which we coexist will never occur again. The importance of understanding biodiversity and of managing it properly for human benefit and of conserving it cannot be overstated.

9. RECOMMENDATIONS

Summarizing the preceding discussion, we recommend the following actions as most significant

1. Draw to the attention of national and international leaders the overwhelming importance of the extinction of biodiversity, so as to bring about its acceptance as the aspect of global change that has the greatest significance for human prospects in the future and is moving the most rapidly. The loss of species is completely irreversible, unlike most other aspects of global change.

2. *Pursue the following research priorities*
   The principal aim of the research effort is to reach a greater understanding of the dimensions of biodiversity, to quantify it in terms of its genetic and functional uniqueness, and to develop a generalized theory of biodiversity as a background for the conservation of organisms, sustainable development, restoration ecology, and the promotion of global stability. For this:

   (a) Much more research is needed into the function of biodiversity, from genes to ecosystems, along the lines of the IUBS/SCOPE/UNESCO Programme of Research on Biodiversity (Solbrig, 1991a, b; di Castri, 1991; di Castri and Younès, 1990). This includes the role that biodiversity plays in ecosystems, the importance of genetic biodiversity, and the role of species diversity in communities.

   (b) Efforts need to be undertaken to improve our knowledge of the diversity of organisms, recognizing that a complete inventory of all organisms is not possible within a reasonable time. Special priority should be given to well-known groups, such as plants and vertebrate animals, and to gaining some appreciation of poorly known ones, such as fungi, bacteria, mites, and nematodes, as well as to freshwater ecosystems, which are highly endangered, and to the coastal seas. Checklists, which may be organized on an international basis, often form an especially useful first approach to the biodiversity of an individual group of organisms or of a region.

   (c) Conduct systematic investigations of the economic value of organisms, and of the ways in which sustainable systems can be based on these properties. Studies of the ways in which economic theory can be modified to take into account the consumption of often-irreplaceable natural resources that are of key economic and ecological importance should also be undertaken.

   (d) Strengthen, on a global basis, research into the scientific basis of conservation, including both *in situ* and *ex situ* strategies and especially restoration ecology.

3. *Institutional arrangements*
   We strongly support the establishment of National Biological Resource Centers (National Biodiversity Institutes) to assess the species richness of individual nations and their economic value, and to manage it for the common benefit, including conserving it. This is a major strategy that we believe will contribute substantially to the development of a stable Earth. Local and regional needs and characteristics must be taken into account in developing national and international policies for dealing with biodiversity.

4. *Traditional uses of biodiversity*
   We recommend the acquisition of as wide as possible an array of information about the traditional use of ecosystems and of biodiversity. This knowledge should be valued so that the principles developed over millenia can be appreciated and applied to the development of regional sustainability.

5. *Training aspects*
   We recommend training an adequate number of geneticists, taxonomists, and ecologists specialized in the study of biodiversity especially in countries where there is a shortage of scientists and a high level of biodiversity, such as tropical areas. The scarcity of taxonomists severely limits the study of all aspects of biodiversity throughout the world. In addition, specialists should be trained to deal with the taxonomy of groups of organisms for which few qualified people exist now.

6. *Conservation*
   Funds are required for the elaboration and continuing support of a world-wide network of protected areas, selected to protect as much as possible of global biodiversity, and of networks of botanical gardens, seed banks, stock culture centers, and similar organizations. These funds will need to come primarily from the wealthy, industrialized nations and flow to the poorer parts of the world; at least 80% of global biodiversity occurs in developing countries.

ACKNOWLEDGMENTS

We are particularly grateful to Otto T. Solbrig, Harvard

University, for his outstanding comments on this chapter, and to Rita R. Colwell, University of Maryland; Penelope L. Firth, US National Science Foundation; Vernon Heywood, Botanic Gardens Conservation International; William M. Lewis, Jr., University of Colorado at Boulder; Thomas Lovejoy, Smithsonian Institution; Norman Myers, Oxford; Gordon Orians, University of Washington; Timothy O'Riordan, University of East Anglia; Walter V. Reid, World Resources Institute; Patricia Werner, National Science Foundation; George M. Woodwell, The Woods Hole Research Center, for their useful suggestions and for the provision of pertinent information.

REFERENCES

Allen, J.D., Flecker, A.S. and Smith, G.R. 1992. Biodiversity conservation in running waters. (Manuscript).

Arroyo, M.T.K., Armesto, J., Squeo, F., and Gutiérrez, J., 1992. Global change: The flora and vegetation of Chile. In Mooney, H., Fuentes, E., Kronberg, B., and Fuenzalida, F., *Northern and Southern Hemisphere Responses to Global Change* (in press), Academic Press, New York.

Barbier, E.B., J.C. Burgess, and Markandya, A. 1991. The economics of tropical deforestation. *Ambio*, **20** (2), 55-58.

Brown, L.R., *et al.* (9 additional authors). 1990. *State of the World 1990*, W.W. Norton and Company, New York.

Colwell, R.R. 1983. Biotechnology in the marine sciences. *Science*, **222**, 19-24.

Colwell, R.R. 1986. The potential of biotechnology for developed and developing countries. *MIRCEN Journal*, **2**, 5-17.

di Castri, F. 1991. Ecosystem evolution and global change. In: O.T. Solbrig and G. Nicolis (eds.), *Perspectives in Biological Complexity*, pp. 189-218. IUBS, Paris.

di Castri, F. and Younès, T. 1990. Ecosystem function of biological diversity. *Biology International*, Special Issue 22. IUBS, Paris.

Dodson, C.H. and Gentry, A.H. 1991 Biological extinction in western Ecuador. *Annals of the Missouri Botanical Garden*, **78**, 273-295.

Ehrlich, P.R. and Wilson, E.O. 1991. Biodiversity studies: Science and policy. *Science*, **253**, 758-762.

Eisner, T. 1990. Prospecting for nature's chemical riches. *Issues in Science and Technology*, **6**(2), 31-34.

Fearnside, P.M. 1990. The rate and extent of deforestation in Brazilian Amazonia. *Environmental Conservation*, **17**, 213-226.

Gaston, K.J. 1991. The magnitude of global insect species richness. *Conservation Biology*, **5**, 283-296.

GESAMP. 1990. *The State of the Marine Environment, UNEP Regional Seas Reports and Studies* No. 115. United Nations Environment Programme, Nairobi.

Grassle, J.F., Lasserre, P., McIntyre, A.D. and Ray, G.C. 1991. Marine biodiversity and ecosystem function: A proposal for an international program of research. *Biology International*, Special Issue 23. IUBS, Paris.

Hawksworth, D.L. 1991. The fungal dimension of biodiversity: Magnitude, significance, and conservation. *Mycological Research*, **95**, 641-655.

Hurst, P. 1990. *Rainforest Politics. Ecological Destruction in South-East Asia*, Zed Books Ltd., London.

Jaenike, J. 1991. Mass extinction of European fungi. *Trends in Ecology and Evolution*, **6**, 174-175.

Lowe-McConnell, R.H. 1987. *Ecological Studies in Tropical Fish Communities*, Cambridge University Press, Cambridge.

MacArthur, R.M., and Wilson, E.O. 1967. *The Theory of Island Biogeography. Monographs in Population Biology*. Princeton University Press, Princeton, New Jersey.

McNeely, J.A. 1988. *Economics and Biological Diversity: Developing and Using Economic Incentives to Conserve Biological Resources*, International Union for the Conservation of Nature and Natural Resources, Gland, Switzerland.

McNeely, J.A., Miller, K.R., Reid, W.V., Mittermeier, R.A. and Werner, T.B. 1990. *Conserving the World's Biological Diversity*, International Union for the Conservation of Nature and Natural Resources, World Resources Institute, Conservation International, World Wildlife Fund-US, and the World Bank, Gland, Baltimore, and Philadelphia.

May, R.M. 1988. How many species are there on Earth? *Science*, **241**, 1441-1449.

Myers, N. 1979. The Sinking Ark. *A New Look at the Problem of Disappearing Species*. Pergamon Press, Oxford.

Myers, N. 1983. *A Wealth of Wild Species: Storehouse for Human Welfare*. Westview Press, Boulder, Colorado.

Myers, N. 1988. Tropical forests and their species. Going, going...? In: *Biodiversity* (ed.) E.O. Wilson, pp.28-35, National Academy Press, Washington, D.C.

Myers, N. 1989. *Deforestation Rates in Tropical Forests and Their Climatic Implications*, Friends of the Earth, London.

Myers, N. 1990. The biodiversity challenge: expanded hot-spots analysis. *The Environmentalist*, **10**(4), 1-14.

National Research Council. 1980. *Research Priorities in Tropical Biology*, National Academy of Sciences, Washington, DC.

Oldfield, M.L. 1984. *The Values of Conserving Genetic Resources*. US Department of the Interior, National Park Service, Washington, DC.

OTA (US Congress, Office of Technology Assesment). 1987. *Technologies to Maintain Biological Diversity*. US Government Printing Office, Washington, DC.

Prescott-Allen, R., and Prescott-Allen, C. 1990. How many plants feed the world? *Conservation Biology*, **4**, 365-374.

Raven, P.H. 1987. Biological resources and global stability. In: *Evolution and Coadaptation in Biotic Communities*. Kawano, S., Connell, J.H. and Hideaka, T., (ed.) pp. 3-27. University of Tokyo Press.

Ray, G. C. 1988. Ecological diversity in coastal zones and oceans. In: *Biodiversity* E.O. Wilson, (ed.). pp. 36-50. National Academy Press, Washington, DC.

Reid, W. V. 1992. How many species will there be? In *Tropical Deforestation and Extinction of Species*. Sayer, J. and Whitmore, T. (ed.) in press. IUCN Tropical Forests Programme Publication Series. IUCN, Gland, Switzerland.

Sagasti, F. 1990. Co-operation in a fractured global order. *New Scientist*, **127**, 18.

Salam, A. 1991. *Science, Technology and Science Education in the Development of the South*. The Third World Academy of Sciences, Trieste, Italy.

Simberloff, D. 1986. Are we on the verge of a mass extinction in tropical rain forests? In *Dynamics of Extinction*. Elliott, D.K., (ed.) pp. 165-180. John Wiley and Sons, Inc., New York.

Solbrig, O.T. 1991a. Biodiversity: Specific issues and collaborative research proposals, *MAB Digest*, **9**, 1-77. UNESCO, Paris.

Solbrig, O.T. 1991b. *From Genes to Ecosystems: A Research Agenda for Biodiversity*. Report of a IUBS-SCOPE- UNESCO Workshop. International Union of Biological Sciences, Paris.

Soulé, M.E. 1991. Conservation: Tactics for a constant crisis. *Science*, **253**, 744-750.

Thorne-Miller, B.L. and Catena, J. 1991. *The Living Ocean. Understanding and Protecting Marine Biodiversity*. The Oceanic Society of Friends of the Earth, Island Press, Washington, DC.

Vitousek, P.M., Ehrlich, P.R., Ehrlich, A.H. and Matson, P.M. 1986. Human appropriation of the products of photosynthesis. *BioScience*, **36**, 368-373.

Wilson, E.O., ed. 1988a. *Biodiversity*. National Academy Press, Washington, DC.

Wilson, E.O. 1988b. The current state of biological diversity. In *Biodiversity* E.O. Wilson, (ed.) pp. 3-18. National Academy Press, Washington, DC.

Woodwell, G.M., ed. 1990. *The Earth in Transition. Patterns and Processes of Biotic Impoverishment*. Cambridge University Press, Cambridge.

World Resources Institute. 1990. *World Resources 1990-91*. Oxford University Press, Oxford.

World Resources Institute, The World Conservation Union, United Nations Environment Programme (WRI, IUCN, and UNEP). 1992. Global Biodiversity Strategy. WRI/IUCN/ UNEP.

III. Responses and Strategies

Introduction

T. O'Riordan

T. O'Riordan

Setting the Scene

The historian Arnold Toynbee, introducing his monumental work *Mankind and Mother Earth* (Toynbee, 1976), contemplated the great enigmas about humanity. Humanity is both good and evil; it destroys but can recreate and restore; humans have rational minds and emotional souls. Alone amongst all living things, humans have a conscience, so are capable of reflective judgment. Throughout all history, humanity has been aware of its capability to render the world uninhabitable. Knowingly to pursue such an outcome would not only be folly; it would be supremely wicked.

The clash between the material and the spiritual qualities of humanity has never been so profound, yet apparently so insolvable, as it is today. As Toynbee ruefully concludes

> It looks as though man will not be able to save himself from the nemesis of his demonic material power and greed unless he allows himself to undertake a change of heart that will move him to abandon his present objective and espouse the contrary ideal (Toynbee, 1973, p. 20).

That ideal is the realization that the gift of life carries with it a responsibility to sustain life on Earth in all its diversity and majesty. That ideal is the recognition that peaceful coexistence is an essential counterpoint to successful competition and spiritual advancement. That ideal is the commitment to a planetary future that promotes the cause of fairness and respect for all the forces that amount to creation and evolution. History and futurity combine into spaceless timelessness. The concepts of sacrifice and self-interest become enveloped in principles of continuance and companionship. This is the essence of sustainable development.

The role of science

How can science contribute towards the attainment of Toynbee's ideal?

Previous sections of this book have pointed to the need of bringing science to bear on the problems of environment and development and on acquiring understanding of the Earth System. In this Section, which deals with the development of responses and strategies, answers are sought to questions such as:

(a) To what extent can science in its broadest sense provide indices of sustainable and non-sustainable development in the quality of life for all people alive and soon to be born?

(b) What is the capability for science and for scientists to co-operate in national, regional and global audits of environment and development that will become the litmus paper for assessing progress of future patterns of development?

(c) What changes need to be put in place to ensure that those future patterns of development will be environmentally tolerable, economically efficient and socially acceptable? This question applies to the transfer of technology, to the application of appropriate prices and quality of life indicators, and to the laws and conventions that regulate individual and collective action.

For dealing effectively with such questions evidently intensive co-operation is required between scientists from natural social, engineering and health sciences and science is already evolving in this direction for which global change is forcing the pace. Similarly much more interaction between the body politic and scientists is required and this can only be successful if this includes the public at large: if sustainable development is to be achieved, people will have to reassess their role and purpose on this globe, what are their priorities, and how they must relate to those already in distress and the countless millions whose plight is bound to get worse unless radical changes in economics, government and international relations take place. These awesome challenges cannot be undertaken by science acting in isolation. People of every nationality and social status will be drawn into the process of renegotiating the character of sustainable growth and the role of science within it. It is only through the understanding, the consent and the commitment of the people that the necessary adjustments in human purpose can occur through peaceful and democratic means. The alternative is much misery, even more oppression and a limitless period of environmental conscription where all humanity will be chained to codes of practice compelling them to save the Earth. To avoid such a dismal future science must play its part. It must be both liberating and listening, flexible and adaptable, humble yet visionary.

Thus besides interdisciplinarity, partnerships will be required between scientists and representatives of the people (not just politicians, but a host of non-governmental interests) so as to prepare options and provide meaningful images of likely future states of sustainability, well-being and vulnerability under various pathways of transition from here to there. Clearly this places a whole new emphasis on the relationship of science in its broadest sense both to the policy-making process as well as to the public at large.

The chapters on response strategies

The introductory remarks made above have an obvious and pervasive significance for the chapters that follow.

This is particularly clear for Chapter 12 on Quality of Life for which Table 1 shows how many disciplines have to be involved and how closely interaction is necessary with the population concerned within their socio-economic–cultural setting. It was concluded that an important task for science was to develop meaningful indicators of the changing state of the quality of life for all peoples. Any satisfactory measure must not only address basic human needs: it should also encompass measures of civil rights, the protection of rights of nature, and the legitimate interests of those yet to be born, all enshrined in national environmental charters. The working group made a special plea for research on quality of life in the urban environment in view of the rapid increase of the proportion of people living in cities. Furthermore, attention was drawn to the effect of migrations on the quality of life of those concerned.

The working group on Public Awareness (Chapter 13) concluded that the interaction between science and the public leaves much to be desired and that major efforts be undertaken to remedy this, i.e. through action by ICSU and the ESF, further research to understand the present deficiencies in interactions, and formation of partnerships with the media. Special attention should be given to the diversity of cultures, traditions and aspirations in generating a common new world ethic.

Chapter 14 on Capacity Building and the working group discussion concerned pointed out that especially in the developing countries where a critical mass of scientists and

Table 1

Common components in Quality of Life definitions

1. BASIC NEEDS	4. COMMUNITY CHARACTERISTICS
– human survival	– shared value system and
– health	norms of behavior
– warmth-energy	– social support network
– shelter	– participation in group
– water	decision making
– nutrition	and action
– personal security	
2. BASIC SOCIAL SERVICES	**5. ENVIRONMENTAL QUALITY**
– education	– sustainable natural resource
– medicine	use
	– pollution and degradation
	– means for recreation
	– survival of other species
	– vicarious satisfaction
3. ECONOMIC ASPECTS	**6. TIME AND SPACE DIMENSIONS**
	Time
– means of livelihood	– positive view of future
– employment	– knowledge of individual and
– income	collective past history
– material wealth	*Space*
	– identify with home and
	territory
	– freedom of movement

science-educators is lacking, not only capacity building for economic development in the traditional sense is required, but also needed are special strategies for sustainable development. Such strategies for building up new scientific infrastructure which might provide an opportunity for "leap frogging" include societal interactivity from the beginning.

Chapter 15 on Technology Policy emphasizes the need for trans-national corporations to act in a societally co-operative manner and be sensitive to the needs of recipient countries when introducing new technologies. This holds true for governments and other development agencies. Indeed there is even a case for some sort of performance bond to be set aside in advance of technology transfer to ensure that victims are adequately compensated if there are adverse outcomes as a result of technological interchange. The working group discussion emphasized that not only transfer and use of technology is important but also the generation of new technologies for which each country should develop a capacity in harmony with its own sustainable development.

The final Chapter 16 dealing with Institutional Arrangements stresses the need for incorporating the natural system in economic accounts. The working group underlined the necessity expressed already in the paper that "a strong support for interdisciplinary scientific work incorporating the natural and social sciences" was needed, reviewed the capacity of existing institutions to carry out performance reviews and also considered the creation of various new arrangements, such as an environmental ombudsman, a proper scientific forum at the highest levels of international government for NGOs and some sort of collective environmental security arrangement to ensure that defaulters to international agreements behave responsibly. In addition it was considered vital to develop economic accounting patterns, possibly via carefully conducted experiments, to try to separate development from environmental and social degradation. The close link between environmental destruction and social and economic advancement is not inevitable, but to date we have developed no proper accounting and governing measures to ensure that the two are not forever interlinked. One outcome of UNCED may be carefully monitored experimental projects in various countries to see how far this can be achieved. It was further agreed that new legal research was needed in close linkage to the need for adjustment of economics to reflect the "right price".

REFERENCE

Toynbee, A. 1976. *Mankind and Mother Earth*. Oxford University Press, Oxford.

Chapter 12: Quality of Life

G. Gallopin and S. Öberg

EXECUTIVE SUMMARY

The ultimate goal of development and of our concern for the environment is to increase the quality of life (QOL). The performance of all global systems, human as well as environmental, should thus be monitored direct or indirect by QOL-indicators.

Quality of life is by definition a subjective concept dependent on cultural perspectives and values. QOL depends on factors essential for a good life like adequate shelter, security of the person, assured access to food, water, and medical care. There are many attempts to measure QOL and a large literature discussing the concept. Some sophisticated measures are developed to address culture specific conditions. However, cross-country comparisons have to use very simple measures, like the Gross National Product (GNP) or the Human Development Index (HDI). The latter provides a more reliable way of measuring QOL. Science can help by objectively defining and measuring these factors. Science can also clarify perspectives on less tractable aspects of Quality of Life, identify further areas of consensus, and devise ways to measure them.

Measures of QOL should avoid ethnographic attitudes but moral judgments could be made on lifestyles that support or limit the quality of life of others. Improving the QOL for one group could degrade the QOL for another and as a consequence different lifestyles emerge and change. Furthermore there is a need to link lifestyles to knowledge on how they are more or less dependent on the use of material and energy.

This chapter is intended to link some of these concepts to problems dealt with during and after the United Nations Conference on Environment and Development (UNCED) in Rio de Janeiro in 1992. The basic question we address is what researchers can do to contribute to the improvement of the quality of life. As a background we discuss present knowledge on both a conceptual and a practical level. What do we mean by quality of life? How do we measure it?

1. BACKGROUND

One topic of the Conference in Rio is the environment. Both social and physical environment are of importance but it is often the physical processes which are focussed upon. Different cultures have different depths of attachment to their natural surroundings, but for most people a good Quality of Life (QOL) is inconceivable without access to a productive and healthy environment. Regardless of cultural values, QOL suffers in an absolute sense when the environment is seriously degraded, as when heavy air pollution causes health problems. The need to rethink old attitudes towards the environment, in which nature and resources were sacrificed in the pursuit of economic growth and material wealth, lies at the heart of the Conference.

It is also important to broaden the definition of environment to include the man-made environment, both social and physical, in the discussions. So far natural sciences have been dominating the environmental debate. Since every second person on the globe lives in a city, engineers, medical experts and social scientists must be more active in creating a livable environment for daily life in cities.

The other Conference topic is economic development, which, rightly or wrongly, is generally regarded as the key to a better QOL. This is especially true for the less developed countries (LDCs) where the goal of improving QOL is inseparable from the goal of increasing material wealth. However, the relation between economic growth and human development is not simple, either in less or more developed countries. This is where a discussion of the concept "Quality of Life" (QOL) becomes important in the UNCED process.

There is a fundamental difference between a "hard" measure like economic growth which to a large extent is oriented towards material consumption, and a "soft" measure of a good life, not included in economic growth, such as leisure time, a sense of security, social participation and other qualities that are values in any society. It is a cliché, but also a true statement, that a good life for people, in all countries, today and tomorrow, is the ultimate goal of the Conference. An improvement in the QOL of the global population includes economic development but with two main restrictions: first, that economic growth, which means increased production and consumption, on the global scale is made sustainable in the long run; second, that global wealth be distributed more evenly.

1.1 Social and economic inequality

Inequality between individuals in different positions, social strata or social groups, is characteristic of all nations. Individuals and groups enjoy differential access to socially valuable rewards of wealth, power and prestige. All complex social systems are hierarchically organized and leaders can, if they want, use their power to benefit themselves, their families, and their friends. Men and women usually have different roles and thus prestige. Inequality may be accepted or not, but it is a fact. The major social science views on the nature of inequality are to be found in functional and conflict theories. As we will show later, in some measures of QOL, inequality is neutral to the measure; in others an increased degree of inequality will lower the average for the population as a whole. Statistical and empirical studies both show highly uneven patterns of growth and distribution of material wealth; reduction of poverty and efforts to satisfy basic human needs are inadequate in most countries. In many countries, both more and less developed, the income gap between the rich and the poor is widening.

Inequality has an important role in the UNCED process; humanitarian arguments have long been used to address starvation and equity arguments applied to questions of development in the LDCs. Added to this is a new rationale for less inequality among individuals and between the North and the South. According to many scientists, a large number of poor people will use natural resources in a non-sustainable way. The contribution of the rich to environmental degradation is well known; less well known is the fact that the poor are forced to do the same. Often they will do this at the expense of their immediate surrounding environment, thus later being forced to migrate to other areas and increase the population pressure in them. In the long run, the rich too will suffer from over-population and poverty.

1.2 Poverty and the need for development

Poor people are often hungry and unhealthy. The number of poor is increasing and the relative gap between the rich and the poor is widening, so poverty will become more visible. Only five or six generations ago there were less than 1 billion inhabitants on Earth. Today, out of the global population of nearly five and a half billion, more than one billion consume for less than one US dollar per day. Around 25,000 children die every day from starvation or from easy-to-cure diseases. One billion people are illiterate and thus lack an important personal resource. In southern Asia and in sub-Saharan Africa, one out of two inhabitants is poor[1].

1.3 Consumption and the need for environmental protection

Poor people, in spite of their low consumption levels, decrease opportunities for future generations. Inhabitants of rain forests use fire to clear land, consuming the forest more rapidly than it can be renewed. Marginal land, vulnerable to erosion and drought, is destroyed for centuries to keep hunger away for a few months.

Every government on Earth hopes to improve QOL within its territory, a hope that is usually perceived in terms of economic growth. This is also true for governments in the more developed countries (MDCs). More than one billion inhabitants in North America, Europe, Japan and Australia have an average daily income of more than 40 US dollars[2].

The effects of these high consumption patterns are well known. There is ample evidence that it is already causing profound changes in the Earth's life-support systems on all scales. The damage to soil and biota caused by logging of rain forests is one example. Another is the use of the atmosphere as a garbage can. The fact that we are changing the chemical balance of the atmosphere, with its complex role as a creator of a livable planet, makes many afraid of sudden large changes in the global heat balance to a new and very different equilibrium.

The present lifestyle in industrialized countries is the goal of many inhabitants in the developing countries. But this lifestyle is based on non-sustainable consumption patterns and production systems. The planet cannot sustain present levels of consumption let alone an increase.

The distinction between rich and poor countries simplifies reality in the sense that there are rich as well as poor inhabitants in both types of countries. Some researchers argue that the number of poor in rich countries is increasing as well as the number of rich. And of course rich people in poor countries can use as much energy and material resources as other rich people and thereby pollute for example the atmosphere.

1.4 Sustainable development

Human behavior and human development inevitably change the environment. Environmental changes can be consistent with an improvement in the quality of life, provided they are sustainable in the long run.

Sustainable development is said to exist if each member of each generation inherits an equally valuable stock of capital – man-made and natural – as the members of the

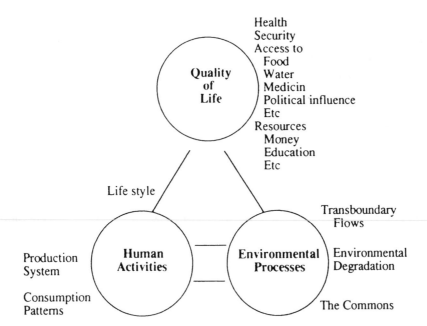

Figure 1: The performance of global systems, human as well as environmental, should preferably be monitored by Quality of Life indicator.

earlier generation (WCED, 1987). This definition raises two problems. One is how to define "valuable". The other is to define what qualifies as "capital". To be clearer we can define these concepts but in doing so we soon find that any attempt to be more specific about sustainability brings in value judgments on what is more or less important for a good QOL. It also brings in good cross-generational values. How much should the present generation be concerned with the future?

It is important to stress this conclusion. Sustainable development is a value judgment which is related to how we define a good quality of life.

It is, however, possible to say something on the balance between material and non-material consumption without making a value statement. Sustainable development will ultimately require not only the stabilization of the total material consumption (for ecological sustainability), but also the approaching of a decent level of material consumption for all people (for social and political sustainability). Population growth and material economic growth will eventually have to stabilize. However, non-material economic growth has no inherent physical limits.

2. EXISTING IDEAS ON QUALITY OF LIFE

There is no absolute or universal consensus on how to define and measure quality of life. On a very general level it is possible to define some dimensions of a good life for individuals or for societies. There is less agreement on ways to measure QOL or policies to improve it. Nevertheless, the concept QOL has been extensively discussed in the literature. Usually authors try to obtain some generalizations across populations and cultures (Fig. 1).

2.1 Objective and subjective perspectives on quality

In social science there is a clear distinction between objective measures and subjective ones. Objective measures are often defined in physical quantities or relations. An objective measure is often defined as a measure of a need or a norm, while a subjective measure measures attitudes, what people want or what they are satisfied with. Subjective measures are more important for the political processes and objective measures are more important in policy implementation.

Earlier attempts to define QOL involve partially overlapping concepts like "happiness", "well-being", "welfare", "standard of living", "living conditions", "way of life", etc. (Framhein, 1975). Usually the concepts address both the objective conditions under which a person lives, including his or her health, and the person's perception, satisfaction or subjective evaluation of his or her life (Mallmann et al., 1978).

Satisfaction is the evaluation of the degree of fulfilment of the desires and aspirations, and is determined subjectively by the person him/herself. Satisfaction is perceived fulfilment of desires and aspirations, based on an individual's perception and knowledge about adequate satisfiers and their availability, and by the person's internal processes and structures. They may therefore in some cases express false, neurotic, or displaced needs (Maslow, 1976).

Desires and aspirations are defined as the concrete forms in which a person seeks to satisfy the perceived needs, specifying the particular satisfier required; desires are immediate, aspirations are mediate. By satisfier we mean any element (material or non-material) whose use or consumption determines the fulfilment or satisfaction or a need, desire or aspiration. Most satisfiers are obtained from the person's environment, and some from the person's internal processes and structures.

As pointed out earlier, measurement of QOL usually includes data on both the objective conditions (frequency of use and/or consumption of satisfiers, their quantity and effectiveness regarding each need), and subjective evaluations. The latter in particular involves various deep methodological problems related to the operation of different psychological mechanisms, and it requires further research, particularly in the case of cross-cultural comparisons (Hankiss, 1976; Hankiss *et al.*, 1978).

Quality of life is a concept of universal applicability, to all societies at all times, although in terms of which dimensions to emphasize and which standards to aim at, it may show variations along the social, cultural, and developmental dimensions. The concept as used here is not restricted, as it is sometimes implied, to amenities or conditions of comfort or well-being, but it embraces the characterization of situations of extreme deprivation[3].

2.2 Health is a central indicator...

The health of a person is important in most definitions. It is often understood as positive physical, mental and social well-being[4]. Health can be conceived as resulting from the fulfilment of human needs, and from the person's internal structure and processes, including, e.g. age, genetic pattern, psychological, perceptual and cultural backgrounds.

2.3 Environment is another...

There are many definitions of an environment. Each person has his or her environment, physical and social. The environment of the person influences the chance of satisfaction of the human needs, desires and aspirations and hence the quality of life. Each society has its own environment, consisting of the physical (natural and man-made) environment enclosed within the territory occupied by the society, plus the external environment, both social (international) and physical. The state of the societal environment influences the functioning of the society, ultimately reflecting on the quality of life of the persons of the society[5].

2.4 Needs and requirements...

The phrase human needs is here used to describe the needs of an individual, while the needs of supraindividual human systems will be called requirements[6]. The following classification of needs[7] together with some of the satisfiers coming from the person's environment, will give an idea of how needs are usually treated in the literature.

Needs for existence or identity (needs whose non-satisfaction results in the annihilation of the system) consists of three groups:
(a) Maintenance: access to food, earnings, shelter and clothing, social and physical habitability, etc.

(b) Protection: access to health services and to health protection; legal defense and protection against violence and repression; prevention and protection against disasters.
(c) Love: ease of interpersonal contacts, attitude or mood of people in general (hostility, distrust, friendship); access to means to keep a family.

Needs for integration or completeness (needs whose non-satisfaction results in the system's inability to perform some of its functions) are here grouped into two sets:
(a) Understanding: access to education and culture; access to information and communication; freedom to exchange ideas.
(b) Self-reliance: possibility for participating in decisions; lack of manipulation, marginalization or repression, fulfilment of individual rights and freedoms.

Needs for optimal functioning (needs whose non-satisfaction results in disturbances in the system's performance of some of its functions) consists of:
(a) Recreation: access to recreational services, and to leisure time.
(b) Creation: access to creative work, and to individual and collective creative activities.

Needs for perfectibility or improvement (needs whose non-satisfaction inhibits the adaptive modification of the system's structure and functioning) are:
(a) Meaning: access to religious, cultural, ideological, and political groups, activities, and freedoms.
(b) Synergy: social encouragement of altruism, generosity, equanimity, solidarity, etc.; trust in persons and social institutions; access to beautiful cities and natural environments; social discouragement of aggressiveness, savage competition, discrimination, etc.

Some human needs are material (i.e. their satisfiers are material things like food, buildings, clothes, etc.) and some are non-material. The concept of basic needs should of course never be restricted to a subset of the material needs. While some needs necessarily require some degree of material consumption, some can be satisfied through non-material means. There is often considerable scope to increase quality of life while stabilizing or even reducing per capita material consumption.

2.5 Global processes affecting the quality of life

It is common in the literature to refer to processes important for changes in QOL. Practically all processes in societies and in nature could be said to have substantial influences on QOL depending on the definition used and the author's disciplinary bias. Some examples of more commonly mentioned ones are:

(a) Changes in settlement patterns, including urbanization.

(b) Ethnic tension and consequent displacements of populations.

(c) Increasing environmental perturbations which result in natural disasters.

(d) Shortage of capital for investments including debt crises in many countries which in turn usually result in declining investments in areas like education and health services.

3. EXISTING MEASURES

In the literature, QOL is usually seen as a holistic measure, but still it should be possible, for analytical purposes, to measure different components adding up to the whole. The first two parts of the concept are usually (i) a value basket which is mainly made up of non-material culture goods, and (ii) a material basket which is made up of the material elements necessary for the sustenance of life[8]. The quality of life for individuals could be measured by "objective" or "subjective" data, but usually objective information is preferred. It is common to classify different measures after their importance, starting with basic needs (food, health, etc.)[9] and ending up with more sophisticated needs (like satisfaction with work, etc.).

Indicators should preferably measure in simple quantitative terms how systems performance over time is related to goals. These goals could be minimum levels or visions. The indicator and the performance should be easy to understand. The primary function of an indicator is simplification. It should, as is indicated by its name, give a clear indication of system performance and trends to policy-makers.

While giving QOL a more precise definition, it is clear that it is possible to draw upon the rich literature on related concepts like welfare, standard of living, living conditions, social indicators or human development indicators[10]. Some of these concepts have a very long tradition, but the more lively scientific debate during the 1960s and 1970s could be seen as a reaction against a use of simple economic measures, especially GNP/capita which was often misused as a welfare measure.

There are several well-known drawbacks with an economic measure as a welfare measure or as a QOL indicator. Let us mention only two of them: Its simplicity and its failure to consider negative contributions. Because it is simple, easy to explain, and easy to find in statistical publications, it is bound to be used (and misused). However, as will be explained later, it has a low correlation to more sophisticated QOL measures. Another drawback with its simplicity is that internal conditions within countries do not show in the figures. Also countries with a fast growing GNP could have large population groups with a decreasing economic standard.

A second problem is its inability to identify economic activity that offsets the negative effects of production. For example, a factory built to produce scrubbers for smoke-stacks is essentially offsetting negative effects of other income-generating, or better welfare-generating, production. There is an on-going discussion on how to consider this drawback in national accounting procedures. The first elaborated effort to construct a QOL-oriented GNP-measure was made by the Economic Council of Japan[11]. They measured the net national welfare index, NNW, in which negative costs, like environmental protection costs, were deducted from GNP and positive benefits, like the value of leisure time, were added.

3.1 Group measures addressing QOL for individuals...

A well-known and simple measure for comparisons of "Quality of Life" in different countries, MDCs as well as LDCs, is the Human Development Index (HDI)[12]. It considers only three aspects: life expectancy, literacy and income, which makes it possible to find adequate comparable data[13]. For the first two indicators, a scale between 0 and 1 is constructed for the lowest and highest values. Life expectancy is today varying between 42 years (Afghanistan, Ethiopia, and Sierra Leone) and 78 years (Japan). Literacy among adults varies between 12% (Somalia) and nearly 100% (many countries). The third indicator, income, is a purchasing-power-adjusted GDP estimate. It varies between US$ 220 (Zaire) and US$ 4861 (nine industrial countries). A scale between 0 and 1 for the logarithm of the indicator is also used here. Averaging the three scales[14] gives the HDI.

3.2 Their social environment...

Individuals living in social groups or societies supporting their lifestyle are better off than others. The same is valid for social and ethnic groups having harmonious relations with other more powerful groups. It is possible to measure the degree of shared value systems, social and political participation, and other community characteristics in QOL-measures.

Relations between groups are more or less problematic depending on, for example, group scale, degree of conflict and its historical context. Some pressures can unite families or ethnical groups, while large conflicts like persecution or civil wars, of course, will lower QOL for most people involved. In the large-scale social environment, some main political problems are related to unsolved fights between ethnical groups over power and territorial control. This is true for situations in several continents, including Africa, with a well-known difference between governmentally controlled territories, which are a colonial heritage, and territories that are the living space for different ethnic groups. Military expenditures in LDCs, on average around 200 billion dollars[15] per year,

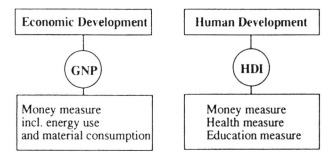

Figure 2: The main international comparisons between countries are today made by the Gross National Product (GNP) or by the Human Development Index (HDI). Quality of life is better measured by the HDI.

are more than that for education and health combined. Technological development of atomic, biological and chemical weapons makes it easier every year for powerful leaders in all parts of the world to threaten others. One QOL measures related to social tension and armed conflicts is the teacher/soldier ratio[16]. This measure, like the HDI, compares the situation in countries and does not give any information on the distribution of costs and benefits of social conflicts among individuals within countries.

In the small-scale, social security for individuals will depend on several factors, including family ties and many other cultural, social and economic traditions. Information on public safety is usually given in statistics on violent crime and accidents. Lately, the frequency of drug crime has become both a threat and a cost for large groups. According to statistics, the value of trade in illegal drugs exceeds that of trade in oil.

3.3 And their physical environment
A high-quality environment, both natural and man-made, is an important objective in all societies. Practical measures on the natural environment could include a suitable topography and climate, clean air and water, low noise level, good soil for food production and a variety and amount of wildlife[17]. For the rural population in LDCs, access to farm land is an essential part of a good QOL. The man-made environment mainly includes so-called physical infrastructures, like dwellings and other buildings, roads and other transportation networks, telecables and other communication networks, utilities for electricity, gas, water, sewage disposal, and solid waste disposal, etc[18]. There is a scientific literature on how to measure access to these environmental utilities. Physical distance and so-called queuing principles, like money, queuing time or membership, will allocate these resources to the population[19].

3.4 Measures addressing equality
It is easy to see how average measures on QOL for large groups sometimes need complementary information on

internal dispersion to make sense, e.g. an increased real total income in a country could hide increasing disparities. If one-fifth of the population earns ten times more than the rest (*per capita*), a total increase of, e.g. 4% could be consistent with a 10% increase among the upper fifth and a 10% decrease among the majority.

An Equal-Weight-Index (EWI) has been jointly constructed by the Institute of Development Studies at Sussex and the World Bank[20] to address the equality question. The index considers both that a majority is more important than a minority (like in democratic systems) and that an increase among the poor is more important than among the rich.

The well-known gender inequality is addressed by many researchers but there are few possibilities to measure and compare the QOL for women in different parts of the world because of data limitations. However, the United Nations Development Programme (UNDP) has constructed two gender specific HDIs on the basis of existing and estimated data[21]. The UNDP concludes that, as the QOL improves in countries, there is a tendency for the situation of women to improve and, eventually, become better than that of men. Among countries with a low or medium human development there is enormous variation in the female–male QOL disparity.

3.5 Measures for individuals
Individuals endowed with high levels of personal "resources," like health and education, are better off than others. In MDCs where most people have their basic needs satisfied, income and economic wealth are usually added to these personal assets. In LDCs income and land ownership are some of the important additions. In all countries, the security dimension is one of the most important. First, security adds to QOL. Second, only secure access to food, medical help and basic knowledge will enable people to think about a shift of consumption patterns from energy and material wasting ones to more environmental sustainable ones.

Most of the scientific literature on QOL indicators deals with conceptual frameworks. Different scientific disciplines specialize on different aspects of life and they are thus more familiar with some indicators: economists on economic growth, geographers on physical access, political scientists on democracy and participation, sociologists on social networks, demographers on infant mortality, etc. Depending on how individuals are treated in theories, different indicators become more central. Starting with a theory on economic growth, individuals are instruments for further production and their abilities to produce (health, education) become important. A welfare approach would concentrate more on perceived and real satisfaction with life. A basic need approach makes no allowance for how countries are governed, e.g. the degree of democracy, and

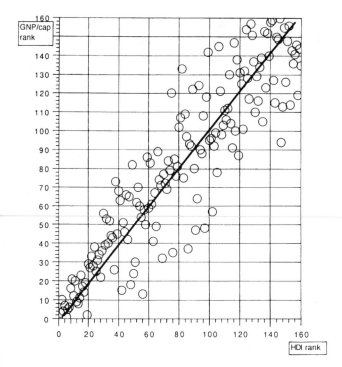

Figure 3: 160 countries ranked according to Gross National Product (vertical scale) and Human Development Index (horizontal scale). Countries with good conditions (low rank in GNP and HDI) in these respects are circles in the lower left hand corner. These countries are often industrialized. Among poor countries (measured by GNP) there is a large variation in the quality of life (measured by HDI). These less developed countries are at the top of the figure.

could therefore more easily be used in an international context than a political science approach.

In the same way, economic indicators like GNP/capita and simple QOL indicators like HDI, are easier to accept than more elaborate measures trying to capture many dimensions of human life. Experience also shows that it is easier to accept international co-operation on indicators of unwanted conditions, like high rates of crime or diseases, than to agree on positive indicators of the good life. To define a good life, the meaning with everything we do, goes too deep into our personal value systems.

There have been several attempts by researchers to develop interdisciplinary conceptual frameworks. Some of them, like the Scandinavian tradition of measuring QOL from a resource perspective and the Anglo-Saxon tradition with social indicators, have been implemented in large surveys in more developed countries.

3.6 Relation between development and income measures

A simple correlation between GNP *per capita* and HDI shows that rich countries usually have a higher QOL than poor. However, the quality dimension measured by HDI (life length, literacy, and income) can vary to a large extent between countries with the same level of production *per*

capita. Some countries chose to translate larger parts of their income than others into an increased quality of life for their people. Countries like Sri Lanka and China do far better on their human development than on their income (see Fig. 3).

4.   THE UNCED DIMENSION

The lifestyle and quality of life in one country has become more and more related to the quality of life in others. Interpersonal and international resource allocation are topics concerning the environment and development nexus. The question of how present resources, especially the commons, are or should be allocated is definitely on the UN agenda. The lifestyle in one part of the world is no longer only a local concern. Complicating the debate about the interterritorial distribution of rights and duties is the question of links between present and future generations. The concept of sustainable development, how to make it easier for every new generation to meet its needs, is targeted towards intergenerational resource allocation. Added to this there are arguments that fairer intragenerational allocation of resources is of importance to social and political stability and sustainability.

4.1 People and their numbers

A commonly used argument to explain environmental degradation, and slow or absent development, is that we are using too much energy and material *per capita*, and that there are too many of us on the planet. Resources like water, soil, land, forestry, and air are scarce or limited, and present population growth, 250,000 new individuals per day, will lead to an absolute shortage. This view is often implicit or explicit in both natural science studies and classical economics.

LDCs tend to emphasize that the lifestyle and *per capita* consumption in MDCs are the main cause of environmental degradation. A citizen in the US consumes 200 times more fossil fuel than his counterpart in Ethiopia. From this, it is not possible to conclude that two children in the US are increasing the CO_2 pollution of the atmosphere as much as 400 children in Ethiopia, but one can argue that birth control or birth restrictions should first be encouraged in rich countries.

In MDCs it is often claimed that family planning programs are needed in poor countries to break through the well-known "poverty trap." Nearly all population increases are now taking place in LDCs. Historically, large families are a new social institution. Thus there could be no old traditions or cultural bounds as a rationale for making this new type of family (with many children surviving) a permanent institution. According to many governments, family planning is a necessity in LDCs to overcome poverty and environmental degradation.

Population growth has also been interpreted as a neutral or positive factor behind economic growth. Some neo-classical economists argue that the basic problem with starvation, poverty, and over consumption is not the number of inhabitants on the globe but inefficient conversion markets. The problem is that markets do not self-adjust to take account of the problems. A consequence of these ideas is that more governmental and international control and interference is needed to correct non-market behavior and to adjust price levels so that they include external effects of consumption patterns. There is thus no need for family planning programs. Every newborn child is looked upon as an investment for the future. Markets would allocate resources (and thus QOL) in an efficient way. Resource scarcity would be met by technological innovations, shifting consumption patterns and substitution of scarce resources with others.

Other perspectives on population growth include discussions on its effects on ethnic and social tension. Population growth is then often seen as dependent on other QOL factors like illiteracy, poverty, and inequality.

4.2 Trans-boundary effects of human behavior

All environmental problems on the UNCED agenda are caused by human lifestyle in a broad sense, including our choice and use of technology. Many of them have to do with trans-boundary movements, e.g.:

(a) CO_2 and other greenhouse gases.
(b) Freons, etc. (ozone depletion).
(c) SO2 and NO_x (acid rain).
(d) Hazardous waste in the air (such as radioactive pollution).
(e) Hazardous waste in the waters.

Trans-boundary flows of chemicals are especially changing the QOL in densely populated, rich regions, like in Europe, where acidification and accumulation of heavy metals in the soil is a problem. Countries which do not install cleaning devices for atmospheric pollution and thus earn money on exporting environmental problems to others seem less eager to put a price on the damage. Why not use one of the commons as a free garbage can when others pay the costs?

Let us once again repeat the main problem with present QOL measures in this perspective. A QOL measure in one country could not be taken seriously if it does not consider the negative effects it will cause in other countries.

Among the trans-boundary effects of human behavior or misbehavior is also international migration. The background behind many contemporary mass migration streams is poverty, starvation or civil wars between ethnic groups. The present number of registered refugees, 15 million, could increase dramatically depending on future changes in human development including ethnic conflict levels, redistribution of resources, including food and economic wealth. In the future migration due to different standards of living will probably increase. Migration from South to North is already taking place in small numbers.

Other trans-boundary flows could be physical, like of arms, drugs and oil, financial or in the form of information. Through world trade, countries usually improve their economic performance but they can also find themselves moving into structural conditions which will lower the QOL for their inhabitants in the long run.

4.3 The commons

The air and the oceans are capital stocks owned jointly by mankind. Any positive flow to the stock (like freshwater from the rivers to the oceans) or negative flow (like greenhouse gases from industrial processes to the atmosphere) change the value of the commons and could thus change QOL.

If one accepts the idea that there is a limited carrying capacity of or ecological capacity for deleterious chemicals, e.g. a specific amount per decade, then it is possible to discuss who should have the right to use this amount. Which group of actors or which nations should have the right to pollute? The rich? The poor? The powerful? The nations who used to pollute? The nations who have not harmed the commons so far? The ones with a good or bad QOL?

One idea would be to give all inhabitants the same right to pollute. Nations with many inhabitants would then, irrespective of their military power, have more permits than small nations. The basic idea, one person, one polluting permit, is the same as in democracies. Individuals or nations with a good QOL – including a large consumption *per capita* – could then buy permits to pollute the commons from those who have (or choose to have) a lower consumption *per capita*.

4.4 Sharing

The lifestyle of a person or a group of people is only a fraction of a larger system of customs, values, beliefs, attitudes, and physical flows. Measures of lifestyle or quality of life for some cannot be understood in isolation from others.

Measures of lifestyle should avoid ethnocentric attitudes. However, transcultural impacts can and should be measured. Moral judgments could be made on lifestyles that support or limit the quality of life of others. Improving the quality of life for one group could degrade the quality of life for another. As a consequence, differing lifestyles are one of the basic sources of conflict in the world. Another consequence is that a person's lifestyle only partly depends on his or her own choice and value system.

Man is social and co-operation is the cornerstone of survival. The need to co-operate has fascinated philosophers throughout the ages, so many of the ideas about how power should be shared between individuals and the collective have a very long history. Let us remind ourselves of two alternative streams of ideas in the scientific literature.

Some would argue that sharing power, which of course is a necessity in the real world, in the long run always leads to totalitarian societies. If we accept "social contracts" in order to avoid violence and anarchy, we support the creation of control systems and superpower that gradually become totalitarian. Others would argue that co-operation and interaction produce prosperity. We should therefore forego a certain amount of freedom for the benefits of living in a civilized and secure society. There one enjoys freedom from hunger and fear, a greater freedom than in a society where every individual is free but insecure and afraid. With this view, societies are seen as advantageous social contracts in which there is a balance between rights and obligations.

Ideas on co-operation and on how power should be shared between individuals and governments have up to now mainly been discussed in their historical, social, and economic context on a local scale, within nations. Public participation is a key concept in some countries. The discussion now has to be extended to a global scale – forced by the physical and economic trans-boundary processes which tend to ignore national boundaries and territorial sovereignty.

4.5 QOL Measures addressing individuals in a sustainable global society

The consumption of energy and materials has reached a level where environmental effects no longer are limited to the immediate surroundings of consumer groups. Costs and benefits have to be shared among people in space and time in a way that needs monitoring and approval. A sharing-perspective has to be added to the traditional wisdom of QOL measures.

In the UN system there is a new emphasis on QOL-related strategy for the coming decade. The UNDP[22], cites three objectives: acceleration of economic growth, reduction of absolute poverty, and the prevention of further deterioration in the physical environment. The departure from earlier development strategies lies in clustering all these objectives around the central goal of enlarging the set of choices available to people. Many will agree on these formulations. When it comes to deciding who should pay the costs there will be less agreement. Still, however, QOL-related questions will be directly or indirectly on the agenda both for the UN Conference in Rio de Janeiro and for future work.

5. POLICY OPTIONS

In the drafting of any policy or declaration regarding management of environmental resources, including the UNCED process, quality of life aspects should be considered explicitly, including their interterritorial (spatial) and intergenerational (time) aspects. In general, considerations of QOL should have a more explicit role in policy formation.

5.1 What can researchers do?

The question is then what researchers can do in this process. Seen from a pessimistic point, we already know more about human needs and sustainability than the present world order can make use of. There is often a lack of will among the rich to increase the quality of life for others. Even with good intentions, there is no organization with the necessary power to change present unwanted conditions. Seen from a more optimistic view, the UNCED-perspective, there are many ways in which researchers can contribute to change the quality of life of individuals.

5.2 New intra- and interdisciplinary knowledge

It goes without saying that policies intended to improve the quality of life must be based on solid scientific knowledge. More knowledge in general about nature and culture, physical flows of material and chemicals, allocation mechanisms on global (and other) scales, human values, and consumption patterns is essential for an understanding of how sustainability could be obtained. The need for more research within most scientific disciplines cannot be over-emphasized.

Research on quality of life and human needs must often have an interdisciplinary approach. Needs do not respect disciplinary boundaries. Often it could be of importance to encourage co-operation between existing disciplines within or outside present universities and research institutes.

5.2.1 How lifestyles emerge and change

One of the most important topics within this context is to understand how lifestyles emerge and change. This type of knowledge could be used, like all knowledge, for "good" and "bad" purposes. Businesses pay large amounts of money to understand how their products would sell better. For world leaders an understanding of lifestyles could be used to influence behavior in order to promote consumption patterns which are consistent with both increases in the quality of life and a long run sustainability. Both "top-down" and "bottom-up" actions are needed to encourage sustainable behavior. Government measures as well as popular movements could be based on this type of knowledge.

5.2.2 *How social values emerge and change*

Behind human actions we find values systems. We have already stressed that ideas on sustainable development are based on value judgments which are related to how we define a good quality of life. Thus the concept of sustainability is founded in social values. It is self-evident that more knowledge on how social values emerge and change is important for future actions within the environment-development nexus.

5.2.3 *Material and non-material consumption patterns*

Present consumption patterns in the rich countries cannot be extended to the global population without placing severe strains on the environment, including the commons. The need for increased material consumption in less-developed countries is evident but unfortunately their plans to increase production are generally based on large increases in energy use, often by using fossil fuel. Technology transfer can reduce some of the pollution but the environment, including the atmosphere, will have to take new large amounts of CO_2 and other chemicals. A realistic scenario is that the present polluters, the rich countries, reduce their pollution to create space for others to pollute. Reduced material consumption and an increase of non-material consumption in rich countries could allow for improvement of the quality of life. A better understanding is needed of how different lifestyles are more or less dependent on the use of material and energy.

Granted that a continuous improvement in the quality of life of the human being is the central goal of sustainable development, the improvement in quality of life could in principle be obtained by alternative mixes of satisfiers of needs, desires and aspirations, even implying the possibility of a continuous increase in economic (non-material) growth. The trend towards the de-materialization of the industrial economy – the use of lower amounts of energy and materials to produce a comparable product – is an expression of this possibility. Deliberate searches for alternative, less material-intensive satisfiers, as well as research into alternative social projects and socio-ecological configurations, should be an important part of future research agendas.

5.2.4 *Nutrition and human health*

In the quality of life discussion there are many topics and research areas which could argue that they should be given priority. Nutrition is one of them and since the composition of our food is so important for human health, this area is one of the most self-evident candidates.

5.2.5 *New development and economic theories*

Progress on measures of development is evident during the last decades. The need for new theories on development is often stated in the literature. What are the factors behind changing living conditions? Why can some countries during some decades develop in several dimensions while others do not?

Also there is a need for theories on how to allocate scarce resources in the real world, a world of violence, crime and bribery, a world in which ethnic groups hate each other and nations are often unwilling to co-operate on trans-boundary phenomena. Present economic theories have problems dealing with the real world and are thus used mainly as a background information by economists working with applied economics.

5.3 Monitoring quality of life

For international comparisons, the measures of quality of life are fairly well developed. The lack of statistical information is more problematic. An improved coverage of known measures is essential for both research and policy-making aimed at improving the quality of life. Since urban life seems to be dominant in the future, new aspects on the relation between humans and their environment could be included in QOL measures.

There is also room for research on how the behavior of individuals in one place influences the behavior of others elsewhere. Measures of this dilemma must in the long run be integrated in the concept of quality of life. More elaborated measures could link the quality of life discussion to ideas on how international relations should be organized in order to increase the sustainability.

5.4 New research tools for impact analysis

Conventional disciplinary research on dependencies between individuals in time and space should be supplemented with a development of scientific tools for synthesizing knowledge. While ordinary research is specializing and analyzing reality in more detail, politicians working with environment and development issues need overviews and integrated knowledge. They must know how different pieces of information are linked to each other in a coherent framework. Here, science should devote more resources to the development of time–space-specific conceptual framework models and tools for impact analysis. These models and tools would also often be suited for making best use of existing knowledge within different scientific fields. With these tools it will be easier to allocate resources where QOL can be most effectively improved with limited resources.

5.5 Advisory groups

Finally, scientific organizations, like ICSU, could arrange independent scientific advisory groups, which could work on the international scene with important QOL questions. Monitoring change and policy impacts would be part of their tasks. The idea is to get the large investments right. A shortage of capital will dominate the international economic arena during the next decade. Investments in the MDCs, including Eastern Europe, will compete with investments in LDCs.

The advisory groups should be specialized in environment and development issues. They could, for example, evaluate the overall results of aid programs to different LDCs. Efficiency improving quality of life, including economic growth and other values should be a guide-line for allocation of investment resources. The group could be used by the World Bank and other organizations to evaluate impacts on the quality of life of existing large investment projects.

In the MDCs there is a need for science to advise on how quality of life could be improved or sustained with less use of energy and material. What mixture of economic measures, such as pricing behavior, regulations, and restrictions on toxic chemicals, will be concordant with both improved quality of life and sustainability? How could we increase consumption in a qualitative way without increasing the quantitative use of energy and materials?

END NOTES

1   See McHale and McHale (1979) or Durning (1989) for a further overview of poverty in the world.
2   GNP per capita in OECD countries in 1987 according to OECD National Accounts.
3   Including those leading to death (that is, to zero quality of life).
4   This is the way health was defined originally by the World Health Organization.
5   The human environment is the environment of a human system (system can be defined at different levels of aggregation, from the individual person to the whole of mankind); it is therefore a complement of the considered human system, and meaningfully defined only in relation to the system (Gallopin, 1981a, 1981b)
6   The justification of needs for groups is given in relation to the fulfilment of the needs of the persons that are members of the group.
7   Needs could be expressed for different levels of aggregation. A characterization of the human needs and requirements combining Sicinski's generalized concept (Sicinski, 1978) with Mallmann *et al.*, (1978) classification of needs, adapting the latter to the needs and requirements was proposed (Gallopin, 1981) for a person, a society and mankind.

8   UNESCO 1978.
9   The basic needs approach has become an important idea in the UN-system.
10 There are hundreds of books on the topic Quality of Life and several authors refer to related concepts, for overviews see, e.g. UNESCO (1978) or Leipert, C. and Simonis, U.E. (1981).
11 See Economic Council of Japan (1973).
12 This index was developed in a collaborative effort by the UN Statistical Office, the World Bank, EUROSTAT, OECD, EEC, ESCAP and USAID.
13 The ideas behind this index were described earlier, see Morris et al (1979), under the name PQLI, the physical quality of life index.
14 UNDP 1990.
15 UNDP 1990.
16 For further information on this index see Human Development Report UNDP (1990)
17 UNESCO 1978.
18 A list of indicators and a proposed methodology for measuring them in order to assist in making public policy can be found in an article by L.W. Milbrath: "Indicators of Environmental Quality" (UNESCO 1978).
19 It is possible to combine physical access measures with availability in terms of supply and demand, see Öberg (1976).
20 See Chenery *et al.*, 1974.
21 UNDP 1990.
22 UNDP 1990.

REFERENCES

Ahmad, Y.J. and Serafy, S.E. and Lutz, E. 1988. Environmental Accounting for Sustainable Development, World Bank, Washington, DC.

Allardt , E. 1976. Dimensions of Welfare in a Comparative Scandinavian Study, Acta Sociologia, 19, pp. 227-240.

Chenery, H.B. et al., 1974. Redistribution with Growth. London. Oxford University Press. Oxford.

Drewnowski, J. 1974. On Measuring and Planning the Quality of Life. Paris: Humanities.

Dréze, J. and Sen, A. 1989. Hunger and Public Action, Oxford, Clarendon Press.

Durning, A. 1989. World Poverty, World Watch Institute.

Economic Council of Japan 1973. Measuring Net National Welfare of Japan. Tokyo: Government printer.

Erikson, R. et al., 1987. The Scandinavian Model, Welfare States and Welfare Research, (eds.) London, M.E. Sharpe.

Framhein, G. 1975. Round Table Meeting on Indicators of the Quality of Life. Summary Report. Organized for UNESCO by the Vienna Center. Budapest, 14-15 November 1975.

Gallopín, G.C. 1981a. Human systems: needs, requirements, environments and quality of life. pp. 124-128 in: G.E. Lasker (ed.). Applied Systems and Cybernetics. Vol. I. The Quality of Life: Systems Approaches. Pergamon Press.

Gallopín, G.C. 1981b. Planning methods and the human environment. Socio-Economic Studies 4, UNESCO, Paris.

Hankiss, E., R. Manchin and L. Füsfös. 1978. Cross-Cultural Quality of Life Research. Center for Quality of Life Research, Budapest.

Hankiss, E. 1976. Quality of Life Models (Hungarian Experience in QOL Research). Meeting of Experts on Indicators of the Quality of Life and the Environment. UNESCO, Paris. SS.76/CONF. 629/5.

Kuik, O. and Verbryggen, H. 1991. In Search of Indicators of Sustainable Development, Kluwer Academic Publ., Dordrecht, The Netherlands.

Leipert, C. and Simonis, U.E. 1981. Social Indicators and Development Planning. In Economics, 24: pp. 47-65.

Mallmann, C.A., M.A. Max-Neef and O. Nudler. 1978. Notes on meaningful and practical measures of health, satisfaction and quality of life. Fundación Bariloche, Argentina.

Maslow, A.H. 1976. Motivation and Personality. 2nd. (ed.) Harper and Row, New York.

McHale, J. and McHale, M. 1979. Basic Human Needs. A Framework for Action. New Brunswick, NJ. Transaction Books.

Morris, M.D. et al. 1979. Measuring the Condition of the World's Poor. The Physical Quality of Life Index. Oxford, Washington, DC., Pergamon Press.

Öberg, S. 1976. Methods of Describing Physical Access to Supply Points, Lund studies of Geography, Lund. Gleerup.

OECD 1976. Measuring Social Well-being. A Progress Report on the Development of Social Indicators. Paris.

OECD. 1982. The OECD List of Social Indicators, Paris.

Pearce, D.W. 1988. Economics, Equity and Sustainable Development. Futures, 20, 6, pp. 598-605

Sachs, I. 1982. Environment and development revisited: ten years after the Stockholm Conference. In: Alternatives 8, pp. 369-378.

Sicinski, A. 1978. The concepts of "need" and "value" in the light of the systems approach. Social Science Information, 17(1), pp. 71-91.

Simonis, U.E. 1990. Beyond Growth. Elements of Sustainable Development, Wissenschaftcentrum Berlin, Edition Sigma.

Titmuss, R.M. 1968. Commitment to Welfare, London, Allen and Unwin.

Todaro, M.P. 1977. Economic Development in the Third World: An Introduction to problems and Policies in a Global Perspective. London, Longman.

UNDP 1990. Human Development Report 1990. New York, Oxford. Oxford University Press.

UNESCO 1978. Indicators of Environmental Quality and Quality of Life, Reports and Papers in the Social Sciences, No. 38.

United Nations 1978. Social indicators: Preliminary Guidelines and Illustrative Series. In Statistical Papers, Ser. M 63.

World Commission on Environment and Development, WCED 1987. Our Common Future, Oxford. Oxford University Press.

Chapter 13: Public Awareness, Science and the Environment

R. Grove-White, S. Kapitza* and V. Shiva

1. INTRODUCTION AND ABSTRACT

Addressing global environmental problems now requires immense political will. But this is likely only to emerge from the political institutions of industrialized countries if there is first widespread public recognition and awareness of the need for appropriate action. So a central question prompting this chapter addresses is: *what contribution can and should science make towards the development of public awareness of environmental problems?*

The chapter refers to some of the current strengths and limitations of scientific knowledge and the scientific research community in this connection. It argues that, if science is to play a role commensurate with its prestige and resources in modern societies, its institutions will themselves have to become more aware of and more responsive to new public anxieties and ambivalences about the way in which their role has developed in recent decades. The evidence suggests that an opportunity now exists for a reconstitution of public trust in scientific understanding of problems, with considerable potential benefits for mankind – but only if the basis of a new concordat with the wider public can rapidly be generated.

Hence, Section I of the chapter considers the prevailing problematic relationship between science and society in many countries. Section II then addresses some of the ways in which new configurations of relationship are beginning to emerge. This leads, in Section III, to a discussion of some of the ways in which public trust in scientific insight might be further reconstituted. It is argued that the immense resources of the scientific community, if renegotiated in a socially sensitive fashion, can be brought to bear with fuller effectiveness to assist public understanding of what is at stake.

In discussing these issues, it is important first to recognize that there are widely diverging conceptualizations of science itself in different cultures around the world. There is a large body of scholarship which disputes the widespread normative assumption in many of the most dominant western scientific institutions, that science is essentially an ethically neutral, reductionist experimental set of practices aimed at advancing mankind's understanding of physical reality (Mulkay, 1979; Lyotard, 1984; Barnes and Edge, 1984). There are a variety of other cultural understandings of science which place the same kinds of practices and aspirations within conceptual frameworks emphazising the religious or cultural (in the widest anthropological sense) underpinnings of such activities (Merchant, 1982; Berman, 1984; Sardar, 1984).

These different conceptualizations – and the clues they provide to possible means for the reconstitution of public confidence – become relevant in the discussions in sections II and III of recent developments in the science/public relationship. However, it is necessary first to consider the present position.

Section I

2. PRESENT TENSIONS BETWEEN SCIENCE AND SOCIETY

In recent decades, the relationship between the publics of both developed and developing countries on the one hand, and scientists and scientific institutions on the other, has become increasingly problematic.

There is little doubt that modern societies have developed in such a way that scientific research and understanding are now crucial to the identification, analysis and rectification of many of their problems – be they medical, defence, transport, food or energy production, or environmental. Moreover, the role of science in both pure and applied forms in helping create the modern world goes beyond simply its physical manifestations in a thousand and one technological forms

---

* Three writers were invited to collaborate on this topic. In the event, the collaboration evolved as a dialogue. Hence, what follows is a main text written by two contributors (Grove-White and Shiva) – and an annex, implicitly a response, by a third (Kapitza).

and patterns of consumption. It has also provided a dominant epistemology (philosophy of knowledge), and indeed an ontology (philosophy of being), for modern secular man (Merchant, 1982). Science has exerted a hugely dominant and pervasive influence on the ways in which 20th Century men and women have come to understand the world and their places in it – an influence which has been greatly intensified in recent decades with the growth of global intellectual networks and shared specialist languages in many of the specific disciplines embraced by the "natural" and "social" sciences. Moreover, the marriage of scientific research with governments and industry in furtherance of international trade and geopolitical competition has led to still more heightening of its prestige and perceived significance at politically influential levels in most modern societies.

Against this background, and recognizing the formidable record of success that scientific discovery and innovation have played in this century in addressing problems (of disease, of industrial production, of infrastructure provision) that had disfigured mankind's history previously, it is not surprising that the role of science in relation to public awareness of environmental problems has frequently been assumed by many in public agencies to be unambiguous; a largely ignorant public, so the doctrine runs, needs to be "educated" by the scientific community on 'the facts' of environmental problems and the solutions necessary for addressing them.

But such an assumption begs many questions. The fact is, notwithstanding the cultural significance of modern science in contemporary societies, public attitudes towards it and towards scientists themselves are now profoundly ambivalent. This has been shown repeatedly in surveys in Europe and North America (e.g. Wynne, 1990). It has been prefigured in critiques of industrial society since the early 19th Century – through "prophetic" writings by influential authors as various as Mary Shelley (Frankenstein), Carlyle, Nietzsche, Spengler, Jaspers, Mumford and Ellul (Winner, 1977). It is now acknowledged increasingly in the 1990s by scientists themselves.

There are several elements to such ambivalence. In the first place, the very centrality of science and technology, in helping create the trajectories of modern industrial society, has implicated them also in such societies' negative features. This can be seen in the stereotyped characterization of "the expert" (social or natural scientist) in popular Western culture – ingenious and powerful, but frequently socially naive and hence dangerous (Winner, 1977). More empirically, the challenges to then- prevailing expert scientific judgments implicit in much of the activity of popular environmental movements – through non-governmental organizations (NGO) – in the 1970s and 1980s, can be read as in part a focussing of public anxiety concerning an important range of human, social and

cultural variables marginalized by the narrow application of scientific methodology to specific environmentally contentious developments (Grove-White, 1991; Eyerman and Jamieson, 1991). There are grounds for understanding many of the public resonances struck by environmental NGOs around the world in the 1970s and 1980s to have reflected the latter's attunement to public unease at the previous dominance of social as well as natural scientific positivism applied in the public arena (Wynne, 1987).

A second, less generalized focus of recent public ambivalence has been the reality that, far from flourishing predominantly in the idealized circumstances of academic independence and detachment, most scientists in contemporary society are employed by particular institutions which have very particular ends in view. A high preponderance of scientists around the world work in private or public sector industry, and only a relatively small proportion in universities or independent research institutions. What is more, funding for the latter is directed to a greater or lesser extent by governments, which can thus exert an influence over the directions in which even publicly funded knowledge is encouraged to develop. Thus, whilst individual researchers may feel themselves to be operating under the principles of detached inquiry, they tend nevertheless to be embedded in more "committed" frameworks. The ideal of scientific inquiry proceeding in splendid isolation from the pressures and contingencies of contemporary political or industrial exigency – an ideal strikingly espoused in learned scientific societies the world over (e.g. Porter, 1990) – needs to be seen against the realities of this background. There is a growing body of sociological evidence suggesting that tacit awareness of these institutional realities of much modern scientific inquiry – that, in effect, he who pays the piper tends to call the tune – is now a significant element in the public ambivalence under discussion.

A third relevant reality emerges from the first two. Even when scientific research is conducted in independent contexts the dominant models of "good science" – high-precision, reductionist, controlled experiments – may tend, cumulatively, to ensure certain emphases in society's understanding of the parameters of particular problems. Often this is healthy. But frequently it is not – as in cases where the social authority of scientific inquiry is invoked by governments or particular powerful sectoral interests (in public or private sectors) to provide reassurance in the face of pollution incidents or other controversial commitments. A number of influential contemporary technologies (pesticides, nuclear power, the internal combustion engine, etc.) have become diffused through society without serious prior analysis of their potential secondary impacts and social costs, or adequate mechanisms for ensuring feedback about unacceptable or regressive social changes engendered by them. (Krohn and Weyer, 1990). The political reliance on reductionist research epistemologies in such circumstances, to provide social reassurance, may well have tended to magnify already existing public anxieties (Wynne, 1987).

To summarise, what the three above observations suggest is that the institutionalized ways in which science has been conducted in industrial societies in recent decades, and the powerful position of scientific epistemology in contemporary public discourse, have themselves contributed to the mixed feelings revealed both in public attitude surveys and in more general sociological commentaries on contemporary "post-modern" culture (Jonas, 1984; Beck, 1986; Giddens, 1990). Recent sociological research confirms that, while in the abstract people believe in science and scientists, in real cases affecting their lives they usually encounter good grounds for mistrust, scepticism or plain lack of interest (Wynne, 1989a). At the very time when we need a new public consensus and trust in the capacity of science – currently, whatever its limitations, mankind's most comprehensively acknowledged code of intellectual communication – to assist in addressing new collective problems, public confidence and trust in that code is fragile and unreliable.

However, the picture is far from being a static one. As the next section argues, there are promising developments emerging.

Section II

3. ENVIRONMENTAL CONTROVERSY AND THE DEVELOPMENT OF SCIENCE

To develop a clearer picture of how science may be able to assist future developments in public awareness of environment and development problems, a more realistic understanding is needed of the interplay between scientific investigation and social contingency on recent such issues. In this section, we consider some examples and their possible implications.

Two of the most significant global environmental problems identified in recent years have been the depletion of the ozone layer through CFCs, and the ways in which cumulative CO_2 emissions appear to be contributing to changes in the global climate. In each case, scientists have played a major role in identifying the problems and in providing reasons for action by government (Lunde, 1991; Warr, 1990). However, the role of social and cultural forces in encouraging the emergence of scientific recognition of the problems has also been crucial.

In the ozone case, the scientific controversy was kept alive in the 1970s and 1980s by environmental lobbies interacting with scientists in the United States. The resulting pressures appear to have contributed to that country's chemical companies coming to favour the development of non-CFC aerosols, with ripple effects in the European Community by rival manufacturers anxiously to maintain their opportunities in US markets. Boasted by Farman et al's confirmation in May 1985 of the existence of an ozone hole over Antarctica, agreement in principle amongst the relatively small numbers of countries and manufacturers responsible for CFC production subsequently proved relatively easy to achieve, in the Montreal Protocol of The Vienna Convention signed in 1987, a fact which should in no way diminish its landmark political significance. In other words, even in this apparently obvious instance – where scientific research is seen as having generated society's understanding of a major problem – the influence of social forces in shaping both the scientists' own preoccupations and the significance attached in particular historical contexts by influential social forces to what they were saying was considerable – and certainly of greater significance than subsequent official reports (e.g. UK Government, 1990) have tended to imply.

The case of global climate change raises similar issues. No one can doubt the major role of the world's scientific community between 1985 and 1990 in alerting the world to the implications of the greenhouse effect and of increasing CO_2 emissions on the global climate. The sequence of events from the Villach conference in 1985, the Inter-Governmental Panel on Climate Change (1989-90), and now the Geneva Convention negotiations, has been underpinned by an impressive consensus amongst scientists in relevant fields concerning the broad mechanisms and processes involved. But again, important questions about social and cultural processes lurk at the very heart of these developments. The field is dominated by uncertainty. Indeed, what is striking to many observers is that, whilst "hard" scientific proof of cause and effect (between given levels of CO_2 emission and measurable changes in climate) is still as elusive as ever, the social and cultural significance attached to the uncertainties involved has escalated rapidly since the mid-1980s. It is this, rather than new scientific evidence, which has helped propel policy communities in many countries (under the aegis of the UN) towards the possibility of new global agreements on CO_2 reductions. As in the ozone case, despite appearances, it has not been a straightforward case of scientific evidence convincing a sceptical public. Rather, there appears to have been a complex interplay between social and cultural forces and the processes of scientific debate and research on the climate issue – an interplay which has both shaped and reflected society's crystallizing concerns about global environmental change, altering its view of the scientific understanding at its disposal, quite as much as the reverse.

What these two important cases suggest therefore is that

the relationships between science and social processes are in practice more complex than many scientists and politicians have understood them to be. Social attitudes and shifting cultural values exert a major, if largely unacknowledged, influence on the ways in which scientists conceive of what they are doing and how they interpret indeterminacies in their work. Seen in this light, the broad consensus about how to view the uncertainties involved in global climatic change reflected in the 1990 IPCC report cannot be viewed in isolation from the concern of growing numbers of individual scientists to engage more directly with society's broader "cultural" anxieties about the pace of uncontrolled environmental change. Such a hypothesis can only be explored from within a social scientific perspective. Nevertheless, when set against the background of widespread public ambivalence about science highlighted in Section I, developments of this kind should be seen as encouraging. Though on the face of it they imply a more ambiguous, less robust, relationship between scientific understanding and public policy than has frequently been assumed in the past, nevertheless they hint at ways in which new, more constructive social partnerships can be forged with respect to societies' need for environmental awareness.

The ozone and global climate cases are far from the only environmental issues that illustrate the point, though they are, of course, amongst the most graphic, since they are global in their implications. But at more regional, national and local levels, similar patterns can be discerned. As indicated above in Section I, much of the popular appeal of environmental movements – and in particular of NGOs like Greenpeace, Friends of the Earth, and many more nationally based groups – has lain in their success in opening up previously uncontroversial areas of society's commitments to new insights, in the process triggering new research understandings. The impetus towards improved scientific understanding of the deleterious consequences of SO_2 emissions (acid rain), of pesticides, of many toxic wastes, of escalating private car ownership, of lead in petrol, of low-level radiation hazards, and of dozens of similar issues, has been driven more by popular intuition and concern crystallized by NGOs in many countries than by the scientific community alone. Repeatedly, it has been environmental movements outside the framework of official institutions which have drawn attention to and exploited new – frequently marginal – scientific hypotheses or studies, drawing them into the mainstream of official attention. Subsequently, such issues have then become reflected in the preoccupations of official agencies and in new official research programs. One can see these processes now reflected graphically in the wave of well- funded new official research programs, social and natural scientific, on environmental issues in North America and the European Community (European

Commission, 1991; IACGEC, 1991; HDGC, 1989). What were seen a decade ago as largely marginal or minority concerns are now being directed to the center of the research community's attention.

Similarly, at more local levels, particularly (though not only) in developing countries, there are an increasing number of cases where indigenous non-formal knowledge is now acting to temper and redirect environmentally hazardous "modernization" projects to wider social advantage. The case of the Chipko "embrace-the-tree" movement in northern India in the late 1970s and early 1980s provides a graphic example. For forestry specialists, the Himalayas have been seen primarily as a timber "mine" generating forestry revenues. For local women, by contrast, the same forests are sources of water, fodder for cattle and fertilizer for agricultural purposes. The impacts of deforestation were first felt by village communities, who started the movement to stop logging in the Himalayas, by embracing trees marked to be felled. This local knowledge led to a shift in forest policy within a decade; logging was banned, and the primary function of the hill forests recognized scientifically as watershed protection. Numerous similar cases of local populations mobilizing against development proposals underwritten by prevailing intellectual orthodoxy in developed and developing countries alike, could be cited. They suggest that, in many circumstances, public awareness of the dynamics of environmental problems is far greater than scientific or policy elites may recognize, perhaps because the latter tend to recognize it only in their own (in this context, rather parochial) idioms of scientific/instrumental rationality, whereas "ordinary people" may express it in other more indirect or colloquial idioms. Be that as it may, popular movements questioning the scientific assumptions underpinning power station or hydro-electric proposals, the efficacy of particular air and water quality regulatory régimes, or habitat protection standards have helped, repeatedly in recent years, to foster new scientific perspectives.

In all of these ways, environmental controversy itself may well have begun to play a role for society, as a sort of non-institutionalized "arena", in which scientists are finding themselves renegotiating their relationships with other "informal" social actors, outside the matrix of official institutions in complex modern states. These processes are poorly understood, not least because the positivistic understanding of science, as a range of activities detached from such cultural dynamics, is so deeply entrenched in much of the self-understanding of the scientific community. The sociology of scientific knowledge is not yet a subject which commands the serious attention of most researchers. Nevertheless, such processes are taking place, with implications which are important for the issue of "Public Awareness" under discussion here.

The most important of these implications is that there are now conflicting pulls within governments and amongst scientists themselves about what assumptions should be made about the most appropriate scientific approaches to environment and development problems.

On the one hand, there appears to be a continuing faith in the appropriateness (and capabilities) of forms of scientific positivism for establishing unambiguous ground-rules for social and political action. This is reflected tacitly in the language and ambitions of growing numbers of new, officially funded research programs, with aspirations to achieving "prediction" or "management" of the state of the global environment. Implicit in such approaches is what might be called a "task force" conception of the institutionalized mobilization of scientific understanding – analogous to the organization of research communities in the Second World War or more recent programs of space exploration. As history shows, the organization of scientists in such dirigist frameworks has achieved significant successes on occasion, both direct and indirect. When research programs are framed on these lines, relevant institutions dependent on public research funds tend to ensure their intellectual resources are channelled to take advantage of the new opportunities. Moreover, private sector R and D capabilities adjust to reflect (and shape) the related emerging trajectories of public policy. Indeed the private sector, being alert to new or burgeoning market opportunities, has been frequently more prescient and flexible about new developments than its public sector counterparts – as can be seen from the current wave of commercially-driven innovation in a number of OECD countries in the pollution control, energy efficiency and "clean technology" spheres. Overall then, one highly conspicuous form of response by the scientific research community to the post-Brundtland focus on environment and development problems has been to lay claim to a major role in establishing "the facts", as a contribution to helping solve such problems, by mobilizing its resources on lines which have proved productive in the past.

On the other hand, political reality is exerting pulls in a sharply contrasting direction. The increasingly international, indeed global, character of environment and development problems means there is now new and urgent emphasis on the significance of international negotiations and agreements. In such contexts, the prevailing state of scientific knowledge – or more accurately the extent of inevitable human ignorance – becomes itself a variable of great significance in the negotiations. Not only do national negotiators – in contexts as various as the World Climate Convention, the North Sea Convention, or (toxic wastes) – operate within a context of ramifying problems of *technical* uncertainty, in the form of our endemic ignorance of the proliferation of potential feedbacks, synergies, and ecological relationships at stake in any

particular set of phenomena. They are also affected by the more radical (though even less well recognized) indeterminacies of the social world. These may frustrate fundamentally the ability to predict the quanta of pollutants at stake, since it is generally impossible to make meaningful predictions about the potential decisions or behaviors of the large number of other social actors who may contribute to the future scale of a particular phenomenon. Still more fundamentally, experience (for example in the toxic waste sphere) shows that even the ways in which pollutants are to be *defined* may vary from one cultural context to another, thus compounding chronically even the basic problem of agreeing workable definitions of the pollutants at stake (Wynne, 1989b).

As one observer has put it, where in the past the relationship between science and politics "was predicated on the belief that science provided decision-makers with objective "hard" facts on which to base their "soft", value-ridden policies", nowadays "the environment is but one area where we now find scientists delivering only "soft", uncertain "facts" to decision-makers facing "hard" decisions" (Newby, 1991). Over the past decade, this reality has led to increasing pressures for "precautionary" action in the making of environmental agreements and legislation. As politicians, pressed to respond to public concerns, find their scientific advisers unable to give them unambiguous answers to an increasingly well-publicized range of potential environmental hazards, they are being forced to act to control particular pollutants ahead of scientific "proof". Variants of the "Precautionary Principle" first framed by the West German government in the early 1970s (Von Moltke, 1985) have already found their way into a number of environmental negotiating fora, such as the North Sea Convention and EC Council of Ministers, though the problems involved are acute.

Strikingly, however, application of precautionary doctrines in global negotiations has proved even more problematic. Moves towards a global climate Convention provide a case in point. Given the immense differential political and economic domestic implications of action to curb fossil fuel consumption in countries at sharply different stages of economic development, the "softness" in scientific understanding of the physical processes involved offers openings for continuing arguments about the appropriate distribution of burdens. Disagreements about the political weight to be attributed to the pervasive scientific uncertainties have been a feature of even US policy towards CO_2 controls – a development which many observers see as reflecting the influence of domestic interests which stand to lose in the short term from more stringent global policies.

In the context of this chapter, however, the importance of such contingencies lies in their implications for an understanding of the actual or potential role of science.

The assumption of many scientists is that it is only scientific evidence which can generate the social authority for political action on complex environmental problems. But what this and other cases suggest is rather the reverse – that the extent to which scientific analyses of particular environmental problems can gain wider authority may itself depend on *the prior existence of "cultural" consensus about the scope and parameters of the problem(s) at issue*. Thus, within the framework of the North Sea Convention or the EC Council of Ministers, where there is already loose consensus about the character and boundaries of particular pollution problems, and a shared political interest in mitigating them, the "softness" in scientific understanding gives rise to relatively few (though admittedly still significant) difficulties. Indeed, it is no coincidence that it is within the European Community, whose member states have developed habits of continuing collaboration in a range of policy fields backed by binding legislative frameworks, that the precautionary approach has begun to gain purchase. Where, by contrast, such prior "cultural" consensus does not exist, as is strikingly the case amongst many of the countries involved in the North-South negotiations of greatest significance for global environmental change, the same "softnesses" become apertures for elaborating a bottomless pit of disagreements (on questions of global equity for example) of much more intractable social and political varieties.

In the latter contexts, even the claimed neutrality, detachment and reductionist approaches of western scientific practice may become seen as further facets of Northern political and economic hegemony. The overwhelming concentration of the world's scientists in Northern countries, and the extent to which even independent research institutions are officially funded in ways which generate still more "expert" knowledge which *de facto* reinforces Northern perspectives (however unwittingly), may also have the same effect.

Drawing these threads together, it becomes plain that the political context in which science has now to be conducted and applied on urgent global environmental questions is generating quite new strains for our understanding of science itself. The "task force" conception outlined earlier, with its positivistic assumption that research can be so organized as to reveal, potentially, a body of incontestable fact about global environmental processes, thus fostering the emergence of universally shared perspectives on the structure of the problems to be solved, is in sharp contrast to the more prosaic reality of real world international negotiation on global questions whose full shape and implications turn out to be radically indeterminate and culturally contingent. As we have already argued, the critical uncertainties on such questions are turning out to be less uncertainties of a physical, technical kind, than

social, cultural and institutional. In this context, the "task force" approach to science's role begins to appear naive and even irrelevant – other than in relation to specific problems on whose dimensions and significance there is already prior and widespread international agreement. The contrasting picture of science as a set of conventions and practices embedded in socially and culturally determined contingencies makes more sense of the ways in which negotiations on such questions as global climatic change have actually been developing. It also has immediate implications for the research community. It challenges natural scientists to begin to develop a greater awareness of the cultural dimensions of science itself, and a more appropriate sense of their need for more negotiated interaction with diverse publics. This being so, it also reveals a potentially important new role for *social* scientific research.

Hitherto, the social sciences – principally such disciplines as law, economics, geography and sociology – have begun to be incorporated into international and national environmental research programs as subordinate to the natural sciences. The implication has been that since the natural sciences can be relied on to reveal progressively more of the character and dimensions of environmental problems, the social sciences are needed to assist in the development of mechanisms – fiscal, legal, cultural – to enable strong environmental policies and agreements to be achieved world-wide. There is undoubtedly just such a role for the social sciences. However, the analysis in this section suggests that there is now an additional, perhaps still more significant contribution they should be encouraged to make. This is to aid improvements in society's understanding of the cultural circumstances in which different forms of knowledge – be they scientific: positivistic, mythic, religious, historical – come to achieve social authority in the real world, as opposed to the closed world of the laboratory or particular closed social institutions.

The examples cited in this section suggest that real world controversies surrounding critical environmental problems may be beginning to encourage new relationships between science and the public. It is striking to recall that, as recently as the mid-1970s, many scientists were equating environmental concern with irrationality and "anti- science" (Maddox, 1975). The ecological sciences were seen as somewhat suspect, against the dominant physics-based reductionist models of "good science". The changes in outlook that have been emerging recently within scientific institutions are thus of the greatest significance for the whole of society. Their causes and structure need urgently to be understood. This is a task which should involve social as much as natural scientists. The results should be seen as highly significant for the self-understanding of the scientific community itself, and for the way in which it organizes itself in future.

At the more immediately practical level, to accelerate the emergence of new relationships between communities of "experts" and the rest of society for their collective benefit, there are a number of steps which can be taken. We now consider these, in the section following.

Section III

4. SCIENCE AND PUBLIC AWARENESS – MOVING FORWARD TOGETHER

Our suggestion has been that science and scientists do not stand outside society's understanding of its environmental problems they are far more implicated in them than they have tended to realise. Encouraged by the spectacle of more and more rapid and encompassing technological development reaching into every corner of the globe, modern societies – their publics and their governments – may have assumed till recently that science could be relied on to generate an endless succession of instrumental solutions to the proliferating side-effects of these patterns of development. But such assumptions are less and less credible. The pervasive sense of environmental crisis which now preoccupies a growing, and influential, proportion of the populations of developed and developing countries alike signals scepticism at that conception of science's role. Nevertheless, the capabilities and prestige of scientific communities world-wide are substantial. They need to be seen as resources available to society.

What this suggests in practical terms is that there may now be two broad ways forward for scientists and scientific institutions to contribute to the processes of enlarged understanding by society of the problems it faces. First, there is undoubtedly an educational role of a more or less traditional kind to be played, aimed at helping heighten public awareness of particular problems on which research has thrown light. But secondly, and more problematically, there is a need for a range of new measures and mechanisms to generate more flexible interaction and cross-fertilization between scientists and other social actors, in order to accelerate the process of integrating new cultural perspectives into scientific research priorities and to stimulate amongst scientists themselves a more appropriate sense of the limitations and culturally contingent character of their own very powerful epistemologies. Just as scientists can assist public awareness of the problems society faces, so too scientists need society's help in generating more realistic awareness of their own roles and responsibilities in the new context of environmental "crisis".

At the most familiar educational level, there are already encouraging, if still embryonic, developments in the educational systems and mass media in many countries. Environmental news agencies such as Panos and Earthscan, the emergence of international film and TV production companies committed to environmental education, and the recognition in a growing number of countries of the significance of environmental studies within the curricula of schools and tertiary educational institutions – these are all healthy developments, which can build awareness of new scientific insight in the environmental sphere. Scientific institutions need to integrate themselves more energetically into such developments.

However, it is striking that far and away the most influential, and indeed "educational", use of the communications media for increasing environmental awareness in recent years has been that by the NGOs – Greenpeace and others world-wide. The inventive staging of events in terms attractive to the media, to dramatise particular environmental problems, frequently with a major scientific component, has become the stock-in-trade of such groups, and has undoubtedly had major social and political impacts. These successes are instructive. They have been achieved because groups like Greenpeace have fashioned themselves to operate within a world of instantaneous communication and fleeting imagery. The particular images they create, which have reached so deeply into public awareness over the past decade and a half, are experienced as what might be called condensed moral fables about contemporary society, fables in which human moral values are seen as pitted against impersonal and powerful commercial and technological forces. Some have suggested there is a danger of such media skills -and associated campaign generation – becoming self-sustaining, maintaining controversy for its own sake, rather than simply to communicate "accurate" messages. But that is to miss the point. The true significance of such NGO successes lies in the cultural resonances the fables themselves have struck in a strikingly sympathetic public opinion. These tend to confirm that scientific forces associated with much technological development are now sensed as socially irresponsible and threatening, even amongst prosperous populations whose material dependence on the fruits of such development is inescapable (albeit in a certain sense unchosen).

This points towards the second, and more complex, area in which the scientific community now needs rapidly to foster innovation. Environmental movements can be pictured as symptoms of industrial societies' unease at gaps between the promises of such societies and certain moral and physical realities the latter have spawned. Ironically, the doctrines of "detachment", "moral neutrality", "problem- solving" and reductionism, which have been so integral to the success and institutional momentum of post-war science, may actually have been feeding such gaps. One tacit message of the success of environmental movements world-wide may be that, precisely because scientists have been so successful in conveying a picture of themselves as professionally neutral and outside the social fray, they are at risk of losing their sense of professional connection with widely sensed human needs and ends.

How might this be remedied? In Section II, we have argued that particular environmental "causes célébres" have served as arenas in which new ranges of social and cultural variables have begun to be brought to the attention of science. In most countries, these processes have taken place outside the framework of formal Parliamentary mechanisms. Only rarely in the 1970s and 80s – countries like Sweden and the Federal Republic of Germany may be exceptions – have the processes been drawn into the mainstream of political parties or similar established political institutions. Indeed, in most countries, there have been an uneven series of attempts since the late 1960s to develop new administrative processes for anticipating and debating potential environmental impacts.

Techniques of Technology Assessment, Probabilistic Risk Assessment, and Environmental Impact Assessment, as well as less methodologically structured forms of public inquiry, have been developed, as means of seeking to "manage" the new social concerns. Parallel processes for the evaluation of new and existing toxic chemicals and new patterns of food production have also assumed increasing significance. However, a recurrent source of tension in all of these developments has been the existence of tacit biases in the largely reductionist appraisal philosophies employed – and the extent to which the expertise available for taking advantage of them has been dominated by existing powerful interests. So although, through NGO institutionalization of public involvement in the processes, new social variables have been insinuated into the debates, this has taken place too much on the terms of the epistemological communities whose actions may have been those triggering public concern in the first place.

Nevertheless, the scientific community should now involve itself much more actively in the openings for public debate and challenge provided by these processes. In the environmental field, they have become highly significant arenas for the exploration of important cultural tensions.

With the escalating mainstream political focus on environmental issues in OECD countries since the mid-1980s, and the accompanying recognition that outright avoidance of pollution is more desirable (and potentially better business) than its post facto regulation, new cultural perspectives – such as those being explored by the Stockholm Environment Institute in the field of waste reduction (SEI, 1991) – are beginning to attract attention. Again, there is scope here for much more active scientific involvement, in processes which signal and may encourage important new cultural adjustments.

Fundamental to these new debates is the need for a freer flow of information on all matters relevant to environmental quality, to facilitate more fully informed public engagement. Freedom of Information legislation, regular State of the Environment reports and frank environmental audits are necessary to assist such public discussion. Scientific institutions, learned societies and professional associations need be active in pressing governments and international agencies for legislation requiring them to be made available as a matter of routine. Similarly, representative bodies of scientists would do well to play more active political roles in helping generate more open administrative procedures for public participation in arguments about environmentally significant technologies – for example, on the lines of the statutory "public hearing" procedures under federal Canadian law or the "major public inquiry" mechanisms which have evolved over the past decade and a half in the United Kingdom. As already explained above, the significance of these processes for the future direction of scientific research priorities – the development of what sociologists of science have termed "post normal" or "vernacular" science (Funtowicz and Ravetz, 1990) – is considerable. A corollary of this is the need for clearer, perhaps statutory, ground rules to be developed for "whistle blowers" – that is, for experts employed in public or private bodies who make available information about deceptions or allegedly dangerous practices in environmentally hazardous industries.

Through advocating and participating actively in socially important innovations of these kinds, the scientific community can achieve benefits of two varieties. First, it will assist the emergence of open debates on issues of significance for contemporary society. But second, and equally important in the present context, it will create opportunities for more fluid interaction between scientists and wider social and cultural forces. The relationships that result should aim to produce new sensitivity amongst intellectual specialists towards societies' deepest anxieties (not least about science itself). Similarly they might be expected to engender a new respect and sensitivity towards the significance for locally applied "solutions" of local informal cultural knowledge.

Finally, complementary to all of these prescriptions is a further need, highlighted in the "Capacity-building" (Chapter 14) in the present volume.

The complexity and multifaceted character of the environmental problematique suggest that present boundaries within the natural sciences, as well as between the natural and social sciences, and indeed between reductionist scientific epistemology and other forms of knowledge (vernacular, religious, intuitive, aesthetic, moral) may now be increasingly unhelpful. Such boundaries have evolved for their own, doubtless compelling intellectual reasons, but in present circumstances too rigid an adherence to them runs the risk of perpetuating rival sets of understandings of what is at stake rather than any more integrated understanding. (The contrasting, and largely incompatible, representations of "the environment" in, say, ecological science, economics and the sociology of knowledge provide a graphic illustration). Recognition of this reality poses acute challenges for academia itself. There is now a vital need to create contexts (institutional homes, research funding programs), to facilitate the emergence of cross-disciplinary understandings and interactions. These in turn need to be translated into teaching programs for a new generation of students in higher education institutions, and into critical research programs for policy communities. There are embryonic signs that such syntheses may be emerging in certain western countries. These developments are discussed further in Chapter 16 (Lang *et al.*) in this volume. For the reasons given in the present chapter, learned societies and scientific associations have a long-term interest in accelerating their emergence. No single initiative could have greater implications for society's improved understanding of environmental problems.

To conclude – it is an obvious implication of this chapter that no group of "experts" can hold "the answers" to society's environmental problems. But the scientific research community has a vital role to play in contributing to public awareness of key dimensions of many of the problems.

The extent to which science's contribution will be experienced as truly useful will depend on the ability of its practitioners to develop new sympathy for and interaction with the insights and perspectives of other, less formalized bodies of knowledge within society at large. Such processes will be helped if "expert" communities can develop a richer epistemological self-awareness, recognizing the contingent character and human limitations of their own intellectual techniques and disciplines. This applies as much to such instrumental social scientific disciplines as economics, as it does within the natural sciences.

The environment and development problematique is confronting society with a sharp new sense of the limits of human mastery and understanding. It poses new ethical and cultural challenges (CSEC, 1991). The richer and more realistic integration of science into these shared concerns will help mankind in the difficulties that lie ahead.

REFERENCES

Barnes, B. and Edge, D. (eds) 1984. *Science in Context.* Oxford, Open University Press.

Beck, U. 1986. Risikogesellschaft (The Risk Society).

Berman, M. 1984. *The Re-enchantment of the World.* Bantam Books.

CSEC 1991. *The Emerging Ethical Mood on Environmental Ethics in the UK.* Report to World Wide Fund for Nature (UK Center for the Study of Environmental Change, Lancaster University.

European Commission. 1991. *Environment 1991-1994. Third Framework Research Programme.* DGXII, Brussels.

Eyerman, E. and Jamieson, T. 1991. *Social Movements: A Cognitive Approach.* Macmillan.

Funtowicz, S. and Ravetz, J. 1990. *Global Environmental Issues and Post-Normal Science.* Council for Science and Society, London.

Giddens, A. 1990. The Consequences of Modernity. Polity Press.

Gilbert, G. and Mulkay, M. 1984. *Opening Pandora's Box: A Sociological Analysis of Scientists' Discourse.* Cambridge University Press.

Grove-White, R. 1991. The emerging shape of environmental conflict in the 1990s. *Journal of Royal Society of the Arts* No. 5419, pp. 437-447.

HDGC. 1989. Human Dimensions of Global Change: An International Programme on Human Interactions with the Earth. HGDC, Toronto, Canada.

IACGEC. 1991. *The UK Research Framework. Inter-Agency Committee on Global Environmental Change*, NERC, UK.

Jonas, H. 1984. *The Imperative of Responsibility: In Search for an Ethics for the Technological Age.* University of Chicago Press.

Krohn, W. and Weyer, J. 1990. Society as Laboratory: The Production of Risks by Experimental Research. Unpublished MS.

Latour, B. 1987. *Science in Action.* Open University Press.

Lunde, L. 1991. forthcoming. Science and Politics in the Global Greenhouse - a Study of the IPCC Process. Fridtjof Nansen Institute, Norway.

Lyotard, J-F. 1984. *The Postmodern Condition: A Report on Knowledge.* Manchester University Press.

Maddox, J. 1975. The Doomsday Syndrome.

Merchant, C. 1982. *The Death of Nature.* Wildwood House.

Moltke, K. Von. 1985. *The Precautionary Principle.* Institute for European Environmental Policy, London.

Mulkay, M. 1979. *Science and the Sociology of Knowledge*, Allen and Unwin.

Newby, H. 1991. One world, two cultures: sociology and the environment. *British Sociological Association*, Network No. 50.

Lord Porter 1990. Anniversary Address by The President. *Royal Society News* (Supplement). Vol. 5, No. 12.

Sardar, Z. (ed.) 1984. *The Touch of Midas - Science, Values and the Environment in Islam and the West.* Manchester University Press.

SEI 1991. Towards a Clean Society. Stockholm Environment Institute/Center for the Study of Environmental Change: Conference Papers.

UK Government. 1990. This Common Inheritance (White Paper). HMSO.

Warr, K. 1990. Ozone: the Burden of Proof. *New Scientist*, **128**, 36-40.

Winner, L. 1977. *Autonomous Technology.* MIT Press.

Wynne, B. 1987. *Risk Management and Hazardous Waste.* Springer-Verlag-22.

Wynne, B. 1989a. Sheepfarming after Chernobyl: A Case Study in Communicating Scientific Information. *Environment,* **31,** (2), 11-39.

Wynne, B. 1989b. The Toxic Waste Trade. Third World Quarterly June 1989, pp. 120-146.

Wynne, B. 1990. Knowledges in Context: Background Paper for Public Understanding of Science Project. Science Policy Support Group, UK (unpublished ms).

Annex to Chapter 13

S. Kapitza

Global environmental issues now loom on our common horizon. After world population growth the environment certainly is the most important of all issues. Our habitat, the world where all people live and multiply, work and enjoy life has an overriding importance for all of us. The environment sums up all human experience. From towns and villages, to roads and waterways, to industry and agriculture, the environment on a planetary scale is the summary result of our activities. The human being has become a factor that independently and powerfully is changing the surrounding world to which he himself belongs. Today we have not only to understand what is happening, but we all of us have to act. Understanding can be provided by science. To act we have to have political will and an industry that can execute the decisions made.

This sequence of connections is too simplistic. Science has advised industry in its relentless development. Industry itself has put pressure on governments to foster development at all costs. On the other hand, have not the governments directed huge resources to devastate the environment by the instruments of war? For modern war, waged by weapons of mass destruction is a war against the environment. By fire or explosions, by poison or radioactivity we can and do ruin the environment for human life on a scale not known before. Society faces all these issues, with people often lost in the rapid developments of our modern world, a world where relentless progress is often misguided and poorly planned, a world that has finally become unproductive if not a menace to our common future. The divisions of our world, the widening gap between the haves and the have nots, these social and economic strains are then multiplied by mass communication, travel and trade enhancing differences, that may lead to profound instabilities and great social upheavals.

In this complex world, guidance was expected from science. But has science delivered? A child of enlightenment and humanism born by reason and having developed into a world-wide cultural and intellectual phenomenon, can science really give advice and answer the pressing issues of our time? Can and will society, through its public institutions, and the people themselves, follow the advice given? Can science really deliver to those in power the changes sweeping our planet, or are powers of our own making too great to be controlled and abated by the very forces that made them possible? These are the questions we have to face.

First we will look into what science is and what can we expect from science and scientists. Next we shall look at society, the people and their public attitudes on these issues and finally suggest recommendations for future action.

Action can be taken in the field of research of science. We have to look into the way society is affected by education and the mass media and how people now react. We will have to take into account both the disparities of our divided world and the common factors on which our destiny depends globally. Both existing actors in industry and government have to take notice of these developments. Finally, new instruments for action will have to be conceived, new organizations having the political will and power to operate on a global scale will have to be built and means found for them to deliver.

SCIENCE

Science in our modern world has become both the agent of progress and a crucial component of development. At the same time it has become a culprit of its own power. Science for the English-speaking world means hard science. But we must remember that Nauka – science in Russian, like Wissenschaft in German – encompasses also the humanities, soft sciences in the parlance of the natural scientist, expressing perhaps the arrogance and sense of superiority of the hard scientist.

This dichotomy is more than a linguistic or semantic difference. We are dealing with two components of science and are looking at an important division splitting its body into two halves, much like our brain is split into two hemispheres. In fact, in the anatomy of our mind this division is a most significant phenomenon.

The hard sciences, calling themselves natural in the first place, physics and chemistry, then biology and geology have developed mainly through a persistent pursuit of analyzing nature into its components, reducing the whole into its parts and then taking to pieces those parts and going still further in search of the ultimate cause. This reductionist approach has proven to be remarkably powerful and intellectually satisfying. In no field has it been so successful as in fundamental studies of nature, motivated by the desire to know, to understand. Thus a remarkable model of nature has been created. By using this knowledge, no less remarkable applications have been invented and engineered by industry.

On the other hand, the development of knowledge, of our understanding, has proceeded also in an opposite direction. From the simple and elementary, the scientists can go to the complex. At each level of complexity a new degree of insight is provided in comprehending the whole, the system of things or ideas. With each step in generalization of our vision we get closer to the real world and discover new concepts penetrating the complexity, introducing new intellectual entities and laws. This synthetic approach is all the more important in that it is in this way we manage to get insight into social phenomena, into ecology and finally into the environment.

But progress in a holistic, integrative understanding of the real world is intellectually much more demanding than the analytic and reductionist approach. Perhaps it could be said that to pursue analytic work all one has to be is clever. To generalize, to explore and summarize, to understand and explain the complexity of the real world, one has to be wise.

Unfortunately, most issues of social relevance, especially of the environment demand an integrative treatment. It is this multidisciplinary approach that makes it so difficult to proceed. For no sum of expertise, no number of specialists with profound insights into their special field of knowledge can by mechanical addition produce the vision of the image of the whole. Intellectually this is one of the great challenges facing modern science, science traditionally dominated by reductionist reasoning.

Even in its organization modern science is not equipped to face these realities. The great institutions of Science, the Academies and Universities are at best a collection of scientists of different disciplines, divided into departments of their own and interacting on a small scale, busy with protecting their own turf in institutes primarily designed to be universal in their mission.

The extent to which science is poorly organized can be seen by the great difficulty in obtaining say a PhD in interdisciplinary studies. Few universities grant such degrees and this recognition can only come at a later stage in a career of a scientist pursuing integrative studies.

This has all led to a certain crisis of identity for modern science and contributed to its alienation from the pressing needs of society. In fact, it may be seen as a difficulty of matching the individualistic attitudes of a modern scientist born into an individualistically minded society, expressing in a socially relevant setting the conflict between the whole and its part. Today, with the predominance of individualistic attitudes, of rights over responsibilities it is not surprising that in the relevance of society towards science a misunderstanding of the past has developed into a conflict of the present.

SOCIETY

This conflict of the whole and the part has now led to a substantial loss of rapport between society and science to the extent that alternative systems of reference, if not knowledge, are professed. In examining the attitudes of Society to science we have to bear in mind the dynamics of current social developments. The present critical conditions in some societies, in the first place in the Soviet Union, have led to manifest distrust of science, of its institutions as part of the establishment (Kapitza, 1991). In other countries expressions of the public distrust and apprehension of science are also seen when the authority of science is challenged. On many occasions this public criticism is proper and signals that science is not beyond reproach, as in fact any social institutions are. Democratic and free enquiry is what can and should help, although in science the answers are based not on public unanimity but by applying scientific reasoning.

Science itself has also contributed by internal difficulties of its own, when the integrity, if not professional honesty of scientists has been publicly questioned. On many occasions science has manifestly not delivered: the promise of safe and abundant nuclear energy, the deliverance of a cure against many common deseases, the promise of machine translation given some 30 years ago is still not with us, space science has certainly provided remarkable results, but also some unfortunate and misguided developments. Is this a temporary folly of development or is science really in trouble? On a number of occasions science has been oversold and the public has lost some of its confidence in scientists, if not in science itself. It must be said that science has been not at its best in delivering its message. The unfortunate separation of research and teaching has contributed to the alienation of science, just as the subservience of science to the military has not on many occasions improved its image in the public eye.

The message of science is delivered to the public by the mass media. Co-operation between scientists and journalists has not contributed to a better public understanding of science. Apart from sensationalizing scientifically important news, the media have undoubtedly contributed to a certain distortion of the public awareness of science. We must also note that academia has mostly been unco-operative in its interaction with the media, leaving much of the reporting to journalists with a lack of scientific training and hardly ever supporting the scientist engaged in the media. The scientific community will denigrate and scorn at a scientist talking to millions on TV demanding much public exposure, when teaching to a dozen students is always considered to be quite proper. Only recently under the pressure of current developments the academic world is gradually changing its mind on these matters, as can be seen in the publication of journals devoted to the public understanding of science and recognizing the social relevance of these activities.

Of special concern should be the responsibilities of those who express their attitudes through the media. In the first place, for whom is a scientist speaking? Is he speaking in his private capacity or is he voluntarily or not expressing the attitudes of the scientific community, of the visible or invisible college to which he belongs? To what extent a scientific controversy can be brought out to the public, a controversy where the specialists themselves are still indecisive? In this case the public will often follow not the one who will prove to be scientifically right, but the speaker or writer who is more literate, eloquent and persuasive. Thus professional ethics have to set the rules and establish certain norms of conduct. One has to bear in mind that the public has to, is even forced by the persuasive power of the media to, believe the expert, for no proof can be offered as it is done in the process of teaching. A student is offered both a proof of a statement and has the capacity and opportunity to question his teacher. All who passively listen to or even read the media in one way or the other are prone to take most statements on trust, or being misled lose all confidence in science.

Finally, today with the advent of the global electronic media – of satellite TV – we have a unique opportunity to literally speak to the world. Unfortunately this remarkable hardware is not used to propagate the global message, to develop public attitudes on the great issues facing us all. This is both a pressing demand and a challenge for the media and the world scientific community.

In the first place we have to be sure of the message to be conveyed to the world. Current experience shows that on many pressing issues it is often very difficult to give unequivocal answers. Hesitancy and indecisiveness follow, not improving the image of science. In this case it is the responsibility of the experts, of the scientists who can best express what is generally taken to be known without sacrificing the truth.

If we are to develop public awareness on global matters, a systematic and long-term effort should be launched, to develop an attitude towards our planetary environment and hopefully instil a system of values. We really have to

invest in the future, delivering a systematic message to humanity on the global environmental imperative.

Scientists are facing an exciting and challenging possibility. Again as in much of modern socio-scientific issues the hardware is here, waiting to be properly used as most of our effort has to go into developing the software. Apart from working out the message itself, perhaps the greatest difficulty is in delivering the common message into different cultures even different worlds, the first and third in particular, where the divisions and differences are so pronounced. This is not so much a question of language or style of presentation. It is a more profound demand to match the universal message into different cultural environments.

The attitudes of different cultures and civilizations are probably the most diverse and fundamental issues we are facing when we have to deliver a message, a message of science. In a more profound way the problem is really of the basic difference in values, the deeply instilled system of beliefs and customs that we may be not only challenging, but even trying to change, and thus questioning something very profound and personal to everyone involved.

In discussing the imminent problem of population growth the number of children comes up, touching on the very basis of social and family life. We have to deal with attitudes towards women and the old or of an indigenous culture and tradition as exposed to, and being vulnerable to, the pressure of the so-called Western civilization. Abstract values are expressed in concrete things and definite actions, illustrated by examples that may or may not be generally applicable, although they come from a universal and generally sound origin.

This universality is one of the issues impeding the way the message of science, of world science may be broadcast. In this case we once again encounter the arrogance of science and of the scientist, who often finds it difficult to deliver his message in a socially agreeable fashion. In this case being right is not all, one has to be understood in the right way.

Public attitudes are to a great extent formed by the effect of impact of often catastrophic events. Chernobyl and Bhopal, the Three Mile Island and the fires on the Kuwait oil fields – these events capture the public mind, as they are focussed and magnified by the media. Often they become the origin of new departures in social awareness with significant social consequences. This, for example, has led to the present public attitude towards nuclear energy in the Soviet Union.

A gradually evolving ecological disaster unfortunately rarely attracts public attention. Only when it has grown out of all proportion does it become, like the Aral Sea in the Soviet Union, a catastrophic event. Only then does the public belatedly become aware of what has happened, in most cases alerted by the media and powerful imagery, justly stirring public emotions. In most cases this happens when it is too late, although occasionally a gradually developing ecological disaster can be caught in its beginning, as it happened in the case of the Lake Baikal.

On a greater scale this is what is now happening to the global environment. The persistent accumulation of CO_2 in the atmosphere in the long run is to change the climate. Should we dramatize this expectant change, or will society simply evolve with the changing environment? How are we to prepare the world and different regions for this eventuality or is this the challenge facing us now, demanding immediate action? At present what should be our concern is the public awareness on these matters in developing attitudes towards this development.

Should we focus on the imminent or not so imminent danger of, say, a rising sea-level for Bangladesh that may be a menace to tens of millions of people in some decades, or should we think about the shift of grain production further to the North in Canada or Siberia?

Is progressive desertification and the movement of Sahara to the South really the greatest issue for Africa today? Often we find these issues to be at the center of public discussion. With no answers provided deep emotions are excited. Could it be that in this case an alarmist attitude is counterproductive? Take for example Bangladesh. It is obvious that it is much easier to protect by dams a country of tens of millions of people where the ratio of population to the size of the dam would be much less than say for Holland that over the last centuries has reclaimed land from the sea and at present has large areas some meters below sea-level. How in the past has the human race reacted towards climate changes? Did the Neanderthalians have conferences on the recession of landbound glaciers or on what happened with the Arabian peninsula before the forest disappeared in the days of the Queen of Sheba when a great irrigation system worked and the land prospered?

Probably some history may teach us more than apocalyptic images devised to scare, rather than enlighten, as so often is happening now in the presentation of global climatic changes. Or do we have to scare people into action? This is a political dilemma in which modern science has been caught, but one of the complex problems scientists have to face.

Once again, no prescription will work, and again we have to go back to the fundamental principles of science, of rational reasoning. The history of science and society has relentlessly taught us that emotional reactions however powerful they may be can only then be productive when they are backed by basically sound and honestly expressed scientific arguments. On the other hand, science, modern science and world science are not omniscient. The limitation of scientific enquiry should be recognized both by the public, the governments, the political bodies and the scientists themselves. The ultimate responsibility lies in recognizing the limits of knowledge. These limits exist both in time and in space, in the depth of our understanding. Perhaps it is here that the measure of wisdom is to be found.

Scientists are but part of humanity and we cannot expect them to be both superhuman or excuse any inhumanity. The public should be also aware of these limitations. Only then can we expect that a proper understanding of our duties will be achieved.

Human conduct is ultimately expressed by values and this applies also to scientists who apart from universal human values are also exposed to the spirit of the scientific endeavor. For many scientists this is a very significant and powerful experience, but on no account can it replace or overrun his or her human objectives. Scientists are often prone to extend their mandate to expertise beyond the immediate field of knowledge. Can a very clever mathematician be as good in public matters as he excels in science? Is a Nobel Prize always a mark of wisdom? Here again perhaps exaggerating the issue we have a socio-scientific problem of relevance.

GLOBAL GOVERNANCE

We have not yet referred to the subject of global governance. Here the world is in the very beginning making the first and hesitant steps after the end of the Cold War. Only recently the UN has managed to really act globally in military matters, acting in fact also on an environmentally relevant issue. Perhaps it is no accident that one of the results of the Gulf War was a great environmental disaster, for modern war more than ever is war against the environment, be it the oil fields or the city of Kuwait.

On the other hand, the unseen war on the global environment is already with us, for there is no other way of describing the summary activities of the human race against its own environment.

Unfortunately here global action is taken slowly and indecisively. For example, does world science have a say at the UN? Today the influence and significance of scientific expertise is such that not to take it into account is seriously to limit the credibility of a world organization. Does industry have to react to decisions that would be made for the cause of the future of our planet. Can the necessary political will be mustered and funds provided for action? At present we can at best pose these questions.

After many years of deliberations and lengthy negotiations we have the Law of the Sea, although it has yet to be properly implemented. Our common heritage is also the atmosphere and the first steps have been made toward its protection. The Montreal Protocol on the Ozone layer is an essential step, although the ban on Nuclear testing in the atmosphere should be considered as an international measure protecting the atmosphere and human life from radiation, once again showing how intimately linked are environmental and military matters.

In the future the international dimension and the role of the UN in acting on environmental matters can only increase. The expectant outcome of the 1992 UNCED meeting in Rio should help to make these developments decisive. With this in view the following recommendations are made:

(a) The world of science should be duly represented at the UN, to voice systematically the opinion of scientists on crucial matters facing the international community.

(b) The UN, UNEP and the organizations responsible for global scientific projects should give all necessary support for international interdisciplinary projects, of which Global Change is currently the most significant, for studying the global environment.

(c) National Academies of Science, international scientific institutions and universities should encourage interdisciplinary research and find ways of establishing high standards of scholarship relevant to global issues and environmental problems, on no account sacrificing the quality of research for its quantity.

(d) The academic community and the media should find ways and means systematically to propagate the global message of science to the world.

(e) The message on global issues should not only be delivered world-wide, where due use should be made of the global TV networks, but greater attention should also be given to matching the message to diverse cultures, taking into account national traditions and aspirations in generating a common new world ethic.

REFERENCE

Kapitza, S. 1991. Anti-scientific and anti-technological trends in the Soviet Union, *Scientific American*, August 1991.

Chapter 14: Capacity Building

D.A. Bekoe and L. Prage

EXECUTIVE SUMMARY

Capacity building is defined as the acquisition of the skills, knowledge and institutions and the generation of motivation to enable certain specific ends to be achieved. The nature of the capacity required for implementing Agenda 21 and the additional scientific agenda of the scientific community into the 21st Century is outlined. Capacity building applies to people as well as to institutions. It requires education, training and the facilities for learning to enable people to cope with their changing worlds. The transition to sustainable development has to be made by everybody. For successful implementation of the anticipated UNCED decisions, i.e. a radical change towards sustainable development practices, capacity building will be necessary in all countries but much more so in the developing countries of the South, and even more so in Africa.

It is suggested that a successful transition will depend on a number of factors which include an orientation of existing capacity in the North, development and strengthening of capacity in the South, the integration of local knowledge with the results of research, and effective action programs which satisfy local needs as well as those related to the global commons, i.e. the atmosphere, the oceans and coastal systems. Interdisciplinarity in approaches to studies of the Earth System and to problem-solving should be encouraged.

Some of the problems of scientific capacity building in the South and strategies for their minimization are discussed.

It is recommended that:

(a) knowledge of the relationship between the environment and development be made available to all citizens through education, training and learning facilities; special attention is to be given to the needs of women, decision-makers and opinion-leaders;

(b) interdisciplinary skills and approaches be cultivated and strengthened;

(c) strategies for capacity building be carefully planned taking into account the specificities of regions and countries;

(d) an environment and development system be established at the national level to make decisions and ensure that all parts of the system are playing their expected roles;

(e) one Regional Environment and Development Center be established in each region to facilitate training and implementation of regional programs;

(f) a capacity audit for each country be carried out to find out the extent of preparedness at all levels to implement UNCED decisions;

(g) the attention of scientists from the industrialized countries be focussed on the environment and development agenda to enable them to fill in gaps in trained manpower and make the orientation necessary for successful implementation of the transition to sustainable development;

(h) the necessary financial resources for capacity building be found at the national level and through appropriate assistance programs.

1. INTRODUCTION

Capacity building in the sense in which the term is used in this chapter applies to people, to institutions and to the processes of scientific learning and effective dissemination and utilization of knowledge. It therefore implies adequate provision for both training and learning. Training comes in the form of preparing all kinds of citizens for the developmental challenges that lie ahead if anything like sustainable existence is to be achieved. Learning is a process of both self-discovery, and of understanding how people hope and fear, acquire, impart and apply knowledge, and cope with their changing worlds. Capacity building therefore implies the extension of the power of education throughout the citizenry as well as to scientific specialists so as to facilitate a creative dialogue. In this sense therefore, capacity building is closely related to, and forms an essential component of public awareness and participation in pathways to sustainable futures. But capacity building as applied to institutions, in the sense of setting up and strengthening the appropriate scientific, educational and policy-oriented institutions to study phenomena, acquire, impart and apply knowledge for the purpose of achieving sustainable human existence, is just as important. Without such institutions, the education of the citizen – the training and learning processes – will be incomplete and will lack the dynamism that goes with the continuous improvement in the state of human knowledge. In this regard, many references have been made in other chapters of this book to gaps in our knowledge of the Earth System and the impact on it of development activities.

Capacity building in the context of the United Nations Conference on Environment and Development refers to the ability of individuals, communities, countries, regions and the whole world to implement Agenda 21 and other decisions and conventions of the Conference. The agenda of the scientific community and its collaborators in the social and engineering sciences requires, however, that capacity building be viewed in an even wider context.

No society has developed a scientific capability which can be considered adequate for its needs for the foreseeable future. So capacity building is by no means confined to the developing world, though the overwhelming imbalance of scientific creativity and training in favor of the rich North requires that the poorer South be given very special treatment and emphasis. This asymmetry of capacity provision is so serious as to threaten the self-generating abilities of many developing countries to research and plan their way towards sustainable futures.

Progress in capacity building in the South rests partly on reducing the exodus of trained scientists, partly on establishing special aid programs that are associated with national education, research and policy environments, and partly on ensuring that local knowledge of environmental adaptation is integrated with the results of research and translated into effective action.

Any agenda for capacity building should include strengthening the role of education generally for people of all ages, ensuring that environment and development linkages are conveyed in meaningful and purposeful ways to all, enabling science in its widest sense to be more integrated and intelligible for all, and using the media – traditional, modern and the non-traditional – in a constructive role to facilitate this process. Above all, people must be induced by legal and other instruments to pursue their development activities in a sustainable way.

In this chapter the situation in Africa has been used to illustrate various aspects of capacity building. There are good reasons for this emphasis. The African situation is significantly worse than in other regions; it is also typical for many of the least developed countries on the other continents. It is necessary, however, that specificity of circumstances and needs be carefully considered in any action programs.

2. CAPACITY BUILDING – DEFINITION AND NEED

For the UNCED process to work effectively and successfully, there must be the necessary skills, knowledge, supporting institutions and motivation in all countries of the world. Where these do not exist in adequate measure, they must be created or extended and strengthened. This process of capacity building is the principal theme of this chapter. It may be seen as embracing the following:

(a) Development of individual, group and institutional capacity for self-sustained learning, generation of technology and implementation of developmental or scientific activities for the solution of a defined range of problems;

(b) Development of the capacities and capabilities of indigenous people for the solution of problems that affect their lives;

(c) Development of the infrastructure to make the above possible; such infrastructure will include political commitment, availability of minimal human resources, favorable conditions for professional work and institutional arrangements for effecting change;

(d) In essence, the development of the ability to cope with environmental stress in the context of the search for a better life.

Accumulating the necessary knowledge about global natural resources so as to understand and explain the processes which are vital to sustainable use and development of these resources is a task which requires the involvement of scientists from a wide range of disciplines. It also requires the participation and engagement of scientists from all regions of the world, who are able to see and understand the local and regional context of the problems under study. Such scientists are also in the best position to help implement the recommendations arising from such research.

The capability for the collection and analysis of data in a scientific way is, however, heavily concentrated in the industrialized countries. This is not to say that this capacity in the industrialized countries is organized in the best possible way for dealing with the problems of the environment and development; existing capacity in these countries will require conscious adjustment and orientation in this direction. Although there are some developing countries which have relatively well-established scientific capacities and infrastructure, most of the countries in Africa, Asia, Latin America and the Caribbean lag far behind the industrialized countries. For example, for the developed market economy countries of the United States, Japan and Europe, the number of scientists and engineers per million of population was, towards the end of the 1980s, about 2800; for the developing countries of Africa, Middle East, Asia, Latin America and the Caribbean, the corresponding number was about 160 (Salam, 1991). A similar picture emerges when expenditures on Research and Development (R and D) and on Development are examined (Mullin, 1991):

2.1 Expenditures on R and D and on development

Category of Countries/Organization	US\$ 10^9
Total Expenditures by all OECD Member States	309
Total Expenditures on R and D by the USA in 1991	157
Expenditure on Development Assistance by OECD and Arab States	55
R and D Budgets of top ten US Corporations for 1991	23
Total R and D Expenditures of all Third World Countries	20
Total R and D budget of General Motors Corporation for 1991	5.4
OECD/DAC countries' support to R and D in all developing countries	2

To focus on just one comparison, the total R and D expenditures of all Third World countries including OECD/DAC (Development Assistance Committee) countries' support to R and D in those countries is about the same as the top ten United States corporations spend on R and D.

Furthermore, in spite of its vast resources and potential, Africa has only about 0.4% of all the researchers of the world, and they account for 0.3% of the internationally accessible scientific publications (Gaillard and West, 1988). A further index of the disparity is the amount of resources spent on salaries as a percentage of the total budget of a developing country research institution. This is often of the order of 90–95%, leaving only 5–10% for all research activities. This sort of ratio is a major constraint on the institution's productivity.

At the national level, there are several tasks for which developing countries cannot afford to depend on scientific leadership and expertise from industrialized countries. Some of these tasks, which require a combination of indigenous knowledge and expertise with scientific training and education in all the relevant sciences, include:

(a) Addressing problems and collecting and analyzing information specific to the needs and conditions of a country or region;

(b) Modifying and adapting international research results and ideas to local and regional needs;

(c) Assisting and advising national authorities in the application of scientific knowledge, thus providing a base for development founded on national resources. The use of technical assistance for this purpose is costly, often does not take local circumstances into account and might not be available when it is needed;

(d) Providing locally relevant competencies in discussions and negotiations with international authorities and organizations, thus balancing the often authoritative role of foreign expertise as well as giving qualified input to the decisions;

(e) Participating in education by supervising students and highlighting scientific issues at different educational levels as well as providing relevant scientific information to the public in national and regional fora and disseminating research results and scientifically based advice to the society.

Besides playing an important role on a national or regional level indigenous scientific capacity will also contribute to "global knowledge".

Within this broad context, actual programs of capacity building will need to meet specific requirements. Capacity building must be purposeful; it must focus on specific problem areas. In the field of agricultural research in Africa for example, the World Bank pointed out in its report on sustained development in Sub-Saharan Africa (World Bank, 1984) a number of weaknesses. It stated:

These weaknesses of agricultural research are not simply, or even mainly, a matter of staff and money. Between 1970 and 1980 the number of research scientists increased from about 1600 to over 4000 – a growth rate of 9.6% a year, compared with a growth rate of just under 9% a year for South and East Asia.

There are now almost half as many research scientists in Africa as in Asia. Expenditure per scientist is also high compared with those Asian and Latin American countries that are regarded as having effective agricultural programs.

The major problem is to develop a more effective use of the existing research capacity. At the national level in many priority areas, research results have not been adequately disseminated. Even local researchers are often underemployed because programs lack focus, continuity and coherence; research management is weak and its status low. In most African countries, researchers are isolated from farmers and extension workers, so nobody can see the direct relevance of research.

In Africa, in spite of the frustrations of natural disasters (such as drought, pestilence and epidemics) and man-made disasters (wars and political conflicts), ways must be found to build up human capital and institutions in a wide range of fields. The common phenomenon of an emphasis on crises and the short-term must be changed. There is a consensus within development circles that Africa, especially sub-Saharan Africa stands in greatest need. The twin issues of human and institutional capacity building have emerged as the central development challenges of the highest priority facing the continent. As pointed out by Jaycox (1990), the development experience of Asia and Latin America shows a gradual build up and investment in their own human capital and institutions. Donor assistance strategies supported this kind of capacity building through substantial and long-term investment.

But Africa has been treated differently. It has been mainly a case of crisis management all along with crisis response by donors, short-term single shot projects, and technical assistance. Although Jaycox refers in this context to economic policy and economic management, his description may be generalized to many other, if not all fields. He continues: "Can we accept the presence of 100,000 expatriate technical advisers in the region (sub-Saharan Africa) – a larger number than at the time of independence? The trends are in the wrong direction!" Very similar comments can be made about policy and management in any other field, including those of particular relevance to environment and development.

It is important to point out that in general national research systems in developing countries are very *young, with small scientific communities*. In the early 1980s there were about 1000 scientists in both Costa Rica and Senegal, and just over 5000 in Thailand (Gaillard, 1990). Even Costa Rica, which had about twice as many scientists compared to the two other countries, as a percentage of the whole population, had 20 times fewer scientists than France and 30 times fewer than the USA. Although these figures are likely to have improved somewhat during the last decade and are better for a few other developing countries, they give an indication of the wide gap in scientific manpower capacity between the Third World and the industrialized countries.

In some countries no comprehensive assessment of capacity needs has been carried out in so far as scientific requirements for pathways to sustainable development are concerned. Such capacity audit would have to extend to indigenous expertise, notably coping strategies for adjustment to environmental stress. It appears that shortfalls in capacity building together with inadequate information on what scientific skills exist that could be tapped in various communities are adversely affecting any meaningful transition towards sustainable futures.

3. MOBILIZING AND STRENGTHENING MANPOWER FOR SUSTAINABLE DEVELOPMENT

Viable and productive research can take place only where there are competent and dedicated scientists who can address problems through an appropriate scientific method. Investments in scientific projects, including facilities and equipment, are worthless if nobody can use them in a competent and productive way.

In spite of the small numbers of qualified scientists in most developing countries, it should be realized that there are *well-trained and highly qualified scientists working in some Third World institutions*. In many cases these scientists have succeeded in building up productive and internationally recognized research groups. It is important that these groups are supported to stimulate and engage colleagues and students through a combination of good training, productive use of intellectual capacity, and dedication to scientific work. Supporting selected existing groups is probably a quicker and more cost-effective way of mobilizing and strengthening indigenous scientific capacities in developing countries than spreading the support thin with regard to geographical distribution and scientific topic. Although these groups might not cover all relevant fields of research they are likely to have an impact on science development in general within their countries. Through the support of such small and already existing centers of competence they are likely to achieve the necessary critical mass and play a catalytic and leading role in developing research in other and related scientific fields. Furthermore, viable research groups may complement each other on a regional basis to cover a wider scientific range.

It would also be proper to organize, as far as possible, scientific courses and other training activities around these selected centers of competence, rather than arranging such activities in the industrialized countries.

A major obstacle for establishing successful research environments in the developing countries is the continuous "brain-drain" from these countries. The example of the Makerere University in Uganda is illustrative. During the period 1986–89 the university lost 34 Ugandans with PhD degrees, 18 of them professors, leaving behind debilitating vacancies in critical areas in university research and teaching (Kajubi, 1990). This situation is not unique. A major factor contributing to such situations is the erosion of professional salaries in many poor countries. In Africa, for example, the real value of researchers' salaries has fallen by a factor of ten in the last 10–15 years.

It is obvious that incentives to keep the scientific talents and entrepreneurs are not sufficient. Consequently it is important that initiatives are taken on a national as well as international level to provide incentives for some of the most promising and best educated scientists from developing countries to stay at research institutions in their home countries.

These initiatives include:

(a) Improving salaries, so that the key scientists at least can afford to do full-time research and teaching, instead of having to look for extra jobs in which they make no use of their scientific skills which represent considerable investments by local governments or donor organizations.

(b) Providing the proper facilities – equipment, supplies, etc. – to do research. Without the scope for pursuing research few, if any, serious scientists are likely to stay in their own countries.

(c) Making it possible for scientists to meet and communicate with other scientists both in the region and internationally, allowing them to participate in international scientific activities in their field of interest.

(d) Engaging local scientists in internationally and regionally sponsored research projects which have to be executed at least partly in their home institutions; such projects should be designed so that they are integrated into the national research and education programs, thus keeping the available manpower to sustain national capacity building.

In recognition of the state of the economy in most developing countries, such incentives obviously have to be introduced slowly but steadily to gradually improve the present situation. It is important, however, that they are considered immediately in all programs which aim to strengthen scientific capacities in developing countries.

Scientists in developing countries must also be encouraged and supported to make contact with and participate in appropriate existing scientific networks, and to establish new networks where these are needed. Support to networking must, however, be conditional to make sure that the networks will be productive. The most successful networks have been found to be those which are confined to a limited topic, having relatively few members who are all actively contributing to the network activities.

4. APPLIED AND BASIC RESEARCH

When research in the Third World is discussed, the issue of applied versus basic research often arises; there are two opposing points of view.

(a) Those who believe that resources for basic research should not be allocated for developing countries, since basic research is more appropriate for the industrialized countries where such research can be sustained through industry and governments. The

results from basic research should be transferred to developing countries when needed;

(b) Those who favor that Third World countries engage in basic research, albeit at a more reduced level and within fewer scientific fields than in the industrialized countries.

A third point of view maintains that it is not meaningful to draw a line between basic and applied research, since the two categories are interlinked. All applied research is founded on, or derived from, basic research. Furthermore, no serious scientist can undertake research projects or stimulate and teach students without being familiar with basic science within his/her field of interest. Moreover the gap between new discovery and application has been closing very rapidly over the years.

There are quite a few examples of groups doing basic research in a variety of Third World countries, managing to act as stimulators and centers of competence on a national and international level. Where such groups or individuals exist, it would seem very restrictive to limit their role to merely applying knowledge from the existing pool and not make any contribution to that pool.

Competently designed basic research in Third World countries should therefore be encouraged as an integral part of the national/regional scientific infrastructure and higher education system. Doing so will also provide a stronger base for applied research and the capabilities to implement scientific results in the pursuit of sustainable development.

5. NATURE OF CAPACITY REQUIRED

From a careful analysis of the above concerns, the following objectives must be met in the design of capacity building efforts aimed at implementing Agenda 21 and the additional scientific agenda of the scientific community into the 21st Century.

(a) *Development of human resources.* This must encompass basic scientific training in a wide range of scientific disciplines for:

(i) the collection, analysis and synthesis of basic data: physical, biological, marine, terrestrial, atmospheric;

(ii) the monitoring of environmental parameters: on the spot and through remote sensing especially in areas outside North temperate zones. It must also include increased training in the engineering and development-related social sciences as well as new technologies and their application. Training should also enable the trainees to interact easily across disciplines and should include an appreciation of interdisciplinary skills.

(b) *Orientation of current skills and acquisition of new*

skills to give a development and environment dimension to the application of those skills;

(c) *The strengthening of research institutions to provide relevant new knowledge* e.g. in atmospheric chemistry, biodiversity and applications of solar and other non-conventional forms of energy. Research institutions should have the capacity to take an interdisciplinary approach to problems and be able to tackle policy and management issues, development/environment problems such as those related to water quality, land fertility, the soil/plant/water system especially in forest, salinized and other degraded areas, modeling and the verification of hypotheses through the collection and use of improved data;

(d) *Strengthening and maintaining scientific infrastructure.* Included here is the maintenance of scientific equipment and the availability of documents and information in the form of books, journals and electronic media;

(e) *Data needs and systems.* Apart from researchers, the various players in the environment and development field require easy and ready access to relevant data and information. The capacity to design dynamic information systems and the collection of data to facilitate the processing, packaging, dissemination and use of such data is of critical importance at a time when the world is overwhelmed by the sheer volume of data in any field. Information should be readily available to individuals, communities, institutions including non-governmental organizations, policy-making bodies, governments and intergovernmental organizations. Facilities for training users of the information systems should also be made available;

(f) *The management of the implementation of the science and technology agenda for environment and development.* The implementation of the agenda needs to be managed. Management at the community, industry, national and international levels will be necessary to ensure that a change of direction will be made and the new direction purposefully pursued. It will involve the following:

(i) Decision-making organs at the various levels;

(ii) Determination of priorities and financial provision;

(iii) Policy-making institutions dealing with such issues as training and use of human resources; new directions in the management of resources taking into account the emerging themes of ecological economics and indigenous adaptive strategies; technology policies covering the acquisition of, and capacity for adaptation of new technologies (e.g. in the areas of energy and agriculture, chemicals and agriculture, biotechnology, efficiency of energy use);

linkages among institutions and users of research results and the productive sector generally; positive discrimination towards women; and empowerment of institutions and communities through support for community organizations;

(iv) Training in regulatory skills to oversee compliance with conventions and monitoring of levels of exploitation of resources;

(v) Development of the input of ecological economics; including valuation of natural resource processes, the social consequences of environmental change and training for development financing;

(vi) Need for international co-operation and the capacity for participation in global programs including new participatory research and training schemes in internationally linked inter-disciplinary centers.

6. INFRASTRUCTURE FOR SCIENTIFIC RESEARCH

Scientific research cannot be done without access to proper and functioning facilities, equipment, and supplies needed to collect and analyze data scientifically. Maintenance of scientific equipment, supply of appropriate documents and information in the form of books, journals and electronic media are key components in a productive research environment. Without the appropriate scientific "tools" no experimental scientist can achieve results of adequate quality, however qualified and dedicated he or she may be. Unfortunately, few scientists in developing countries work under conditions where the appropriate prerequisites to do research are at hand. In the industrialized countries scientists have the support of students, technicians and other back-up staff who can help with the research work, and the operation and maintenance of scientific equipment and other research facilities. These possibilities seldom exist in developing countries.

In an enquiry about research conditions, covering about 500 scientists in 67 developing countries, scientists were asked to identify the major factors limiting their research work. The difficulty of maintaining equipment in good working order was identified as the second most important constraint, after chronic lack of funds (Gaillard and Quattar, 1988). The problem of ill-maintained and non-functioning scientific equipment in developing countries is generally recognized by organizations donating equipment as well as by the national authorities responsible for research (Siamwiza, 1985). Various attempts have been made to alleviate this serious constraint (Prage 1989, 1990). Most of the efforts have, however, been on an *ad hoc* basis, with almost no attempts at co-ordination. Unfortunately the equipment problem is seldom seen and addressed from the point of view of its users, i.e. the researchers, students and technicians in the developing countries. It is important that the existing capacities to operate and maintain research facilities and equipment in developing countries are mobilized and strengthened.

Whenever support is given to scientific projects which include equipment, arrangements should be made to ensure that the equipment can be kept functioning during most of its normal lifetime. Attitudes of preventive maintenance should be promoted, and rehabilitating existing equipment should be given preference over acquiring new ones. Training of technicians and scientists to operate and maintain their equipment should be strengthened through hands-on practical experience and the concept of "training of trainers". As much as possible of the training and instruction should be done in the institutions, preferably as "hands-on" training on existing equipment.

Instead of continuing to procure or deliver vast amounts of scientific equipment and supplies without taking precautions to make sure that these items would come to proper use over a reasonable period of time, national and international organizations should make every effort to develop new mechanisms and to design equipment support programs that contribute to the building of indigenous capacities. Such improved mechanisms will undoubtedly result in large budget savings, when it is considered that 30–50% of the potentially useful scientific equipment acquired by Third World countries through purchase or donation are not in use because of inadequate maintenance.

The skills needed to manage a "research environment" are seldom at hand in developing countries. There are numerous examples where proper research facilities and manpower resources have failed to produce the expected results because of lack of managerial skills. It is therefore essential that increased efforts are made to train the staff of research institutions in the management of their resources.

7. EDUCATION FOR ENVIRONMENT AND DEVELOPMENT

Capacity building for environment and development involves both persons and institutions. The commitment of persons of all ages can only be secured through a larger social process in which education must play a key role. At all levels, education should incorporate the objective of fostering an understanding of the role of environment and development issues in the society, emphasizing that they are two sides of the same coin.

Education is a process of teaching as well as learning. The greatest educational skill is to be able to accumulate information, organize it in meaningful ways and impart its significance to others. Such skills should be shared by all and not confined to "teachers". In view of the urgency of capacity building for environment and development

through education, it is necessary to select from a wide range of educational methods, including new and innovative ones, such as the use of story-telling, drama and song in non-literate communities, and to target specific sections of society: schools, colleges, informal study groups, and all adults with special emphasis on women. The saying that "when you educate a man, you educate an individual but when you educate a woman, you educate a nation" is apposite. For it is through mothers that all children receive not only their first education but also their attitude to learning. The new methods and approaches must facilitate the creation of new attitudes, values and relations that should guide our individual and group relationships with the environment and development.

At present, environmental education in the sense of preparing citizens for the transition towards a more sustainable future is still in an embryonic state, even in the more advanced countries. Environmental education should not just address a more integrated and applied environmental science. That is vital, but it is not a sufficient condition for successful capacity building. Environmental education has to be accompanied by a radical change in social and political attitudes concerning the ability of all people to be properly informed about environmental change and their individual and collective roles in the transition to a sustainable future. Such information requires access to appropriate knowledge about the state of the environment, about rates of change and what they mean for human health as well as the health of the environment, the economy and individual well-being, and about the political factors that inhibit or promote the cause of sustainable development. This is the capacity that is generally missing in almost all countries.

The role of scientists in the education process is to provide appropriate information and to demonstrate that science and technology are linked to both environmental and developmental issues. To do this effectively, scientists need to develop better communication skills and should try to reach out to, and influence decision-makers and opinion-leaders, i.e. politicians, government officials, industrialists and journalists. Only then could a political/public climate which is sympathetic to capacity building in science and technology be created, ultimately leading to governmental commitment to development based on environmental considerations.

Scientists also have the responsibility to explain to the citizen the importance of sustainable development to the well-being of the individual and the future of the society. Environmental processes and their various local and regional manifestations have to be explained in terms of the ways they will threaten or benefit individual livelihoods and opportunities. Such explanations should provide the motivation for individuals, communities and nations to make the transition to sustainable development.

8. RECOMMENDATIONS

1. At the very basic level, *knowledge of the relationship between the environment (the Earth System) and development should be made available to everybody.* This can be achieved through the formal school system and through facilities for learning provided outside school and targeted to all identifiable groups, *especially mothers*; on whom much of initial learning of all children depends. *Decision-makers, politicians and other opinion-leaders are also to be specially targeted* to provide new emphasis and support for sustainable development. New and innovative educational methods should be sought.

2. Both at the higher education and community levels, *facilities should be provided for a problem approach to learning, the orientation of skills* and the acquisition of new skills for the achievement of sustainable development. In recognition of the highly interdisciplinary nature of the subject, efforts must be made to develop, strengthen and encourage interdisciplinary studies at all levels.

3. *Strategies for capacity building* should have the following elements:
   (a) influencing the decision-making process through a program of education targeting decision-makers and opinion-leaders (politicians, government officials, industrialists and journalists);
   (b) invoking and enhancing self-interest and motivation of individuals and communities for sustainable development;
   (c) education for environment and development for all citizens;
   (d) promoting the use of existing capacity;
   (e) creating a critical mass of relevant scientists and managers with the appropriate skills and providing the necessary infrastructure in all countries and regions;
   (f) promoting as models existing success stories, stressing the circumstances and factors leading to success, the necessary management techniques and the need for co-operation among countries;
   (g) the need for specificity in capacity building from region to region and from country to country;
   (h) promoting the use of existing capacity.

4. In each country, there should be a *national environment and development system* made of:
   (a) a policy-and decision-making body,
   (b) a system of monitoring institutions,
   (c) education and training institutions,
   (d) productive sector representation, and
   (e) schools, centers or institutes for interdisciplinary and problem-oriented studies and research.

Within the system, there should be facilities for providing support for sustainable development activities, such as monitoring, provision of scientific services and information, research, policy studies and general and specialised support to industry.

5. There should be in each region of the world at least one *Regional Environment and Development Center*; to facilitate capacity building at the national level through training and through the implementation of regional programs. These institutions should also be used to support regional co-operation and South-South co-operation, with attention being given to region and country specificity of needs and circumstances. Similar institutions which aim at strengthening scientific communities in developing countries and promote the development of these countries through interaction of their scientists with those of other countries and with industry should be supported.

6. *The capacity built through the above recommendations should be actively used to facilitate the implementation of the decisions* of UNCED through participation in the science agenda. In this connection, one of the Scientific Sessions and Symposia of the Third World Academy held in Caracas, Venezuela, 15-19 October 1990 had recommended that:

   > studies of the environment should be initiated by well-trained scientific and engineering personnel for the collection of basic data on the Third World Environment as a system, and for the monitoring of change in space and time. Such a program will permit the Third World to participate substantially in the current debate on global environment change (TWAS, 1990).

7. The international scientific community should launch a *"capacity audit"* to assess the present state of capacity for environmentally tolerant development in every country. It is recognized that some audits of capacity for specific needs have been carried out in many countries in co-operation with some United Nations agencies. This audit, however, should be more comprehensive, should include indigenous expertise, should extend to all the relevant sciences, and should cover people as well as institutions and organizations. It should address in a meaningful way the state of science as it is practised in environmental preservation and the transition towards sustainable development. The results of the audit should indicate what orientation of existing capacity needs to be made and the additional capacity required.

8. The United Nations and its agencies, national governments and other agencies should ensure that *financial resources are available for the*

implementation of these recommendations. Serious consideration should be given to the restructuring of national budgets to make funds available.

REFERENCES

Gaillard, J. and Quattar, S. 1988. Purchase, use and maintenance of scientific equipment in developing countries. *Interciencia,* **13**, 2, 65-70.

Gaillard, J. and West, R. 1988. Scientific research in Africa. *Afrique Contemporaine*, **148**, 3-29.

Gaillard, J. 1990. Science in the developing world: foreign aid and national policies at a crossroad. *AMBIO* **XIX**:8, 348-353.

Jaycox, E. V. K. 1990. *Capacity Building in Africa: Challenge of the Decade in Capacity Building and Human Resource Development in Africa.* Kwapong, A. A. and Lesser, B. (eds.) Lester Pearson Institute, Dalhousie University, pp.113-116.

Kajubi, W. S. 1990. The university and development in sub-Saharan Africa. The case of Makerere in Uganda. *The Courier*, **123**, 61-64.

Mullin, J. 1991. Private communication based on information and estimates compiled from a variety of sources.

Prage, L. 1989. Operation and maintenance of scientific equipment in developing countries. *Food Laboratory News*, **16**, 15-19.

Prage, L. 1987. Procurement, operation and maintenance of scientific equipment in developing countries. Part I. Preliminary report. SAREC.

Salam, M. A. 1991. Science, Technology and Science Education in the Development of the Society, *Third World Academy of Sciences*, pp. 57-58.

Siamwiza, M. N. 1985. Procurement and maintenance of scientific equipment in SADCC countries (National Council for Scientific Research Zambia) and International Development Research Center, pp. 12-31.

Third World Academy of Sciences, 1990. Third General Conference, Caracas, Venezuela. Key recommendations, B.6, p.3.

World Bank, 1984. *Toward Sustained Development in Sub-Saharan Africa*, World Bank, Washington DC, pp. 31-32.

Chapter 15: Policies for Technology

C. Juma and M. Sagoff

EXECUTIVE SUMMARY

This chapter calls for new approaches in science and technology policy to respond to the growing importance of environmental considerations in economic development. We recommend metaphors that represent socio-economic systems as complex, dynamic and non-linear. We argue for institutional reforms central to sustainable development. The chapter notes that the use of environmentally-friendly technologies will require policies based on evolutionary approaches to development and which emphasize technological diversity and access to scientific and technological information. The chapter stresses that sustainable development will depend largely on the ability of the world community to redirect the trajectory of technological change (development, transfer, diffusion and adoption) towards environmentally sound paths.

Important Facts
The current concern over environmental degradation has helped to highlight the importance of technological change in economic development. The public perceives technology as a principal cause of environmental degradation. Accordingly, the world community asks for the transfer and diffusion of environmentally friendly technologies in developing countries.

These perceptions emerge as the world undergoes major geopolitical transformation. Improving East–West relations, new regional marketing blocs, growing poverty in the developing countries, especially Africa, and the demand for democratic reform in most of the world create new conditions for policy reform. The preparatory process for the United Nations Conference on Environment and Development (UNCED) to be held in June 1992 in Brazil adds further pressure on the world community to think in concrete terms about solving environmental problems.

The climate for research is also changing. Scholars emphasize the role of the market both in causing and in solving social and ecological problems (Wolf, 1990). On the one hand, many students of development see free and open markets as the key to prosperity – to increasing employment, controlling inflation, and alleviating poverty. On the other hand, if markets produce wealth by running down capital stocks and by destroying ecological support systems, whatever prosperity they create must be short-lived, and the long-term costs may not be worth the immediate gains. Accordingly, researchers emphasize productive systems that, to the extent feasible, maximize economic growth but minimize throughput, i.e. the removal of low entropy resources from nature and the return of high entropy wastes into it.

These changes are taking place at a time of unprecedented generation of scientific information. The information, which is essential in solving persistent and emerging ecological problems is, however, unevenly distributed. Developing countries have limited access to this global fund of knowledge and also lack the technological capacity to transform the information they acquire into practical applications that can assist in the transition towards sustainable development.

Research Themes
The field of technology policy research is still nascent. It is only in the last decade that technology policy came to be recognized as a discipline. Much of what is referred to as "technology policy studies" originated from sociological critiques of science. Only a few institutions have focussed on the role of technological innovation in economic change. The emergence of environmental concerns will place new pressure on technology policy studies to incorporate ecological questions into research programs.

Of particular importance in such attempts is the need for alternative epistemological approaches to the analysis of technological change, for example, the use of ecological and biological metaphors. Other areas that will need research attention include an assessment of the modes of technology flow and the implications of the pervasiveness of science-intensive technologies for technology transfer.

Urgent Actions

There are at least three urgent matters that need to be addressed by the international community in applying technology to sustainable development. The first is the introduction of policy measures that promote the creation and conservation of technological diversity. The second includes policies and institutions that promote access to (public domain and proprietary) scientific and technological information necessary to develop environmentally friendly systems of production. The third comprises measures aimed at enhancing the technological capacity of developing countries. These measures include training, redirection of development assistance, and national policies that promote creativity and innovativeness.

1.   INTRODUCTION

"Nowhere has liberal philosophy failed so conspicuously," Karl Polanyi wrote in *The Great Transformation* (p. 33), "as in its understanding of the problem of change." The transformation of Europe from feudalism to industrial capitalism brought "an almost miraculous improvement in the tools of production, which was accompanied by a catastrophic dislocation of the lives of the common people." Polanyi includes in "liberal philosophy" the tendency to judge events from an economic rather than from a cultural or social point of view. This philosophical approach applauds the "miraculous improvement in the tools of production" but passes over the "catastrophic dislocation of the lives of the common people" as well as the degradation of the environment.

Today, the growing concern over environmental degradation has heightened the need to understand and reshape the direction of socio-economic transformation. The transformation of Europe from traditional to industrial and then to post-industrial societies has taken centuries to accomplish and has resulted from cumulative changes over time (Rosenberg and Birdzell, 1986; Landes, 1990). On other continents, for example, in Africa, this transformation may be measured in decades and it occurring at a time when minor perturbations may result in major changes over short time scales. The problem we face is to achieve the economic benefits of development at lower human and ecological costs. The solution lies in the transition to sustainable development as broadly defined by the World Commission on Environment and Development (WCED). This chapter argues that sustainable development will require policies

for technology that are responsive to social and environmental imperatives.

2.   TECHNOLOGY AND SUSTAINABLE DEVELOPMENT

2.1. Technology and the changing world order

The current environmental crisis has brought into sharp focus the role of science and technology in shaping the future. The current global situation is marked by crises, discontinuity and innovation. A new global order is in the making heralded by changes in East-West relations as well as by the emergency of democratic governments in Eastern Europe, Africa, Asia and Latin America.

Amid these sweeping global geopolitical changes, environmental awareness has begun to shape the direction of technological innovation and to provide new guidelines for market behavior, at a time when humankind has access to unprecedented amounts of scientific and technological knowledge (Heaton *et al.*, 1991; UN, 1991; OECD, 1990; Dosi, 1988). These new conditions promote policies that apply technology to support sustainable development.

The starting point for these new policies lies in the conception of socio-economic systems as the result of an evolutionary process driven by innovation. The analogy between biological and technological evolution helps to explain this idea, especially in so far as one takes the appropriate units of selection to be organizational and informational structures rather than individual firms.

We emphasize two metaphors drawn from this analogy, namely, the idea that evolution proposes solutions to local problems and that these solutions rely on co-operation at least as much as on competition. We shall argue that technology policy ought to help communities to meet short-run challenges and thus, if they wish, to maintain their composition and boundaries in the face of changing and therefore challenging economic conditions. We argue against policies that promise great economic success in the distant future at the cost of human misery in the short run.

The findings of the WCED, chaired by Norwegian Prime Minister Gro Harlem Brundtland, highlight the waning of a world view that treated nature as a collection of reducible entities existing in reversible time and oscillating in equilibrium. This worldview has left indelible marks on the face of the Earth and in the minds of people. The Brundtland Report dealt more with the marks on the face of the Earth and less with the grip of these old-fashioned epistemologies on human minds. It called into question, however, the global market uniformity in which firms compete for success by developing newer and often larger technological systems without regard to the effects of these technologies on sustainable cultural and ecological relationships. A discussion of technology policies, seen here as choices about the articulation of human creativity in socio-economic evolution, must start with our perception of the world in general, and technology in particular.

The Brundtland Report has provided a basis for dealing with epistemological issues by emphasizing the moral context of the notion of sustainable development. Issues of equity – both in time and space – are central to the Report. For example, the Report asserts that a "world in which poverty is endemic will always be prone to ecological and other catastrophes" (WCED, p.8). But reducing poverty requires the introduction of new knowledge and organizational forms into the economic system. Technology and related institutional reform thus represent a key element in achieving sustainable development (Israel, 1989; Matthews, 1986; North, 1990). The concept of sustainable development, according the Report, implies limits "imposed by the present state of technology and social organization on environmental resources and by the ability of the biosphere to absorb the effects of human activities" (WCED, p. 8).

The Report further stresses that a new era of economic sustainable development is essential for meeting basic needs and to assure that the poor get a fair share of the resources they require. But such equity cannot be achieved in the absence of "political systems that secure effective citizen participation in decision-making and by greater democracy in international decision-making" (WCED, p.8). Not only has the Brundtland Report made new demands on the way the world treats access to resources, but it has also called for the recognition of the fact that the kinds of innovations necessary to facilitate economic renewal can be effectively achieved only in open political systems. Economically enlightened dictatorships can no longer be justified.

The Brundtland Report (WCED, p.9) sets forth an agenda for change that is admittedly uncertain and entails painful choices:

> Yet in the end, sustainable development is not a fixed state of harmony, but rather a process of change in which the exploitation of resources, the direction of investments, the orientation of technological development, and institutional change are made consistent with future and well as present needs. We do not pretend that the process is easy or straightforward. Painful choices have to be made. This, in the final analysis, sustainable development must rest on political will.

The notion of choice is a critical element in technological and economic evolution. It suggests that change is not linear and that societies determine their own future. Indeed, political activities are often associated with the choice of future directions under conditions of uncertainty.

2.2 Mechanistic versus evolutionary perspectives

The Brundtland Report highlights the inadequacies of a world view that treats nature as a reducible reversible system that may be analyzed, understood, and anticipated by identifying and classifying its components and the causal links between them. This mechanistic model was so successful in classical physics that people "believed (erroneously) that analogous ideas must apply in the domain of biology, ecology and human sciences, and particularly of course, economics" (Allen, 1991).

This world view emphasized the universal (the laws of nature common to all systems) and the necessary (events we may predict on the basis of the functioning of those laws). An evolutionary perspective, in contrast, emphasizes the particular and the contingent. It seeks to explain events by reference to diversity, innovation and contingency – as well as to values that guide human action.

The choice with respect to technology transfer is as clear as it is compelling. We may continue the current practice of transferring technological structures as whole blocks to benefit particular factions in less developed nations. These factions, in return, profit by supplying commodities to be consumed in developed countries. We may instead work incrementally in local and regional contexts to increase the capacity of the population of undeveloped nations to solve their immediate problems and provide for their own needs.

A concern with sustainability leads us to take an evolutionary approach, i.e. to understand how individuals at the local level adapt means-at-hand to ends-in-view and thus solve the immediate problems posed by their contiguous environment. The evolutionary perspective concentrates on the local, the incremental, and the short run, but bears in mind the global and the long run. It attempts to form the future from materials borrowed from the past and picked up from the immediate environment. This evolutionary approach presents the development process as a long effort in social learning (Lall, 1987). It relies on experimentation and innovation and on biological, technological and cultural diversity.

2.3 Towards a techno-ecological paradigm

The success of an evolutionary process depends largely on the diversity of its functional units. In biological, as well as in technological and social processes, diversity is essential for the creation of new systems. Diversity, be it biological, technological or cultural, is the principal condition of innovation, adaptation and change. The conservation of diversity is therefore an important metaphor with wide-ranging policy implications.

Let us consider the analogy between biological and technological change. Evolutionary biologists today disparage the view associated with Herbert Spencer and before him Thomas Hobbes that identifies a competitive struggle among individuals as the principal source of evolutionary change. Biologists are likely to put as much emphasis on co-operation as on competition, on symbiosis as on singularity, on co-evolutionary strategies as on individual selection (Gadgil and Thapar, 1990). In many instances, the unit of selection is not the individual but a symbiotic system characterized by its ability to maintain itself in the face of a changing environment. And the environment may be itself local: success may depend more on one's adaptation to immediate circumstances, i.e. to define and create a niche, than on one's survival in a war of each against all.

The view of global competition presents us with this kind of war: an atomistic or granular competition between individuals or firms each striving to achieve an advantage one over the other. While we recognize the role of competition in fostering efficiency and innovation, we also note that when firms or individuals are confronted with major external challenges, they rely on co-operation as a means of survival. The current ecological threats to the world community demand co-operation often in the place of competition.

Markets often reward those whose virtues and values succeed in contexts of competition rather than community. An evolutionary model of social development, in contrast, creates niches wherever it can – in virtually any space, local, regional, or global. It makes use of co-operative, communitarian, and symbiotic strategies to follow many paths – not just one path – to improve the human prospect. It builds on diversity and creates a nested hierarchy of structures referred to as environments. Moments of crisis create new conditions for novelty: new technological systems, institutional arrangements and cultural values.

The problem in starkest terms is this: to succeed in a global market, one must achieve an advantage over all others with respect to the same conditions. The idea of adaptation, in contrast, allows one to define a niche that bypasses that kind of competition. We see little reason to believe that the less developed world has competitive advantages that will serve it well in global markets, except in very rare circumstances or in situations where they are acting collectively. Rather, the opportunities to seize lie in local and regional co-operation (Saasa, 1992).

We may then take the unit of selection in economic development to be the design of a technological and informational system in relation to a local and regional – rather than a global – economic and ecological setting (Clark and Juma, 1987, p.117). We may think of these systems as structuring niches – i.e. as evolutionary strategies which, for all their differences, can co-exist and even co-evolve and build on ecological or biological principles (Moser, 1991).

Evolutionary strategies for technological development may be sustained (and can be sustainable), moreover, because of their relationship to local and indigenous institutions, practices, capacities, resources, and opportunities as well as ecological sensitivity. In that sense, technological evolution like biological evolution does not have a theory or an essence but a history. We should not try to derive laws but to understand cases and examples. The variety, diversity, and contingency of evolutionary paths – biological and technological – are the keys to adaptation.

As a result, we should caution – as many commentators have done – against the idea that technological solutions can be applied to, or imposed on, local problems exogenously. An evolutionary approach supports the common wisdom that successful technology transfer – including cleaner technologies – depends on efforts to improve the capacity of developing countries to choose, recreate, adapt, and in that way assimilate technical knowledge to meet their own patterns, practices, interests, and needs. This must originate from within.

The problem, then, is to enhance the capacity of societies to integrate novel technologies into extant systems and to avoid the imposition of particular technologies where no infrastructure exists to assimilate them (Kakazu, 1990). Local initiatives and priorities provide the essential basis for external assistance; successful programs of foreign assistance muster and concentrate the experience and expertise of communities and in that sense import technology as dictated by local needs.

To some extent, local communities can rise to the challenge of a new technology; thus technology transfer is properly understood as an interactive process or epistemological dialogue. How the process works depends on the details of the institutional, economic, social and ecological context. Thus, it is not sufficient that a new technology pass the tests of efficiency and environmental soundness. Sustainability also depends on practicability, i.e. on the extent to which a transfer makes sense given local experience, institutions, economic structures, and knowledge. Ecological criteria can later be brought to bear on the development of the technology as it continues to undergo adaptation and modification.

2.4 Technology and progress

The history of this century and the prospects we face in the next make the following conclusion inescapable – that technological evolution, like biological evolution, shows no built-in or automatic tendency to make the world better or to improve the lot of humanity over the long run. As evidence of this, one may note that periods of greatest technological advance were often periods of the greatest social decline, e.g. the fall to the Roman and the Ottoman empires. At other times, during the Renaissance, for example, technological advance appears to follow social improvement as a consequence rather than to proceed it as a cause.

Technological evolution like biological evolution has no inherent structure, purpose, goal, telos, or direction. While biological evolution may be said to be purposive in so far as it tends to produce systems that are designed more or less well in respect of local environmental conditions, it cannot be said to have a purpose in relation to a goal or an ideal. As S. J. Gould (1977, p.45) writes: "Natural selection is a theory of local adaptation to the changing environments. It proposes no perfecting principles, no guarantee of general improvement in short, no reason . . . favouring innate progress in nature".

Our tendency to identify or at least to conflate technological advance with human development inhibits us from consciously shaping the former to enhance the latter. This tendency follows from a nearly habitual belief in progress – a commitment to the proposition (as expressed, for example, by the young Adam Smith) that in the long run forces of economic and technological change will produce the best of all possible worlds if allowed to run their course.

Even if we could control technology well enough to subjugate it to a conception of progress, moreover, it would be one conception of progress, which might seem less attractive to many people and societies, including those for whom that technology may be prescribed. We should respect cultural and ethical pluralism or diversity – "that is, the conception that there are many different ends that men may seek and still be fully rational, fully men, capable of understanding each other and sympathizing and deriving light from each other" (Berlin, 1991, p.11).

The recognition of cultural diversity must not be confused with relativism, the absurd belief that nothing is good or evil in itself but only in the eyes of particular cultures. The great ills we face today – war, hunger, disease, and ecological degradation – plainly are evils and it is easy to see why. It should also be easy to see that we should seek technologies that can alleviate these ills in the short and long-term in ways that respect cultural diversity. We should also be suspicious of technologies that may make things worse in the near term but, according to some theory, will eventually produce Utopian results.

2.5 The treadmill and the trampoline

Two contrasting metaphors help describe two opposing models of technology transfer. The first metaphor, the "treadmill," sets technology primarily in the context of economic competition among individual firms. Technology in that context is proprietary; it must be efficient; it substitutes for rather than builds upon earlier productive systems. The second metaphor, that of the trampoline, sets technology in the context of large and flexible social-support systems. The emphasis lies on networking, information exchange, co-operation, and a shared basis for individual and social development.

The functioning of the technological treadmill in agriculture is well understood. The first firms to adopt new technologies gain a competitive advantage in world markets. Firms engaged in agricultural biotechnology, for example, are successfully developing industrial methods to produce many of the key imports – cocoa would be an example – the wealthier countries now buy from less developed nations. Nearly all high-value crops have become targets of displacement through biotechnology research (Juma, 1989, p.135).

Many commentators see the likely result of biotechnological advance in agriculture to be the disappropriation, as it were, of nations from the commodities of which they historically have been the source. Through "bioreplacement", biorelocation, and "biosubstitution", technological evolution overcomes biological evolution, for example, by enabling industrialized nations to "grow" cocoa butter from rapeseed oil and thus to underprice West African producers. Many observers support the contention that "production is being relocated from third world farms to industrialized country laboratories" (Busch et al., 1991; Rogoff and Rawlins, 1987).

The functioning of the technological treadmill in agriculture – the introduction of innovations the early users of which, by underpricing their competitors, profit while increasing surpluses that depress prices still further – by allowing industrialized nations to underprice less industrialized nations in markets for agricultural commodities, seems poised to aggravate rising levels commodity surpluses and therefore, ironically, of impoverishment. Rather than tending by its logic to make everyone better off, technological advance may redistribute wealth to the rich, i.e. make the rich countries richer while taking from the poor countries even the little that they have.

This result, however, is hardly inevitable. A new technology, such as biotechnology, has no predetermined role to play in human affairs; how it is used remains up to us. Biotechnology offers vast possibilities for the production of safe healthful food, the adaptation of farming to varying conditions, the degradation of wastes, the creation of new pharmaceuticals and materials, and so on and on. To seize the many promethean opportunities for human good that this technology (among others) offers, we must resist the attitude of fatalism. We must reject the idea that forces beyond our control ordain social outcomes, whether for good or evil, over the long run.

We should guide the transfer of technology not with the image of a treadmill but that of a interconnected net or trampoline in mind. A broadly based indigenous capacity to adapt technology to local needs and conditions – rather than surrendering to the exigencies of the global marketplace – would provide a spring board for economic activity ranging from subsistence farming to entrepreneurship for export. Often capacities for technology assimilation and development are limited when taken one by one; they can have a synergistic effect, however, when combined. The institutional networking of knowledge, experience, and expertise available within a region, then, would seem to be a more urgent prerequisite for successful innovation than the transfer of the latest technology by which a firm in the developing world might gain an ephemeral advantage in global competition.

3. TECHNOLOGICAL DIVERSITY AND IMPLICATIONS

3.1 Technological recombination and adaptation

Studies of innovation tend to focus on "breakthrough" and emerging technologies, for example, microelectronics, biotechnology, and new materials. These studies create the impression that new technologies sweep out old ones by rendering them obsolete – as in the story of Aladdin's lamp. While this approach helps researchers to map the frontiers of technological innovation and to identify new market opportunities, it tends to hide a major characteristic of technological change: the recombination or blending of "new" technologies with established ones to broaden the range of technological options. Moreover, the analysis treats products as discrete entities and not as combinations of different technologies or functional units.

As new inventions come in use, existing patents fall in the public domain. Yesterday's technology, moreover, can often solve todays problems. As technical information becomes commonplace, the prospects for technological adaptation and recombination increase. Few if any technologies arise solely from new information and inventions; most use existing information in new ways. Indeed, the kind of innovation most wanted involves the application of existing – not the development of new – technology.

Even in frontier technologies, recombinations of available applications of knowledge play an essential role. In photovoltaics, for example, advanced technology is used only in the design of the cells while the rest of the system relies on traditional power storage as well as metal technologies. There is also a synergistic relationship between different bodies of scientific knowledge. Advances in biotechnology, for example, depend on advances in information processing and instrumentation. Developments in new materials depend on information technology and to a certain extent on advances in biotechnology. Japan's advanced materials research, for example, builds on developments already accomplished in information technology and biotechnology (Lastres, 1990; Bowonder and Miyake, 1990).

"If we ask how inventions come into being," an observer comments, "a sound operating rule is to examine the flows

of information that converged . . . when the new combinations came into existence" (Aitken, 1985, p.15). The policy challenges of the future will thus involve increasing access to a wide range of technical information, much or all of which may be in the public domain, in order to generate new technological options.

This view departs from current approaches that emphasize the revolutionary nature of novel technologies. While we recognize the far-reaching impact of these advances, their greatest importance may lie in their recombination with conventional technologies to form technological hybrids. A good rule of thumb, then, is to think in terms of the integration or improvement not the replacement of one technological system by advances in another. One may argue that technological diversity and the ability to combine technologies are critical to innovation consistent with sustainable development. Frontier technologies are useful to the developing countries only when viewed as ways of improving – not replacing – prior and familiar ways of doing things.

3.2 Technological innovation and environmental quality

Theorists measure technological innovation generally in the context of increased productivity and competitiveness. This view is reinforced by analytical models that either treat natural resources as free or undervalue them. This situation is starting to change as regulatory bodies begin to require firms to take notice of their environmental impacts. Environmental regulations affect technology choice and lead to modifications. Regulations will profoundly affect the process of technological innovation over the long run (Kemp and Soete, 1990; Porter, 1991).

The 21st Century will place greater attention on environmental and ecological considerations in technology development, e.g. in renewable energy and environment-friendly materials. The closing of industrial cycles to increase the recovery of resources (recycling) presents another incentive for increased innovative activity (Krupp, 1990). Technologies that were seen as separate will be brought together to create new options and market opportunities. Interest in measuring and monitoring environmental quality will continue to stimulate the generation of sensing technologies. Biosensors, for example, will become a new growth area.

Technical solutions or "fixes", however, may receive – and do receive – too much attention. Industry will introduce incremental innovations to improve on the environmental standing of conventional technologies. However environmentally benign these technologies may become, they may support patterns of activity that threaten in the aggregate to overwhelm ecological support systems. The protection of the environment and the sustainability of development require the reorganization of social structures and expectations – ways of doing work, seeking entertainment, communicating, and so on. Better pollution controls on refineries, for example, will not help societies to build transportation systems less dependent on gasoline.

Social innovation offers as much promise as technological innovation for sustainable development and resource use. The design on new urban areas, for example, could take into account long-term energy and environmental considerations. Developing nations offer considerable opportunities for re-designing urban transport systems. Areas that are not overcommitted to traditional technologies and energy paths may find a quicker road to ecologically sensible infrastructure design. African countries, for example, are able to adapt to new technologies more easily than many industrialized nations because they are not too heavily committed to the old technological systems.

In discussing these issues, one has to take account of the co-evolution of technological and social and political systems. The genuine realization of sustainable development, associated with diffusion of relevant technologies, will go hand in hand with the integration of environmental considerations into political processes, for it is here that the broad trajectories of socio-economic evolution are shaped and opportunities are seized or lost (Krupp, 1990; Bijker *et al.*, 1987).

3.3 Local participation
The growing realization that long-term economic development depends on the integrity of the environment represents one of the most significant changes in public awareness in recent years. The rapid growth in environmental awareness has by far outstripped the capacity of existing economic theories to integrate environmental considerations into the planning process and national accounts. Prospects for long-term growth are now viewed as dependent on effective environmental management. This is important, especially for countries that rely on the natural resource base for their livelihood.

While "green" political parties have emerged in industrialized countries, political life in developing countries turns primarily on meeting basic needs by distributing access to natural resources. In this respect, matters pertaining to environmental management immediately become political without necessarily being labeled "green". Environmental issues will thus be part of the growing demands for greatest democracy. Indeed, effective environmental management requires the same kind of openness that is consistent with democratic rule.

Access to technologies necessary for the sustainable use of natural resources is increasingly important, especially when governments become major economic actors. The economic liberalization and regulation under way in many countries will give local populations more control over

resources. Enterprises will need to make peace with their neighbors by being responsive to local environmental concerns as well as follow to government regulations.

But even more importantly, local communities and non-governmental organizations (NGOs) are innovators in their own right. In this respect the access of local communities and NGOs to technological and environmental information is essential. Research institutions must take into account the needs of these communities and NGOs in designing technological systems. For example, the successful introduction and diffusion of energy-efficient charcoal stoves (*jiko*) in East Africa resulted from collaboration between NGOs research institutions and local communities. Non-governmental organizations are thus key actors in promoting cleaner technologies.

So far, issues of equity and gender have not been important in the design of technological systems. But local control requires that equity and gender considerations inform the design and choice of technologies. The issue in the long run concerns not just the ability of local communities to control technology but their role as participants in the process of technological innovation.

3.4 Technology transfer and acquisition
Judging from the record of technological performance in Africa, the relevant policies and institutional arrangements for supporting technological transformation are still nascent (Goka and Mihyo, 1990). The tendency in developing countries is still to rely on "technology transfer", a term that needs to be defined in the context of policy and practice.

Technology transfer has come to mean "the transfer of systematic knowledge for the manufacture of a product, for the application of a process or for the rendering of a service and does not extend to the transactions involving the mere sale... of goods" (UNCTAD, 1990, p.48) Technology transfer, on this conception, resembles heat transfer – in the absence of impediments it just flows from those with more to those with less. If barriers are removed, according to this definition, the flow of technology to developing countries will increase.

This assumption does not hold in practice. Relaxation of intellectual property and licensing restrictions, for example, will not necessarily lead to greater technology transfer. The failure of developing countries to use fully technological information in the public domain – e.g. patents that have expired – illustrates the problem. The ability to assimilate technology – not barriers to transfer – is the primary impediment to technological development in developing countries. Those who cannot use public domain technologies are unlikely to find patented technologies any more useful. In fact, the ability to utilize new technological options depends largely on how much of the prior knowledge has already been assimilated.

While in the 1960s and 1970s, technology transfer was largely effected through international investment and technical assistance, this is starting to change as a result of changes in the character of technology. In the field of biotechnology, for example, technology flows largely through training and access to information and not necessarily the movement of large quantities of equipment or machinery between countries. However, discussions on this subject still rely on the images of the 1960s when the movement of hardware was closely linked to the notion of technology transfer. The challenge facing the world community is how to enhance the diffusion of, and the access to, technology and information about technology while raising the capacity of developing countries to search, choose, adapt, and develop their own technologies according to their local needs and conditions.

As one commentator has said, technology is a game for the rich, a dream for the poor, and a key for the wise; it is the master key of development. The wisdom of a country is generally deposited in its policies and practices. The ultimate indicator of a country's wisdom is the status of its policies and practices concerning technology. In Africa, many of the policy elements necessary for making use of the information available are still missing, and there is no clear recognition of the links between technological change and economic transformation. This inhibits genuine partnership on technology from taking place.

A genuine and effective technological partnership will require willingness to transfer the relevant technology. But this is only one aspect of the process – the one that has been the focus of technology transfer discussions for a long time. The other aspect is the effort of those acquiring the technology. The effectiveness of this effort is reflected in the assimilation of technology as well as the development of local technological capability. It is useless to discuss the prospects for partnership in technological development without paying special attention to the ability of the developing countries to acquire technology effectively.

Partnerships in technological development will not arise automatically as a result of funding, the relaxation of restrictions, the introduction of lending facilities, or the signing of conventions. To a large degree, it will take a measure of wisdom and effort. Most African countries, for example, have not yet improved the ability of the research and business communities to have access to information banks or databases in the industrialized countries. This not only limits access to technology, but it also reduces the prospects for commercial and technological partnerships.

4.   TOWARDS TECHNOLOGICAL
         CO-OPERATION

Dominant economic theories of technological development emphasize international competitiveness (Porter, 1980;

Stegmann, 1989). While this model can be justified in the sense that firms do have a certain measure of loyalty to their country of origin, these firms often enhance their competitiveness by sharing between nations the skills, knowledge, and competence at their disposal (Levy and Samuels, 1991). This co-operative approach exhibits the collaboration required to achieve sustainable development. The recent signing of a collaborative agreement between the International Business Machines (IBM) and Apple Computers, traditional rivals in the personal computer industry, illustrates the growing importance of co-operation, especially at the pre-market stage. It is important to distinguish the stages of the innovation process at which co-operation and competition occur (Katz and Ordover, 1990).

One area in which such a collaborative approach can succeed is the development of biotechnology and the conservation of biological diversity. Currently, the question is cast in the framework of North–South relations and reciprocity at the bilateral national level. That genetic resources and biotechnology are unevenly distributed, however, provide a basis for research collaboration. This will not necessarily result from commitments in the form of conventions, but through specific research programs which are designed to take advantage of the differing endowments of countries. Research institutions and universities could play a key role in fostering new partnerships that include the development of new biotechnology products as well as the transfer of technology to developing countries through training programs.

The recent agreement between Merck, the largest pharmaceutical firm in the world with an annual research budget of US$600 million, and the Costa Rican Instituto Nacional de Biodiversidad (INBio), is an example. Under the agreement, Merck has provided the institute US$1 million for non-exclusive bio-products explorations over two years. The two institutions will share royalties from the sale of products resulting for the deal. It is notable that this agreement relies not on any international convention on the matter but reflects the interest of the two parties to derive benefits from biodiversity.

Conventions and multilateral agreements, however, could help by harmonizing practices and by indicating directions for institutional reform. The role of conventions in shaping the direction of innovation would therefore vary considerably from technology to technology. While some conventions may provide direct compensation for discontinued technological practices (as in the case of ozone-depleting substances), others may call for reforms to facilitate collaborative research, for example, in the area of biotechnology and biodiversity.

But collaboration alone will not improve the performance of technically underdeveloped firms or nations. More is needed, specifically the development of expertise within these firms and nations to utilize the results of externally performed research (Mowery and Rosenberg, 1989, p.290). This will require greater information sharing – and greater attention to information transfer.

4.1 Information sharing
Information is the currency of change (Clark and Juma, 1987). Sustainable development will need to start with partnerships to generate and disseminate information. While the industrialized countries have entered the information age, less developed nations have yet to do so. Information is probably the most expensive and the scarcest resource in the developing world. The current decay of institutions – the social mechanisms for managing information – may indicate that there is not much left for them to manage.

The massive amount of information available in the industrialized countries is not reaching developing ones, and whatever is reaching them is likely to be treated as proprietary. The few institutions that manage to maintain access to the global pool information, that is to say, tend to "privatize" what they collect. These circumstances prevent the transition towards sustainable development. An urgent need exists, then, for new forms of partnership to make information available to the developing world.

One of the priority areas in this effort is to establish direct communication links between developing country institutions and their counterparts in the industrialized countries for purposes of exchanging information. These links have two major uses. First, they will offer information on sustainable development otherwise unavailable to researchers in developing countries. Second, these links would inform those researchers about the kinds of information they should collect in their own countries (OTA, 1990).

These two purposes suggest different kinds of institutional arrangements. The first can be implemented through links with institutions whose aim is to disseminate information per se. The second would be related more specifically to research institutions. The problem, however, is that it is not possible to determine in advance the manner that information is likely to be utilized, and it would be unwise at this stage to restrict information to certain kinds of institutions. The priority should be to promote measures that allow for the widest range of access. Institutions such as the Network for Environmental Technology Transfer (NETT) based in Brussels and the International Environmental Bureau (IEB) of the International Chamber of Commerce (ICC) could extend their operations to the developing countries and facilitate access to the information on their databases.

There are many options for improving the dissemination of information ranging from on-line access to publications and other media. All these need to be explored. We should begin by establishing direct links with the many information facilities in the industrialized countries. Where information is unavailable, expectations must be modest. Sustainable development, owing to its complexity and breadth, requires developing nations to have far more access to information.

Partnerships in information dissemination should also include the acquisition of specific information facilities such as books, journals, magazines and databases, including information that is collected or generated by UN agencies and other international organizations. Most UN agencies are commercializing some of their publications to generate revenue, putting these publications out of the reach of many readers in developing countries. Depository centers in the developing countries as well as the waiving of royalties payments for UN publications re-issued by developing countries institutions would alleviate this problem.

4.2 Research and development
The role of science and technology in sustainable development, which has been identified as a major area for international co-operation, presents many opportunities for partnership. These include joint research programs, limited research and development collaborations, subcontracting arrangements, long-term institutional collaboration, and other measures. In the past, research programs in the developing world would have followed an agenda set by developed nations. The times are changing, however, and opportunities for real collaboration are starting to emerge (NRC, 1991). Environmental problems specific to developing countries but that reply on global action – the protection of rain forests, for example – require equitable partnerships. In addition, the capacity of developing countries to define research projects has increased.

4.3 Policy research
Pronouncements of a few people familiar with the problems of developing nations are often confused with policy research, which is a discipline in its own right. This confusion arises because the discipline of policy research is still nascent in developed countries and virtually absent in most developing ones. The problem is compounded because scientists and economists often believe that they should speak to power directly and that what they have to say constitutes the rational basis for policy making.

Theories of rational decision-making, however, have little applicability to its practice anywhere; politics is politics not "rational choice" in every part of the world. Policy choices depend, therefore, on a wide range of factors which should be the subject of policy analysis. The area of empirical policy research offers new possibilities for partnership between industrialized and developing countries. If policies are treated as experiments, as they should be, the developing world may be seen as a laboratory. Partnerships in building and interpreting these policy experiments will enhance information sharing as well as policy research.

4.4 Institution-building and training

"Both theory and experience", Bates (1989 p.4) writes, "underscore the significance of appropriate institutions" for sustainable development. Economic historians and developmental economists have long stressed that institutions are as important as markets in facilitating social and economic improvement (North, 1981; Hayami and Ruttan, 1985). Institutions are the arrangements through which people manage change. More precisely, institutions offer ways people may cope with risks and therefore ways they may take risks – which is the necessary first step to manage change.

Recent evidence on the history of technological transformation, especially in newly industrialized countries (NICs), shows clearly the role of institutional change in development. South Korea, for example, has over the last 30 years evolved a complex network of local institutions to promote technological change. The major institutional reforms introduced in Korea to promote technological development include the Korea Ministry of Science and Technology (MOST) in 1967, Korea Institute of Science and Technology (KIST) in 1966, Korea Advanced Institute of Science (KAIS) in 1971. Other major new institutions include the Korea Science and Technology Information Center (KORSTIC) whose main task was to serve as an information clearing house for the private sector (Choi, 1991). The country has also established the Presidential Commission on Science and Technology which serves as a link between these institutions and central government at the presidential level. These research and information institutions were also accompanied by new organizations which provide financial support for research and development. These include the Korea Technology Advancement Corporation (K-TAC), Korea Technology Development Corporation (KTDC), Korea Development Investment Corporation (KDIC) and Korea Technology Finance Corporation (KTFC).

Accompanying these institutional arrangements are legal and tax reforms aimed at promoting technological development, for example, tax exemptions for foreign engineers, research institutes, technology sales, technology-intensive enterprise initiators and projects using new technology. Other measures include tax credit and accelerated depreciation for research and vocational training equipment, tax credit for expenditures on research and training, tax reduction (65–70%) on imports of research equipment, tariff reduction (30–55%) on imports

of high technology industrial equipment, and deductions for research and development (Lee *et al.*, 1990).

In many developing nations institutions are often created for economic reasons: to reduce or spread the risks of investment, to accumulate capital, to organize production, and to regulate opportunistic behavior. Many non-market institutions – political bureaucracies, for example – take on economic functions, with respect both to productivity and redistribution. Political institutions also often invigorate interest-groups that cut across and religious rivalries, and this serves both a social and an economic purpose.

To a large extent, donor agencies have encouraged institution-formation by providing funds for infrastructure development as well as by endowing institutions, e.g. banks, import-export organizations, and industrial and agricultural co-operatives. Investment in people – in the capacity or competence for leadership – should also be a priority. So far, training programs have been designed to academic specifications. This kind of training produce academics and researchers; institution builders, however, are also needed.

Joint training programs that utilize the relative advantages of local and foreign institutions would be particularly helpful. (The Rockefeller Foundation has begun an important initiative in this area). A relaxation of residency requirements in the interest of competence building would assist collaborative training programs to gain stronger foothold in the developing world. Flexible training arrangements should also allow trainers from the industrialized countries to learn more about the developing world and thus improve their competence in, and understanding of, local problems.

As we come to see sustainable development as a dialectical or co-evolutionary process among economic, political, and social institutions, the theoretical basis for technology transfer will improve. One commentator (Bates, 1989, p.153) observes: "Economy and polity thus interact, generating a process of change. But the process of change is path-dependent; the course of the path is shaped by the initial institutional endowment. In this way, each society generates its own history."

4.5 Collaborative research outposts

Industrialized countries have established a wide range of research institutes in the developing world. Many of these were established at the time when raw materials shortages were a threat to the economies of the west. The original competence of these institutes, therefore, lay in prospecting for these materials. Developing countries turn to these institutes today, however, for the technological information and knowledge needed to process and market raw materials. As the developing world seeks to establish sustainable economies, institutions conceived to identify and process materials are now needed to become conduits, producers, and suppliers of much broader kinds of information.

Institutions whose main mandate is to provide technological information in the developing world, such as patent offices, may not exist in the developing world and, when they do, they are often marginal to the process of development. Indeed, some policy-makers and activists oppose the establishment of patent officers on the argument that adherence to intellectual property rights will impede development.

While information-gathering agencies are sometimes found on the local level, they are not yet active enough to enable countries to cope with their current informational needs. International research organizations make a contribution of course, when they locate branches in developing countries. What would help most, however, would be for research institutes established by developing countries to open branches in developed countries. This would open up new possibilities for collaboration and new mechanisms for sharing the global fund of knowledge. For example, the Nairobi based African Center for Technology Studies (ACTS) has established its Biopolicy Institute at Maastricht in the Netherlands. By waiting for international and other agencies to pass on technological information to them, developing countries leave their future to fate.

In a world dominated by multilateral agencies, interest in supply-side approaches to information are likely to become dominant, for example, the so-called "information clearing-houses" in industrialized countries. Instead of seeking to strengthen institutions locally, the tendency is to extend the activities of the clearing houses to the developing countries. The assumption here is that these agencies know what the developing countries want and are therefore ready to provide the relevant information. This smacks of institutional arrogance. The field of information is highly unpredictable and one does not really know what is useful until the need for it arises.

4.6 Development assistance

One of the most important forms of international co-operation lies in development assistance. So far, such assistance has been measured largely by the levels of aid transferred to developing countries. This assistance matters most when it strengthens scientific and technological capabilities in developing countries. In addition to the measures that are currently undertaken to support technological development and training, development assistance could also support collaborative research and technological co-operation between institutions in the donor and recipient countries. Priority could be given to the transfer of environmentally friendly technologies to the developing countries.

Assistance could also support programs aimed at adjusting or modifying existing technological systems in developing countries to meet new environmental standards. Donor agencies could help with the acquisition of useful proprietary technologies. Where venture capital is lacking to develop biorational technologies, development assistance could be used (Gerrits and van Latum, 1991). Development assistance would complement efforts in developing countries to reduce the participation of the government in private enterprise. Financial resources previously provided to public enterprises could be converted to venture capital.

4.7 Energy: an illustration

The development of technology for supplying the energy needs of developing nations illustrates how the general principles we have described here might be applied. First, transferred large-scale technologies to be successful must take root in indigenous knowledge, institutions, and practices. According to the World Bank (1988), the performance of imported energy utilities in the developing world remains poor especially when those utilities are vulnerable to political patronage and controls. The disadvantages outweigh the advantages of whole-system transfer. To be sure, foreign suppliers offer reliability, while local suppliers may lower quality and increase delivery time. However, increasing reliance on local supply enhances indigenous technological capability.

The countries of the world differ greatly in their energy resources. In producing petroleum for export, Nigeria burns off enough natural gas to serve the energy needs of most of Africa. Other nations may be poor in petroleum but rich in hydro resources. The resource base is rarely the limiting factor. Rather the limiting factor is usually the ability to bring traditional and imported technologies together in a domestic technological array that responds to local technological capacities and energy needs. It is notable that a viable energy technology policy will need to take into account the principles of technological diversity. Renewable energy technologies, particularly solar and biomass, need to be stressed as means of increasing diversity in energy sources and stability of supply (Scurlock, Rosenschein and Hall, 1991). The pressure to make a transition away from fossil fuels in light of global climatic change has increased the need to develop alternative energy technologies.

At present, sub-Saharan Africa with the exception of South Africa uses less electricity than New York City. The absence of electricity in rural areas serves as a barrier to the introduction of other technologies, for example, computers. Accordingly, there are "limitations on the extent to which market forces can be relied upon to develop appropriate solutions . . ." (IEA, 1985). There seems to be no alternative to government policies in organizing an appropriate response to the energy needs of developing countries.

Nevertheless, since fossil fuel based energy systems are not as entrenched in the developing as in the developed world, developing nations may be more amenable to sustainable energy technologies. This may include installation of solar, wind-powered, and hydroelectric systems in traditional sectors that rely primarily on fuelwood and animals for energy. It may also include a move away from fossil fuels insofar as practicable in modern sectors.

The success of energy production in developing countries depends far less on access to large-scale energy systems than on attention to institutional arrangements, appropriateness to local needs and resources, the ability to make adjustments incremental and an often minor adjustments, and the ability of local technicians to adapt to social and technological change. Thus, it becomes more important to think in terms of broad policies of adaptive technology assessment than to try to duplicate in any way approaches to energy production used in developed nations.

5.   SYSTEMIC TECHNOLOGY POLICY MEASURES: EXAMPLES

We have argued that adjustments in institutional environments must accompany advances in the application of science and technology. We have now to identify the needed kinds of political, institutional, and legal reforms. Our recommendations for technology development, acquisition, transfer, and diffusion include systematic technology policy measures (STEMs) (see Table 1) which provide guidance for decisions under conditions of limited resources, ecological imperatives institutional fragmentation, and pressure to show results.

We shall describe a sample of systematic technology policy measures (STEMs) by which developing nations may promote and stimulate indigenous technology. These policy measures build on existing circumstances: they organize the economic and institutional terrain with the minimum in investment, administration, staff, and infrastructure. The chief factors in all STEMs are information flow, technical content, and institutional networking. The STEM approach differs from other strategies of policy formation in that it emphasizes synergistic links between sectors, a systems approach rather than a linear cause-and-effect structure.

The long-standing emphasis on technology transfer as a tool for international competitiveness makes it difficult for African nations to apply emerging technologies to solve local problems and achieve self-sufficiency. These technologies, in so far as they speed up the treadmill of innovation, larger surpluses, falling prices, and further innovation, undermine rather than enhance the global competitiveness of economies dependent on agriculture

and natural resources. The industrialization of agriculture forces these economies to emphasize local rather than export markets and self-sufficiency over competitiveness.

If corporations can produce substitutes or equivalents for chocolate, for example, in Europe as easily as in Africa – if industrial bioprocessing of vegetable oils into cocoa butter can be sited in New Jersey as well as in the Ivory Coast – then there is reason to believe that Africa will become marginalized in the world economy. Technology transfers may not change this trend, but they can enable local economies to provide for themselves by converting obsolete export industries to production for local and regional needs.

A number of policy measures have been used with varying success to facilitate technological innovation (Table 1). The feasibility of these measures depends largely on the current technological level of the country, access to the global fund of scientific knowledge, existing institutional arrangements and flexibility, and internal capacity to implement those policies.

To take an example: the independence of public sector industries from political patronage and manipulation may be the single important factor in predicting their efficiency; it is the notable difference, for example, between public utilities that produce electric power efficiently and those that do not. Measures related to direct assistance for research and development would be appropriate in public sector industries insulated from political control to maintain professional competence. These measures might not be appropriate in other contexts, given the current pressure on developing countries to reduce their public expenditure and rationalize public sector operations.

5.1 Public and private procurement
Since governments are major purchasers in Africa, public procurement policies could encourage indigenous technological innovation and development as well as redirect technological change towards environmentally sound paths. At present, most African countries operate on a "lowest bidder" principle in order to reduce costs. Large-volume suppliers, who can take advantage of economies of scale, tend to edge out or discourage local producers who must charge higher prices. If purchasing agents considered the difference between the lowest bid and the price of a local product as an investment in research and development, however, they might be able to direct and improve the pace of technological innovation.

Purchasing agencies would need to secure the services of engineers, designers, and other technologists to make the case for accepting a local higher bid; agencies may also have to relax standards, thus encouraging gradual improvement. Such a procurement system would also be linked to an intellectual property régime which would

Table 1 : Government innovation policy measures

Measure	Examples
Procurement	Central and local government purchases and contracts, public corporations, R and D contracts, prototype purchases, setting of design criteria, choice of priority technologies
International trade	Trade agreements, technology acquisition agreements, tariffs, foreign exchange regulations, export compensation, import subsidies, licensing
Public enterprise	Innovation by publicly owned industries, setting up of new industries, pioneering use of new techniques by public corporation, participating in private enterprises
Scientific and technical	Research laboratories, support of research associations, learned societies, professional associations, research grants
Education	General education, universities, technical education, retraining
Information	Information networks and centers, libraries, advisory and consultancy services, databases, technology monitoring, liaison services, public awareness
Financial	Grants, loans, subsidies, financial sharing arrangements, venture capital, R and D limited partnerships, provision of equipment, buildings or services, loan guarantees, duty and customs remissions, export credits
Taxation	Company, personal, indirect and payroll taxation, tax allowances, tax exemption for private foundations
Legal and regulatory	Patents, utility models, plant breeders rights, environmental and health regulations, contractual arrangements, conventions, inspectorates, monopoly regulations
Political	Planning, regional policies, honours or awards for innovation, encouragement of mergers or consortia, public consultation, creation of new institutions, setting up of research funds, initiating legal reforms
Public services	Purchases, maintenance, supervision and innovation in health service, public building, construction, transport, telecommunications, infrastructure
External assistance	External aid, technical assistance, local and external training
International relations	Sales organizations, trade and diplomatic missions (science attachés), technical co-operation, research representatives

Source: African Center for Technology Studies, Nairobi.

protect innovations that lack novelty (in the legal sense of the word) or build on information in the public domain. An appropriate intellectual property system might encourage the reverse-engineering and the re-invention, as it were, of needed items. We shall return to this possibility presently.

As the developing countries restructure their economies to promote private sector activities, governments will need to introduce regulations and incentives that promote the use of cleaner process and product technologies.

5.2 Foreign exchange management

Even in situations in which the government is not the major buyer, it may nevertheless influence technology development through foreign exchange management. The government of Zimbabwe (under settler rule) required investors to recover in a set period the foreign exchange they spent in setting up a venture. This measure forced investors to buy from local producers whenever possible to conserve foreign exchange. As a result a local network of suppliers sprang up to meet this demand.

This lesson can be applied to other nations. In Kenya, for example, occasional restrictions on the allocation of foreign exchange – although instituted primarily for other reasons – has worked to encourage local industrial inputs.

Foreign exchange regulations might be coupled with technology acquisition criteria that emphasize the role of learning at the level of the firm. This would discourage the importation of "turn-key" or "push-button" plants.

5.3 Financial local research

Local research efforts require support from whatever sources can be found; usually, governments offer some funding, but more must be done to encourage private sector and individual support. In many developing countries (including Brazil, Argentina, Philippines, and Nigeria), local entrepreneurs provided support for local research. Yet private investors tend to avoid risk and, thus, prefer short-term or quick turn-around research to long-term development.

Government in developing countries could use tax law – i.e "deductions" – to encourage private sector support for research, the establishment of private foundations, and set-asides for research and development by firms. Developing countries use tax codes in this way to encourage research and development. In the US, for example, there are over 22,000 private foundations which support a wide range of social activities including research. State incentives make it possible to establish such civil institutions. Incentives of this kind are particularly hard to monitor, however, since it is easy to mask ordinary expenses as investments in research. We thus strongly recommend policies that promote the establishment of local foundations that support the development of cleaner technologies.

The lack of venture capital to support research and development remains a major obstacle to innovation. African banks minimize the risks they take; accordingly, they tend to support only proven innovations. This is particularly discouraging since technological development in western countries such as the US depends greatly on the willingness of bankers to back new ideas. The close link between finance and innovation does not exist in Africa, and bankers are interested in financing sellable products not ideas. Alas, this attitude presumes that all the economically worthwhile ideas will come in from elsewhere and that therefore there is less reason to invest in research.

The structure of the banking system in developing countries – including the role of central banks – needs to be re-examined in relation to its responsibility to support innovation. Laws dealing with public finance, commercial and financial institutions, and international financial agreements could be amended to enable local banking institutions to support local research.

5.4 Intellectual property protection

The international patent system affects innovations largely at the frontiers of technology. If developing countries wished to hitch their wagons to the star of global competitiveness, then, indeed, the patent system might thwart their efforts. We have mentioned that advances in agricultural biotechnology, for example, seem poised to replace agricultural with industrial processes for producing commodities like cocoa, vanilla, specially oils, and so on. If developing nations wished to get on the technological treadmill – if they wished to compete in the industrial fabrication of these products – then patents might be one of a thousand obstacles in their way.

We see no competitive advantage or other economic reason, however, for developing nations to adopt this strategy. If novel technologies were free to all comers, we could except enormous surpluses to flood markets, ensuring that none but the very most efficient and best capitalized ventures would survive round after round of falling prices.

Advances in technology, for example, biotechnology, can assist African nations to satisfy the needs of their own people if not improve their competitive position in serving the affluent in other nations. Research and development in biotechnology would have to be directed, however, to use by small-scale and even semi-subsistence farmers. To do this "biotechnologists should focus on the problems of these farmers; secondly, there should be enough

information about their farming systems to generate ideas for appropriate biotechnological research" (Bunders, 1988, p.179).

To some extent, there have been advances in biotechnology appropriate to the needs of developing nations: Monsanto Corporation is developing a drought-resistant cassava and many experimenters are breeding small ruminants that can be herded for meat and milk. To realize the immense contribution biotechnology can make to sustainable development in developing countries, those countries must themselves have the capacity to improve and apply technology in relation to local needs. Evidence from different parts of the world illustrates the potential for blending traditional technologies with innovations in microelectronics, energy production, and biotechnology (Katz, 1987). The debate over patents is misplaced insofar as it ignores the possibility of permissible technological hybrids or recombinants that could assist industries that serve the common people in developing countries. The main concern in technology development is not barriers patents pose but the ability of people in developing nations to blend existing with imported technologies.

A large body of immensely useful technological knowledge lies in the public domain and is unaffected by the international patent system. Developing nations can meet many or most of their requirements by tapping this pool of knowledge and recombining it with local innovations to generate alternative technologies. This is precisely the basis for innovation in the informal sectors of the world. If this is not happening, it is because the informal sector has very limited access to technological knowledge. Where information exists, it is as likely to be kept as trade secrets as shared. Developing countries need an intellectual property system that would promote innovations in the public domain as well as the local production of appropriate or intermediate technologies.

Patent offices in Africa are therefore more useful as databases of technological diversity than as centers for granting patents to local inventors. The Kenyan Industrial Property Act, for example, puts more emphasis on the promotion of technological innovativeness than on granting patents.

Patent laws in Africa do not adequately address the need to stimulate and protect "appropriate" or "intermediate" technologies such as cook-stoves, wind-mills, water pumps, grain stores, oil presses, and plows. The definition of "patentable invention" excludes a large number of socially useful technologies from protection. These kinds of inventions could be encouraged and protected through utility certificates or "petty patents" of the kind Germany and Japan used extensively to upgrade their innovative capacity.

The Japanese introduced such utility certificates in 1905 to protect booming household industries and other manufacturers in the informal sector. This sector was largely based on practical but novel contrivances which combined existing technologies and therefore might not be patented in the usual way. Innovation in the Japanese textile industry, for example, was stimulated through utility certificates granted to handlooms that combined traditional technology with new knowledge gained from reverse-engineering textile machinery imported from Europe.

Reverse engineering can be a portal through which appropriate technology enters the economy of developing nations. In Japan, managers and engineers were hardly cowed by the patent on a foreign invention. They reverse-engineered and either improved it or applied it to something else. They became used to thinking about technology not as a black box but as integrated system of knowledge. They used the factory as a laboratory (Freeman, 1987). Reverse-engineering helped promote communication and co-operation across industries. Similar processing of knowledge and collaborative learning is found in the informal sector of developing countries. For that reason, the 1989 Industrial Property Act of Kenya provides for utility certificates.

In a world where sustainable development depends on co-operation and technological diversity, intellectual property laws need to encourage the wide use of knowledge already in the public domain. In addition, it is necessary to expand the variety of intellectual property protection instead of arguing for uniformity. New régimes of protection, for example, are needed for encouraging the conservation of biological diversity at the local level.

6. OUTLOOK AND ACTIONS

While the 1970s and 1980s could be characterized as periods of "natural resource wars," the 21st Century is likely to see "information wars." Information is the ultimate currency of change. It is the resource from which innovations grow. The same technologies that have made it possible to disseminate information so widely can also be used to control its distribution. It is this factor that makes information and its technology sources of power and a potential battlefield between institutions in the next century. The current debate over intellectual property protection is possibly the beginning of a long struggle over access to information in general. Nowhere is this matter more obvious than in the controversy over an international convention on biological diversity. The developing countries are arguing that they will make genetic resources available to industrialized countries if these countries agree to the transfer of biotechnology. In both cases, what is at stake is information: biological and technological.

Efforts to achieve sustainable development will not yield results unless countries, especially in the developing world, have access to the global body of scientific and technological knowledge and build the relevant capacity to apply this knowledge. In this respect, it is important that measures that make technological information available also improve the capacity of the developing nations to solve environmental and economic problems using technological options.

For policy formulation, it is important to distinguish between *technological co-operation* at the research level and *market competition* at the product level. The emerging world order reflects both an interest in competition while at the same time new and more complex patterns of co-operation and partnership are starting to emerge. The simplistic perceptions of "pure market competition" are being replaced by more sophisticated views which recognize the shaping of market institutions to reflect social values and ecological responsibility.

In the words of the Brundtland Report: "Knowledge shared globally would assure greater mutual understanding and create greater willingness to share global resources equitably" (WCED, 1987, p.11). The issue of sharing the global fund of knowledge and information requires urgent consideration. It not only involves access to information, but it includes the provision of infrastructure for information sharing. Access to information is not adequate unless there are individuals to translate "information into a form intelligible to participants" (Aitken, 1985, p.17). Establishing information sharing programs needs to start with public domain information.

The developing countries, on the other hand, will need to introduce policies and institutional reforms that facilitate innovation and change. These policies will need to encourage the spirit of free enquiry as well as the creation of civil institutions. This, in our view, is the beginning point for the democratic transition necessary to generate and harness innovation for sustainable development.

The waning of the mechanistic world view must now be accompanied by creating and institutionalizing alternative approaches based on biological and ecological metaphors. The transition towards sustainable development will be to a large extent a revolution in the way we view the world – it is an epistemological transformation which articulates itself in the design of new technologies and institutions.

REFERENCES

Aiken, H. 1985. *The Continuous Wave: Technology and American Radio, 1990–1932.* Princeton, NJ, Princeton University Press.

Allen, P. 1991. *Evolution and Leaning in Economic Theory.* Fifth Hayek Symposium, Freiburg.

Bates, R.H. 1989. *Beyond the Miracle of the Market.* New York, Cambridge University Press.

Berlin, I. 1991. *The Crooked Timber of Humanity.* New York, Alfred A. Knopf.

Bijker, W.E. *et al.* 1987. *The Social Construction of Technological Systems, New Directions in the Sociology and History of Technology.* Cambridge, Mass., MIT Press.

Bowonder, B. and Miyake, T. 1990. Technology development and Japanese industrial competitiveness, *Futures,* **22**, 1.

Bunders, J. 1988. Appropriate biotechnology for sustainable agriculture in developing countries, *Trends in Biotechnology,* **6**, 173–180.

Busch et al. 1991. *Plants, Power and Profit: Social, Economic and Ethical Consequences of the New Biotechnology.* Cambridge, Mass.: Blackwell.

Choi, H.S. 1991. Science and technology policies in the industrializations of a developing country: Korean approaches. In M.A. Salam, *Science, Technology and Science Education in the Development of the South.* Trieste: Third World Academy of Sciences.

Clark, N. and Juma, C. 1987. *Long-Run Economics: An Evolutionary Approach to Economic Growth.* London: Pinter Publishers.

Clark, N. and Juma, C. 1991. *Biotechnology for Sustainable Development: Policy Options for Developing Countries.* Nairobi: Acts Press, African Center for Technology Studies.

Dosi, G. 1988. Institutions and markets in a dynamic world, *The Manchester School,* **LVI**, no. 2.

Freeman, C. 1987. *Technology Policy and Economic Performance: Lessons from Japan.* London: Pinter Publishers.

Gadgil, M. and Thapar, R. 1990. Human Ecology in India: Some Historical Perspectives. In: M.A. Salam, *Science, Technology and Science Education in the Development of the South.* Trieste: Third World Academy of Sciences.

Gerrits, R. and van Latum, E.B.J. 1991. *Bio-pesticides in Developing Countries: Prospects and Research Priorities,* Biopolicy International 1. Nairobi: Acts Press, African Center for Technology Studies.

Goka, A.M. and Mihyo, B. 1990. *Technology Policy Institutions in Selected African Countries.* Ottawa: International Development Research Center.

Gould, S.J. 1977. *Ever Since Darwin.* W.W. Norton: New York.

Hayami, Y. and Ruttan, V. 1985. *Agricultural Development: An International Perspective.* Baltimore: Johns Hopkins University Press.

Heaton, G. et al. 1991. *Transforming Technology: An Agenda for Environmentally Sustainable Growth in the 21st Century.* Washington, DC: World Resources Institute.

IEA (International Energy Agency). 1985. *Energy Technological Policy.* Paris: Organization for Economic Co-operation and Development.

Israel, A. 1989. *Institutional Development: Incentives to Performance.* Baltimore: Johns Hopkins University Press.

Juma, C.1989. *The Gene Hunters: Biotechnology and the Scramble for Seeds,* Princeton NJ: Princeton University Press.

Kakazu, H. 1990. *Industrial Technology Capabilities and Policies in Selected Asian Developing Countries.* Manila: Asian Development Bank.

Katz, J. 1987. *Technology Generation in Latin America Manufacturing Industries.* London: Macmillan.

Katz, M. and Ordover, J.A. 1990. R and D co-operation and competition, Brookings Papers: *Microeconomics.* Washington,

DC: Brookings Institution.

Kemp, R. and Soete, L. 1990. Inside the 'Green Box': on the economics of technological change and the environment. In: C. Freeman and Luc Soete (eds). *New Explorations in the Economics of Technological Change.* London: Pinter Publishers.

Krupp, H. 1990. A Vision of S and T Policy in a Resource-conscious Society, In: Jon Sigurdson (ed.) *Measuring the Dynamics of Technological Change.* London: Pinter Publishers.

Lall, S. 1987. *Learning to Industrialize: The Acquisition of Technological Capabilities by India.* London: Macmillan.

Landes, D. 1990. Why are we so rich and they so poor?, *American Economic Review,* **80**, 2.

Lastres, H.M.M. 1990. *The Impact of Advanced Materials on World Development,* United Nations Industrial Development Organization, Vienna, Austria.

Lee, K.R. et al. 1990. Interim Evaluation of Localization Policy and Ways to Improve. Research Report No. 196. Seoul: Korea Institute for Economy and Technology.

Levy, J.D. and Samuels, R.J. 1991. Institutions and Innovation: Research Collaboration and Technology Strategy in Japan. In: L.K. Mytelka, (ed.) *Strategic Partnerships and the World Economy: States, Firms and International Competition.* London: Pinter Publishers.

Matthews, R.C.O. 1986. The economics of institutions and the sources of growth, *Economic Journal,* **96**, 384.

Moser, A. 1991. Ecologically Sustainable Technology: The New Technology Paradigm. Vienna: Chapter Presented at the International Conference on An Agenda of Science for Environment and Development into the 21st Century. November 24–29.

Mowery, D. and Rosenberg, N. 1989. *Technology and the Pursuit of Economic Change.* Cambridge: Cambridge University Press.

North, D.C. 1981. *Structure and Change in Economic History,* New York. Norton.

North, D.C. 1990. *Institutions, Institutional Change and Economic Performance.* Cambridge: Cambridge University Press.

NRC (National Research Council). 1991. *Towards Sustainability: Soil and Research Priorities for Developing Countries.* Washington, DC: National Academy of Sciences.

OECD. 1990. *Economic Instruments for Environmental Protection.* Paris: Organization for Economic Co-operation and Development.

OTA. 1990. *Critical Connections: Communication for the Future.* Washington, DC: Office of Technology Assessment.

Polanyi, M. 1944. *The Great Transformation.* Boston: Beacon Press.

Porter, M.E. 1980. *Competitive Strategy: Techniques for Analyzing Industries and Competitors.* Free Press.

Porter, M.E. 1991. America's green strategy, *Scientific American.* April.

Rogoff, M. and Rawlins, S. 1987. Food security: a technological alternative, *Biosciences,* **37**, 800–807.

Rosenberg, N. and Birdzell, L.E. 1986. *How the West Grew Rich,* I.B. Tauris, London. UK.

Saasa, O. (ed.) 1992. *Joining the Future: Economic Integration in*

Africa. Nairobi: Acts Press, African Center for Technology Studies.

Scurlock, J., Rosenschein A. and Hall, D.O. 1991. *Fuelling the Future: Power Alcohol in Zimbabwe*. Nairobi: Acts Press, African Center for Technology Studies.

Stegmann, K. 1989. Policy rivalry among industrial states, *International Organizations*, **43**, 1,73–100.

UN (United Nations). 1991. Utilization of Economic Instruments. Geneva: Preparatory Committee for the United Nations Conference on Environment and Development.

UNCTAD (United Nations Conference on Trade and Development). 1990. *Transfer and Development of Technology in Developing Countries: A Compendium of Policy Issues*. New York: United Nations.

WCED (World Commission on Environment and Development). 1987. *Our Common Future*. Oxford: Oxford University Press.

Wolf, Jr. C. 1990. *Markets or Governments: Choosing Between Imperfect Alternatives*. Cambridge, Mass.: MIT Press.

World Bank, 1988. *Energy Issues in Developing Countries*. Washington, DC: World Bank.

Chapter 16: Institutional Arrangements

W. Lang, J. B. Opschoor and C. Perrings

EXECUTIVE SUMMARY

This chapter addresses the problem of institutions and environmental change from a very broad perspective. Institutions are taken to mean not only organizations but all arrangements, social conventions and codes of conduct governing human behavior. It concentrates on two aspects of the problem: economic aspects on the one side, and legal and political aspects on the other.

The authors present a number of recommendations. These fall into two main areas: a research agenda and scientific institutions. The research agenda was divided into the following three categories:
1. The environmental impacts of institutional change.
2. The scope for, and barriers to, a transition to sustainable development strategies.
3. The institutions, policies and instruments required by a sustainable development strategy.

These stress a number of issues:
(a) the growing importance of decentralized decision-making.
(b) the importance of "delinking" economic growth and environmental degradation.
(c) the welfare gains from promoting efficiency in the use of environmental resources, "getting prices right".
(d) the need to protect ecological systems through the use of safe minimum standards.
(e) the importance of institutions and mechanisms for linking development and environment.

The scientific institutional recommendations included:
(a) The strengthening of co-operation between the natural and social sciences.
(b) The need to establish networks of (regional) multidisciplinary environmental research centers.
(c) The importance of countering the asymmetry in access to information – largely a North–South problem.
(d) The establishment of mechanisms to improve the science input into multilateral decision-making.

1. INTRODUCTION

The notion of "institutions" appears sometimes to be used in the narrow sense of "organizations". Here we use the term "institutions" in the scientific sense, originally due to Veblen, by which it covers all "arrangements" (formal and informal ones), social conventions, and codes of conduct governing human behavior. In this sense, institutions include both the organizations that give force to social conventions, and the less tangible social structures governing particular aspects of human behavior: markets and market regulating structures, property rights, social ethics and so on. The focal point of "organization" is still important for this discussion, however. For it is around organizations that the cultural norms of rules, behavior, values and expectations revolve. This is the political ecology of institutional change.

Institutions would accordingly include, for example, at the one extreme IGOs, national and federal governments, and all the other informal grass-roots associations of resource users – grazing associations or borehole syndicates for example.

This chapter is prepared within the context of two major global trends, each subject to its own dynamics but both closely related. The first is the process of global environmental change and environmental degradation. It is this trend and the adverse effects it is having on human welfare that has motivated both this Conference and the broader processes of which it is a part. The second trend is, however, equally significant. It is the trend towards the democratization of institutes and the decentralization of decision-making, not only in Eastern Europe and the former USSR, but through much of the Third World.

The implication of the first trend is both that the stock of natural assets that forms the inheritance of each generation is declining in value, and that the set of options available to successive generations is contracting. The implication of the second trend is that increasingly the state of the environment, and so the well-being of future generations, is going to rest on the private decision of those having access to the resource base.

Our starting point is therefore the fact that the health of the environment is increasingly going to rest in the independent decisions of billions of individual users of environmental resources. The underlying causes of environmental change due to human behavior will increasingly come to lie in the determinants of those individual decisions: the available information on the environmental effects of resource use, preferences of consumers and the technology available to producers, the rate at which users discount the future effects of current actions, the property rights that define their endowments, the set of relative prices that determine market opportunities associated with those endowments, the cultural, religious and legal restrictions on individual behavior that prescribe the range of admissible actions, and so on. The general problem addressed by this Conference derives from the fact that, while the decisions of resource users may be privately rational, they appear to be socially damaging – compromising the interests of present as well as future generations. The problem addressed in this chapter derives from the fact that, many of the determinants of socially damaging individual decisions reflect the institutions, as we have defined the term, that govern human behavior.

We are concerned to identify both the institutional causes of environmentally unsustainable behavior, and the scope for institutional reform. There will be barriers in the way of the adoption and implementation of these improvements. Some of these barriers are frictional as when institutional rearrangements are associated with reshuffling of costs and benefits, others are structural (when causes of unsustainability are inadequately addressed/curbed).

2. ECONOMIC ASPECTS

2.1 Institutional causes of unsustainability

Unsustainability can be related to a range of societal processes and elements. Traditionally, it has been linked to economic growth and population expansion, or to the combination of these two as outlined in Chapter 1 by Arizpe *et al.* Other factors mentioned include short-sighted technological developments and poverty (for the latter, see notably The Brundtland Report – WCED, 1987). Behind these forces, there are some more deep-seated mechanisms such as the creeping dominance of the Western cultural world views (including views of humankind-nature relationships; see, e.g. O'Riordan 1981; Opschoor 1989; Arizpe *et al.*, 1991) that are firmly associated with economic expansion over the globe.

It is conventional to refer to the "failure" of institutions wherever they lead to socially damaging decisions. The most frequently cited institutional failures are government

failures and market failures (see OECD, 1990). However, while these are the most visible of failures, it is important to appreciate that they are often the effect of another source of failure – what may be termed informational failure. This can often be traced back to the way that scientific research and development is organized, and scientific results are disseminated. Just as it is now widely recognized that environmental systems and economic valuation of the role and integrity of such systems should be merged (WCED, 1987, 62ff), so there is a growing understanding of the need to integrate the sciences both conceptually and structurally. This point is outlined in the report of the Bergen Conference (NAVF, 1990, 329-342).

Government failures can be divided into policy failures and administrative failures. The term "policy failure" refers to the range of regulatory instruments, fiscal, exchange rate, monetary, price, income and other policies which so distort the private cost of environmental resource use as to make it privately rational to damage the social heritage. It is common to speak of policy failure in three rather different cases.

One speaks of policy failures in cases where prevalent policies are based on past decisions in which ecological or environmental considerations were given insufficient weight. This very often is the case with sectoral policies where sectoral interests and powers have predominated over, or excluded ecological considerations, or with policies dating back to periods of time when environmental problems were not yet perceived fully. The evidence is everywhere, particularly in the fields of agriculture, energy, transportation and financial accounting. In every case policies designed to promote a particular sectoral interest, where environmental consequences have not been weighed or costed, turn out to erode sustainable ecological systems.

Secondly, one may speak of policy failure when policy is directed at enhancing economic growth per se, that is, disregarding social and ecological repercussions of the growth paths that these policies imply. This is sometimes known as the equity dimension and is evident in the treatment of cultural minorities when their habitats and ways of life are destroyed, as well as in the increasing vulnerability of the very poor when forced on to marginal lands to keep alive.

Thirdly, one may speak of policy failures when national or international policy is unable or unwilling to develop adequate institutional checks on market failures (see below).

The notion of administrative failures refers to a range of problems within the organization of government at the various levels, leading to inadequate policy formulation and/or inadequate policy implementation. Examples include rigidities due to entrenched traditional divisions of labour within administrative organizations (very often along sectoral lines), high time preference even within governmental organizations, insufficient integration between agencies and departments, instruments or powers insufficient to achieve policy objectives, and lack of instruments or powers to ensure policy implementation within the economic processes.

In the following, we indicate some environmentally relevant failures in both macro-economic and sectoral policies. Economic growth in itself may be beneficial for many obvious reasons.

Economic growth may also be consistent with the maintenance of environmental quality, and is certainly feasible in the foreseeable future. However, given a finite resource base, and finite waste absorptive capacity, a growth oriented strategy is unlikely to be sustainable in the very long run (Perrings, 1987). The policy failure in current growth strategies lies in the fact that long-term environmental effects tend to be ignored for the sake of short-term income gains. The risk is that we will collectively be confronted with environmental costs that are either irreversible (e.g. species extinction) or very costly to redress (e.g. the impact of pollution in Northern Bohemia).

This failure is in part a matter of social preferences: one task for science is to make it clear the likely consequences of continuing to foster such preferences and how to reorder them by informed consent and on a socially just basis.

In order to understand the determinants of the growth process, it is conventional to point at the predominant value structure according to which more is preferred to less by economic agents. This is a far from complete analysis; it fails to come to terms with some other basic "push" factors behind growth such as:

(a) a tendency to bridge wealth gaps both within and between economies, the aspiration levels always being set at the levels of those that are materially "advanced" while the latter's material welfare continuously expands;

(b) the desire to ensure continuity in a competitive and uncertain context (especially at the level of firms) entailing a demand for growth, profit, market control, etc. This is an economic system-inherent growth tendency;

(c) the attempts by governments to maintain political security and continuity by providing dissident and emancipatory groups with more absolute and relative wealth and income, not by taking this from the privileged in absolute terms, but by redistributing from a growing national product;

(d) technological development as we have seen recently, has tended to be labor-replacing. In societies favoring full employment this implies a substantive push for continued economic growth at a rate beyond the technology induced rate of growth of productivity.

This suggests how formidable a task will lie ahead of our political organs if ultimately the levels of economic activities will have to be controlled and redirected. This might be the case if changing the technologies or locations of our activities would not be sufficient from an ecological perspective. In most (if not all) market economies, however, societies and their governments are not equipped to redirect and control activity levels. And if they are, very often effective policies would require a much higher degree of international co-operation in environmental and economic politics than the prevalent one.

Also, efforts to reorient economic processes in developing countries will require rates of transfers of capital, technology and skills that exceed the current levels of availability or absorbability. As long as these circumstances remain unaltered, the necessary conditions to achieve sustainable economic development will not be met.

One possible way forward for science is to model pathways to future sustainable macro-economic futures, by manipulating prices, enhancing the capability of indigenous science, and reviewing the compensatory measures necessary to put in place. This could be done via models, or through detailed case studies of particular cultures and economies.

2.2 Sectoral policy

Very often past sectoral policies (e.g. in the field of energy, agriculture and transportation) in many countries and regions at least in the North, have been decided upon primarily with the sectoral interests in mind, at best with consideration for tradeoffs vis-à-vis other established sectors. Environmental concerns have not been appropriately internalized. Apart from this, not only private sector but also public sector decision-makers have limited time horizons and/or discount future consequences of present decisions. Thus, policy formation may suffer from biases towards stronger (in terms of economic and political power or significance) sectors and against interests that cannot manifest themselves on markets and in the political arena, such as future generations' interests.

Sectoral policy failure may often result in the subsidization of sectoral activities so that prices no longer reflect even appropriate private costs. In resource-related sectors such as agriculture, water, timber and energy, this leads to artificially low resource prices. On top of this (environmental) externalities are often ignored so that private costs in themselves were distorted reflections of social costs. In such cases, users of the products of these sectors are paying less than the social costs their use gives rise to; they thus are induced to consume more than would be the case were the price corrected for social costs. Prices then give the wrong signals and the sector may expand to levels beyond what is socially desirable.

2.3 Price distorting policies

Perhaps the most environmentally significant of the policy failures are those which drive a wedge between the true social cost of resource use, and the cost to the individual user – the private cost of resource use. There exists a range of fiscal, exchange rate, price and incomes policies which have the effect of encouraging the overutilization of environmental resources. Administered prices in agricultural markets, destumping subsidies in agriculture and stumpage fees or royalties in forestry, have encouraged deforestation at excessive rates both in terms of rates of felling and the clearance of ever more marginal land for agricultural purposes (Warford, 1987; Pearce *et al.*, 1988; Repetto, 1989). Subsidies designed to promote cash cropping as a means of increasing export revenue have resulted in leaching, soil acidification, and loss of soil nutrients, and to the reduction in the resilience of key ecosystems (Grainger, 1990).

One task for science is systematically to examine these distortions for a range of countries and environmental/social conditions to examine what would be the environmental consequences of retaining or removing them, who would gain and lose thereby, and what compensatory and regulatory institutions should be put in place to achieve the best result.

The same set of policies has frequently had the effect of depressing incomes of rural dwellers, with similarly adverse effects on the environment. Indeed, the Brundtland Report recognized the close relationship between poverty and environmental degradation. Because what matters is consumption today, people in poverty tend to discount the future costs of resource use at a very high rate (Perrings, 1990). In meeting their immediate consumption needs, they are compelled to ignore the potential future consequences of the use they are making of environmental resources.

A second cause of poverty is the distribution of assets, whether marketed or non-marketed. The very poor also tend to be those with access to few productive assets, and are frequently the most likely to degrade those assets. There has been a marked and continuing tendency for the distribution of both assets and income to widen over time in many of the low income countries, reflecting both the erosion of traditional rights of access to the resource base and the increasing human population pressure. Gender is an important factor in this trend. Female headed households typically have access to a much smaller asset base than male headed households, and it is not coincidental that relative poverty in the sense of relative deprivation is reckoned to bear most heavily on women (UNDP, 1990).

2.4 Market failures

Market failures denote the failure of the market, as an institution, to allocate resources in the best interest of society. The main cause of such failure is the absence of markets – particularly for the environmental effects of economic activity. This gives rise to what are known as external effects. It is useful to distinguish between two types of externality. Reciprocal externalities are those in which all parties having rights of access to a resource are able to impose costs on each other (as often happens with resources in common property). Unidirectional externalities are those in which processes in the common environment ensure that the short run external environmental costs or benefits of resource use are "one way" (for example, deforestation by the users of an upper watershed inflicts damage on the users of the lower watershed). Both types of externality give rise to "cost shifting" or "displacement of costs" (Kapp, 1950; Opschoor, 1989; Pearce and Turner, 1990). This means that part of the adverse consequences of one actor's decision are passed on to others to bear. Economic activities lead to effects that are external to those who decide over these activities in the first place. In other words, economic activities lead to social costs (including the costs of environmental degradation) that are not fully translated into private costs, or internalized into the private decision-making mechanism.

This practice of cost shifting is facilitated by what could be labeled as: the "distance factor". The consequences or effects of environmental degradation in relation to economic activities manifest themselves at often large distances from the source or agent causing them. This may be a distance both in terms of space and time (e.g. DDT in polar ice caps, chemical time bombs, and climate change). Effects of environmental degradation are thereby shifted on to other people, to future people and even to other species. There is a third type of distance involved, namely that between the level of one's individual influence and the level at which a problem must be addressed for its solution. One could refer to this as distance in decision-level. Single actors in a multi-actor context may face situations where their privately optimal behavior may lead to socially or collectively undesirable overall outcomes (the "prisoners" dilemma in the case of very few actors, or the "tragedy of the commons" in the case of many actors). Examples include countries sharing a common resource, individual fishermen exploiting a shared fish population, etc. In many cases, the absence of control and intervention by national or international authorities, leads to an irrational exploitation of a shared or common property resource, to ongoing pollution, etc.

Distance between cause and effect is an obvious problem. But the dilemma is worsened if scientific linkage is lacking or inadequate. Furthermore, if such distances facilitate cost shifting, then what is optimal from an individual perspective may not at all be optimal from a social or collective perspective. Where such distance factors prevails and the party on which the burden is shifted cannot counteract this distance by pressing his interest, government intervention is needed. This is quite obviously the case with a range of environmental problems.

The reasons why these "external" interests are not adequately internalized, include:

(a) absence of legally based "property rights" protecting the damaged party, or of liability/accountability regulations enforceable upon the causal agent;

(b) absence of means to exert "countervailing power" through the political system (lack of voting power as in cases of trans-boundary cost shifting, or intertemporal cost shifting, or cost shifting onto other species), or through the market place (i.e. lack of purchasing power).

Reasons why this situation is not easily changed by installing more appropriate institutions or legislation, include the filtering process applied by any political system in responding to claims for systemic changes: the filters (again) of time preference (whereby future effects and future interests are discounted away), and of present purchasing or voting power, both heavily biased in favour of the predominant economic and political forces.

2.5 Asymmetrical access to information and information failures

It is clear that asymmetry in the access to information has the tendency to widen the income and consumption gaps between the developed and the developing countries. Policies are being discussed aimed at achieving higher volumes of transfers of technology from North to South. Fundamental issues with respect to intellectual property rights are involved. These should preferably be dealt with through international negotiations.

The international research community also has a responsibility to seek ways to enhance fairness in the distribution of access to knowledge and knowhow. Apart from conventions and codes addressing these issues, research institutions and (national and international) science organizations could devote part of their budgets to financing such effective transfers. This point is developed by Bekoe and Prage in Chapter 14 in this Volume.

Informational failures have occurred where science and technology have set incorrect priorities and launched inappropriate research and development programs. This, in part, has been a consequence of the way scientific and technological discovery and development have traditionally been organized, especially in so far as this existing organizational structure stood in the way of a correct understanding and even perception of environmental problems.

There has been a persistent "monodisciplinary" bias in the analysis, prediction and mitigation or prevention of environmental change in the sense that there has been insufficient dialogue between the natural and the social sciences. Problems of environmental changes have thus been studied often without adequate attention for social causes of degradation, social response to degradation and ways of addressing these social causes.

Many existing research institutions have responded to the emerging environmental problems by at best extending their capacities and instruments to some of these problems and often rather late. In changing this, science has to take at heart the implications of the saying: "If you only have a hammer you look at every problem as though it is a nail". What is needed is a strategy based on a meta-approach whereby one seeks new arrangements cutting across disciplinary and paradigmatic boundaries. One may have to go beyond existing scientific institutions and infrastructures and ask what structures and networks are needed to deal with the problem of ensuring sustainability, how they should be linked etc., how they should be financed and with what sort of multi-purpose toolkit they ought to be provided. The conclusions of Group 3 of the conference: "Sustainable Development, Science and Policy (NAVF Bergen, 1990: 329-342) are supportive of this position."

2.6 Institutional reform towards sustainable economic development

The main lessons from the above are that there will be a need for managing growth (at least where it has become ecologically inviable) and for redirecting it by new institutions and instruments capable of achieving a rebalancing of rights and powers.

Below, we attempt to present a general introduction to institutional changes that current experience offers. In the subsequent Section we pay special attention to conceivable international institutional implications.

Managing economic growth in the sense of curbing it, may be needed when in the longer run growth would otherwise lead to an overall environmental pressure beyond what the ecological buffers can stand. That implies that these limits be determined and translated into politically accepted threshold values below which the economy has to remain. Given that growth-induced environmental change may bring the biosphere to states that have been hitherto unknown (e.g. likely overall atmospheric temperature rises at rates beyond those in the past), these thresholds may have to be set at safe levels, in accordance with a precautionary approach including the notion of "safe minimum standards", namely applying brakes to the scale of resource extraction in the face of ignorance about future availability.

Economic activities likely to bring societies near or beyond these threshold values would have to be scrutinized

288   *An Agenda of Science for Environment and Development into the 21st Century*

for their environmental impacts. If these are unacceptable, and if there is no scope for reducing them by technological innovation, then it may be necessary to restrict the level of economic activity. Given the very considerable and fundamental uncertainty associated with the future environmental effects of current activities, the potential for "overshooting" ecological thresholds where optimal levels of private resource use are determined on the basis of (current) market prices, and the irreversibility of environmental effects such as species extinction, the best means of ensuring the ecological sustainability of economic activity is likely to lie in the application of safe minimum standards of environmental access. Indeed, in systems that are neither controllable nor observable through the price mechanism, and where many environmental assets are in the nature of public goods, safe minimum standards will be an essential part of a strategy of sustainable development. This may ultimately result in changes of the overall patterns of production.

In other cases (where the danger of exceeding critical ecological thresholds is less) it may be sufficient to rely on economic instruments. That is, governments may wish to intervene in the market by changing relative costs to the users of resources (by shifting environmental costs back on to the causing agents). This would indirectly change decisions on environmentally undesirable investments, raw materials, and final products. This would mean that the polluter is made to pay (analogous to the user pays-approach advocated above). However, there will always be cases where direct regulation of economic behavior will be warranted or even preferable, e.g. from an efficiency or effectiveness point of view. Environmental policy thus inevitably extends the powers of the state into areas (e.g. economic planning, pricing policy etc.) from where it is, in fact, very often withdrawing.

We note, in passing, that the approach advocated above implies a decision by society on its attitude *vis-à-vis* risk and especially uncertainty. In accordance with the Scientific Conference preceding the ECE Conference on Sustainable Development (NAVF, 1990), a plea may be made to governments officially to adopt the Precautionary Principle in situations of likely irreversible changes and/or extensions of the state of the biosphere into domains beyond past experience.

Policies of curbing economic growth described so far, do not affect the basic forces underlying growth; they merely mitigate the effects of these forces. Depending on how effective these growth-curbing policies are, more basic strategies might be needed. These could be referred to as preventing non-viable levels of economic growth.

First, society can stimulate further technological innovation oriented towards reducing the environmental burden of economic activities, enhancing the environment's capacity to generate economic inputs and

improving the environment's buffering capacities. The current scientific and technological research programs and the institutional arrangements between the organizations responsible for drawing up and executing these programs, must be reviewed.

In the second place, poverty alleviation at the global level would both directly and indirectly (through its impact on population size) reduce long-term environmental pressure. However, this will come about only via economic development and growth, implying additional environmental burdens in the short run. Poverty alleviation without changing the quality of economic growth, is a cul-de-sac; such quality changes in East and South will only come about in so far as the consumption patterns in the West will manifestly reflect new environmental values and if there is an explicit willingness on the side of the richer countries to assist the poorer countries in achieving this reorientation in their production patterns whilst guaranteeing them rising material *per capita* welfare levels.

Thirdly, the most profound policy to prevent growth would be that of reducing (world) market insecurity and competition. As this comes close to the very essence of our (world) economic system, one cannot but hope that the environmental crisis can be resolved without having to consider changes as fundamental as these.

Redirecting economic growth is the second avenue. From a structural perspective, society needs to prevent or reduce cost shifting tendencies so that prices reflect (marginal) social costs and thereby provide appropriate and correct signals to decision-makers in the economic and political process. This requires an institutional reduction of the impact of distances between cause and effect in space, time and decision level.

Distance in space may lead to redistributions of environmental impacts either through transportation through air and water, or via redistributions of sources of environmental impacts in relation to changing spatial patterns of trade and investment. Proper pricing policies may make far away environmental repercussions of economic behavior count in decision-making. As we suggested above, the use of safe minimum standards will be an essential part of any environmental policy supporting sustainable development. But there will be many cases where it will be sufficient to introduce new or extended legal arrangements for liability and accountability by changing the structure of property rights in environmental resources and environmental effects. This may be particularly appropriate where the problem is one of local degradation as a consequence of unidirectional externality.

In cases of internationally shared environments or environmental resources in common property, new legal and administrative institutions may be needed to facilitate the negotiation of acceptable outcomes.

Distance in time needs to be overcome by extending time preference to favour future gains from present day "sacrifices" and by altering the preoccupation in the public sector with matters of immediate urgency. In concrete terms, the concept of "sustainable use of the environment" needs to be adopted by individual states, and to be made operational. One form this could take is the adoption of some type of "legacy principle", possibly enshrined in the forthcoming Earth Charter, whereby countries agree to pass on to the next generation an environmental quality and environmental resource stock at least as large as the one they found. Institutionally, this would have to be complemented by installing some authority or body to represent and defend future generations' interests, e.g. an Ombudsman-type organization for this specific purpose.

The problems created by the distance between decision levels (the prisoners' dilemma) can be overcome by creating platforms or authorities at levels high enough to cope with the problem at hand; that is at least to discuss it and exchange information, and preferably to have some authority over the joint resource or environment and their uses. Many of these problems are manifest at the international level. One task for science here would be the compilation and standardization of state of the environment reports for all countries.

Sectoral policies, both national and international, will have to be reviewed in the light of the user pays and polluter pays principles in order to present the full price of the involved goods and services to those who enjoy them. In calculating these full social costs, governments may have to apply social discount rates lower than the private ones, in order to take societally acceptable time preference into account.

There is a need to review taxation policies at its foundations. In so far as scarcity is one justification for choosing tax bases, most current systems turn labour into an over-expensive factor of production reinforcing tendencies towards unemployment. Ecotaxes are based on differences in environmental pressure or resource claims, and may be a very useful addition to the set of fiscal instruments.

They are an example of fiscal instruments with a potentially very powerful economic incentive impact towards environmentally friendly behavior in consumption and production (see Opschoor and Vos, 1989). Basically what has been recommended above, amounts to an alteration of rights such that environmental pressure is recognized as a new type of claim on livelihood, existence rights (of species), etc., to be compensated for by those laying that claim, to those on whom the claim is being laid. Compensation will have to follow new regulations on accountability and liability. This could be in the form of an extension of the polluter pays principle to not only the measures prescribed by environmental policy, but to all

types damage costs. Another extension might be that non-compliance with agreed or prescribed practices be punishable much more heavily than is currently done, or that some *ex ante* "performance bond" be made possible to be returned upon behavior according to agreement, as advocated by Arizpe *et al.*, in Chapter 1 in this Volume. Such fundamental reversals in legal status of polluters *vis-à-vis* pollutees, will not easily come about and may need political mobilization and coalition formation between various non-governmental organizations and economic interest groups.

3. POLITICAL AND LEGAL ASPECTS

3.1 Dual challenge

Discussing institutional arrangements in the context of the contribution of sciences to "environment and development" implies that two different tasks be considered:

(a) How to organize science as a whole in the most appropiate way in order to attain a maximum yield for "environment and development"?

(b) How to conceive and realize institutions (organizations) at the various levels (national, regional, global) on the basis of know-how accumulated in the political and legal sciences?

3.2 Societies in need of institutional change

Reviewing the present situation, one becomes aware that institutions are an issue mainly in societies "in the making", societies which need further perfection in order to meet the expectations of their members:

(a) In most developing countries their lagging behind as regards environmental standards or the implementation of environmental laws is largely attributed to the lack or non-performance of institutions fully geared to the objectives of "environment and development". Thus the demand for institution-building or the improvement of existing institutions.

(b) In the international society co-operation required to meet trans-boundary or global challenges in respect of "environment and development" is suffering from the fragmented and decentralized nature of that society. The system of sovereign nation-states frequently runs counter to the need for comprehensive, rapid and enduring actions.

Both types of society call for improvement, an improvement to which the political and legal sciences could contribute.

Previously Communist societies in Central and Eastern Europe are not "societies in the making" but "societies in transition". Nevertheless their problems closely resemble difficulties encountered by developing countries thus, numerous recipes and suggestions related to developing countries may to some extent also be applied to these societies moving from command-economy to free market-economy.

3.3 Institutions in general

As regards institutions (organizations) in general their behavior can be characterized by the following points:

(a) Their structures as well as their mandate are constantly sensing and adapting themselves to changing circumstances (new demands of their clients or supporters, new insights, reappraisal of previous views); the creation of special departments (ministries) for environmental protection in most developed countries in the mid-1970s is a case in point.

(b) They have a collective mind-set, which is the outgrowth of demands (expectations of clients) addressed to them. This "collective will" is also influenced by the values dominant in the respective organizations. Furthermore this mind-set strongly interacts with the organizational structure: the lagging behind of most East European states in environmental matters has to be traced back to the old Communist structures, where environmental concerns were mainly entrusted to ministries in charge of attaining certain production goals (coal, steel, electricity, etc.).

(c) They abide by rules of procedure (guidelines for decision-making) which they have developed themselves in order to ensure a consistent pattern of behavior.

(d) They are likely to have a degree of knowledge about their area of action which includes data-bases of varying degrees of completeness. As a rule, decision-making has to rely on very limited data, but this may be compensated by other factors such as the pressure of public opinion. In 1987, when governments agreed to control-measures for substances likely to deplete the ozone layer, they has no firm proof of the ozone depleting capacity of certain substances they were prohibiting.

(e) They are subject to budget battles and similar constraints. Thus their action – in favour of "environment and development" – is restricted to the immediately attainable. Economic feasibility has become a major constraint in most environmental endeavors. Interest groups representing the economy (business, trade unions) are ready to pursue environmentally oriented policies, only if one can convince them that high costs arise only in the short-term or are likely to be commuted into gains or that costs may be reduced to a level that can be swallowed.

(f) They do not pursue their action in a political vacuum; there do exist structural relationships to the pattern of political power. Thus policies on climate change will be strongly influenced by the automobile industry and associated interests. Countries relying mainly on private (individual) transport will show a high degree of reluctance when they are requested to radically reduce their emissions of carbon dioxide. Organizations which represent consumer interests may wield considerable power if they are able to initiate boycotts of environmentally dangerous products.

3.4 Institutions at the national level

Institutions at the national level perform a vast variety of functions. Such institutions include also those acting at a lower level (grass-roots, local, provincial etc.), though not all functions may be carried out at all levels; some division of labour – well known in the case of federal states – between the various levels of action is common practice. Functions to be fulfilled by national institutions comprise the following ones:

(a) *Education/training and information*; the importance of these functions results from the need to raise the awareness of those degradation occurs quite frequently as a side-effect of legitimate activities devoted to economic development. Institutions involved in this exercise are not only schools or universities but also the media.

(b) *Participation of the public in decision-making*, be it related to environment, be it related to development, is also a tool to raise awareness. As broad participation is the salient feature of the democratic system, it may well be argued that democracy (such as market economy and other factors) contributes positively to the process of environment and development. Social organizations representing sectoral or partial interests play an important role in this context; they mobilize pressure on the government so that it takes action which meets the expectations of their members.

(c) *Rule-making/application of rules/adjudication of competing claims*, constitute the hard core functions of the legal structure necessary to draw the limits beyond which economic activities are likely to damage the environment.

(d) *Monitoring the environment* is a function not to be neglected as decision-making has to be based on reliable data, the validity of which is not challenged by social organizations (citizens' groups, green movements, business lobbies).

(e) *Promotion of science* is another function upon which rests any sound environmental policy, which fully takes into account economic feasibility and social acceptance. Science has a role to play at various intervals: when the threat to the environment is perceived: when society/economy seek substances and procedures less detrimental to the environment; when scientific know-how is required to verify compliance with rules and regulations.

Developing societies frequently have a choice: to import and to adapt institutions working well elsewhere (e.g. environment impact assessment) to local conditions or to establish institutions fully rooted in their traditional cultures. Most probably a mix of these two approaches is the likely outcome. A broad process of awareness-raising is of special importance in these societies, to ensure that the twin-goal of "environment and development" is not understood as a political objective imposed from outside ("ecological imperialism").

3.5 Institutions at the regional level
Institutions (organizations) at the regional level perform some, but certainly not all, functions which are covered by national institutions. At this level, institutions yield a maximum of results, if they are part of a broad integration process. In spite of certain drawbacks of the EEC, as regards its environmental policies, it constitutes the most advanced example of common environmental action above the level of the nation-state. This environmental action is, however, subject to the principle of subsidiarity; this implies that common action takes only place in those instances, where national actions do not suffice. Thus regional action is not going to replace national action but to complement it; centralization is to be kept to a minimum. In this context it should also be recalled that the EEC does not only create environmental law among and within its member-states but negotiates also on behalf of its member-states in global fora devoted to the elaboration of international environmental law.

3.6 Institutions at the global level
Global institutions (organizations) are called upon whenever a challenge to the global commons has to be met. Here the interests of the greatest possible number of states have to be taken into account. As these interests vary according to the level of economic development, global treaties and conventions, in order to obtain the maximum number of contracting parties, have to differentiate between rights and duties of developing countries and those of other countries (see the Montreal Protocol on Substances that deplete the Ozone Layer).

At this global level institutions may be easily established and are likely to work if they do not interfere with the traditional prerogatives of states and governments. Thus functions such as information exchange, monitoring, assistance and counseling (see below: Science Council) are

likely to be well performed. Difficulties are to be encountered by institutions which perform more intrusive functions such as the adjudication of environmental claims (see below the Ombudsman) or the application of sanctions in case of non-compliance (see below the System of Collective Environmental Security).

3.7 Interdependence of levels
The above scale of action-levels does not indicate that actions do occur in a strictly separate way. Quite to the contrary, there exists interdependence between these levels, as already was indicated when referring to the role of the EEC as a negotiator of global treaties. Sometimes global treaties need to be complemented by regional or bilateral treaties in order to become fully operational (see the Convention on Assistance in Case of Nuclear Accidents). This interdependence of action-levels should be added to the already well known interdependence of economies and the so-called interdependence of issues.

3.8 Involvement of sciences
The political and legal sciences, as well as other social sciences, may play an important role in the process of "environment and development", leading humanity far into the 21st Century. Their first task should be to support institutional change at the various levels so that institutions become fully responsive to the new needs of "environment and development". Where these sciences are of the view that mere adaptations do not suffice they will have to propose models for new institutions. Such models could be proposed for developing societies as well as for the international society at large. As regards the former, model-laws and model-institutions should correspond to the economic, social and ecological conditions prevailing in most developing countries. As regards the latter, these sciences should assist diplomats and treaty-makers when they are drafting new rules of international law and when they are reorganizing existing institutions or establishing new ones. These sciences could also assist in removing traditional obstacles to international co-operation, by for instance redefining the concept of sovereignty. Furthermore in view of the growing difficulties to arrive at international treaties which successfully meet the requirements of "environment and development", negotiating techniques will have to be improved. Such improvement needs inputs from a broad range of social sciences (political science, social psychology, etc.), which could throw more light on the characteristics of environmental negotiations.

3.9 International environmental framework
The political and legal sciences could develop the main elements of an "international environmental framework". These elements may serve as a source of inspiration for

politicians and diplomats called upon to create post-UNCED-structures. Within such "framework" may figure the following elements; the order in which these elements are listed corresponds to considerations of short-term and long-term feasibility:

(a) As the *Earth Charter* to be adopted by the Rio-Summit is likely to be a rather concise document, in order to obtain a maximum of support, it would be advisable to start soon after its adoption a broad exercise of exploring its implications, of refining its content, of highlighting its possibilities of practical application, etc. Another consequence would be the rethinking of traditional principles (polluter pays principle, precautionary principle, sovereign equality of states etc.), in the light of the specific requirements of "environment and development";

(b) As the *Agenda 21*, also to be adopted by the Rio-Summit, will be a comprehensive catalogue of measures, it seems advisable to charge a formal or informal body with the continuous review of its realization; here the so-called Commission for Sustainable Development (as suggested by the Aspen Institute) may perform useful functions;

(c) As *Agenda 21* is to be realized by numerous international organizations, it would be advisable to establish a *Co-ordination Board*, which should be empowered to ensure that no duplication of work already done elsewhere occurs. In view of the limited availability of resources this institution should be construed as strong as possible;

(d) A *Global Watch Mechansim*, linking existing national and regional institutions together into one, efficiently performing network seems equally advisable; this would require that standards of measuring and respective methods are fully harmonized, in order to ensure the comparability of data. Already in this context problems may arise in relation to the confidentiality of data and the readiness of some governments to make them accessible to a broader public;

(e) As the task of translating scientific insights into concrete action related to "environment and development" is insufficiently performed at the various levels, it seems advisable to establish *Regional Science Councils* at each of these levels. To establish such Councils at the regional level (in close association with the respective regional commission of the UN) would be a first step. These Councils should inter alia be entitled to give advice to governments: without request, if development-projects are likely to have trans-boundary environmental consequences; upon request, if development-projects do not affect the environment beyond the area of national jurisdiction. Furthermore these Councils should encourage the broadest possible exchange of information and mobilize all available resources of science for the benefit of "environment and development". They would also be involved in the transfer of knowledge and know-how across borders. The impact of these Councils on decision-making will to a large extent depend on the quality and impartiality of the advice they are able to give. This advice given by scientists will be accepted by policy-makers much more easily, if scientists are ready to assume responsibility for the implications of their findings;

(f) Equally at the regional level (regional commissions of the UN) an *Environmental Ombudsman* could be installed; this institution would receive complaints of individuals or NGOs concerned, if the national government (or similar bodies at the national level) has not sufficiently taken care of the respective environmental concern. The respective government will be expected to respond to this complaint. This procedure would to some extent follow the model of international practice in the field of human rights. The real powers of this institution would vary from one region to the other in accordance with the degree of values shared by the governments of the region. But even in regions where governments are not yet ready to accept a certain external control of their policies the mere existence of such an institution would be helpful because it would serve as a sounding board reflecting the actual level of environmental awareness in government and the broad public;

(g) As international environmental law needs coherent and systematic procedures and structures of rule-making it may be advisable to include in this framework a *UN-Commission on International Environmental Law*. Such body could be established along the lines of existing legal bodies (ILC, UNCITRAL); its task would include the drafting of legal instruments as well as the reviewing of legal texts emanating from other international organizations. A primary function of this Commission could be the elaboration of principles of international environmental law (responsibility and liability, etc.). Such code of principles would facilitate the pulling-together of existing "régimes" devoted to various sources of environmental degradation (air pollution, ozone depletion, etc.). By means of such an institution the legal science would have an immediate impact on the codification and progressive development of international environmental law;

(h) The ultimate achievement of the environmental framework would be a *System of Environmental Collective Security*. It would follow the model of enforcement action as recently pursued by the Security Council in the Gulf conflict. A government which is guilty of most seriously violating its obligations under international environmental law or which is guilty of a large-scale degradation of the environment beyond its national borders, would be sanctioned by an international body functioning along the lines of the Security Council. Such sanctions would include trade embargoes, embargoes on transfer of technology etc. Any collective action against the wrong-doer or law-breaker should be more acceptable to the international society ("a society in the making") than sanctioning by individual states which feel affected by that behavior (self-help).

3.10 Non-governmental organizations
Finally, special attention should be paid to non-governmental organizations. Although representing sectoral interests their action cannot be neglected. They contribute to the early and late phases of environmental action: discovery of threats, awareness-raising, control of governments as regard compliance with existing duties (watch-dog-function). As these NGOs represent the germs of world public opinion it may be advisable to establish some kind of NGO-forum, which would be either of a general nature or would be assigned to each and every "régime" (air pollution, ozone depletion, waste transport and management etc.). This would realign the basis of community power and indicate that scientific endeavor needs to be integrated vertically, i.e. by liaising with people on a direct basis, as well as laterally, i.e. by linking the disciplines. This will especially be the case when North-South trade results in unsustainable patterns of production in developing countries, or where developing countries are (e.g. as a consequence of international debt servicing obligations) are forced to sell out natural resources on the world market. In fact, the environmental crisis is so linked with other aspects of international relationships (trade, debts, etc.) that the debate on a new international economic order needs to be fundamentally amended with another "basket" for ecological issues. Institutionally, this would have to be complemented by installing some authority or body to represent and defend future generations' interests, e.g. an Ombudsman-type organization as outlined above.

4. RECOMMENDATIONS

4.1 On economic aspects
Our recommendations fall into two main categories: those concerning the items on an agenda for research, and those concerning the enabling institutions.

4.2 Research agenda
We have identified three broad areas in which research is urgently required: the environmental impacts of

institutional change: the scope for, and barriers to, a transition to sustainable development strategies; and the institutions, policies and instruments required by a sustainable development strategy.

4.3 The environmental impacts of institutional change
Given the trend towards a rapid decentralization of research allocation decision, particularly in Eastern Europe and the USSR but also in may other parts of the world, it is expected that the environmental effects of human activity will increasingly be the consequence of private decision. For this reason we consider that it will be increasingly important to understand the determinants of private decisions on the use of environmental resources, and the factors leading private individuals to ignore the environmental costs these decisions have for others now and in the future. This implies a need for research in the following areas:

(a) the contradictory effects of institutional and environmental change on the decision-making process: many current institutional trends have the effect of increasing the opportunities open to individuals, many of the current environmental trends having the reverse effect;
(b) the determinants of private decisions with respect to the allocation of environmental resources paying particular attention to the interdependence of income, uncertainty, impatience or myopia and the preferences of consumers;
(c) the private and social costs of environmental goods and services, paying particular attention to the causes of the divergence of private and social costs;
(d) the net environmental implications of (a) to (c);
(e) special attention must be given to decision-making and alternative options for women in their environmentally based activities and to the impacts on decision-making of poverty and insecure access to environmental resources.

In all cases it is recommended that research be at a geographical or spatial level that is compatible with the environmental effects of interest. It is important that research be at a sufficiently disaggregated level to identify the often highly localized institutional and other causes of environmental degradation.

4.4 The scope for and barriers to a transition to sustainable development strategies
The second set of research issues identified concern the potential for making an adjustment towards a set of more environmentally sustainable development strategies, given the dynamics of both environmental and social processes:
(a) given the dynamics of the biogeochemical cycles, and the future implications of changes that have already

taken place, the extent of socio-economic adjustments already taking place and expected to be required;

(b) the potential for making economic growth more environmentally benign (delinking growth and environmented degradation) through change in technology, patterns of consumption, geographical location of economic activity, and through the enhancement of the productivity and assimilative capacity of the environment;

(c) the range of potential "transition pathways" available, given the particular features of local and regional resource endowments, institutions and levels of development, etc. Pathways must be identified that do not lead to economic crisis in the North spilling over through declining exports into the South;

(d) the impediments or barriers to the adoption of sustainable development strategies, due to the cultural, institutional informational and economic conditions currently obtaining, paying special attention to the origins of price distortions, policy failures, problems in law enforcement, poverty and inequality, and the restrictions on the mobility of human population and resources imposed by current political arrangements (especially the impermeability of national boundaries to human mobility).

4.5 The institutions, policies and instruments required by a sustainable development strategy

While some progress has already been made towards the identification of the necessary conditions for an environmentally sustainable development strategy, much remains to be done. Most recent research has concentrated on the efficiency issues (getting prices right) but it is not at all clear how far efficiency in the allocation of environmental resources is a condition for sustainability. Six high priority areas of research are identified:

(a) the scope for improving the efficiency of environmental resource allocation through the reallocation and redefintion of property and access rights, paying special attention to issues of intellectual property, liability and accountability;

(b) the scope for improving the efficiency of environmental resource allocation through economic policies to reduce the wedge between the private and social cost, (applying the user pays principale) paying special attention to taxes, subsidies, price, monetary and exchange rate policy;

(c) the scope for improving the sustainability of resource use through the transfer of income (to counter the negative effects of poverty on environmental resource allocation decisions);

(d) the scope for improving the sustainability of resource use by bounding economic activities using regulatory restrictions, safe minimum standards and other application of the precautionary principle;

(e) the scope for improving the sustainability of resource use by changing consumption preference and technology (through, e.g. education) without negative effects on growth in the South, and finally

(f) the scope for addressing both efficiency and the sustainability of environmental resource use through reform of the institutions charged with the management of environmental resources in common property at the local, national, regional, and international levels.

4.6 Enabling scientific institutions

Under this heading we propose the following points, ensuing from para. 2.5:

4.6.1 Multidisciplinary co-operation

There has to be continued and more intensive co-operation between the social sciences and the natural sciences in two ways:

Firstly, the traditional disciplines and existing science organizations should co-operate in addressing, such issues on the society–environment interface that merit concerted activities. This may not in itself lead to a better understanding of what is happening in environmental systems, but it will definitely enhance the likehood that science based recommendations can be made that address environmental problems at their (societal) source: human decision-making.

Secondly, networks are needed, of dedicated institutes that are capable of working creatively and in an inter- or at least multidisciplinary mode on the causes and remedies of environmental degradation and unsustainability and related issues such as poverty, population growth, environmental insecurity. This chapter presents an argument for especially strengthening the social science and economics components of such institutes and networks. There are many impediments to progress in multidisciplinary work, but a network of international centers of excellence geared to a proper research program should help to overcome these.

4.6.2 Dissemination of knowledge

Access to information is unevenly distributed. There has to be a better exchange of knowledge and knowhow particularly between North and South, but also South–South and East–West. This can already be undertaken on a voluntary basis by scientific institutions. In some countries institutes have committed a fixed part of their budget for this purpose.

Similarly, there must be a mechanism to support capacity building so that science in the South need no longer to rely on external sources of data.

4.7 Enabling political and legal structure

As the present political and legal structure – mainly in developing societies and the international society as a whole – does not fully meet the requirements of "environment and development" a major overhaul of institutions has to take place. Political and legal sciences are called upon to assist decision-makers in this exercise, which should take place at the national, regional and global levels:

(a) Functions referred to in 3.4. should be entrusted to existing institutions, if they perform well. Otherwise institutions should be improved or replaced by new ones;

(b) Regional institutions should be entrusted with tasks provided functions are better fulfilled at this level than in the national context;

(c) Global institutions should be in charge wherever global challenges have to be met.

ICSU or other international bodies (International Council of Environmental Law, (ICEL), International Institute for Applied Systems Analysis (IIASA etc.) should mobilize political and legal sciences in order to trigger and/or accelerate institutional change. The draft for an "international environmental framework" (see 3.9) constitutes not only a broad agenda for research but also a broad program for political action.

REFERENCES

Arizpe, L. et al. 1991. Primary Factors Affecting Population and Natural Resource Use. This volume.

Aspen Institute 1991. Report of the Working Group on International Environment and Development Policy.

Conference 1990. *Sustainable Development, Science and Policy.* NAVF, Berger, 329-342.

Grainger, A. 1990. *The Threatening Desert-Controlling Desertification.* London, Earthscan.

Kapp, K.W. 1950. *Social Costs of Private Enterprise.* (paperback ed 1971 Schocken Books Inc, New York. NAVF. 1990. The Conference Report. Conference: "Sustainable Development, Science and Policy". NAVF, Bergen 1990: 329-342.

Kiss, A., Shelton, D. 1991. *International Environmental Law,* London, Graham and Trotman.

Koning de, H.W., ed 1987. Setting environmental standards, Guidelines for Decision-making, WHO, Geneva.

Lang, W. 1989. *Internationaler Umweltschutz, Volkerrecht und Aussenpolitik zwischen Okonomie und Okologie,* Vienna, Orac.

Lang, W. 1991. Negotiations on the Environment. In: Kremenyuk Victor, *International Negotiation,* San Francisco-Oxford, Jossey Bass.

Lang, W. 1991. Is the Ozone Depletion Régime a Model for an Emerging Régime on Global Warming? *UCLA Journal of Environmental Law and Policy,* **9**/2.

NAVF (Norwegian Research Council for Sciences and the Humanities). 1990. Sustainable Development, Science and Policy. Conference Report, Bergen 8-12 May 1990. Oslo.

OECD 1990. *The Economics of Sustainable Development*: a Progress Report. OECD, Paris.

Opschoor, J.B. 1989. *Na ons geen zondvloed: voorwaarden voor een duurzaam milieugebruik.* Kampen: Kok-Agora.

Opschoor, J.B. and Vos H.B. 1989. *Economic Instruments for Environmental Protection.* OECD, Paris.

O'Riordan, 1981. *Environmentalism.* Pion Ltd, London.

Pearce, D.W., E.B. Barbier and A. Markandya 1988. *Environmental Economics and Decision-making in Sub-Saharan Africa.* LEEC-paper 88-01. London.

Pearce, D.W. and Turner R. K. 1990. *Economics of Natural Resources and the Environment.* Harverster Wheatsheaf, New York/London.

Perrings, Ch. 1987. *Economy and Environment.* Cambridge, Cambridge University Press.

Perrings, 1990. Economic growth and sustainable development. In: Conference 1990, above.

Repetto, R. 1989. Economic incentives for sustainable production. In: Schramme G. and Warford J.J. (eds). *Environmental Management and Economic Development.* Baltimore, Johns Hopkins for the World Bank, 69-86.

Sand, P. 1990. *Lessons learned in Global Environmental Governance,* World Resources Institute.

Thacher, P. *Global Security and Risk Management, Background to Institutional Options for Management of the Global Environment and Commons,* World Federation of United Nations Associations.

UNCTAD, 1990. *The Least Developed Countries*: 1989 Report, Geneva Report No. TD/B/1248.

UNDP 1990. *Human Development Report 1990.* Oxford, Oxford University Press.

Warford, J.J. 1987. *Environment and Development.* World Bank/IMF Development Committee, Washington, DC.

WCED (World Commission on Environment and Development) 1987. *Our Common Future.* Oxford University Press, Oxford.

List of Participants

Al-Awadi, Dr Abdul Rahman
Regional Organization for the
Protection of the Marine
Environment, ROPME
P.O. Box 26388
13124 Safat, Kuwait
Tel: (965) 531 2140
Fax: (965) 532 4172/531 2144
Tlx: 44591 ROPME

Anstee, Ms Margaret J.
United Nations Office
Vienna International Center
P.O. Box 500
1400 Vienna , Austria
Tel: (43 222) 21131

Arizpe, Prof. Lourdes
Instituto de Investigaciones
Antropologicas, U.N.A.M.
Ciudad Universitaria
Delegacion Coyoacan
04510 Mexico, D.F., Mexico
Tel: (52 5) 548 3419
Tel + Fax: (52 5) 548 3667
Fax: (52 73) 17 59 81

Arroyo, Dr M.T. Kalin
Departamento de Biologia
Facultad de Ciencias
Universidad de Chile
Casilla 653, Santiago, Chile
Tel + Fax: (56 2) 271 5464
Fax: (56 2) 2712983
Tlx: 240230 BOOTH CL

Arunin, Dr Somsri
Soil Salinity Research Section
Land Development Dept.
Pahonyotin Road, Bangkhen
Bangkok 10900, Thailand
Tel: (66 2) 579 5546
Tel: (66 2) 579 0111
Fax: (66 2) 561 1230
Tlx: 21505 IBSRAM TH

Ayensu, Prof. E.S.
Panafrican Union for Science
& Technology
P.O. Box 16525 Airport
Accra, Ghana
Tel: (233 21) 77 55 54
Fax: (233 21) 77 31 16
Tlx: 2685 CRAMER

Ayibotele, Prof. N.B.
Water Resources Research
Institute, P.O. Box M.32
Accra, Ghana
Tel: (233 21) 775 351
Fax: (233 21) 773 068

Ayres, Dr R.U.
E.P.P.
Carnegie-Mellon University
5000 Forbes Ave
Pittsburgh, PA 15213, USA
Tel: (1 412) 268 2678
Fax: (1 412) 268 3757

Ba, Dr Mariline
c/o Dr E.S. Diop
UNESCO - B.P. 3311
Avenue Roumé, Dakar
Senegal
Tel: (221) 23 50 82
Fax: (221) 23 83 93
Tlx: 51410 UNESCO SG

Bahmani Fard, Ms. Tish
ICSU Administrative Assistant
51, Bld de Montmorency
75016 - Paris, France
Tel: (33 1) 45 25 03 29
Fax: (33 1) 42 88 94 31
Tlx: 645 554 ICSU
E.mail:
Omnet: ICSU.Paris
Telecom Gold 10075: DBIO 126

Barberan, Dr José
Instituto de Ciencias del
Mar y Limnologia, UNAM
Ap. Postal 70-305,CU,
04510 Mexico D.F.
Mexico
Tel: (52 5) 550 5873
Fax: (52 5) 548 2582
Tlx: 1774523 UNAMME

Barrère, Dr Martine
23 rue Jean Brunet
92190 - Meudon, France
Tel: (33 1) 45 34 58 48
Fax: (33 1) 46 34 39 40
(Association Descartes)

Batisse, Dr Michel
UNESCO
7, Place de Fontenoy
75700 Paris, France
Tel: (33 1) 45 68 40 51
Fax: (33 1) 40 65 9897

Bauer, Prof. S. J.
Universität Graz
Institute für Meteorologie
und Geophysik
Halbärthgasse 1, 8010 Graz
Austria
Tel: (43 316) 380 5255/56
Fax: (43 316) 35566
Tlx: 311662 UBGRZ A

Beckers, Dr H. L.
Balnor, Hook Heath Road
Woking, Surrey GU22 OLF, UK
Tel.+ Fax:(44 483) 772501
Or
Benoordenhoutseweg 28
2596 BB Den Haag
The Netherlands
Tel: (31 70) 324 66 53

Bekoe, Prof. D.A.
IDRC
P.O. Box 62084
Nairobi, Kenya
Tel: (254 2) 33 0850
Fax: (254 2) 21 4583

Benedick, Dr R. E.
World Wildlife Fund
The Conservation Foundation
1250 Twentyfourth Street NW
Washington DC 20037, USA
Tel: (1 202) 778 9704
Fax: (1 202) 293 9211
Tlx: 64505 PANDA

Berger, Prof. S.
Institute of Human Nutrition
Warsaw Agricultural University
166 Nowoursynowska
02766 Warsaw, Poland
Tel: (48 22) 43 90 41 (61, 81)
Fax: (48 22)47 15 62
Tlx: 81 62 38 SGGW PL

Bolin, Prof. Bert
Kvarnasvagen 6
18451 Osterkar, Sweden
Tel: (46 8) 15 7731
Fax: (46 8) 15 7185
Tlx: 15959 MISU S

Bonkoungou, Dr E.G.
c/o OAU/SAFGRAD
B.P. 1783, Ouagadougou
Burkina Faso
Tel: (226) 306071

Bonnin-Roncerel, Dr Annie
Climate Network - Europe
46, rue du Taciturne
1040 Brussels, Belgium
Tel: (32 2) 231 0180
Fax: (32 2) 230 5713
Tel: (32 2) 539 30 01 home

Botero, Dr Margarita Marino de
El Colegio Verde de Villa de Leyva
Carrera 3a. N°10-92
Santa Fe de Bogotà, D.C., Colombia
Tel: (57 1) 341 0459
Fax: (57 1) 342 3237
Home:
Carrera 5, N° 28-12, Apt. 601
Santa Fe de Bogotà, Colombia
Tel: (57 1) 243 8545

Bourdeau, Dr Ph.
CEC, (DG XII) directorate E
Square du Meeus 8
1049 Brussels, Belgium
Tel: (32 2) 236 88 14
Fax: (32 2) 236 88 24
Tlx: 21877 COMEUR B

Bourenane, Prof. Nacer
Research Fellow(CREAD)
Hai Mohamadia Baty
Appart.137, El Harrach 16130
Algiers, Algeria
Tlx: 61520 CREAD DZ

Branis, Dr Martin
Institute for Environmental Studies
Charles University, Benatska 2
128 01 Prague 2,Czecoslovakia
Tel: (42 2) 29 79 41 ext.395
Fax: (42 2) 291958 or 296084

Brennan, Ms Maureen
ICSU Administrative Assistant
51, Bld de Montmorency
75016 - Paris, France
Tel: (33 1) 45 25 03 29
Fax: (33 1) 42 88 94 31
Tlx: 645 554 ICSU
E.mail: Omnet: ICSU.Paris
Telecom Gold 10075: DBIO 126

Bruce, Dr James P.
1875 Juno Ave.
Ottawa, Ontario K1H 6S6, Canada
Tel: (1 613) 731 5929
Fax: (1 613) 731 3509

Buat-Ménard, Dr Patrick
Center des Faibles Radioactivités
Parc du CNRS ,91198 Gif sur Yvette
Cedex,
France
Tel: (33 1) 69 82 35 40
Fax: (33 1) 69 82 35 68

Bugge, Dr Hans Chr.
Dept. of Public & International
Law, University of Oslo
Karl Johans gt. 47
0162 Oslo, Norway
Tel: (47 2) 41 15 20
Fax: (47 2) 33 63 07(office)

Burnett, Sir John
Dept. of Plant Sciences
University of Oxford
South Parks Road
Oxford OX1 3RB, UK
Tel: (44 865) 270880
Fax: (44 865) 513324

Cano, Dr Guillermo J.
Fundacion Ambiente y Recursos
Naturales, Monroe 2142
1428 Buenos Aires, Argentina
Tel: (54 1) 781 6115/9171
Fax: (54 1) 781 6115
Tlx: 22088 CARTE AR
or:
Arenales 2040 - 7° B
1124 Buenos Aires, Argentina

Cetto, Prof. Ana Maria
Institute of Physics
Universidad Nacional Autonoma
de Mexico, Apdo Postal 20-364
Mexico 20, D.F., Mexico
Tel: (52 5) 550 5939
Fax: (52 5) 548 3111
Fax: (52 5) 548 8186

Chadwick, Prof. M.J.
Stockholm Environment Institute
Box 2142
10314 Stockholm, Sweden
Tel: (46 8) 723 02 60
Fax: (46 8) 723 03 48
Tlx: 19580 SEI S

Colombo, Dr Umberto
E.N.E.A.
Viale Regina Margherita 125
00198 Rome, Italy
Tel: (39 6) 8528 2214
Fax: (39 6) 8528 5804
Tlx: 610183/610167 ENEA

Concepcion, Dr Mercedes B.
Population Institute
CSSP, University of The Philippines
Palma Hall, Room 236
Diliman, Quezon City, Philippines
1100
Tel: (63 2) 976061 loc. 6997
Fax: (63 2) 992863 or 991929

Conway, Dr R.A.
Union Carbide Chemicals
& Plastics Company Inc.
P.O. Box 8361, South Charleston
WVA 25303, USA
Tel: (1 304) 747 4021
Fax: (1 304) 747 5430

Cordani, Prof. Umberto G.
Instituto de Geociencias-USP
Cx. Postal 20899
01498 Sao Paulo, Brazil
Tel: (55 11) 211 2847
Fax: (55 11) 210 4958
Tlx: 1182564 UVSI BR

Corell, Dr Robert
National Science Foundation
Room 510 - 1800 "G" Street
N.W.; Washington D.C. 20550
USA.
Tel: (1 202) 357 9715
Fax: (1 202) 357 9629
Tlx: 892438

Cunha, Dr Luis Da
Nato Scientific Affairs Division
North Atlantic Treaty Organization
1110 Brussels, Belgium
Tel: (32) 7284111
Fax: (32) 7284232
Tlx: 23 867 NATOHQ

Dagne, Dr Ermias
Natural Products Research Network
for East & Central Africa
(NAPRECA)
Chemistry Dept.
Addis Ababa University
Box 1176, Addis Ababa
Ethiopia
Fax: (251 1) 550911

Daniel, Prof. R.R.
COSTED Secretariat
24 Gandhi Mandap Road, Guindy
Madras 600 025, India
Tel: (91 44) 41 9466/4543
Fax: (91 44) 41 4543 or 411589
Tlx: 4121014 CLRI IN

de Janosi, Dr Peter E.
Director
International Institute for
Applied Systems Analysis
IIASA
2361 Laxenburg, Austria
Tel: (43-2236) 71 521 599
Fax: (43-2236) 71313
Tlx: 79137 IIASA A

Delsol, Dr Frédéric
World Meteorological Organization
Case Postale 2300
1211 Geneva 20, Switzerland
Tel: (41 22) 7308212
Fax: (41 22) 7342326

Dente, Dr Bruno
Instituto per la Ricerca Sociale
Via XX Septembre 24
20123 Milano, Italy
Tel: (39 2) 4815653
Fax: (39 2) 48008495

Di Castri, Prof. Francesco
UNESCO
7 Place de Fontenoy
75007 Paris, France
Tel: (33 1) 45 68 41 57
Fax: (33 1) 45 66 90 96

Diop, Dr E.S.
12, avenue Roume
c/o Unesco Breda
P.O. Box 3311, Dakar, Sénégal
Tel: (221) 23 50 82
Fax: (221) 23 83 93
Tlx: 51410 or 21735 SG

Diop-Mar, Dr Ibrahima
Faculty of Medicine & Pharmacy
University of Dakar
Dakar-Fann, Senegal
Tel: (221) 210346 or 210360
and:
Clinique Fann-Hock
Rue 70 x 55, Dakar, Senegal
Tel: (221) 21 03 46
Fax: (221) 25 29 52

Dooge, Prof. J.C.I.
Center for Water Resources
Research
Civil Engineering Dept.
University College Dublin
Earlsfort Terrace, Dublin 2
Ireland
Tel: (353 1) 693244
Fax: (3531) 754568
Tlx: 32693 UCD EI

Döös, Dr Bo
International Institute for
Applied Systems Analysis
IIASA
2361 Laxenburg, Austria
Tel: (43-2236) 71 521 599
Fax: (43-2236) 71313
Tlx: 79137 IIASA A

Dotto, Ms Lydia
2650 Marsdale Drive #303
Peterborough, Ontario K9L 1Y1
Canada
Tel + Fax: (1 705) 741 1476

Dudal, Prof. Raoul
Faculty of Agricultural Sciences
Laboratory of Land Management
92 Kardinaal Mercierlaan
3001 Leuven (Heverlee), Belgium
Tel: (32) 16 23 13 81
Fax: (32) 16 20 50 32
Tlx: 25941 ELEKUL

El-Lakany, Prof. Hosny
Desert Development Center
The American University in Cairo
113, Sharia Kasr El Aini
P.O. Box 2511, Cairo, Egypt
Tel: (20 2) 354 2969
Fax: (20 2) 355 7565
Tlx: 92224 AUCAI UN

Elshout, Dr A.J.
K.E.M.A., Utrechtseweg 310
6812 AR Arnhem
The Netherlands
Tel: (31 85) 56 23 81
Fax: (31 85) 51 50 22
Tlx: 45016 KEMA NL

Epstein, Sir Anthony
University of Oxford
Nuffield Dept. of Clinical Medicine
John Radcliffe Hospital
Headington
Oxford OX3 9DU, UK
Tel: (44 865) 221334
Fax: (44 865) 750506

Falkenmark, Prof. Malin
Swedish Natural Science Council
P.O. Box 6711
113 85 Stockholm, Sweden
Tel: (46 8) 610 07 07
Fax: (46 8) 610 07 40
Tlx: 13599 RESCOUN S

Fall Ndiaye, Dr Khady
Ministère de la Femme, de
l'Enfant et de la Famille
3, rue Béranger-Ferraud
Angle Rue des Essarts
B.P. 4050, Dakar, Sénégal
Tel: (221) 23 89 85
Fax: (221) 23 46 22

Fenstad, Prof. J.E.
Institute of Mathematics
University of Oslo
Blindern, P.O. Box 1053
0316 Oslo 3, Norway
Fax: (47 2) 85 43 49

Field-Juma, Ms Alison
Initiatives Limited
P.O. Box 69313
Nairobi, Kenya
Tel: (254 2) 744047 or 744095
Fax: (254 2) 743995
Tlx: 22448

Fleissner, Prof. Peter
Technische Universität Wien
Institut für Gestaltungs-und
Wirkungsforschung
Abteilung für Sozialkybernetik
Möllwaldplatz 5
1040 Wien, Austria
Tel: (43 222) 504 11 86
Fax: (43 222) 504 11 88

Freney, Prof. J.
CSIRO Division of Plant Industry
G.P.O. Box 1600l
Canberra ACT 2601, Australia
Tel: (61 6) 246 5442
Fax: (61 6) 246 5000
Tlx: 62351

Fuchs, Prof. R.G.
The United Nations University
Toho Seimei Building
15-1, Shibuya 2-chome
Shibuya-ku, Tokyo 150, Japan
Tel: (81 3) 3499 2811
Fax: (81 3) 3499 2828
Tlx: J25442 UNATUNIV

Gabaldon, Dr Arnaldo
Centro Cuidad Comercial
Tamanaco, Torre A, Piso 6
Oficina 608, Chuao
Caracas, Venezuela
Fax: (58 2) 9593161

Gallopin, Dr Gilberto C.
International Institute for
Sustainable Development
212 McDermot Avenue
Winnipeg, Manitoba, Canada
R3B 0S3
Tel: (1 204) 958 7700
Fax: (1 204) 958 7710

Gandhi, Ms Maneka
A/4 Maharani Bagh
110 065 New Delhi, India
Tel: (91 11) 6840402 or 6847576

Giesecke, Dr A.A.
Centro Regional de Sismologia,
Apartado 14-0363, Lima, Peru
Tel + Fax: (51 14) 336750
Fax: (51 14) 338934
Tlx: (36) 20053 PE PB LIMTC

Giovannini, Dr Bernard
Académie Internationale de
l'Environnement
4, Chemin de Conches
1231 Conches, Switzerland
Tel: (41 22) 789 13 11
Fax: (41 22) 789 25 38

Glaser, Dr Gisbert
UNESCO
7, Place de Fontenoy
75700 - Paris, France
Tel: (33 1) 45 68 40 53
Fax: (33 1) 45 66 90 96

Glatzel, Prof. Gerhard
Universität für Bodenkultur
Institut für Forstökologie
Peter Jordanstrasse 82
1190 Wien, Austria
Tel: (43 222) 342500/528
Fax: (43 222) 477896

Golitsyn, Prof. G.S.
Institute of Atmospheric Physics
USSR Academy of Sciences
3 Pyzhevsky, 109017 Moscow
USSR
Tel: (7 095) 231 55 65
Tel: (7 095) 331 32 69 home
Fax: (7 095) 292 65 11
(attn: IFA box 109.52))
Tlx: 411700 PTB SU
(attn: IFA box 109.52)

Golubev, Dr Genady N.
At the time of the Conference:
Faculty of Geography
Moscow State University
Leninskie Gory
119899 Moscow, USSR
Tel: (7 095) 939 3962 office
Tel: (7 095) 336 2353 home
Fax: (7 095) 939 2114
Tlx: 411483 MGU SU
Subsequently:
IUCN-The World Conservation
Union
Avenue du Mont-Blanc
1196 Gland, Switzerland
Tel: (41 22) 64 9114
Fax: (41 22) 64 2926
Tlx: 419605 IUCN CH

Goodman, Dr Gordon T.
20 Dunstal Field
Cottenham
Cambridge CB4 4UH, UK
Tel: (44 954) 522 59
Fax: (44 954) 504 25

Gordon, Dr W.E.
1400 Hermann, Unit 7-H
Houston 77004, Texas, USA
Tel: (1 713) 527 4939
(Rice U. office)
Fax: (1 202) 334 3094
(NAS office)
Fax: (1 713) 285 5143
(Rice U. office)

Grabow, Dr W.O.K
Dept. of Medical Virology
Institute of Pathology
University of Pretoria
P.O. Box 2034
Pretoria 0001, South Africa
Tel: (27 12) 319 23 51
Fax: (27 12) 324 48 86
Tlx: 320912

Grove-White, Mr R.
Center for Science Studies &
Science Policies
University of Lancaster
Lancaster, LAI 4YN, UK
Tel: (44 524) 63806
Fax: (44 524) 843 934 or 846339

Gutman, Dr Pablo
CEUR, Av. Corrientes 2835, p.7
1193 Buenos Aires, Argentina
Tel: (54 1) 961 0309/2268
Fax: (54 1) 542 2719
home: Melian 2215
1430 Buenos Aires, Argentina
Tel + Fax: (54 1) 542 2719

Hall, Dr David O.
Division of Biosphere Sciences
University of London
Kings College London
Campden Hill Road
London W8 7AH, UK
Tel: (44) 71 333 4317
Fax: (44) 71 937 7783
Tlx: 8954102 BBSLON G

Hassan, Prof. M.H.A.
c/o ICTP, P.O. Box 586
34100 Trieste, Italy
Tel: (39 40) 2240 328
Fax: (39 40) 224 559
Tlx: 460392 ICTP I
Cable:CENTRATOM
E. Mail: TWAS@ITSICTP.Bitnet

Hefny, Dr Kamal
Research Institute for
Ground-water
4 Wady Street Heliopolis
Cairo, Egypt
Tel: (202) 662095 or 954948
Fax: (202)958729

Herrera, Dr R.
Instituto Venezolano de
Investigaciones Cientificas (IVIC)
Carretera Panamerica km.11
Apartado 21827
Caracas 1020A, Venezuela
Tel. (58 2) 501 1014
Fax (58 2) 571 3164
Tlx: 21657 VC

Higuchi, Prof. Keiji
College of International Studies
Chubu University
Kasugai, Aichi 487, Japan
Tel: (81) 568 51 1111 ext. 2765
Fax: (81) 568 52 1325

Holligan, Dr P.M.
National Environmental
Research Council
Plymouth Marine Laboratory
Prospect Place
Plymouth PLI 3DH, UK
Tel: (44 752) 222 772
Fax: (44 752) 670 637

Houghton, Sir John
Meteorological Office
London Road
Bracknell RG12 2SZ
Berkshire, UK
Tel: (44) 344 854600
Fax: (44) 344 856909
Tlx: 849801 WEABKA

Htun, Dr Nay
Director of Programmes &
Special Adviser UNCED
P.O. Box 80, 1231 Conches
Switzerland
Tel: (41 22) 789 35 35 or 1676
Fax: (41 22) 789 3536

Hughes, Ms Janet
TWAS Administrative Assistant
c/o ICTP, P.O. Box 586
34100 Trieste, Italy
Tel: (39 40) 224 0328
Fax: (39 40) 224 559
Tlx: 460392 ICTP I
Cable:CENTRATOM
E. Mail: TWAS@ITSICTP.Bitnet

Isebor, Dr Catherine
Nigerian Institute for
Oceanography & Marine Research
P.M.B. 12729, Victoria Island,
Lagos, Nigeria

Jellali, Dr Mohammed
Direction de la Recherche et
de la Planification de l'Eau
Administration de l'Hydraulique
B.P. Rabat-Chellah, Rabat
Morocco
Tel: (212 7) 77 90 16
Fax: (212 7) 77 60 81
Tlx: 360 82 PLANEAU

Juma, Dr C.
African Center for Technology
Studies, P.O. Box 45917
Nairobi, Kenya
Tel: (254 2) 744047 or 744095
Fax: (254 2) 743995

Kaarhus, Dr Randi
Norwegian Institute for
Urban & Regional Planning
P.O. Box 44 Blindern
0313 Oslo, Norway
Tel: (47 2) 95 88 00
Fax: (47 2) 60 77 74

Kapitza, Prof. S.P.
Institute for Physical Problems
USSR Academy of Sciences
ul. Kosygina, 2
117334 Moscow, USSR
Tel: (7 095) 137 6577 office
Tel: (7 095) 237 4007 home
Fax: (7 095) 938 2030

Kasperson, Dr Roger
Center for Technology,
Environment & Development
Clark University
950 Main Street
Worcester MA 01610-1477
USA
Tel: (1 508) 751 4605
Fax: (1 508) 751 4600

Keckes, Dr Stjepan
9, Chemin Taverney
Grand Saconnex/Genève
Switzerland
Tel: (41 22) 798 89 45 home
Fax: c/o C. Cameron/IRPTC/Unep
 (41 22) 733 2673

Kempe, Dr Stephan
Universität Hamburg
Institut für Biogeochemie und
Meereschemie
Bundestrasse 55
2000 Hamburg 13, FRG
Tel: (49 40) 4123 5234
Fax: (49 40) 4123 5270
Tlx: 214732 UNIHH D

Khandakar, Dr Khurshida
Irrigation Support Project
for Asia & The North East
(ISPAN) House N° 26
Road N° 34 Gulshan Model Town
Dhaka, Bangladesh
Tel: (880 2) 881 570-2
Tel: 329747 (home)
Fax: (880 2) 883 097 c/o ISPAN
Tlx: 671215 BRIGHTWAY

Khoshoo, Dr T.N.
Tata Energy Research Vihar
7, Jor Bagh
New Delhi 110 003, India
Tel: (91 11) 4623935 (office)
Tel: (91 11) 6442784 (home)
Fax: (91 11) 4621770
Tlx: 31-61593 TERI IN

Khosla, Dr Ashok
Development Alternatives
B-32, Institutional Area
Tara Crescent
New Mehrauli Road
New Delhi 110 016, India
Tel: (91 11) 665370 or 657938
Fax: (91 11) 686 6031
Tlx: (31) 73216 DALT IN

Kinane, Ms Margaret
ESF Administrative Assistant
1, Quai Lezay-Marnésia
67080 - Strasbourg Cedex
France
Tel: (33) 88 76 71 31
Fax: (33) 88 37 05 32
Tlx: 890440

Kindler, Dr J.
Institute of Environmental
Engineering
Warshaw Technical University
Ul. Nowowiejska 20,
OO-653 Warsaw, Poland
Tel + Fax: (48 22) 21 89 93
Tlx: 813307 PW PL

Kneucker, Dr Raoul F.
Bundesministerium für
Wissenschaft und Forschung
Sektion IV, Freyung 1/403
1014 Vienna, Austria
Tel: (43 222) 53 120 2170
Fax: (43 222) 53 120 2223

Koptyug, Prof. V.A.
USSR Academy of Sciences
Leninskii Prospekt 14
117901 Moscow V-71, USSR
Tel: (7 095) 234 2549
Fax: (7 095) 230 2630
Fax: (7 095) 230 2043
Tlx: 411964 ANS SU

Kreutzer, Dr Frantz
ORF- Club 2
Würzburggasse 30
1136 Vienna, Austria
Tel: (43 222) 87878/2148
Fax: (43 222) 87878/2215

Krichagin, Dr V. I.
Institute for Social & Economic
Problems of Population
Krasikova 27
117218 Moscow, USSR
Tel: (7 095) 465 67 31
Fax: (7 095) 253 90 98
attn. TYNOROY

Krippl, Ms Elisabeth
International Institute for
Applied Systems Analysis
IIASA
2361 Laxenburg, Austria
Tel: (43-2236) 71 521 599
Fax: (43-2236) 71313
Tlx: 79137 IIASA A

Kulasooriya, Prof. S.A.
Dept. of Botany
University of Peradeniya
Peradeniya, Sri Lanka
Tel: (94) 888693 ext.206
Fax: (94) 888151

Kullenberg Dr Gunnar
UNESCO
7, Place de Fontenoy
75700 Paris, France
Tel: (33 1) 56 48 39 83 or 85
Fax: (33 1) 40 56 93 16
Tlx: 204461

Lago, Dr Regina C.A.
CTAA/EMBRAPA
Av. das Americas 29.501
23020 Rio de Janeiro, Brazil
Tel: (55 21) 410 1353 ext. 177/180
Fax: (55 21) 410 1090
Tlx: 33267 EBPA

Lal, Prof. D.
Geological Research Division
Scripps Institution of Oceanography
Mail Stop GRD-0220
University of Calif. San Diego
La Jolla, CA 92093-0220, USA
Tel: (1 619) 587 1535
Tel: (1 619) 534 1829
Fax: (1 619) 534 0784
Tlx: 188929

Lang, Dr Winfried
Mission Permanente de l'Autriche
9-11 rue de Varembé
1211 Genève 20, Switzerland
Tel: (41 22) 733 7750
Fax: (41 22) 734 4591

Lang, Prof. Istvan
Hungarian Academy of Sciences
Roosevelt tér 9
Budapest 1051, Hungary
Tel: (36 1) 111 9812
Fax: (36 1) 112 8483
Tlx: 224 139

Lasserre, Prof. Pierre
Observatoire Océanologique de
Roscoff
Station Biologique de Roscoff
Place G. Teissier,
29680 Roscoff, France
Tel: (33) 98 29 23 00
Fax: (33) 98 29 23 24

Lee, Dr Hahn Been
Korea Institute of Science &
Technology
P.O. Box 131 Cheongryang
Seoul 130-650, Korea
Tel: (82 2) 962 4612
Fax: (82 2) 963 4013
Tlx: 27380 KISTROK

Lichem, Dr Walther
c/o Federal Ministry for
Foreign Affairs
Ballhausplatz 2
1014 Wien, Austria
Tel: (43 222) 53115
Fax: (43 222) 53115 3621
Fax: (43 222) 5354530

Lie, Prof. Ulf
University of Bergen
Center for Studies of
Environment & Resources
Allegt. 36, 5007 Bergen, Norway
Tel: (47 2) 60 50 90
Fax: (47 2) 96 30 50

Lindahl-Kiessling, Prof. Kerstin
Uppsala University
Dept. of Zoophysiology
Box 560, 751 22 Uppsala
Sweden
Tel: (46 18) 182616
Fax: (46 18) 518843
Home:
Ripvägen 14,
75653 Uppsala, Sweden
Tel: (46 18) 324497
Fax: (46 18) 324497

Liss, Prof. Peter S.
University of East Anglia
School of Environmental Sciences
Norwich NR4 7TJ, UK
Tel: (44 603) 59 2563
Fax: (44 603) 50 7719
Tlx: 975197

Liu, Dr Jingyi
c/o Hao Qian, UNCED
P.O. Box 80
1231 Conches, Switzerland
Tel: (41 22) 789 1676
Fax: (41 22) 466 815

Löffler, Prof. Heinz
Universität Wien
Institut für Zoologie Abteilung
Limnologie
Althanstrasse 14
1091 Wien, Austria
Tel: (43 222) 313 36
Fax: (43 222) 313 36700

Lorius, Dr Claude
Laboratoire de Glaciologie &
Géophysique de l'Environnement
B.P. 96
38402 Saint-Martin d'Hères Cedex
France
Tel: (33) 76 82 42 79
Fax: (33) 76 82 42 01
Tlx: 80131 LGGE

Loucks, Prof. Daniel P.
School of Civil &
Environmental Engineering
Cornell University, Hollister Hall
Ithaca, NY 14853-3501, USA
Tel: (1 607) 255 4896
Fax: (1 607) 255 9004
Tlx: WUI 6713054

Lutz, Dr Wolfgang
International Institute for
Applied Systems Analysis
IIASA
2361 Laxenburg, Austria
Tel: (43 2236) 715 21
Fax: (43 2236) 713 13
Tlx: 79137 IIASA A

MacNeill, Dr J.
Institute for Research on
Public Policy
275 Slater Street, 5th Floor
Ottawa, Ontario K1P 5H9
Canada
Tel: (1 613) 238 2296 office
Tel: (1 613) 749 8681 home
Fax: (1 613) 235 8237

Magadza, Prof. Chris
University of Zimbabwe
Lake Kariba Research Station
P.O. Box 48, Kariba
Zimbabwe
Tel: (263) 22312
Fax: (263 4) 732828
Tlx: 26580 UNIVZ ZW

Mahtab, Dr F. U.
Planning & Development
Services Ltd.
31, Bijoy Nagar
North South Road
Dhaka-1000, Bangladesh
Tel: (880 2) 405064
Fax: (880 2) 863325
Tlx: 642546 CKC BJ

Marton-Lefèvre, Ms Julia
ICSU Executive Director
51, Bld de Montmorency
75016 - Paris, France
Tel: (33 1) 45 25 03 29
Fax: (33 1) 42 88 94 31
Tlx: 645 554 ICSU
E.mail:

Omnet: ICSU.Paris
Telecom Gold 10075: DBIO 126
Mascarenhas, Dr Adolfo
Institute of Resource Assessment
University of Dar es Salaam
P.O. Box 35102, Dar es Salaam
Tanzania
Tel: (255 51) 49192
Tel: (255 51) 48573 home
Fax: (255 51) 48224/48409
Tlx: 41327 UNISCIE
Tlx: 41561 UNIVIP

Mascarenhas, Dr Ophelia
University of Dar es Salaam
P.O. Box 35097, Dar es Salaam
Tanzania
Tel: (255 51) 49039/48235
Tel: (255 51) 48573 home
Fax: (255 51) 48224/48409

Matsuno, Prof. Taroh
Center for Climate System
Research, University of Tokyo
Bunkyo-ku, Tokyo 113, Japan
Tel: (81 3) 3812 2111 ext.2680
Fax: (81 3) 5800 6893
Tlx: 2722126 UTGAB

Mattos de Lemos, Dr Haroldo
Federal University of
Rio de Janeiro
Rua Pereira da Camara, 30
22631 Barra da Tijuca
Rio de Janeiro, RJ, Brazil
Tel: (55 21) 399 5336
Fax: (55 21) 399 8995

Matzner, Prof. Egon
Technische Universität Wien
Institut für Finanzwissenschaft
und Infrastrukturpolitik
Karlsgasse 11/267
1040 Wien, Austria
Tel: (43 222) 588 01
Fax: (43 222) 587 79 72

Maurer, Dr Ludwig
Ludwig Boltzmann Institut
für Biologischen Landbau
Rinnböckstrasse 15
1110 Wien, Austria
Tel: (43 222) 74 36 31/71
Fax: (43 222) 74 33 51

Maynard, Dr Nancy G.
Executive Office of the
President, OSTP
Environmental Division
Washington, D.C. 20506
USA
Tel: (1 202) 456 6202
Fax: (1 202) 395 3719

McBean, Prof. Gordon A.
Atmospheric Science Programme
Dept.of Geography
University of British Columbia
1984 West Mall, Vancouver BC V6T
1Z2
Canada
Tel: (1 604) 822 5940
Fax: (1 604) 822 6150
Tlx: 04-51233 UBC PURCH

McGowan, Mr Alan
Scientists' Institute for
Public Information
355 Lexington Avenue
New York, N.Y. 10017, USA
Tel: (1 212) 661 9110
Fax: (1 212) 599 6432

McLaren, Dr Digby
Royal Society of Canada
P.O. Box 9734
Ottawa, Ontario K1G 5J4
Canada
Tel: (1 613) 992 3468
Fax: (1 613) 992 5021

Medina, Dr Ernesto
Lab. Plant Physiological
Ecology, I.V.I.C
Centro de Ecologia Y Ciencias
Ambientales
Aptdo 21827, Caracas 1020-A
Venezuela
Tel: (58 2) 5011014
Fax: (58 2) 5727446
Tlx: 21657

Menon, Prof. M.G.K.
President ICSU
77 Lodi Estate
New Delhi 110 003, India
Tel: (91 11) 462 00 62
Fax: (91 11) 371 06 18

Merle, Ms Elisabeth
ICSU Administrative Assistant
51, Bld de Montmorency
75016 - Paris, France
Tel: (33 1) 45 25 03 29
Fax: (33 1) 42 88 94 31
Tlx: 645 554 ICSU
E.mail:
Omnet: ICSU.Paris
Telecom Gold 10075: DBIO 126

Messerli, Prof. Bruno
Institute of Geography
University of Berne
Hallerstrasse 12
3012 Berne, Switzerland
Fax: (41 31) 65 85 11

Millward, Mr Michael
ICSU Assistant Executive Director
51, Bld de Montmorency
75016 - Paris, France
Tel: (33 1) 45 25 03 29
Fax: (33 1) 42 88 94 31
Tlx: 645 554 ICSU
E.mail: Omnet: ICSU.Paris
Telecom Gold 10075: DBIO 126

Mitra, Dr A. P.
Bhatnagar Fellow
National Physical Laboratory
Hillside Road
New-Delhi 110012, India
Tel: (91 11) 585298 office
Fax: (91 11) 572 2678
Tlx: 31 77384 RSD IN
Tlx: 31 77099 NPL IN

Moldan, Prof. Bedrich
Geological Survey
Malostranské n.19
118 21 Praha 1, Czecoslovakia
Tel: (42 2) 533641 or 538605
Fax: (42 2) 533564

Mooney, Dr H.A.
Dept. of Biological Sciences
Stanford University
Standford, CA 94305-5020
USA
Tel: (1 415) 723 1179
Fax: (1 415) 723 9253

Moss, Dr Richard H.
Royal Swedish Academy of
Sciences, IGBP Secretariat/HDGECP
Box 50005, 104 05 Stockholm
Sweden
Tel: (46 8) 15 0430
Fax: (46 8) 16 6405
Tlx: 17509

Moura, Dr A.D.
INPE
Av. dos Astronautas 1758
C.P. 515, Sao Jose dos Campos
SP 12201, Brazil
Tel: (55 123) 229977
Fax: (55 123) 218743
Tlx: 1233530

Munn, Dr R.E.
Institute for Environmental
Studies, University of Toronto
Ontario M5S 1A4, Toronto,
Canada
Tel: (1 416) 978 8202
Tel: (1 416) 484 6551 home
Fax: (1 416) 978 3884

Mykletun, Dr J.
Norwegian Research Council for
Science & the Humanities
Sandakervn. 99
0483 Oslo 4, Norway
Tel: (47 2) 15 70 12
Fax: (47 2) 22 55 71

Nador, Mr Balazs
Administrative Assistant
Hungarian Academy of Sciences
Roosevelt tér 9
Budapest 1051, Hungary
Tel: (36 1) 111 9812
Fax: (36 1) 112 8483
Tlx: 224 139

Natvig, Dr Jacob B.
Dept. of Chemical Immunology
Rikshospitalet University Hospital
F. Qvams Gate 1
0172 Oslo 1, Norway
Tel: (47 2) 11 1510
Fax: (47 2) 20 7287

Nkusi, Dr Marie-Thérèse
Université du Burundi
Faculté des Sciences
B.P. 2700, Bujumbura
Burundi
Tel: (257 22) 224854
Fax: (257 22) 3288
Tlx: 5161

Norse, Dr David
At the time of the Conference:
FAO/Agricultural Dept.
Via delle Terme di Caracalla
00100 Rome, Italy
Subsequently:
53 Summer Road
East Molesey
Surrey KT8 9LX, UK
Tel: (44 81) 398 4715
Fax: (44 81) 941 6909

Nowotny, Prof. Helga
Institute for Theory & Social
Studies of Science
University of Vienna
Sensengasse 8
1090 Vienna, Austria
Tel: (43 222) 402 76 01
Tel: (43 222) 402 76 02/ext. 12 or 14
Fax: (43 222) 408 88 38

O'Brien, Prof. James J.
Mesoscale Air-Sea
Interaction Group
B-176, 020 Love,
The Florida State University
Tallahassee, FL 32306-3041
USA
Tel: (1 904) 644 4581
Fax: (1 904) 644 8579
Tlx: 5706017589

Öberg, Dr S.
International Institute for
Applied Systems Analysis
IIASA
2361 Laxenburg, Austria
Tel: (43 2236) 715210
Fax: (43 2236) 71313

Okigbo, Dr Bede Nwoye
Director, United Nations University
Programme on Natural Resources
in Africa (UNU/INRA)
c/o Unesco/ROSTA
P.O. Box 30592
Nairobi, Kenya
Tel + Fax: (254 2) 520043
Tlx: 22275 UNESCO KE

Okonkwo, Prof. S.N.C.
MAIZE Research Programme
IITA, PMB 5320, Ibadan
Nigeria
Tlx: 31417 TROPIB NG

Oleru, Prof. U.G.
Dept. of Community Health
College of Medicine
University of Lagos
P.M.B. 12003, Lagos, Nigeria
Tel: (234 1) 801 500
Tlx: 27636 LUTHCM NG

Opschoor, Prof. J. B.
RMNO, P.O. Box 5306
2280 HH Rijswijk
The Netherlands
Tel: (31 70) 398 5880
Fax: (31 70) 398 5837

O'Riordan, Prof. Timothy
University of East Anglia
School of Environmental
Sciences
Norwich, Norfolk NR4 7TJ
UK
Tel: (44 603) 592 840
Fax: (44 603) 507 719
Tlx: 975197

Pachauri, Prof. R.K.
Tata Energy Research Institute
7 Jor Bagh, New Delhi 110003
India
Tel: (91 11) 462 7651
Fax: (91 11) 462 1770
Tlx: 31-61593 TERI IN

Palz, Dr W.
CCE, DG XII
Square de Meeus 8
1040 Brussels, Belgium
Tel: (32 2) 235 69 22
Fax: (32 2) 236 30 24

Parikh, Dr Kirit S.
Indira Gandhi Institute
of Development Research
Gen. Vaidya Marg, Goregaon (E)
Bombay-400 065, India
Tel: (91 22) 6800918
Fax: (91 22) 6802752
Tlx: 70040 IGI IN

Parry, Prof. Martin
Environmental Change Unit
1a Nansfield Road
Oxford OX1 3TB, UK
Tel: (44 865) 281180
Fax: (44 865) 281181

Patarroyo, Dr Manuel E.
Instituto de Immunologia
Hospital San Juan de Dios
Avenue 1, Number 10-1
Bogota, D.E., Colombia
Tel: (571) 233 9006
Fax: (571) 280 3999
Fax: (571) 280 1616

Pawlik, Prof. Kurt
Psychologisches Institut I
Universität Hamburg
Von-Melle-Park 11,
200 Hamburg 13, FRG
Tel: (49 49) 4123 4722
Fax: (49 40) 4123 6591
Tlx: 214732

Perrings, Prof. Charles
Dept. of Economics
University of California
Riverside, CA 92521, USA
Tel: (1 714) 787 5037
Fax: (1 714) 787 5685

Pescod, Dr M.B.
Dept. of Civil Engineering
University of Newcastle
upon Tyne
Newcastle upon Tyne NEI 7RU
UK
Tel: (44) 91 222 6000
Fax: (44) 91 261 1182
Tlx: 53654 UNINEW G

Petit, Dr Michel
Ministère de l'Equipement,
du Logement, des Transports &
de l'Espace
Délégation Générale à l'Espace
Colline Sud - Plot H
92055 Paris La Défense
Cedex 04, France
Tel: (33 1) 40 81 35 27
Fax: (33 1) 40 81 35 28

Praderie, Ms Françoise
ICSU Science Officer
51, Bld de Montmorency
75016 - Paris, France
Tel: (33 1) 45 25 03 29
Fax: (33 1) 42 88 94 31
Tlx: 645 554 ICSU
Email:Omnet: ICSU.Paris
Telecom Gold 10075: DBIO 126

Prage, Dr Lennart
I.F.S.
Grev Turegatan 19
11438 Stockholm, Sweden
Tel: (46 8) 791 2900
Fax: (46 8) 660 2618

Pravdic, Prof. Velimir
Center for Marine Research
Rudjer Boskovic Institute
P.O. Box 1016
41 001 Zagreb, Croatia
Yugoslavia
Tel: (38 41) 42 53 84
Tel: (38 41) 43 51 11 ext. 1215
Fax: (38 41) 42 04 37

Preining, Prof. Othmar
Universität Wien
Institut für Experimentalphysik
Strudlhofgasse 4
1090 Wien, Austria
Tel: (43 222) 34 26 30
Fax: (43 222) 3102683
Tlx: 116222

Prinn, Prof. Ronald G.
MIT, Room 54-1312,
Cambridge, MA 02139, USA
Tel: (1 617) 253 2452
Fax: (1 617) 253 0354
Tlx: (23) 921 473 MIT CAM

Qassim, Prof. Raad Y.
School of Engineering-UFRJ
P.O. Box 68529
CEP 21945 Rio de Janeiro
Brazil
Tel: (55 21) 280 75 43
Fax: (55 21) 590 48 40
Tlx: 2133817

Qian, Dr Hao
Senior Adviser UNCED
P.O. Box 80
1231 Conches, Switzerland
Tel: (41 22) 789 1676 ext.327
Fax: (41 22) 466 815

Rabinowitch, Dr V.
The John D. & Catherine T.
MacArthur Foundation
Suite 1100
140 South Dearborn Street
Chicago, IL 60603, USA
Tel: (1 312) 726 8000
Fax: (1 312) 917 0200

Rabinowitch, Ms Mary Martha
1212 North Lake Shore Drive
10 AS
Chicago, IL 60610, USA
Tel: (1 312) 943 3197

Ramalingaswami, Prof. V.
Dept. of Pathology
All India Institute of
Medical Sciences
Ansri Nagar
New Delhi 110 029, India
Tel: (91 11) 652 352
Fax: (91 11) 686 24 35

Ramallo, Dr Luis
ISSC Secretary General
1, rue Miollis
75015 - Paris, France
Tel: (33 1) 45 68 25 58
Fax: (33 1) 43 06 87 98

Raven, Dr P. H.
Missouri Botanical Garden
P.O. Box 299, St. Louis
MO 63166-0299, USA
Tel: (1 314) 577 5110
Fax: (1 314) 577 9595

Ravetto, Dr Alicia
SOLTECNICA Energy Consultant
RT3, Box 169
Pittsboro, N.C.27312
USA
Tel: (1 919) 5425361
Fax: (1 919) 8323332

Ripert, Dr Jean
Ministère des Affaires
Etrangères
37, Quai d'Orsay
75007 - Paris, France
Tel: (33 1) 47 53 53 53
Fax: (33 1) 47 53 50 85

la Rivière, Prof. J.W.M.
International Institute for
Hydraulic & Environmental
Engineering, P.O. Box 3015
Oude Delft 95, 2601 DA Delft
The Netherlands
Tel: (31 15) 78 30 60
Fax: (31 15) 12 29 21
Tlx: 38099

Rodda, Dr John C.
Hydrology & Water Resources
Dept.
WMO, Case Postale N°5
1211 Geneva 20, Switzerland
Tel: (41 22) 734 8245
Fax: (41 22) 734 2326

Rosswall, Prof. Th.
IGBP Secretariat
Royal Swedish Academy of
Sciences
Box 50005
104 05 Stockholm, Sweden
Tel: (46 8) 16 6448
Fax: (46 8) 16 6405
Tlx: 17509

Ruchirawat, Dr Mathuros
Chulabhorn Research Institute
c/o Faculty of Science
Mahidol University
Rama 6 Road
Bangkok, Thailand
Tel: (662) 247 1900
Fax: (662) 247 1222

Ruellan, Dr Alain
Programme Environnement
C.N.R.S.
15, Quai Anatole France
75700 - Paris, France
Tel: (33 1) 47 53 13 62
Fax: (33 1) 47 53 12 21
Tlx: 260034

Saavedra, Prof. I.
Physics Dept.
Faculty of Physics &
Mathematics
University of Chile
Casilla 487-3, Santiago
Chile
Tel: (56 2) 710732
Fax: (56 2) 712799
Tlx: 243302 INGEN CL

Salam, Prof. A.
President TWAS
c/o ICTP, P.O.B. 586
34100 Trieste, Italy
Tel: (39 40) 22401
Fax: (39 40) 224163
Tlx: 460302

Sang Soo, Prof. Lee
Korea Advanced Institute of
Science & Technology
P.O. Box 150
Chongyangni, Seoul, Korea
Tel: (82 2) 966 1931
Fax: (82 2) 968 2259
Tlx: 26795 KAISROK

Sanhueza, Dr Eugenio
IVIC, Apartado 21-827
Caracas 1020-A, Venezuela
Tel: (58 2) 501 1414
Fax: (58 2) 571 3164
Tlx: 21657

Schleicher, Prof. Stefan
Universität Graz
Institut für Volkswirtschaftslehre
und -politik
Schubertstrasse 6a
8010 Graz, Austria
Tel: (43 316) 380 3440

Sdasyuk, Dr Galina
Institute of Geography
USSR Academy of Sciences
Staromonetny Perm. 29
109017 Moscow, USSR
Tel: (7 095) 434 51 72
Fax: (7 095) 230 20 90
Tlx: 411781 GLOBE SU

Semesi, Dr Adelaïda K.
University of Dar es Salaam
Botany Dept.
P.O.B. 35060, Dar es Salaam
Tanzania
Tel: (255 51) 49192
Fax: (255 51) 48224
Tlx: 41343 MILLER

Sendov, Prof. Bl.
Bulgarian Academy of Sciences
Center for Informatics &
Computer Technology
"Acad. G. Bontchev"
Str. Bl. 25-A
1113 Sofia, Bulgaria
Tel: (359 2) 70 84 94
Fax: (359 2) 70 72 73
Tlx: 22056 KZIIT BG

Sène, Monsieur Djibril
Député, Président de la
Commission des Affaires
Etrangères à l'Assemblée
Nationale
B.P. 86, Dakar, Sénégal
Tel: (221) 23 55 73
Tlx: 61265 ASNAT SG

Shamir, Prof. Uri
Faculty of Civil Engineering
The Technion, 32000 Haifa
Israel
Tel: (972 4) 292239
Fax: (972 4) 220133
Tlx: 46406 TECON IL

Sharma, Dr Manju
Dept. of Biotechnology
Block N°2, 7th floor
CGO Complex, Lodi Road
New Delhi 110 003, India
Tel: (91 11) 360598 office
Tel: (91 11) 674587 home
Fax: (91 11) 362884
Tlx: 31 74105 BIOT IN

Shechter, Prof. Mordechai
Natural Resources &
Environmental Research Center
University of Haifa
Mt Carmel, Haifa 31905, Israel
Tel: (972) 4 240083
Fax: (972) 4 342101/4

Silver, Prof. Leon T.
Division of Geological &
Planetary Sciences
California Institute of
Technology
Mail code 170-25
Pasadena, California 91125
USA
Tel: (1 202) 334 2807
Fax: (1 202) 334 2231
Tlx: 4900009522

Sinha, Prof. S. K.
Indian Agricultural Research
Institute
New Delhi 110 012, India
Tel: (91 11) 575 4595
Fax: (91 11) 575 2006
Tlx: 3177161 IARI IN

Skinner, Dr Brian J.
Economic Geology
Yale University
91-A, Yale Station
New Haven, CT 06520
USA
Tel: (1 203) 432 3175
Fax: (1 203) 432 9819

Smith, Dr John
European Science Foundation
1, Quai Lezay-Marnésia
67080 - Strasbourg Cedex
France
Tel: (33) 88 76 71 31
Fax: (33) 88 37 05 32
Tlx: 890440

Sokolov, Dr V.
Institute of Evolutionary
Morphology & Ecology of
Animals, Leninsky Prospekt 33
117071 Moscow, USSR
Tel: (7 095) 232 20 88
Fax: (7 095) 129 13 54
Tlx: 411682 MAB SU

Solbrig, Prof. Otto T.
Dept. of Organismic &
Evolutionary Biology
Harvard University Herbaria
22 Divinity Avenue
Cambridge, MA 02138, USA
Tel: (1 617) 495 4302
Fax: (1 617) 495 9484

Sombroek, Dr Wim G.
At the time of the Conference:
International Soil Reference
& Information Center (ISRIC)
P.O. Box 353 - 9, Duivendaal
6700 AJ Wageningen
The Netherlands
Subsequently:
Land & Water Development
Division, F.A.O.
Room B-733
Viale delle Terme di Caracalla
00100 Rome, Italy
Tel: (39 6) 5797 3155
Fax: (39 6) 5782 610
Tlx: 610181 FAO I

Somlyody, Dr L.
International Institute for
Applied Systems Analysis
IIASA, 2361 Laxenburg, Austria
Tel: (43-2236) 71 5210
Fax: (43-2236) 71313
Tlx: 79137 IIASA A

Stacher, Dr Ulrich
Dept. for Co-ordination
Federal Chancellery
Ballhausplatz
1014, Vienna, Austria
Tel: (43 222) 53115
Fax: (43 222) 53115 4227
Tlx: 1370900

Stewart, Dr J.W.B.
College of Agriculture
University of Saskatchewan
Saskatoon, Sask. S7N OWO
Canada
Tel: (1 306) 966 4055
Fax: (1 306) 966 8894

Stewart, Dr R.W.
4249 Thornhill Crescent
Victoria, B. C.
Canada, V8N 3G6
Tel: (1 604) 477 1247
Fax: (1 604) 477 3725

Stigliani, Dr William M.
International Institute for
Applied Systems Analysis
IIASA, 2361 Laxenburg, Austria
Tel: (43 2236) 715210
Fax: (43 2236) 71313
Tlx: 79137 IIASA A

Stoszek, Dr Karl
Institut für Waldbau
Peter Jordanstrasse 70
1190 Vienna, Austria
Tel: (43 222) 342500627
Fax: (43 222) 3691659

Strömberg, Prof. J.O.
Kristineberg Marine Biological
Station, Kristineberg 2130
45034 Fiskebackskil, Sweden
Tel: (46 523) 22 007 home
Fax: (46 523) 22 871
Tlx: 17073 ROYACAD

Strong, Mr Maurice F.
Secretary General UNCED
P.O. Box 80, 1231 Conches
Switzerland
Tel: (41 22) 789 35 35 or 1676
Fax: (41 22) 789 3536

Sun Honglie, Prof.
52 Sanlihe Road
Beijing 100864, China
Tel: (86 1) 8012880
Fax: (86 1) 8011095
Tlx: 22474 ASCHI CN

Sun Zonglu, Prof.
Peking University Hospital
Haidian, Beijing 100871
China
Tel: (86 1) 282471 ext.3063
Fax: (86 1) 2564095
Tlx: 22239 PKUNI CN

Szöllösi-Nagy, Dr S.A.
Division of Water Sciences
Unesco
7 Place de Fontenoy
75700 Paris, France
Tel: (33 1)45 68 10 00
Fax: (33 1)45 67 58 69

Thacher, Mr P.S.
54 Gold Street
Stonington, CT 06378, USA
Tel: (1 203) 535 0633
Fax: (1 203) 535 4787

Thonstad, Prof. Tore
University of Oslo
Dept. of Economics
P.O. Box 1095, 0317 Oslo 3
Norway
Tel: (47 2) 45 51 60
Fax: (47 2) 45 50 35

Tindimubona, Dr A.
African Academy of Sciences
P.O. Box 14798, Nairobi
Kenya
Tel: (254 2) 802176
Fax: (254 2) 802185
Tlx: 25446 AFACS

Tinker, Dr P.B.
Terrestrial & Freshwater Sciences
Natural Environment Research
Council, Polaris House, North Star
Avenue, Swindon SN2 1EU, UK
Tel: (44 793) 411 523
Fax: (44 793) 411 502
Tlx: 444293 ENVRE G

Tundisi, Dr José G.
Universidad de Sao Paulo
Escola de Engenharia de Sao
Carlos, Av. Dr Carlos Botelho 1465
CX Postal 359, Sao Carlos
Brazil
Tel: (55 162) 781144
Fax: (55 162) 715726
Tlx: 162 411 USPO BR

Umaña, Dr Alvaro
Central American Management
Inst. (INCAE)
P.O. Box 960-4050
Alajuela, Costa Rica
Tel: (506) 41 22 55
Fax: (506) 43 91 01
Tlx: 7040 ICAE

Van Lookeren Campagne , Dr N.
Van Bergenlaan 6
Wassenaar 2242 PV
The Netherlands
Fax: (31 10) 4696971 office
Tel: (31 1751) 14085 home
Tel + Fax:(31 1751) 18821 home

Venkataraman, Dr K.
Industrial Technology
Development Division /UNIDO
CIX, P.O. Box 300
1400 Vienna, Austria
Tel: (43 222) 211310
Fax: (43 222) 232 156
Tlx: 135612 UNO A

Vida, Prof. Gabor
Genetics Dept.
Eötvös University
Muzeum Krt. 4/a
Budapest 1088, Hungary
Tel: (36 1) 118 1296
Fax: (36 1) 118 2694

Villegas, Prof. Raimundo
IDEA
Parque Central, Apartado 17606
Caracas 1015A, Venezuela
Tel: (58 2) 962 1601
Fax: (58 2) 962 1602
Tlx: 24593 FIIEA VC

Villevieille, Mr Adelin
c/o UATI, Unesco
1, rue Miollis
75015 - Paris, France
Tel: (33 1) 45 66 94 10
Fax: (33 1) 43 06 29 27

Vranitzky, Dr Franz
Federal Chancellor
Bundeskanzler, Ballausplatz 2
1014 Vienna, Austria

Wasawo, Dr David P.S.
P.O. Box 41024
Nairobi, Kenya

Weish, Dr Peter
Österreichische Akademie
der Wissenschaften
Kommission für Humanökologie
Messepalast 1, Stiege 14
1070 Wien, Austria
Tel: (43 222) 93 64 78/93 73 02

White, Dr Gilbert F.
Institute of Behavioral Science
Campus Box 482
University of Colorado
Boulder, CO 80309-0482
USA
Tel: (1 303) 492 6311
Fax: (1 303) 492 6924
Tlx: 303492

Whyte, Dr Ann
IDRC
Social Sciences Division
P.O. Box 8500, Ottawa
Canada KIG 3H9
Tel: (1 613) 236 6163 ext.2558
Fax: (1 613) 238 7230
Tlx: 053- 3753

Woods, Dr J. D.
Marine & Atmosperic Sciences
NERC
Polaris House, North Star Avenue,
Swindon SN2 1EU, UK
Tel: (44 793) 41 16 37
Fax: (44 793) 41 15 45
Tlx: 444293 ENVRE G

Yankov, Prof. Alexander
Bulgarian Academy of Sciences
7 November Str. N°1
1046 Sofia, Bulgaria
Tel: (359 2) 87 46 24 or 84 141
Fax: (359 2) 88 04 48
Tlx: 224-BAN BG

Zhao Qiguo, Prof.
Nanjing Institute of
Soil Science
71 East Beijing Road
P.O. Box 821
Nanjing 210 008, China

Tel: (86 25) 713781
Fax: (86 25) 712663
Tlx: 34024 ISSAS CN

Zylicz, Prof. Tomasz
Ministry of Environmental
Protection, Economics Dept.
Ul. Wawelska 52/54
00922 Warsaw, Poland
Tel: (48 22) 25 02 67
Fax: (48 22) 25 41 41
Tlx: 817857 PIU PL

LIST OF AUTHORS WHO DID NOT ATTEND THE ASCEND CONFERENCE

Chapter 1: *Population and Natural Resource Use*

Professor R. Costanza
Chesapeake Laboratory
University of Maryland
P.O. Box 38, Solomans
Maryland 20688, USA
Telephone: (1 301) 326 4281
Telefax: (1 301) 326 6342

Chapter 2: *Agriculture, Land Use and Degradation*

Dr Clive James
c/o CIMMYT
APD-Postal 6-641
Lisboa 27, 06600 Mexico D.F.
Telephone: (52 595) 455499/45395 x 22
Telefax: (52 595) 410 69

Chapter 4: *Energy*

Professor J.P. Holdren
Energy & Resources Group
University of California
100 T-4, Berkeley, CA 94720, USA
Telephone: (1 415) 642 1139
Telefax: (1 415) 642 1085

Chapter 5: *Health*

Professor S.K.D. Bergstrom
Karolinska Institute
Nobel Department, Box 60400
10401 Stockholm, Sweden
Telefax: (46 8) 326 888

Chapter 6: *Global Cycles*

Dr R. Victoria
Centro de Energia Nuclear
na Agricultura (CENA)
Caixa Pastal 96
Pivacicaba, S.P. Brazil
Telefax: (55 194) 228 339

Professor M.G. Wolman
Department of Geography
& Environmental Engineering
Johns Hopkins University
Baltimore 21218, MD, USA
Telefax: (1 301) 3387075

Chapter 8: *Marine and Coastal Systems*

Dr P. Bernal, Executive Director
Instituto de Formento Pesquero (IFOP)
Avenue Jose Domingo, Canas 2277
Casilla 1287, Santiago, Chile
Telephone: (56 2) 2256 325
Telefax: (56 2) 2254 362

Chapter 11: *Biodiversity*

Professor J. Sarukhan, Universidad
Nacional, Automoma de Mexico
Torre de Rectoria, 6° Piso
Cludad Universitaria, 04510 Mexico
Telephone: (52 5) 548 4040
Telefax: (52 5) 550 8772

Chapter 13: *Public Awareness, Science and Environment*

Dr. Vandana Shiva
Research Foundation for Science,
Technology & Natural Resource
Policy, 105, Rajpur Road
Dehra dun 248001, India
Telephone: (91 135) 23374
Telefax: (91 135) 283 92

Chapter 15: *Policies for Technology*

Dr M. Sagoff
Institute for Philosophy & Public Policy
University of Maryland,
0123 Woods Hall College Park,
MD 20742, USA
Telephone: (1 301) 405 4762/4759
Telefax: (1 301) 314 9346

List of Observers

IIASA Observers

Clark, Dr M.
International Institute for
Applied Systems Analysis
IIASA
2361 Laxenburg, Austria
Tel: (43-2236) 71 521 599
Fax: (43-2236) 71313
Tlx: 79137 IIASA A

Heilig, Dr Gerhard
Population Programme
International Institute for
Applied Systems Analysis
IIASA, 2361 Laxenburg, Austria
Tel: (43-2236) 71 521 599
Fax: (43-2236) 71313
Tlx: 79137 IIASA A

Kulshreshta, Dr Surendra
Water Resources Project
International Institute for
Applied Systems Analysis
IIASA, 2361 Laxenburg, Austria
Tel: (43-2236) 71 521 599
Fax: (43-2236) 71313
Tlx: 79137 IIASA A

Nakicenovic, Dr Nebojsa
Project Leader
Environmentally Compatible
Energy Strategies
International Institute for
Applied Systems Analysis
IIASA, 2361 Laxenburg, Austria
Tel: (43-2236) 71 521 599
Fax: (43-2236) 71313
Tlx: 79137 IIASA A

Shaw, Dr Roderick
Project Leader
Global Environmental Security
International Institute for
Applied Systems Analysis
IIASA, 2361 Laxenburg, Austria
Tel: (43-2236) 71 521 599
Fax: (43-2236) 71313
Tlx: 79137 IIASA A

Wessels, Dr Jacobus
Methodology of Decision
Analysis Project
International Institute for
Applied Systems Analysis
IIASA, 2361 Laxenburg, Austria
Tel: (43-2236) 71 521 599
Fax: (43-2236) 71313
Tlx: 79137 IIASA A

AUSTRIAN Observers

Deistler, Prof. Manfred
Technical University Vienna

Goldmann, Dkfm. Wilhelmine
Österreichische Arbeiterkammertag
Prinz-Eugen-Strasse 20-22
1040 Wien
Tel: 50 165-2367
Fax: 50 165-2230

Grübler, Dr Arnulf
Internationales Institut
für Systemanalyse
Schlossplatz 1
2361 Laxenburg
Tel: 2236/715210-470
Fax: 2236/71313

Hafner, Prof.
Universität Wien
Institut für Völkerrecht
Universitätsstrasse 2
1090 Wien
Tel: 222/42 92 86-160
Fax: 222/402 79 41

Haider, Prof. Manfred
Universität Wien
Institut für Umwelthygiene
Kinderspitalgasse 15
1095 Wien
Tel: 222/43 15 95 -300
Fax: 222/402 05 10

Hantel, Prof. M.
Universität Wien
Institut für Meteorologie
und Geophysik
Hohe Warte 38
1190 Wien

Hauck, Doz. Helger
Universität Wien
Institut für Umwelthygiene
Kinderspitalgasse 15
1095 Wien
Tel: 222/43 91 15-312
Fax: 222/402 05 10

Heindler, Prof. Manfred
Energieverwertungsagentur
Opernring 1/R/3
1010 Wien
Tel: 222/586 15 24
Fax: 222/56 94 88

Jansen, Prof. Peter-Jörg
Technische Universität Wien
Institut für Energiewirtschaft
Gusshausstrasse 27-29
1040 Wien
Tel: 222/58 801-5200
Fax: 02665 273

Kolb, Doz. Helga
Zentralanstalt für Meteorologie
und Geodynamik
Hohe Warte 38
1190 Wien
Tel: 222/36 44 53-2401
Fax: 222/36 64 570

Kuhn, Prof. Michael
Universität Innsbruck
Institut für Meteorologie
und Geophysik
Innrain 52
6020 Innsbruck
Tel: 512/507-2171
Fax: 512/507-2170

Lang, Mag. Roland
Österreichischer Arbeiterkammertag
Prinz-Eugen-Strasse 20-22
1040 Wien
Tel: 222/501 65/0
Fax: 222/501 65/2230

Meissner-Blau, Ms Fredda
ECEUROPA
Bräuner strasse 10
1010 Wien

Moser, Prof. Anton
Institut für Biotechnologie
Technische Universität Graz
Petersgasse 12, 8010 Graz
Tel: 316/873-8405
Fax: 316/811050

Moser, Prof. Franz
Technische Universität Graz
Institut für Verfahrenstechnik
Inffeldgasse 25, 8010 Graz
Tel: 316/873-7460
Fax: 316/873-7469

Münz, Doz. Rainer
Österreichische Akademie
der Wissenschaften
Institut für Demographie
Hintere Zollamtsstrasse 2b
1033 Wien
Tel: 222/7121284
Fax: 222/712 97 01

Neuwirth, Doz. Fritz
Zentralanstalt für Meteorologie
und Geodynamik
Hohe Warte 38
1190 Wien

Ott, Prof. Jörg
Universität Wien
Institut für Zoologie
Althanstrasse 14
1091 Wien
Tel: 222/313 36-1317
Fax: 222/313 36-700

Raggam, Prof. August
Technische Universität Graz
Forschungsinstitut für
Alternative Energienutzung
Krenngasse 37/V, 8010 Graz
Tel: 316/824 171
Fax: 316/827 685

Rozsenich, SC Dr Norbert
Bundesministerium für
Wissenschaft und Forschung
Sektion II, Freyung 1
1014 Wien
Tel: 222/53 120-2227
Fax: 222/53 120-2217

Stachowitsch, Dr Michael
Universität Wien
Institut für Zoologie
Althanstrasse 14
1091 Wien
Tel: 222/313 36-1851
Fax: 222/313 36-700

Unterbrunner, Doz. Dr Ulrike
Universität Salzburg
Institut für Didaktik der
Naturwissenschaften
Hellbrunner Strasse 34
5020 Salzburg
Tel: 662/8044-0

Wieser, Prof. Wolfgang
Universität Innsbruck
Institut für Zoologie
Technikerstrasse 25
6020 Innsbruck
Tel: 995/7480-5300
Fax: 995/7480-5358

Wimmer, Prof. Dr Norbert
Universität Innsbruck
Institut für öffentliches Recht
und Politikwissenschaft
Innrain 80, 6020 Innsbruck
Tel: 995/5070-2670

Wohlmeyer, Prof. Dr Heinrich
Österreichische Vereinigung
für agrarwissenschaftliche
Forschung, Kleine Sperlgasse 1/37
1020 Wien
Tel: 222/35 00 32, 26 22 42
Fax: 222/21 13 62 987

UNCED Observer

Matsushita, Dr Kazuo
UNCED Secretariat
P.O. Box 80
1231 Conches, Switzerland
Tel: (41 22) 89 1676
Fax: (41 22) 46 68 15

IAEA Observer

Srinivasan, Dr M.
Internal Atomic Energy Agency
Wagramerstrasse 5
P.O. Box 100
1400 Vienna, Austria
Tel: (43 1) 2360
Fax: (43 1) 234564

ILSI Observer

Julkunen, Ms Päivi
International Life Sciences Institute
Risk Science Institute
1126 Sixteenth Street, N.W.,
Washington D.C. 20036, USA.
Tel: (1 202) 659 3306
Fax: (1 202) 659 8654

UNEP Observer

Evtéev, Dr S.
UNEP, P.O. Box 30552
Nairobi, Kenya
Tel: (254 2) 33 3930
Fax: (254 2) 52 0711
Tlx: 22068

UNIDO Observers

UNIDO
Vienna International Center
P.O. Box 300
1400 Vienna, Austria
Tel: (43 222) 211 310
Fax: (43 222) 232 156
Tlx: 135612 UNO A

Burmistrov, Mr Y.
Daniel, Ms M.
Fujita, Mr K
Subrahmanyam, Mr D.
Wiedemann, Mr P.
Williams, Mr R.

Members of the ASCEND Advisory Committee

    M.G.K. Menon, President of ICSU, Chairman   (India)
    U. Colombo, President of ESF, Vice-Chairman   (Italy)
*   J.C.I. Dooge, President-elect of ICSU, Vice-Chairman   (Ireland)
    Abdus Salam, President of TWAS, Vice-Chairman   (Pakistan)

Members from ICSU Executive Board
    U. Cordani   (Brazil)
    M.A. Epstein   (UK)
*   J.W.M. la Rivière   (Netherlands)
*   I. Lang   (Hungary)

Members from ICSU Advisory Committee on the Environment
*   G.T. Goodman   (UK)
    P. Thacher   (USA)
    N. van Lookeren Campagne   (Netherlands)
    G. White   (USA)

Members from ICSU Committee on Science and Technology in Developing Countries
*   D.A. Bekoe   (Ghana)
    Ana Maria Cetto   (Mexico)
    R. Daniel   (India)
*   M.H.A. Hassan   (Sudan)

Representative of Federal Government of Austria
*   U. Stacher

Invited Members

	Lourdes Arizpe	(Mexico)		V.A. Koptyug	(USSR)
	P. Bourdeau	(Belgium)		G. Kullenberg	(Denmark)
	J. Burnett	(UK)		B. Moldan	(Czechoslovakia)
	Rita Colwell	(USA)		J. Mykletun	(NAVF)
	F. Di Castri	(Italy)		Helga Nowotny	(Austria)
*	B. Döös	(IIASA)	*	T. O'Riordan	(ESF)
	R. Fuchs	(USA)	*	J. Smith	(ESF)
	G. Golubev	(USSR)		Sun Honglie	(China)
	J. Houghton	(UK)		M.S. Swaminathan	(India)
*	H. Jacobson	(ISSC)	*	L. Walløe	(Norwegian Academy)
	M. Kassas	(Egypt)			

A **Bureau** consisting of selected members (including representatives of the sponsoring bodies and host country) of the Advisory Committee, was constituted to be responsible for the details of Conference preparations.

* = Bureau Member.

Description of co-sponsors

The International Council of Scientific Unions (ICSU)
ICSU was created in 1931 (succeeding its predecessor the International Research Council) to promote international scientific activity in the different branches of science and their applications for the benefit of humanity. ICSU is a non-governmental organization with national scientific members in 75 countries and scientific unions in 20 disciplines. The combination of these two groups provides a wide spectrum of scientific expertise enabling members to address major international interdisciplinary issues which none of them could handle alone. In addition, ICSU has 29 Scientific Associates. The Council seeks to accomplish its role in a number of ways: by designing and co-ordinating major international interdisciplinary research programs (e.g. the International Geosphere-Biosphere Programme and the World Climate Research Programme – the latter with WMO); by creating interdisciplinary bodies which undertake activities and research programs of interest to several member bodies (e.g. oceanic, space, water, genetic research), and by carrying out activities of common concern to all scientists (e.g. free circulation of scientists, teaching, science and technology in developing countries). The Council also acts as a focus for the exchange of ideas and information and maintains close working relations with a number of other bodies, both inter-governmental and non-governmental.

The Third World Academy of Sciences (TWAS)
The Third World Academy of Sciences was founded in 1983 and officially launched and inaugurated by the Secretary General of the United Nations, Mr Perez de Cuellar in 1985. It has succeeded in uniting the most distinguished scientists from the Third World. Currently, it has 270 members from 52 Third World countries including the living 10 Science Nobel Laureates of Third World origin. The founding President of the Academy is the Nobel Laureate Professor Abdus Salam of Pakistan. The Academy is a non-governmental, non-political and non-profit-making organization whose main objectives are to support scientific excellence and research in the Third World through awarding annual prizes to eminent scientists from the Third World who have made significant contributions in science, providing research grants to young scientists from developing countries and encouraging scientific contacts between research workers in developing countries, providing Fellowships to facilitate contacts between research workers in developing countries and encouraging scientific research on major Third World problems. The Academy was granted official NGO Consultative Status with the United Nations in 1985. It is presently located on the premises of the International Center for Theoretical Physics at Miramare, Trieste, Italy, a Center sponsored by the IAEA and UNESCO. The programs of the Academy are supported by the Italian Government, the Canadian International Development Agency and a number of other donors.

The European Science Foundation (ESF)
The ESF is an association of its 56 member research councils, academies, and institutions devoted to basic scientific research in 20 countries. The ESF brings European scientists together to work on topics of common concern, to co-ordinate the use of expensive facilities, and to discover and define new endeavours that will benefit from a co-operative approach.

The scientific work sponsored by ESF includes basic research in the natural sciences, the medical and bio-sciences, the humanities and the social sciences.

The ESF links scholarship and research supported by its members and adds value by co-operation across national frontiers. Through its function as a co-ordinator, and also by holding workshops and conferences and by enabling researchers to visit and study in laboratories throughout Europe, the ESF works for the advancement of European science.

The International Institute for Applied Systems Analysis (IIASA)
IIASA is an international, non-governmental research institution sponsored by scientific organizations from 15

countries. It was created in 1972 on the initiative of the USA and the former USSR, and its membership now also includes 11 European countries, Canada and Japan. The Institute's overall goal is: to provide objective, authoritative, timely and relevant environmental and societal change by means of international, interdisciplinary, and non-governmental scientific studies for the benefit of the public, the scientific community, and national and international institutions. The program of substantive and methodological research covers global and regional environmental and demographic issues and their inter-relations; technological and economic developments; systems and decision sciences; and international negotiations.

International Institute for Environmental Technology and Management (Stockholm Environmental Institute)

The Stockholm Environmental Institute (SEI) is an international institute for environmental management and technology as applied to problems of the environmentally sustainable development of society. SEI receives substantial core-funding from Sweden but as decided by Parliament, it functions as an independent foundation governed by an international Board. The Board members are from developing and industrialized countries world-wide and act in their personal capacities as distinguished persons of great experience in dealing with issues related to environment and development globally. The scientific and administrative work of the Stockholm Environmental Institute is co-ordinated by a small core staff along with guest scholars at its Head Office in Stockholm. The scientific work is further sustained through collaboration with an international network of scientists, project advisors and field staff. The Institute has two additional branch offices housing senior scientists and support staff in Boston, Mass, USA and York, UK.

The International Social Science Council (ISSC)

The ISSC, founded in 1952, is an international non-governmental organization in category A with Unesco (consultative and associate relations) and in consultative status category B with the Economic and Social Council of the United Nations Organization. The major aim of the ISSC is to advance the social sciences and their application to major contemporary problems by means of co-operation among social scientists and social science organizations at the international and regional level.

The ISSC is the meeting place and co-ordinating body for 14 international disciplinary associations in the social sciences (legal sciences, economics, geography, administrative sciences, law, peace research, political science, sociology, international studies, anthropology and ethnology, psychology, population studies, public opinion research, mental health) and the federation of social

sciences organizations. In addition, ten professional organizations are associate members of the ISSC (economic history, industrial relations, criminology, data organizations, applied psychology, genealogy and heraldy, future studies, life sciences, French speaking sociologists, and scientific editors).

Through its international and multidisciplinary networks the ISSC serves as a co-ordinating body not only within the social science community at large but also in establishing close co-operation and communication with relevant research programs undertaken by natural scientists within IGBP specifically and ICSU in general. Since 1988, among its many other activities, the ISSC has set up the Human Dimensions of Global Environmental Change Programme.

The Norwegian Research Council for Science and the Humanities (NAVF) and the Norwegian Academy of Science and Letters

NAVF was established in 1949. The Council is a strategic research organization bearing a national responsibility for long-term basic research. In addition to the Board, it consists of four sub-councils: the Council for Social Science Research, the Council for Research in the Humanities, the Council for Natural Science Research and the Council for Medical Research. Some of NAVF's tasks are:

(a) to initiate and support research within areas of special importance to society;
(b) to promote the distribution of research results;
(c) to allocate funds for research purposes;
(d) to assist public authorities and private organizations with expert opinions.

The NAVF attaches particular importance to the promotion of high quality interdisciplinary research oriented toward international networking and science collaboration. This objective is of central guidance to the way in which the NAVF seeks to instigate a broadly based scientific follow-up of the World Commission's Report on Environment and Development, a task mandated to it by the Norwegian Government in 1987.

The Norwegian Academy of Science and Letters is a national member of ICSU. The Academy was founded in 1857 and is a non-governmental, nation-wide, interdisciplinary body which in principle embraces all fields of learning. Its members are divided into a science class and a humanities class, and each class is subdivided into groups for the constituent disciplines. The main purpose of the Academy is the advancement of science and scholarship. It provides a national organ of communication within and between the various learned disciplines, and it represents the sciences and humanities of Norway *vis-à-vis* foreign academies and international organizations. The Academy performs these functions by initiating and supporting research projects, arranging meetings and seminars on subjects of current interest, publishing scientific and scholarly works, participating in and nominating representatives for national and international bodies.

Acronyms

ACE	Advisory Committee on the Environment
ACTS	African Centre for Technology Studies
AEZ	Agro-Ecological Zones
AGCM	Atmospheric General Circulation Models
ALE	Atmospheric Lifetime Experiment
ASCEND 21	An Agenda of Science for Environment and Development into the 21st Century
AVHRR	Advanced Very High Resolution Radiometer
BAHC	Biospheric Aspects of the Hydrological Cycle
BAT	Best Available Technology
CEFIC	European Council of Chemical Industry Federations
CGIAR	Consultative Group on International Agricultural Research
CLICOM	Climate Data Management System
COSPAR	Committee on Space Research
COSTED	Committee on Science and Technology in Developing Countries
DAC	Development Assistance Committee
DAWN	Development Alternatives for Women in a New Era
DTM	Digital Terrain Modelling
EEC	European Economic Community
EEZ	Exclusive Economic Zone
EIA	Energy Information Administration
ENSO	El Nino-Southern Oscillation
ENUWAR	Environmental Consequences of Nuclear War
EPA	Environmental Protection Agency
ESCAP	Economic and Social Commission for Asia and the Pacific
ESF	European Science Foundation
EWI	Equal Weight Index
FAO	Food and Agriculture Organization of the United Nations
GAGE	Global Atmospheric Gases Experiment
GARP	Global Atmospheric Research Programme
GAW	Global Atmospheric Watch
GCIP	GEWEX Continental International Project
GCM	General Circulation Models
GCOS	Global Climate Observing System
GCTE	Global Change and Terrestrial Ecosystems
GEMS	Global Environmental Monitoring System
GESAMP	Group of Experts on Scientific Aspects of Marine Pollution
GEWEX	Global Energy and Water Cycle Experiment
GIPME	Global Investigation of Pollution in the Marine Environment
GIS	Geographical Information Systems
GNP	Gross National Product

GOOS	Global Oceans Observing System
GRID	Global Resource Information Database
GWP	Gross World Product
HDGECP	Human Dimensions of Global Environmental Change Programme
HDI	Human Development Index
HWRP	Hydrology and Water Resources Programme
IACGEC	Inter-Agency Committee on Global Environmental Change
IAEA	International Atomic Energy Agency
IAHS	International Association of Hydrological Sciences
IBSRAM	International Board for Soil Research and Management
ICC	International Chamber of Commerce
ICEL	International Council of Environmental Law
ICRAF	International Council for Research in Agroforestry
ICSU	International Council of Scientific Unions
ICTP	International Centre for Theoretical Physics
IDNDR	International Decade for Natural Disasters Reduction
IEA	International Energy Agency
IEB	International Environmental Bureau
IFDC	International Fertilizer Development Center
IGAC	International Global Atmospheric Chemistry Programme
IGBP	International Geosphere–Biosphere Programme
IHP	International Hydrological Programme
IIASA	International Institute for Applied Systems Analysis
INBio	Instituto Nacional de Biodiversidad (Costa Rica)
IOC	Intergovernmental Oceanographic Commission
IPCC	Intergovernmental Panel on Climate Change
IPNS	Integrated Plant Nutrition Systems
IRRI	International Rice Research Institute
ISBI	International Sustainable Biosphere Initiative
ISCRAL	International Scheme for the Conservation and Rehabilitation of African Lands
ISEW	Index of Sustainable Economic Welfare
ISIS	International Soil Information System
ISNAR	International Service for National Agricultural Research
ISRIC	International Soil Reference and Information Center
ISSC	International Social Science Council
ISSS	International Society of Soil Science
IUBS	International Union of Biological Sciences
IUCN	International Union for the Conservation of Nature
JGOFS	Joint Global Ocean Flux Study
K-TAC	Korea Technology Advancement Corporation
KAIS	Korea Advanced Institute of Science
KDIC	Korea Development Investment Corporation
KIST	Korea Institute of Science and Technology
KMOST	Korea Ministry of Science and Technology
KORSTIC	Korea Science and Technology Information Center
KTDC	Korea Technology Development Corporation
KTFC	Korea Technology Finance Corporation
LDCs	Less Developed Countries
LOICZ	Land-Ocean Interactions in the Coastal Zone
MAB	Man and the Biosphere Programme
MDCs	More Developed Countries
NAS	National Academy of Sciences (USA)
NATO	North Atlantic Treaty Organization
NAVF	Norwegian Research Council for Science and the Humanities

NETT	Network for Environmental Technology Transfer
NGO	Non-governmental organizations
NICs	Newly Industrialized Countries
NOAA	National Oceanic and Atmospheric Administration (USA)
NRC	National Research Council (USA)
OECD	Organisation for Economic Co-operation and Development
OSLR	Ocean Science in Relation to Living Resources
OTA	US Congressional Office of Technology Assessment
PAGES	Past Global Changes Project
PCE	Personal Consumption Expenditures
PPP	Polluter Pays Principle
QOL	Quality of Life
RURR	Remaining Ultimately Recoverable Resources
SBI	Sustainable Biosphere Initiative
SCAR	Scientific Committee on Antarctic Research
SCOPE	Scientific Committee on Problems of the Environment
SCOR	Scientific Committee on Oceanic Research
SCOSTEP	Scientific Committee on Solar Terrestrial Physics
SCS	Soil Conservation Service
SEI	Stockholm Environment Institute
SERI	Solar Energy Research Institute
SOTER	World Soils and Terrain Digital Data Base
START	Global Change System for Analysis, Research and Training
STEMS	Systematic Technology Policy Measures
STEP	Solar Terrestrial Energy Programme
SWCC	Second World Climate Conference
TIGER	Terrestrial Initiative in Global Environmental Research
TOGA	Tropical Ocean Global Atmosphere Programme
TROPENBOS	Stimulation Programme for Research in Tropical Forest Areas
TSBF	Tropical Soil Biology and Fertility
TWAS	Third World Academy of Sciences
UK-NERC	UK-Natural Environment Research Council
UNCED	United Nations Conference on Environment and Development
UNEP-GRID	UNEP Global Resource Information Database
UNEP	United Nations Environment Programme
UNESCO	United Nations Educational, Scientific and Cultural Organization
UNFPA	United Nations Fund for Population Activities
UNIDO	United Nations Industrial Development Organization
VIC	Vienna International Center
WASAD	International Action Programme on Water and Sustainable Agricultural Development
WCED	World Commission on Environment and Development
WCRP	World Climate Research Programme
WDC	World Data Centre
WEC	World Energy Conference
WHO	World Health Organization
WMO	World Meteorological Organization
WOCE	World Ocean Circulation Experiment
WRAP	Waste Reduction Always Pays
WRI	World Resources Institute
WWW	World Weather Watch

Index

CIMMYT, 30
civil rights, 224
clean technologies, 96, 243, 271
climate, 131, 142, 144, 158, 165, 168,
    180
  change, 122, 148, 150–153, 192,
      199, 200,
  models, 150
  dry tropical, 80, 189
clouds, 147, 150, 199
co-operation, 279
  on energy, 115
  as a means of survival, 268
coastal,
  aquifers, 195–196
  development, 161
  engineering, 165
  habitats, 164
  oceans, 158, 169
  plains, 158
  sediments, 158, 163
  upwelling, 161
commons, 234, 291
common good, 119
communication skills, 261
competition, 94
  for land, 82
  for water, 82
competitiveness, 270, 272, 278
conservation biology, 212
conservation of biodiversity, 206, 213
  of diversity, 267
  of technological diversity, 266
constant rates scenario, 64
consumption and the need for
    environmental protection, 228
consumption, 62, 119
  control, 69, 75
  non-material, 236
  patterns, 65, 98, 209, 234, 236, 288
continental slopes, 158, 164
contingency, 267
controversies, 244
convention, 243
  on the conservation of biological
      diversity, 213
coral reefs, 163, 165, 207
COSTED, 1
costs, 80
cost shifting, 286
crop, 80, 82
cultural,
  diversity, 269
  evolution, 67

forces, 241, 246
  values, 241, 268
culture, 65, 75
  goods, 231
cycle,
  hydrological, 129, 138
  materials, 97
  product, 95
cycles, biogeochemical, 129, 132,
    138, 141, 150
cyclones, 144

dam, 196
databases, 386
data needs and systems, 260
debt for nature swaps, 72
decentralization of decision-making,
    283
decision-makers, 73, 262, 283, 285,
deforestation, 96, 97, 158, 165, 198,
    208, 209, 210, 286
  in the Tropics, 82
degradation, 79, 82, 174, 224
  ecological, 269
  forest, 103
Delft Declaration on water resources,
    202
demersal fish, 162
democracy, 290
democratization, 283
denitrification, 374
desertification, 83, 96, 98, 252
deposition,
  acidic, 132, 136, 141, 144, 145, 153
developing countries, 82, 201, 266
development,
  assistance, 97, 275
  economic, 61, 227, 260
  environment, 68, 197, 230
  human, 119, 227, 233
  strategies, 235
  sustainable, 266
dimethyl sulphide, 129, 134, 135,
    136, 162, 165
dinoflagellates, 161
disaster,
  ecological, 251
  man-made, 258
  natural, 258
disease, 85, 121, 122, 188, 258, 269
diversity,
  conservation of, 267
  cultural, 269
  genetic, 207, 212

dominant models of "good science",
    240
drainage water, 196
drought, 191, 194
drugs, 234

earth charter, 292
  system, 127, 130, 184, 188–190
east european states, 290
ecological,
  degradation, 269
  disaster, 251
  economics, 260
  sustainability, 96
  thresholds, 288
ecology, 250
economic,
  efficiency, 97
  growth, 61, 227, 284
  restructuring, 98
  theories, 236
  wealth, 232
economic limits to growth, 69
economics, ecological, 260
ecosystems, 150, 151, 158, 161, 162,
    163, 165, 215
  aquatic, 190
  biodiversity, 183
  coastal, 157–159, 164
  manipulation, 199
ecotaxes, 289
education, 74, 75, 245, 261, 277
  and training, 168, 201, 290
efficiency of energy use, 67, 97, 100,
    109, 111, 243
El Niño, 141, 152, 157, 161, 187
electric car, 93
electricity generation, 104
emission,
  anthropogenic, 148
  limits, 97
  permits, 93
  standards, 93, 98
  taxes, 93
energy, 104, 106, 108–111, 230, 270
  alternative strategy, 111
  availability among countries, 104
  balance, 190
  co-operation, 115
  conservation, 100
  efficiency, 67, 97, 100, 109, 111,
      243
  efficient scenario, 103
  future, 109

Printed in the United States
By Bookmasters